Mitigation Strategies and Risk Management for Food Contaminants

Matthew Chidozie Ogwu
Appalachian State University, USA

Sylvester Chibueze Izah
Bayelsa Medical University, Nigeria

IGI Global
Scientific Publishing
Publishing Tomorrow's Research Today

Vice President of Editorial	Melissa Wagner
Director of Acquisitions	Mikaela Felty
Director of Book Development	Jocelynn Hessler
Production Manager	Mike Brehm
Cover Design	Jose Rosado

Published in the United States of America by
IGI Global Scientific Publishing
701 East Chocolate Avenue
Hershey, PA, 17033, USA
Tel: 717-533-8845
Fax: 717-533-7115
Website: https://www.igi-global.com E-mail: cust@igi-global.com

Library of Congress Cataloging-in-Publication Data

LCCN: 2025041423 (CIP Data Pending)
ISBN13: 9798337370521
Isbn13Softcover: 9798337370538
EISBN13: 9798337370545

British Cataloguing in Publication Data
A Cataloguing in Publication record for this book is available from the British Library.

Table of Contents

Detailed Table of Contents

Climate change is emerging as a profound disruptor of food safety, influencing the prevalence, distribution, and behavior of contaminants across global food systems. This chapter provides a comprehensive examination of how climate-induced environmental changes, including rising temperatures, shifting precipitation patterns, sea-level rise, and extreme weather events, exacerbate the risks associated with chemical, biological, and physical food contaminants. Key mechanisms such as the mobilization of heavy metals, increased pesticide use, enhanced fungal mycotoxin production, and altered pathogen ecologies are examined in relation to climate variability. The growing presence of microplastics and other physical contaminants is also analyzed, particularly in regions vulnerable to flooding and drought. The chapter demonstrates how climate change exacerbates food contamination routes, compromising public health and food security, particularly in low- and middle-income nations, by combining interdisciplinary evidence. It further highlights contamination hotspots, socioeconomic vulnerabilities, and implications for global food governance. The chapter concludes with strategic recommendations including climate-resilient agricultural practices, integrated surveillance systems, community engagement, and cross-sectoral policy frameworks such as One Health. Through this lens, the chapter contributes critical insights into the complex intersections of climate change and food contamination, offering a roadmap for mitigation and adaptation in an era of environmental uncertainty.

Emerging chemical contaminants, including microplastics, per- and polyfluoroalkyl substances (PFAS), and pharmaceutical residues, pose escalating, multi-pathway threats to global food security. Their environmental persistence, bioaccumulation potential, and association with significant human health risks demand urgent attention. This chapter synthesizes critical findings on contaminant entry routes into

residues, and environmental impact. In response, novel decontamination technologies have emerged that aim to enhance microbial safety while preserving food quality and aligning with sustainability goals. This chapter explores contemporary physical, chemical, and biological approaches, including ultraviolet-C light, cold plasma, high-pressure processing, organic acids, electrolyzed water, and bacteriophage applications highlighting their mechanisms, effectiveness, and limitations. Additionally, the chapter reviews emerging and hybrid systems such as photocatalytic decontamination, ultrasound-assisted sanitization, supercritical CO_2 treatment, and smart sensor-integrated solutions, which offer synergistic benefits and greater adaptability to modern food processing demands. Despite their promise, challenges such as regulatory inconsistencies, high implementation costs, and consumer skepticism hinder widespread adoption. Nevertheless, increased demand for clean-label, minimally processed, and safe food is driving innovation. This chapter underscores the need for collaborative efforts in research, standardization, and policy support to facilitate the transition from traditional to next-generation decontamination strategies in food safety systems.

Nanotechnology in the food sector offers innovative solutions for mitigating biological, chemical, and physical contaminants across the supply chain. This chapter explores its core principles and diverse applications in enhancing food safety, quality, and sustainability. Key technologies (nanosensors, nano-adsorbents, nano-coatings, and smart packaging) enable real-time detection, neutralization, and removal of contaminants. Nanomaterials like silver nanoparticles, quantum dots, carbon nanotubes, and nano-silica exhibit strong antimicrobial and toxin-adsorbing properties, while nano-encapsulation supports delivery of detoxifying agents and functional compounds. The chapter also examines regulatory, ethical, and public perception issues, highlighting nanotechnology's role in reducing food waste, improving agricultural efficiency, and supporting eco-friendly packaging. Prospects for the field are discussed within the context of interdisciplinary research, digital integration, and consumer trust.

Preserving food flavor, freshness, and nutritional content is possible through careful packaging, which accounts for physical, mechanical, and epidemiological aspects. Unfortunately, the accumulation of non-biodegradable waste is a direct result of the conventional packaging's use of non-renewable fossil fuels, which harms the environment. Therefore, biodegradable films for food packaging are in high demand; they provide a cost-effective solution that meets industry standards while maintaining food safety, and consumers are becoming more conscious of the importance of purchasing sustainably sourced, high-quality food. Learn why biodegradable films are crucial for food safety and how they function in packaging. Biodegradable polymers derived from polysaccharides are the focus of this research, along with their application in multi-criteria food packaging and the success or failure of developing smart biodegradable films.

Food safety has become an increasingly important global issue in recent decades, mainly due to the increasing contamination of food products. This chapter examines the potential of natural compounds derived from plants, microorganisms, and marine organisms as safer and more sustainable alternatives to synthetic chemicals

Preface

Mitigation Strategies and Risk Management for Food Contaminants brings together a timely and diverse collection of contributions aimed at addressing the growing complexity of food safety in the face of emerging contaminants, evolving climate conditions, technological advancements, and socio-cultural dynamics. This volume is designed to serve as a resource for researchers, practitioners, policymakers, and students engaged in food safety, public health, agriculture, and environmental sciences.

The global food system is increasingly challenged by contaminants that originate from a variety of sources including industrial processes, climate-related phenomena, packaging materials, agricultural inputs, and biological vectors. Addressing these challenges requires more than traditional food safety protocols. It demands interdisciplinary collaboration, innovative technological interventions, and the integration of traditional and contemporary knowledge systems.

This edited book is a sequel to earlier explorations into food contaminant detection and monitoring. It focuses on mitigation strategies and risk management frameworks that are critical to strengthening food system resilience and protecting public health. The chapters in this volume are authored by experts from across the globe, reflecting a range of regional experiences, scientific insights, and policy perspectives.

The following chapter overview summarizes the content and focus of each contribution.

CHAPTER OVERVIEW

Chapter 1: Climate Change and Its Influence on Food Contaminant Patterns

This chapter explores how climate change acts as a significant driver in altering the behavior, distribution, and risk of food contaminants. It discusses the implications of shifting weather patterns, rising temperatures, and environmental disruptions on the mobilization of heavy metals, pathogen ecologies, pesticide usage, and mycotoxin production.

Chapter 2: Microplastics, PFAs, and Pharmaceutical Residues as Emerging Contaminants in the Global Food Chain

Focusing on three major categories of emerging chemical contaminants, this chapter critically reviews their sources, pathways into the food chain, toxicological implications, and regulatory challenges. It also evaluates current detection technologies and mitigation strategies to guide informed policy development.

Chapter 3: Risk Assessment Frameworks and Predictive Modeling for Food Safety Decision-Making

This chapter bridges conventional risk assessment methodologies with predictive modeling tools such as machine learning and statistical models. It provides insights into how these tools can enhance food safety risk evaluation and support proactive decision-making in dynamic and data-driven environments.

Chapter 4: Novel Decontamination Technologies and Emerging Approaches in Food Safety

The authors present a range of modern physical, chemical, and biological decontamination techniques, including UV-C light, cold plasma, bacteriophages, and hybrid systems. These emerging technologies are assessed for their efficacy, sustainability, and potential to replace or complement traditional methods.

Chapter 5: Nanotechnology Applications in Food Contaminant Mitigation

This chapter delves into the potential of nanotechnology to transform food safety practices through innovations such as nanosensors, nano-coatings, and smart packaging. It outlines the mechanisms by which nanomaterials can detect, neutralize, or remove contaminants and discusses the regulatory and ethical issues involved.

Chapter 6: Biodegradable Films for Mitigating Food Contaminants through Sustainable Packaging

Focusing on sustainable packaging solutions, this chapter explores the development and application of biodegradable films made from polysaccharides and other natural polymers. It highlights their role in preserving food quality, reducing plastic waste, and aligning with consumer demand for environmentally friendly products.

Chapter 7: Utilization of Natural Compounds as Food Contaminant Mitigation Agents

This chapter reviews the use of plant- and microbe-derived compounds as alternatives to synthetic chemicals in food safety applications. It discusses their antimicrobial, antioxidant, and detoxifying properties, along with practical considerations for stability, regulation, and sensory effects.

Chapter 8: Genetically Modified Crop Risk Assessment and Biosafety

Drawing from a comprehensive meta-analysis, this chapter evaluates the safety of genetically modified crops with respect to contaminant risk. It introduces omics-based approaches for early biomarker detection and proposes a communication and surveillance framework to support biosafety governance.

Chapter 9: Indigenous and Traditional Knowledge Systems in Food Safety and Contaminant Management

This final chapter emphasizes the value of Indigenous and traditional knowledge in maintaining food safety and preventing contamination. It explores a wide range of culturally embedded practices such as fermentation, herbal preservation, and ecological monitoring, and argues for their integration into modern food safety strategies.

CLOSING REMARKS

This book highlights the urgent need for innovative, inclusive, and sustainable approaches to managing food contaminants in a rapidly changing world. By combining cutting-edge technologies with traditional wisdom and systemic risk management, it offers a comprehensive perspective on how to strengthen food safety from farm to fork. The editors express their gratitude to all the contributing authors for their valuable insights and to the institutions and organizations supporting this work.

Matthew Ogwu
Appalachian State University, USA

Sylvester Izah
Bayelsa Medical University, Nigeria

Chapter 1
Climate Change and Its Influence on Food Contaminant Patterns

Matthew Chidozie Ogwu
https://orcid.org/0000-0001-6054-1667
Appalachian State University, USA

Aliu Olugbemiga Ojo
https://orcid.org/0000-0003-4942-7454
Auchi Polytechnic, Nigeria

ABSTRACT

Climate change is emerging as a profound disruptor of food safety, influencing the prevalence, distribution, and behavior of contaminants across global food systems. This chapter provides a comprehensive examination of how climate-induced environmental changes, including rising temperatures, shifting precipitation patterns, sea-level rise, and extreme weather events, exacerbate the risks associated with chemical, biological, and physical food contaminants. Key mechanisms such as the mobilization of heavy metals, increased pesticide use, enhanced fungal mycotoxin production, and altered pathogen ecologies are examined in relation to climate variability. The growing presence of microplastics and other physical contaminants is also analyzed, particularly in regions vulnerable to flooding and drought. The chapter demonstrates how climate change exacerbates food contamination routes, compromising public health and food security, particularly in low- and middle-income nations, by combining interdisciplinary evidence. It further highlights contamination hotspots, socioeconomic vulnerabilities, and implications for global food governance. The chapter concludes with strategic recommendations including climate-resilient agricultural practices, integrated surveillance systems, community

DOI: 10.4018/979-8-3373-7052-1.ch001

engagement, and cross-sectoral policy frameworks such as One Health. Through this lens, the chapter contributes critical insights into the complex intersections of climate change and food contamination, offering a roadmap for mitigation and adaptation in an era of environmental uncertainty.

1. INTRODUCTION

The ongoing phenomenon of climate change has emerged as a significant driver influencing global food security, fundamentally altering agricultural productivity and the broader food supply chain. As articulated by Baldos and Hertel, climate change is projected to affect not only food production but also the nutritional availability, stability, and accessibility of food, which are critical components of food security (Baldos and Hertel, 2014). The increasing frequency and intensity of climate-related events such as floods, droughts, and shifting weather patterns pose significant risks to food systems globally, as demonstrated in studies examining resilience in various countries (Favas et al., 2024). Such vulnerabilities underscore the complexities of maintaining food security while adapting to these changes. Food contaminants, defined as substances that compromise food safety, come in various forms: chemical (including pesticides and heavy metals), biological (such as pathogens and toxins), and physical (like foreign objects) (Assunção et al., 2018). These contaminants not only affect food safety but also pose significant risks to human health, leading to increased morbidity in populations already vulnerable due to poverty and malnutrition (Pickson and Boateng, 2021). As such, understanding the changing patterns of food contaminants is crucial to addressing food security in the context of climate change.

In recent years, numerous studies have documented the interconnectedness of climate change and food systems. For instance, Favas et al. (2024) emphasize that extreme weather events, accelerated by climate change, threaten food supply chains and lead to food price volatility, which in turn directly affects the nutritional quality of available food. In examining the effects on various regions, Sabola notes the potential for climate change to negatively impact food production systems, particularly in emerging economies where agricultural resilience is already precarious (Sabola, 2024). Such findings underscore the urgent need to investigate how these environmental changes contribute to the emergence or exacerbation of food contaminants. Efforts to mitigate the adverse impacts of climate change include adaptive strategies within agricultural practices, aimed at enhancing food security while addressing health concerns posed by food contaminants. Research indicates that climate-smart agriculture incorporates techniques that not only improve agrarian productivity but also prioritize food safety, thereby addressing the dual challenge posed by climate change (Connolly-Boutin and Smit, 2015). This methodology is crucial, as evidence indicates that climate change promotes the proliferation of mycotoxins and other contaminants that may compromise food safety and health outcomes (Medina et al., 2017).

Moreover, studies have shown that the safety of food systems can be significantly compromised due to adverse climatic conditions, which promote the growth of pathogens and toxins (Janevska et al., 2010). For example, rising temperatures and fluctuating moisture levels can lead to increased contamination rates by mycotoxigenic fungi, further complicating the food security landscape (Fels-Klerx et al., 2016). This association highlights a cycle where degraded agricultural environments caused by climate change not only limit food availability but also introduce or increase foodborne health risks, ultimately challenging the nutritional outcomes for affected populations. Understanding these dynamics necessitates a thorough examination of the pathways linking climate factors, food contaminant emergence, and food security outcomes. It requires an interdisciplinary approach, integrating perspectives from environmental science, public health, and agricultural policy to formulate comprehensive strategies that enhance both climate resilience and food safety across diverse contexts.

The intricate interactions between climate conditions and food security must also consider socioeconomic dimensions that influence vulnerability and exposure to food contaminants. For instance, regions already experiencing high levels of food insecurity, such as sub-Saharan Africa, face compounded risks as climate change continues to intensify (Addisu and Wajera, 2024). Studies have increasingly focused on these relationships to highlight the disproportionate effects of climate variability on smaller and marginalized farming communities, identifying both economic and social pathways through which climate change impacts food safety at a household level (Kanu and Onyekwere, 2024).

This chapter aims to critically examine the multifaceted impacts of climate change on the occurrence, distribution, and dynamics of food contaminants across global supply chains. By synthesizing current scientific evidence, the chapter explores how shifting climatic conditions—such as rising temperatures, altered precipitation patterns, and extreme weather events—affect the behavior and prevalence of chemical, biological, and physical contaminants in food systems. It also highlights emerging risks and underexplored pathways through which climate stressors exacerbate food safety challenges, particularly in vulnerable and low-resource contexts. The chapter contributes to knowledge by bridging gaps between climate science and food safety research, offering an integrative framework that links environmental change with public health risks. Furthermore, it highlights the importance of adaptive monitoring systems, climate-resilient agricultural practices, and cross-sectoral policy approaches (e.g., One Health and food security frameworks) in mitigating the evolving threat of contamination under changing climate scenarios. This contribution is especially timely as it informs both academic inquiry and policy innovation in addressing food safety in an era of global environmental change.

2. CLIMATE CHANGE DRIVERS AND ENVIRONMENTAL PATHWAYS

One of the most significant issues facing the world today is climate change, which has a profound impact on various environmental systems and manifests as a complex array of interconnected phenomena. Food safety is greatly impacted by the changes it makes to the environment. The transit, permanence, and concentration of food contaminants are influenced by several key factors, including sea level rise, changing precipitation patterns, rising temperatures, and extreme weather events (Table 1). The driving force behind climate change is rooted in the increase of greenhouse gases, such as carbon dioxide, methane, and nitrous oxides, primarily resulting from human activities. These gases create a greenhouse effect, leading to global temperature rise, which has been documented and understood since the late 20th century. The Intergovernmental Panel on Climate Change (IPCC) and numerous studies highlight that the average global temperature has already risen by approximately 1.4°C since the late 19th century, and this trajectory is expected to continue unless drastic measures are undertaken to curtail emissions (Malhi et al., 2021; Zhou et al., 2023).

The rise in global temperatures is intrinsically linked to changes in precipitation patterns, a phenomenon exacerbated by increased concentrations of greenhouse gases in the atmosphere. These changes manifest not only in precipitation volume but also in the intensity and frequency of extreme weather events. Research indicates that as the climate warms, the moisture-holding capacity of the atmosphere increases, as articulated through the Clausius-Clapeyron relationship, leading to more extreme precipitation events (Kim et al., 2018; Li et al., 2019). For instance, studies have shown a notable shift in precipitation patterns worldwide, with an increase in heavy rainfall incidents combined with longer dry spells, effectively altering both hydrological cycles and agricultural productivity across diverse biomes (Zhang et al., 2013; Verma et al., 2024).

Moreover, the implications of changing precipitation patterns include an increased frequency and intensity of extreme weather events, such as floods and droughts. The consequences reverberate through ecological systems, affecting biodiversity, altering habitat suitability, and causing shifts in species distribution due to the pressure exerted by these changes (Shih et al., 2021; Narayanan et al., 2024). Extreme weather events have been identified not only as immediate threats but also as catalysts for long-term ecological shifts, which may hinder ecosystem functions and services upon which humanity fundamentally depends (Beard et al., 2016; Ogwu et al., 2023). Increased precipitation variability has been linked to detrimental impacts on human health, particularly through the proliferation of waterborne diseases, especially in coastal and estuarine environments that are vulnerable to runoff from heavy rainfall (Shih et al., 2021).

In parallel, ocean acidification —a direct consequence of increasing CO_2 levels —presents another critical environmental pathway affected by climate change. The absorption of excess atmospheric CO_2 by ocean waters leads to a decrease in pH levels, thereby affecting marine ecosystems, particularly coral reefs and shellfish, which are sensitive to changes in acidity (Kroeker et al., 2011; Woesik et al., 2013). Studies reveal that ocean acidification can alter species interactions and ecosystem dynamics, often leading to decreased biodiversity and productivity within marine food webs (Kroeker et al., 2011). The profound effects on calcifying organisms underscore the need for adaptive management strategies to maintain the resilience of marine systems amidst these changes.

Sea level rise, resulting from the thermal expansion of seawater and increased melting of land-based ice, is another critical consequence of climate change. Projections indicate that global sea levels could rise from 75 to 90 centimeters by 2100, significantly threatening coastal ecosystems and human settlements (Woesik et al., 2013). The combination of rising sea levels with extreme weather can exacerbate coastal flooding, leading to socioeconomic disruptions, habitat loss, and increased salinity in estuarine systems, which adversely impact freshwater supplies and agricultural outputs (Mercer et al., 2012).

Simultaneously, climate change exacerbates the transport and transformation of contaminants in both terrestrial and aquatic environments. Variability in temperature and rainfall patterns impacts the bioavailability of pollutants and nutrients, altering their interaction within ecosystems. For example, heavier rainfall events can lead to increased runoff, thereby enhancing the transport of agricultural chemicals into nearby water bodies, subsequently affecting aquatic life and human health through contaminated drinking water supplies (Huang et al., 2020). The influence of climate change on contaminant dynamics necessitates that environmental monitoring integrate climatic variables to better understand and mitigate impacts on water quality and ecosystem health (Loecke et al., 2017).

Furthermore, shifts in climatic patterns can lead to complex interdependencies among species and alter ecological resilience. Specific research indicates that climate change may modify plant-insect interactions, thereby affecting both agricultural production and natural ecosystems. Increased temperatures and changing precipitation patterns can lead to mismatches between the life cycles of plant species and their pollinators or pests, resulting in cascading effects throughout the food web (Jamieson et al., 2012). This is particularly concerning in agricultural contexts where crop yields depend heavily on the stability of these ecological interactions (Verma et al., 2024). The myriad effects of climate change on environmental pathways highlight the intricate interplay between greenhouse gas emissions, global temperature rise, altered precipitation patterns, rising sea levels, ocean acidification, and contaminant dynamics. Each driver contributes to a complex framework of ecological interdependencies that must be understood as global society seeks to mitigate and adapt to the ongoing impacts of climate change. Continuous research is crucial for enabling informed decision-making and effective policy development that supports both human and ecosystem health in a rapidly changing climate.

Table 1. Climate change drivers and environmental pathways affecting food contamination

Climate Change Driver	Environmental Pathway	Associated Contaminants	Impact on Food Systems
Increased Temperature	Enhanced microbial growth and toxin production	Pathogens (e.g., *Salmonella, Listeria*), and Mycotoxins	Elevated spoilage rates, increased foodborne illnesses, and higher aflatoxin risk in cereals
Changes in Precipitation	Flooding or drought alters soil and water dynamics	Heavy metals (e.g., lead, cadmium), Pesticides, and Nitrates	Soil leaching and runoff increase uptake of toxicants by crops; water contamination
Sea Level Rise	Saltwater intrusion into agricultural lands and freshwater sources	Salinity, Arsenic, and Persistent Organic Pollutants (POPs)	Compromised irrigation water quality and food crop viability
Extreme Weather Events	Infrastructure damage, increased runoff, and disruption of cold chains	Pathogens, Physical contaminants (debris, and microplastics)	Greater post-harvest losses, contamination during processing, and distribution
Drought Conditions	Concentration of pollutants in reduced water volumes	Cyanotoxins and Agrochemical residues	Higher contaminant concentration in irrigation sources and animal feed
Altered Ecosystem Dynamics	Expansion of vector ranges and introduction of new pests and diseases	Zoonotic pathogens and Mycotoxins	Increased crop damage, need for more pesticide application, and zoonotic foodborne diseases
Carbon Dioxide Elevation	Altered plant physiology and microbial interactions	Nitrate accumulation and Reduced nutritional quality	Decreased food quality and safety; shifts in microbial populations

3. IMPACT OF CLIMATE CHANGE ON CHEMICAL FOOD CONTAMINANTS

Climate change influences the presence and behavior of chemical contaminants in food systems. Rising temperatures and altered precipitation patterns accelerate the formation, mobility, and bioavailability of hazardous chemicals, including mycotoxins, heavy metals, and pesticide residues. Figure 1 illustrates the key pathways through which climate-induced environmental changes contribute to increased contaminant uptake by crops and accumulation in animal feed, posing growing risks to food safety and public health. Heavy metals are widely distributed in the environment and can accumulate in food products, posing significant health risks. Climate change can exacerbate the mobilization of heavy metals through mechanisms such as increased flooding and soil erosion, which are becoming more frequent due to altered precipitation patterns and extreme weather events (Martin et al., 2024). Specifically, areas prone to flooding may experience increased leaching of metals from contaminated sites, transferring these pollutants directly into aquatic environments and subsequently into the food chain. Furthermore, rising temperatures can enhance the rate of weathering processes, leading to increased bioavailability of heavy metals in soil, which is further aggravated by soil erosion (Leite et al., 2025; Imarhiagbe et al., 2024). Studies have indicated that climate-induced alterations to hydrology can significantly affect the transport and transformation of these metals, with implications for agricultural lands (Steffens et al., 2014).

Figure 1. Influence of climate change on chemical food contaminants

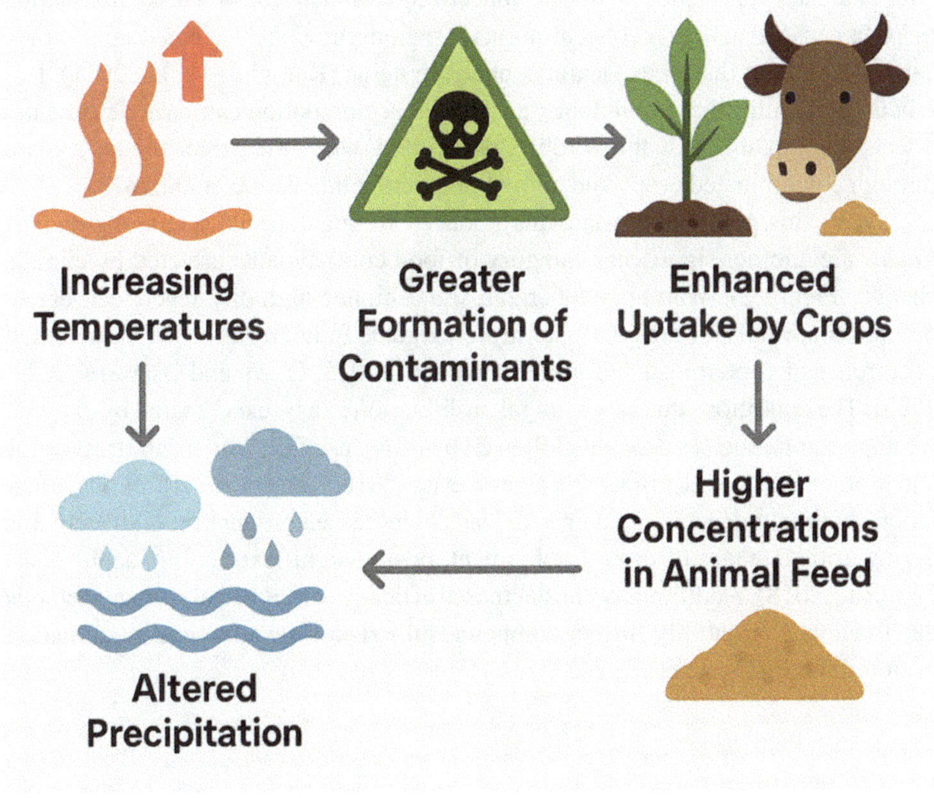

In conjunction with heavy metals, pesticide residues represent another critical pathway of contamination. Climate change is already influencing agricultural practices, leading to an increase in pesticide application due to the proliferation of agrarian pests that thrive in warmer conditions (Gagnon et al., 2015; Agoh et al., 2024). The increased frequency of pests can prompt farmers to resort to intensive pesticide use, thus elevating the levels of pesticide residues in food products. Moreover, climate factors, such as temperature and precipitation, significantly influence pesticide degradation rates and their efficacy. It is well established that higher temperatures can increase the volatility of pesticides, resulting in higher atmospheric concentrations and subsequent deposition on crops (Tudi et al., 2021). Furthermore, climate change can also alter the efficacy of pesticide formulations, necessitating higher application rates, which in turn increases the risk of residue accumulation in food (Martínez-Megías et al., 2022).

Persistent Organic Pollutants (POPs) are another area of concern. These toxic substances, which include a range of industrial chemicals and pesticide by-products, are known to bioaccumulate in the food web. Climate change can exacerbate their spread via changes in temperature and precipitation patterns. Rising temperatures have been linked to increased volatilization and transport of POPs, making their environmental persistence even more concerning (Kannan, 2023). This redistribution of POPs can lead to higher concentrations in regions previously deemed safe, as they can travel greater distances via atmospheric systems (Kattwinkel et al., 2011). It is crucial to monitor these pollutants, as their bioaccumulation can have detrimental effects on human health and wildlife, particularly when they enter the food chain through contaminated crops and animal products (Oldenkamp et al., 2023).

Mycotoxins, particularly aflatoxins produced by fungal species such as *Aspergillus flavus*, are another significant category of food contaminants affected by climate change (Figure 2). Warmer temperatures and higher humidity levels can create optimal conditions for the growth of mycotoxigenic fungi, resulting in an increased occurrence of contaminated crops (Rincón et al., 2015; Ogwu and Osawaru, 2022; 2023). The conditions that favor fungal proliferation can be exacerbated by extreme weather events, such as droughts followed by heavy rainfall, which can stress crops and promote fungal infections (Hasegawa et al., 2018; Yang et al., 2021). The escalation in climate-driven fungal growth leads to increased production of mycotoxins during critical stages of crop development, posing significant public health risks (Wu et al., 2018). Additionally, the decreased efficacy of traditional storage methods due to climate variability further compounds the risk of mycotoxin contamination in stored foods (Ogwu et al., 2023).

Figure 2. Mycotoxin-Affected Food Products. (A) Aspergilus fumigatus, which releases aflatoxin, (B) molds on red corn, (C) molds on maize flour, (D) molds on red corn, (E) molds on soybean flour, (F) molds in rice grains, (G) mold on white maize grains, and (H) molds on white corn

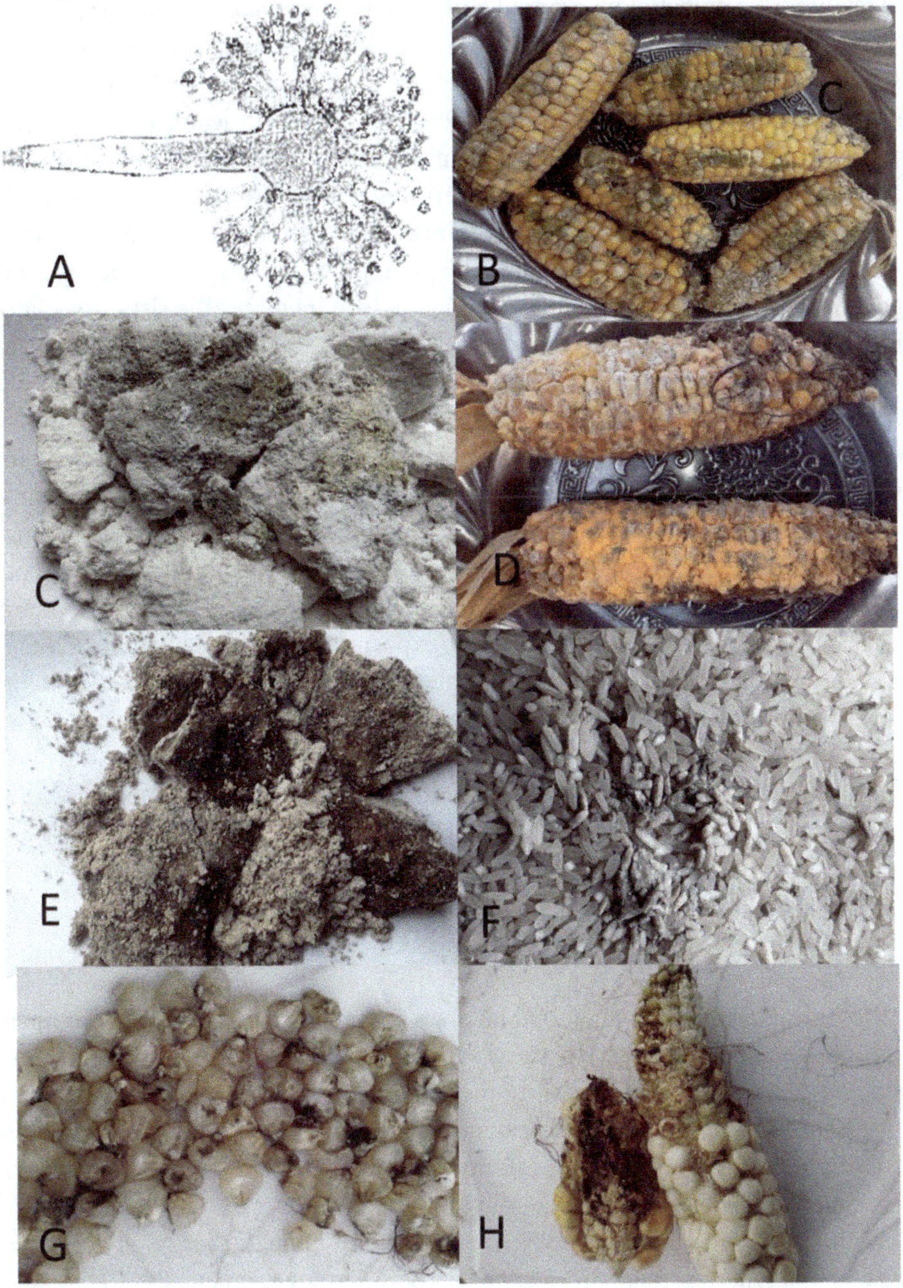

Source: Mafe and Büsselberg (2024)

Moreover, it is essential to consider the interconnectedness of these issues. The destabilization of ecosystems due to climate change can lead to shifts in biodiversity that may alter the dynamics of natural pest control and fungal diseases, creating a feedback loop that increases reliance on chemical solutions. As ecosystems become less resilient, the potential for crops to withstand pest pressures diminishes, leading to even greater pesticide application, which can subsequently contaminate water resources. This highlights the importance of understanding the complex interactions between climate change, agricultural practices, and ecosystem health in managing the risks associated with chemical food contaminants effectively (Steffens et al., 2014).

Additionally, there is a growing concern regarding how climate change could exacerbate food insecurity issues, which are already dangerously linked to agricultural productivity. Warmer and drier climates may compromise crop yields, necessitating a greater reliance on chemical inputs, which in turn increases the risks associated with chemical residues in food products (Wei et al., 2022; Ogwu and Kulkarni, 2024). Mitigation strategies that successfully reduce greenhouse gas emissions can inadvertently lead to increased food prices and supply issues of key agricultural commodities, further impacting food security (Schwiesow et al., 2024). Thus, the response to climate change must be holistically integrated into food systems planning, considering the chemical contaminant pathways and socioeconomic factors that may arise. It is also pertinent to address community health and environmental justice considerations. Vulnerable populations, often residing in regions with high agricultural output, face disproportionate risks related to exposure to these contaminants. Climate change exacerbates these risks through intensified agricultural practices required to maintain productivity in the face of shifting conditions (Imarhiagbe and Ogwu 2022; Martin et al., 2024). A comprehensive understanding of the risks involved can inform better practices and policies geared towards protecting public health.

4. IMPACT OF CLIMATE CHANGE ON BIOLOGICAL FOOD CONTAMINANTS

Biological contamination of food involves the presence of harmful microorganisms or biological agents that can cause illness when ingested. These contaminants, including bacteria, viruses, parasites, fungi, and prions, often enter the food chain through contact with contaminated water, soil, animals, improperly handled or stored food, and food handlers who are infected (Table 2). Biological contamination is one of the leading causes of foodborne diseases globally, posing significant risks to public health and food security. The influence of climate change on biological food contaminants represents a critical nexus between environmental alterations and public health. As global temperatures rise and precipitation patterns fluctuate, we observe a notable expansion of microbial pathogens, with profound implications for food safety. Notably, pathogens such as *Salmonella* and *Listeria* are becoming increasingly prevalent due to the favorable conditions stimulated by climate change, particularly warmer temperatures and higher humidity levels (Smith and Fazil, 2019; Akil et al., 2014). The review by Smith and Fazil (2019) highlights the complexity of these interactions, noting that extreme climate events significantly contribute to the increased risk of foodborne illnesses associated with microbial pathogens. These conditions bolster pathogen proliferation, potentially leading to more frequent outbreaks of foodborne diseases.

Table 2. Some biological sources of food contamination

Contaminant Type	Examples	Primary Sources	Common Food Vehicles	Health Risks
Bacteria	*Salmonella, E. coli, Listeria, Campylobacter*	Contaminated water, soil, animal feces, and undercooked meats	Poultry, beef, dairy, and fresh produce	Food poisoning, diarrhea, and kidney failure
Viruses	Norovirus, Hepatitis A and E, Rotavirus	Infected food handlers, sewage-contaminated water	Shellfish, fresh produce, ready-to-eat foods	Gastroenteritis, liver damage, fever
Parasites	*Toxoplasma gondii, Giardia, Cryptosporidium*	Contaminated water, undercooked meat, feces	Pork, lamb, raw vegetables, untreated water	Neurological effects, dehydration, and intestinal distress
Fungi (Molds and Yeasts)	*Aspergillus, Penicillium, Fusarium*	Poor storage conditions, humidity, and contaminated grains	Cereals, nuts, dried fruits, and dairy	Mycotoxicosis, allergies, and immunosuppression
Protozoa	*Entamoeba histolytica, Cyclospora*	Human or animal waste, untreated water	Berries, leafy greens, and untreated water	Amoebic dysentery, prolonged diarrhea
Helminths (Worms)	Tapeworms, roundworms, flukes	Undercooked meat or fish, poor sanitation	Pork, beef, freshwater fish, and vegetables	Nutritional deficiencies, intestinal blockage
Prions	Misfolded proteins cause BSE (mad cow disease)	Infected animal tissues, nervous system products	Contaminated beef products	Neurological degeneration (e.g., vCJD in humans)

Seasonality and geography of foodborne illness outbreaks are profoundly affected by climate change. Research suggests that warmer temperatures may alter the timing and frequency of these outbreaks, as seen through increased cases of illnesses such as *Campylobacter* and *E. coli*, which are directly associated with higher ambient temperatures (Smith and Fazil, 2019; Smith et al., 2019). The presence of pathogens like Campylobacter in poultry is shown to be strongly correlated with temperature increases, indicating that even minor climate shifts could significantly elevate the risks associated with foodborne pathogens (Semenza et al., 2012). Furthermore, Akil and Ahmad highlight that rising temperatures are correlated with increased incidences of Salmonella infections, demonstrating a clear linkage between climate variables and the dynamics of foodborne diseases (Akil and Ahmad, 2023; Akil et al., 2014). Thus, climate change leads to altered patterns in disease dynamics, with specific seasons becoming more pronounced for certain pathogens.

Moreover, the transmission of vector-borne zoonotic diseases into food systems is exacerbated by climate change, driven by variations in temperature and precipitation, which in turn affect the habitats and behaviors of vectors such as mosquitoes and ticks (Cissé, 2019). The effects of a changing climate on water systems can introduce these pathogens into livestock and crop production, further complicating the food safety landscape. It is projected that climate-induced changes will necessitate modifications across food supply chains, emphasizing the need for integrated approaches to manage biological contaminants in food systems (Akil, 2021; Akil and Ahmad, 2019; Hamaideh et al., 2024). Cissé notes that low- and middle-income countries are particularly vulnerable, as deficiencies in water and sanitation are likely to compound the effects of climate change on food safety (Cissé, 2019).

Potential economic burdens mark the escalation of foodborne diseases resulting from these phenomena. A scoping review conducted in Canada projects an alarming financial implication of approximately US$30-US$62 billion annually by 2050 due to heightened risks of biological food contamination (Zai et al., 2024). Such extensive economic forecasts underscore the need to address the role of climate variability in food safety. In the specific context of fresh produce contamination, climate change has a profound influence. Research suggests that elevated temperatures have a significant impact on the persistence and growth of foodborne pathogens on crops, particularly leafy greens. Samut et al., 2023; Liu et al., 2016). The seasonal variations in pathogen presence further complicate the agricultural practices necessary to mitigate risks, underscoring the need for revised farming strategies to counteract the impacts of climate change (Akil, 2021; Samut et al., 2023). Enhanced moisture levels can facilitate microbial growth, underscoring the crucial role of humidity as a climate factor that warrants further investigation to safeguard food systems (Gao et al., 2014).

Furthermore, the interplay between climate components and foodborne pathogens leads to a complex scenario in which pathogen emergence is compounded by agricultural practices and technological advancements (Lake, 2017; Hamaideh et al., 2024). For instance, the indiscriminate use of antibiotics in livestock may contribute to the emergence of resilient strains of pathogens, which is exacerbated by climate-induced stresses (Bintsis, 2017).

In response to these multifaceted challenges, it's critical to implement adaptive management strategies. This proactive approach can help public health officials anticipate and respond to health risks associated with foodborne diseases triggered by climate change. Ebi (2011) emphasizes the potential for adaptive management frameworks to provide structured decision-making processes, thereby enhancing resilience against health risks induced by climate alterations. The impact of climate change on biological food contaminants is an intricate interplay of warming climates, increased humidity, and altered seasonal patterns, resulting in varied risks for foodborne pathogens. This necessitates a comprehensive understanding of the linkages between environmental changes and public health implications to formulate effective responses to mitigate the increasing burden of foodborne illnesses.

5. IMPACT OF CLIMATE CHANGE ON PHYSICAL FOOD CONTAMINANTS

The interrelationship between climate change and physical food contaminants is an increasingly vital area of study. Physical contamination of food occurs when foreign objects or materials are unintentionally introduced into food products at any stage of the food supply chain. These contaminants can originate from raw materials, equipment, packaging, the processing environment, or even human handling. Although often visible, physical contaminants pose serious health risks, ranging from choking and injuries to long-term exposure to harmful materials, such as microplastics (Table 3). This nexus becomes especially relevant when considering the growing prevalence of microplastics in food systems, both terrestrial and aquatic, and the concurrent increase in contamination from soil particles and pollutants due to extreme weather events. The complexities of these interactions necessitate an investigation into the mechanisms by which climate change exacerbates the presence of these contaminants and the implications for food safety.

Table 3. Some physical sources of food contamination

Source Category	Examples	Typical Entry Point in Food Chain	Potential Health Risks
Foreign Objects	Stones, glass shards, metal fragments, wood chips	Harvesting, processing, and packaging	Choking, internal injuries, and dental damage
Packaging Material	Plastic fragments, staples, foil pieces, and ink residues	Packaging machinery, damaged packaging	Ingestion of harmful materials, toxic exposure
Processing Equipment	Nuts, bolts, screws, blades, plastic guards	Wear and tear or breakage of food processing machines	Physical injury, contamination with lubricants or oils
Construction Materials	Dust, insulation fibers, nails, plaster	Nearby construction activities in food facilities	Respiratory irritation, ingestion of foreign debris
Personal Items	Jewelry, hair, false nails, bandages	Human error during handling, poor hygiene practices	Infection risk, consumer distrust, brand damage
Pest Parts	Insect fragments, rodent hairs, eggshells	Contaminated raw materials, poor storage, and sanitation	Disease transmission, allergic reactions
Microplastics	Synthetic fibers, fragmented plastic particles	Water, seafood, processed food, and environmental contamination	Chronic inflammation, endocrine disruption
Natural Plant Material	Twigs, seeds, and husks are not intended for consumption	Inadequate sorting, threshing, or winnowing processes	Potential choking, reduced food quality
Soil and Sand	Grit, pebbles, and clay particles	Inadequate cleaning of fresh produce or root crops	Tooth damage, digestive irritation

Microplastics have emerged as pervasive contaminants within both aquatic and terrestrial ecosystems, with climate change acting as a significant amplifier of these pollutants. Increased temperatures and altered precipitation patterns promote the fragmentation of larger plastics into microplastics, facilitating their incorporation into food chains (Fenta and Kebede, 2019). Warming oceans, for instance, accelerate the breakdown of plastic debris, while extreme weather events, such as floods, can wash microplastics into water sources used for agriculture and drinking (Fenta and Kebede, 2019). Consequently, the ingestion of contaminated food poses significant health risks, as evidenced by the rise in foodborne illnesses linked to various contaminants, including microplastics (Parise, 2018). Furthermore, the interaction of microplastics with biological and chemical contaminants has been widely documented, influencing their toxicological profiles and potentially amplifying health risks to humans (Assunção et al., 2018; Parise, 2018).

Additionally, climatic stresses exacerbate the contamination of agricultural systems through soil erosion and the dispersal of pollutants. Extreme weather events, particularly those characterized by heavy rainfall or prolonged droughts, can lead to the mobilization of soil particles and pollutants into food production systems. Heavy rains can wash chemical residues, heavy metals, and pathogens from surface runoff into agricultural fields Parise, 2018; Lassa et al., 2018). This situation is exacerbated by the increase in surface flooding, which affects water quality by introducing harmful substances from urban areas and agricultural runoff into crops (Ogwu 2019; Fenta and Kebede, 2019). It is crucial for food safety policies and practices to consider these dynamics, particularly as the resilience of food systems is compromised by climate change. The impact of climate change on mycotoxin contamination further illustrates the connection between environmental stressors and food safety risks. As temperature and humidity levels rise, the proliferation of fungi such as Aspergillus and Fusarium increases, leading to enhanced mycotoxin production on food crops (Nji et al., 2022; Perrone et al., 2020). The relationship between climate variability and mycotoxin contamination is well-documented, with key studies indicating that alterations in climatic conditions can significantly enhance the risk of aflatoxin and other mycotoxin occurrences in food systems (Muharremi et al., 2022; D'Amato and Akdiş, 2020). Furthermore, these changes can lead to broader food security issues as infestations can diminish crop yields and threaten food availability, impacting overall public health (Nji et al., 2022; Lassa et al., 2018).

Additionally, extreme weather phenomena can trigger shifts in ecological dynamics, which can affect the fungal community composition within agricultural systems, increasing the potential for mycotoxins to thrive under new environmental scenarios (Leggieri et al., 2019; Nji et al., 2022; Perrone et al., 2020). The resilience of food systems is therefore contingent on climate adaptation strategies that mitigate the impact of these contaminants. For example, implementing practices that enhance soil health and cultivate climate-resilient crops can limit the spread and occurrence of harmful fungi, thus safeguarding food safety (Dasgupta and Robinson, 2022; Parise, 2018). Moreover, socioeconomic aspects play a crucial role in how communities respond to the consequences of climate-related food contamination. Disadvantaged communities often face heightened risks due to food insecurity and a lack of access to clean water, which is exacerbated by climate-induced disruptions to agricultural production (Zai et al., 2024; Lassa et al., 2018). Emergency policies, such as food reserve strategies highlighted by Lassa et al. (2018), can serve as critical buffers against the dual challenges of climate change and food safety threats like contamination. Considering these interconnections, health professionals must advocate for integrating climate change considerations into public health frameworks, particularly in the context of food safety (Eckelman and Sherman, 2018). Furthermore, sustainable food policy development will facilitate the creation of comprehensive and targeted responses to the multifaceted challenges posed by climate change (Macdiarmid and Whybrow, 2019; Connors et al., 2022; Scharadin et al., 2023).

6. CLIMATE-SENSITIVE FOOD SYSTEMS AND CONTAMINATION HOTSPOTS

The impact of climate change on food systems is profoundly felt across vulnerable supply chains, particularly in fresh produce, dairy, seafood, and grains, where contamination hotspots emerge as significant concerns. In regions characterized by high climate variability, such as Sub-Saharan Africa and South Asia, agricultural practices are heavily influenced by climatic factors, which in turn exacerbate food safety issues and food security challenges. Climate-sensitive food systems are further complicated by the interactions between climate change, supply chain dynamics, and contamination risks, necessitating a comprehensive examination of these elements.

One critical aspect of climate variability that impacts food systems is the heightened risk of food contamination, particularly in vulnerable supply chains. Studies indicate that Sub-Saharan Africa's food safety landscape is fraught with challenges due to inadequate food hygiene practices among food handlers, leading to a high incidence of foodborne diseases (Atalay et al., 2025). The lack of robust food safety regulations creates a fertile ground for contamination across various food categories, underscoring the intimate link between climate variability, agricultural practices, and food safety (Masipa, 2017). This phenomenon is not limited to Sub-Saharan Africa; similar patterns of deterioration in food safety can be observed in South Asia, where climatic fluctuations also disrupt agricultural output and market dynamics (Tesfay, 2024).

The role of climate-induced stressors, such as increased temperatures and erratic rainfall patterns, on food supply chains is evidenced in the observed rise in food-borne pathogens linked to climate change (El-Khishin et al., 2017). For example, the proliferation of mycotoxins, potentially lethal contaminants produced by mold on crops, can significantly increase during periods of climatic stress, resulting in health risks for both humans and livestock (Nleya et al., 2018). Thus, the mapping of contamination hotspots under future climate scenarios reveals a worrying trajectory for food safety in vulnerable agricultural sectors. Such research helps identify specific commodity risks, allowing stakeholders to prioritize interventions and enhance food security (Guo et al., 2020). Geopolitical crises and deteriorating economic conditions can exacerbate contamination hotspots in agricultural regions. The ongoing conflict in Ukraine has significantly disrupted global food supply chains, leading to price volatility that affects food safety and quality, particularly in resource-constrained environments in Africa (Jagtap et al., 2022). As transportation costs surge and the availability of seasonal labor diminishes, the pressures on agricultural production intensify, increasing the likelihood of food contamination incidents throughout the supply chain (Jagtap et al., 2022). Moreover, urbanization rates influence climate change and exacerbate these issues by transforming local food systems (Sugg et al., 2023). The transition towards supermarket-dominated food environments has mixed implications for dietary quality and safety. Supermarkets often implement higher food safety standards. Still, they can also introduce new forms of contamination risks through their supply chains, which may involve complex international sourcing practices (Wanyama et al., 2019; Osawaru and Ogwu, 2020). Consumers in rapidly urbanizing regions face the dual challenge of navigating food environments that may not prioritize safety, while also dealing with the rising costs of nutritious food as climate impacts become more evident (Izah et al., 2022; Obahiagbon and Ogwu, 2023; 2024).

Additionally, as food systems struggle to adapt to changing climatic conditions, both food waste and losses significantly contribute to the overall inefficiency of these systems (Kosoe and Ogwu, 2024; Nizzy and Ogwu, 2024). It is estimated that a considerable proportion of food produced is lost or wasted at various stages of the supply chain due to inadequate handling practices and poor storage facilities, particularly in Sub-Saharan Africa (Oelofse and Nahman, 2012). These losses not only have economic ramifications but also increase the risk of contamination, as perishable products become more susceptible to contamination. Initiatives aimed at mapping food loss and waste hotspots can prove invaluable in identifying where interventions can reduce the risks associated with food safety and enhance overall system resilience (Guo et al., 2020; Egboduku et al., 2024). Research has shown that comprehensive policy approaches are essential in addressing the multi-dimensional impacts of climate change on food security. Implementing targeted agricultural policies that promote climate-resilient practices can significantly enhance both food availability and safety (Masipa, 2017; Wudil et al., 2022). Moreover, investment in agricultural infrastructure, training in food hygiene practices, and extension services can help mitigate the risks associated with food contamination in vulnerable supply chains (Wudil et al., 2022; Akudugu and Ogwu, 2024).

In light of these complexities, the need for robust data collection and utilization in mapping contamination hotspots is paramount. Advanced tools, such as the Rapid Alert Supply Network Extractor (RASNEX), have been developed to track contamination incidents in food supply chains, generating insights that can guide stakeholders in improving their safety management systems, especially in the context of emerging climate threats (Lorenzen et al., 2021). By comprehensively analyzing incidences of contamination and correlating them with climatic data, policymakers can better strategize interventions that are both timely and effective. The intersection of climate variability, food systems, and contamination hotspots is an intricate web that requires targeted research and policy interventions. The vulnerabilities in essential food supply chains in regions such as Sub-Saharan Africa and South Asia necessitate a nuanced approach informed by empirical evidence and realistic assessments of the ongoing and future impacts of climate change on food safety and security. Clarity in these areas is crucial to ensure that food systems can not only survive but thrive amidst an increasingly uncertain climate landscape.

7. IMPLICATIONS OF CLIMATE CHANGE FOR FOOD SAFETY AND PUBLIC HEALTH

Climate change poses significant challenges to food safety and public health, particularly through altered environmental baselines that complicate risk assessment. These changes not only affect agricultural productivity but also introduce new contaminants along with novel exposure routes that were previously unrecognized. Dietary alterations resulting from contaminant-driven food loss create further nutrition-related public health issues. Modifications in temperature, precipitation, and weather extremes affect the presence and behavior of biological, chemical, and physical contaminants in food and water. These changes increase the risk of foodborne illnesses, toxic exposures, and nutritional deficiencies, particularly among vulnerable populations (Table 4).

The challenges presented by shifting environmental baselines are diverse and include the rise of new food safety hazards. According to Spiegel et al. (2012), climate change affects the types of crops cultivated and their geographical origins, which can increase food safety risks due to altered feed sources and production systems within the dairy production chain. Additionally, Battilani et al. (2016) note that climate change is anticipated to exacerbate aflatoxin contamination in maize in Europe, presenting significant concerns for food and feed safety. Their research highlights a correlation between climatic conditions and the prevalence of mycotoxins, establishing a pathway that emphasizes the need for proactive risk assessments for food safety as climate-related factors evolve.

As climate change progresses, the emergence of new pathogens and contaminants in the food supply becomes increasingly likely. Fenta and Kebede (2019) outline potential pathways through which climate-related factors might affect food safety, including altered precipitation and temperature patterns that can increase the frequency and intensity of foodborne disease outbreaks associated with pathogens. This assertion supports the findings of Duchenne-Moutien and Neetoo (2021), whose comprehensive review highlights the direct correlation between the impacts of global warming on weather variability and changes in food safety dynamics. Moreover, Mills et al. (2010) affirm that climate change will exacerbate the occurrence of vector-borne diseases, complicating both public health and food safety issues, especially in agricultural contexts.

The emergence of new contaminants in food systems is influenced by changing climatic variables, altering how existing pathogens interact with food products and their environment. Borgå et al.'s (2010) research demonstrates that climate-induced alterations in marine ecosystems can lead to increased bioaccumulation of organic contaminants, posing risks not only to aquatic life but also to human populations through food chains. Wu et al. (2011) discuss how climatic trends in the U.S. could influence the prevalence of mycotoxins in maize, underscoring the need for comprehensive monitoring of agricultural hazards in a changing climate. Collectively, these studies underscore a pressing need for the evolution of food safety protocols and contamination management systems to address these challenges.

Nutritionally, climate change will induce significant transformations due to contaminant-driven food loss and shifts in dietary patterns. Lake et al. (2012) articulate that as climate change reshapes food security, there is potential for declining nutritional quality amid rising food prices, which may further exacerbate health inequalities. This interplay suggests that populations may become increasingly reliant on lower-quality foods, potentially leading to adverse health outcomes, including obesity and malnutrition. These findings align with those of Chen et al. (2021), who emphasize that systemic failures in food production exacerbate nutritional deficits among vulnerable populations facing climate-related agricultural declines, thereby compounding existing public health challenges.

Moreover, the disconnect between climate awareness and food choices can impede the necessary dietary transitions to mitigate climate impacts. Korkala et al. (2014) indicate that individuals may express concern about climate change, but this awareness does not always translate into healthier nutritional practices that could bolster food sustainability. This highlights the critical need for educational initiatives that encourage the public to take actionable measures to address these pressing issues. Additionally, studies by Lee et al. (2022) support these findings by illustrating variances in attitudes toward climate-related dietary changes among different demographics, underscoring the complexity of creating effective interventions.

In the global context, climate change-induced agricultural losses underscore the urgency for adaptive strategies across regions. As Adeagbo et al. (2021) point out, understanding the factors that influence adaptation strategies is crucial for comprehending how communities respond to climate change challenges, emphasizing the interconnected nature of food security, health, and economic viability in the face of these threats. This highlights the importance of integrating food safety strategies with broader health and nutrition narratives, particularly in regions most severely affected by climate change. The cascading effects of climate change on food safety and public health necessitate a comprehensive, multi-faceted approach to risk assessment, monitoring, and intervention strategies. Understanding the dynamic interactions among environmental factors, food systems, and public health frameworks will be key to mitigating the negative consequences of climate change. Collaborative efforts across sectors will be essential in developing robust policies and innovative strategies that tackle both immediate and long-term health and safety challenges. The effects of climate change are pervasive and warrant urgent attention from policymakers, researchers, and public health officials to safeguard food systems and health outcomes in a warming world.

Table 4. Climate Change and the Emerging Risks to Food Safety and Human Health

Climate Change Impact	Food Safety Implications	Public Health Consequences
Rising Temperatures	Increased growth and survival of pathogens (e.g., *Salmonella, E. coli, Listeria*)	More frequent foodborne illnesses, especially in vulnerable populations
Altered Rainfall Patterns	Contamination of water sources and crops via runoff, leaching, or flooding	Exposure to heavy metals, pesticides, and nitrate contamination in drinking water and food
Extreme Weather Events	Damage to infrastructure, storage, and cold chains	Spoiled food, compromised nutrition, and increased food insecurity
Drought and Water Scarcity	Concentration of contaminants in reduced water volumes; overuse of wastewater	Accumulation of chemical residues in food; risk of long-term toxic exposure
Humidity and Moisture Increases	Elevated fungal growth and mycotoxin production in grains and nuts	Chronic exposure to aflatoxins is linked to liver cancer and immune suppression

continued on following page

Table 4. Continued

Climate Change Impact	Food Safety Implications	Public Health Consequences
Changing Ecosystems and Pest Range	Increased need for pesticide use; introduction of new pathogens and vectors	Residual pesticides in food; emergence of zoonotic and vector-borne diseases
Sea-Level Rise and Salinization	Saltwater intrusion into croplands and freshwater supplies	Increased dietary sodium, bioaccumulation of toxic elements like arsenic
Disruption of Traditional Practices	Loss of indigenous food safety and preservation knowledge	Reduced adaptive capacity and increased reliance on unsafe food practices
Reduced Nutritional Quality	Climate stress on crops alters micronutrient content	Malnutrition, especially in children and pregnant women
Increased Use of Chemical Inputs	Reliance on fertilizers and agrochemicals to compensate for yield losses	Residue-related health risks and antimicrobial resistance development

8. CLIMATE CHANGE MONITORING, MITIGATION, AND ADAPTATION STRATEGIES FOR ADDRESSING FOOD CONTAMINATION

Climate change poses significant challenges to food safety and public health, necessitating innovative monitoring, mitigation, and adaptation strategies to address emerging food contamination issues. Central to this discourse are the development of surveillance systems that integrate climate and contaminant data, the promotion of climate-resilient agricultural practices, particularly integrated pest management (IPM), and the emergence of robust policy frameworks in line with initiatives such as the Codex Alimentarius and the One Health approach.

8.1. Surveillance Systems Integrating Climate and Contaminant Data

The need for effective surveillance systems is exacerbated by the changing climate, which directly influences food safety by altering the dynamics of pathogens and contaminants. Comprehensive surveillance mechanisms document the intertwined effects of climate variables and food safety risks associated with biological and chemical pollutants (Zai et al., 2024). Moreover, Sawatzky et al. emphasize the necessity of integrating various data sources, including environmental and health data, to optimize decision-making related to public health interventions in the context of climate change (Sawatzky et al., 2018). The development of data-driven approaches for monitoring foodborne disease pathogens underlines the importance of real-time data for effective risk assessment and prevention strategies in a climate-affected landscape (Smith et al., 2019). Adequate public health surveillance must also encompass diverse populations and geographic regions, addressing variances in climate impacts on health outcomes (Moulton and Schramm, 2017). For example, the incorporation of big data analytics into surveillance systems can enhance understanding of trends in food contamination over time and space, promoting more tailored and proactive responses to emerging threats (Rodríguez-González et al., 2019). Hence, an integrated, multifaceted surveillance framework is crucial for effectively safeguarding food safety in a fluctuating climate.

8.2. Climate-Resilient Agricultural Practices and Integrated Pest Management (IPM)

The intersection of climate change and agricultural productivity necessitates the adoption of climate-resilient practices that can mitigate both environmental stressors and food contamination risks. Cheng et al. (2022) emphasize that adaptation strategies must be contextually relevant, as strategies that are effective in developed settings may not be applicable in developing contexts where resources are limited. Implementing climate-smart agricultural methods can enhance food security while also reducing the incidence of foodborne diseases. IPM is a critical component of these strategies, focusing on the sustainable management of pests through an integrated approach that combines biological, cultural, physical, and chemical tools (Cheng et al., 2022). Research indicates that effective IPM can significantly reduce the risk of chemical pesticide use and food contamination (Chandorkar, 2023). This is particularly relevant in horticulture, where extreme weather events can increase pest populations and pathogen loads in crops, thereby exacerbating food safety challenges. Recent studies suggest that climate change will likely shift the patterns of foodborne pathogens, making ongoing evaluations of these agricultural practices even more critical, especially in established food systems (Ziska et al., 2016). Thus, integrating climate resilience into agricultural practices through IPM must be prioritized as a key mitigation strategy for food contamination.

8.3. Policy Frameworks: Codex Alimentarius, One Health, and National Strategies

Within the realm of policy frameworks, the Codex Alimentarius provides a vital international standard for food safety that can help mitigate the risks associated with climate change (Ziska et al., 2016). Its development emphasizes multi-sectoral collaboration to enhance food security and public health through comprehensive food safety practices. The adoption of the One Health approach is equally important, as it advocates for a holistic view of human, animal, and environmental health, acknowledging the interconnectedness of these domains (Zinsstag et al., 2018). By addressing health at the intersection of these areas, policymakers can develop interventions that align with ecological balance and health improvement across populations. Additionally, the implementation of national strategies that incorporate global health guidelines and local climate data can facilitate effective responses to food safety risks. For instance, integrating real-time climate data into national food safety frameworks may enhance risk communication among stakeholders, leading to a more synchronized approach in managing food safety during climate-induced disasters (Harris-Glenville and Cloos, 2024). Emphasis on localized data collection and analysis will enable nations to tailor their responses to specific environmental challenges, fostering resilience against climate-induced food contamination.

8.4. Interventions in the Face of Extreme Weather Events

The impacts of extreme weather events on the food supply chain underscore the urgent need for intervention initiatives that can mitigate contamination risks stemming from flooding, droughts, and other climate-related extremes. Flooding can introduce pathogens into food systems, creating serious health risks for affected communities (Jiang et al., 2015). Mitigation practices must therefore be grounded in evidence-based protocols for addressing such contamination risks in real time. Public health agencies must enhance collaboration with meteorological departments to ensure timely access to climate information, which informs food safety protocols. Strategies such as preemptive food safety alerts during anticipated extreme weather events can foster greater community preparedness and resilience (Lake, 2017). Research indicates that systems must evolve to incorporate climate forecasting and food safety risk assessment to serve the best public health agendas (Moulton and Schramm, 2017).

8.5. Role of Community and Stakeholder Engagement

Effective climate change counter-strategies also rely heavily on community and stakeholder engagement. Promoting environmental health literacy can enhance public understanding of the relationships between climate change, food safety, and health outcomes (Marsili et al., 2021). Community-based initiatives that educate individuals about their local ecosystems and related food systems empower them to take proactive measures to safeguard their health and minimize contamination risks (Imarhiagbe et al., 2022; Kosoe et al., 2023). Moreover, implementing participatory approaches in health surveillance can improve the responsiveness of programs to community needs and experiences, significantly improving the effectiveness of interventions tailored to specific local challenges posed by climate change (Siya et al., 2021). These engagements can also ensure that cultural practices and knowledge inform food safety initiatives, advancing towards more sustainable and community-specific solutions. More research is needed to unpack the complexities of climate change impacts on food safety (Evivie et al., 2020). Efforts should be directed toward developing advanced modeling techniques to predict contamination risks associated with climate variability, particularly for under-researched pathogens (Gbashi et al., 2024; Ogwu et al., 2025a;b). Enhanced modeling can help policymakers preemptively address the health implications of these threats. There is a growing need for interdisciplinary collaboration among climatologists, epidemiologists, and food safety experts to generate comprehensive insights on how food systems can adapt to changing climates (Chandorkar, 2023). Cross-sector partnerships can drive innovation in developing effective food safety policies, thereby securing public health against the backdrop of a continuously changing environment.

9. CLIMATE CHANGE RESEARCH GAPS AND FUTURE DIRECTIONS FOR ADDRESSING FOOD CONTAMINATION

The increasing recognition of climate change as a significant driver of food safety challenges underscores the urgent need for interdisciplinary models and frameworks that adequately capture the complex interactions between environmental factors and food safety risks. Research indicates that climate-related phenomena, such as extreme weather events and shifts in temperature and precipitation patterns, are linked to rising levels of pathogens and toxins in the food supply, thereby amplifying global food safety challenges (Duchenne-Moutien and Neetoo, 2021; Ziska et al., 2016; Heidari et al., 2023). This situation prompts calls for innovative approaches to risk assessment and management. These interdisciplinary models should incorporate biological and environmental sciences alongside social sciences and local knowledge systems to comprehensively enhance understanding and mitigation of risks associated with food safety amid climate change (Schattman et al., 2018; Tirado et al., 2010).

The inclusion of indigenous and local knowledge systems is particularly crucial in addressing the complexities of food safety and contamination that arise from climate change. Indigenous communities possess a wealth of localized knowledge on environmental interactions, resource management, and food safety practices that have been developed over generations, which can significantly enrich scientific approaches to understanding food safety risks (Saltzman et al., 2023; Hoover et al., 2015). It is emphasized that situating local context within food safety narratives enhances risk communication and enables more effective community-level interventions (Godde et al., 2021; Schmitt and Sullivan, 2022). Furthermore, interdisciplinary collaboration fosters knowledge sharing, capturing the multifaceted and localized impacts of climate change on food systems, ultimately leading to greater resilience in local food safety management strategies (Siedlok and Hibbert, 2013; Subedi et al., 2018). Innovation in detection technologies constitutes another vital aspect in the fight against food contamination, which is exacerbated by climate change. Advanced detection methods, utilizing technology such as biosensors and molecular techniques, are essential for timely and accurate identification of foodborne pathogens and chemical contaminants that are influenced by fluctuating environmental conditions. Nielsen et al., 2021; Clayton et al., 2016). The development of predictive modeling tools that integrate climate data with food safety metrics could enhance monitoring of food safety hazards and support proactive measures in food supply chains (Fisher and Chen, 2011; Liu et al., 2021). Collaborative efforts within the scientific community can streamline innovations in detection technologies aimed at addressing the evolving challenges posed by climate change on food systems (Rong et al., 2024; Enova et al., 2024).

Risk communication, informed by integrated interdisciplinary models, emerges as a crucial component in addressing food safety concerns under a changing climate. Effective communication strategies that engage various stakeholders, including farmers, distributors, consumers, and policymakers, must prioritize clarity, contextual relevance, and cultural sensitivity (Chandorkar, 2023; Clayton et al., 2016). Establishing partnerships that unite scientists, practitioners, and community representatives will be integral to building trust and ensuring that food safety messages resonate with affected populations (Bohensky et al., 2011; Min and Hong, 2021). Additionally, utilizing technology to develop innovative communication platforms can effectively disseminate critical information on food safety threats and promote community engagement (Griffith et al., 2010; Zhang et al., 2023). Engaging in interdisciplinary research that transcends traditional academic boundaries is paramount for understanding the multifaceted impacts of climate change on food safety. By harnessing methodologies from diverse fields from environmental science to psychology researchers can obtain more nuanced insights into the interactions between climate phenomena and food safety risks, informing policy and practice (Rong et al., 2024; Liu et al., 2021). Additionally, these interdisciplinary approaches may prove essential for addressing the underlying social determinants of health, particularly for communities vulnerable to climate-induced food safety hazards (Schmitt and Sullivan, 2022; Nielsen et al., 2021). Integrating theories from psychology can also enrich risk communication strategies by incorporating behavioral insights into how individuals perceive and respond to food safety risks (Sidikjan et al., 2022; Min and Hong, 2021).

The challenges of food contamination under climate change are intricately linked to equity considerations. Vulnerable populations often face disproportionate risks stemming from environmental contaminants and disrupted food systems due to climate change (Godde et al., 2021; Hoover et al., 2015). Addressing disparities in access to safe food systems necessitates inclusive dialogues and the application of interdisciplinary frameworks that prioritize equity in developing food safety policies (Saltzman et al., 2023; Clayton et al., 2016). Implementing policies that specifically target at-risk communities while also incorporating local knowledge and experiences will be key to reducing the risk of foodborne illnesses and enhancing community resilience amid climate-induced changes (Heidari et al., 2023; Hoover et al., 2015). Acknowledging and addressing existing research gaps regarding climate change and food safety is crucial. The existing literature suggests that the complexity and unpredictability of climate impacts present significant challenges to current food safety paradigms (Tirado et al., 2010; Li et al., 2016). Thus, refining existing risk assessment methods that account for the uncertainties introduced by climate change is essential for developing robust food safety strategies (Rong et al., 2024; Subedi et al., 2018). Future research in this area should not only strive to illuminate these complexities but also promote adaptable frameworks capable of evolving in response to changing environmental contexts (Duchenne-Moutien and Neetoo, 2021; Tirado et al., 2010; Siedlok and Hibbert, 2013).

10. CONCLUSION

Climate change is reshaping the global food safety landscape by altering the occurrence, intensity, and distribution of food contaminants across diverse ecosystems and supply chains. The dynamic interactions between rising temperatures, shifting precipitation patterns, increased humidity, and the frequency of extreme weather events directly influence the pathways and persistence of chemical, biological, and physical hazards in food systems. These changes not only exacerbate existing risks—such as heavy metal mobilization, mycotoxin proliferation, and microbial pathogen expansion—but also introduce novel contamination challenges with potentially serious implications for public health and nutrition.

This chapter highlights the urgent need for anticipatory, integrative strategies that align climate adaptation with food safety objectives. In particular, vulnerable regions with limited infrastructure and regulatory capacity face disproportionate risks, requiring targeted interventions and capacity-building efforts. Advancing interdisciplinary and transdisciplinary research is crucial for developing predictive models, early warning systems, and tailored mitigation measures that accurately reflect the complex realities of climate-contaminant interactions. Equally important is the establishment of robust, adaptive governance frameworks that bridge environmental, agricultural, and health sectors, guided by principles such as One Health and food system resilience. Promoting climate-smart agricultural practices, empowering local communities, and strengthening international cooperation will be essential to safeguarding food security and public well-being in a rapidly changing world. A collaborative, evidence-informed response remains vital for navigating the interconnected challenges of climate change and food contamination.

REFERENCES

Addisu, S., & Wajera, A. (2024). Climate change and variability impacts on food security of smallholder farmers in Assosa district, Ethiopia. DOI: 10.21203/rs.3.rs-4187640/v1

Adeagbo, O., Ojo, T., & Adetoro, A. (2021). Understanding the determinants of climate change adaptation strategies among smallholder maize farmers in southwest Nigeria. *Heliyon, 7*(2), e06231. DOI: 10.1016/j.heliyon.2021.e06231

Agoh, E. C., Ogwu, M. C., Chukwuemeka, O. S., & Ekeledo, P. I. (2024). Environmental and human health effects of cassava processing and processing waste. In *Plant Biology, Sustainability and Climate change, Sustainable Cassava.* Academic Press. DOI: 10.1016/B978-0-443-21747-0.00001-1

Akil, L. (2021). Trends of foodborne diseases in Mississippi: Association with racial and economic disparities. *Diseases (Basel, Switzerland), 9*(4), 83. DOI: 10.3390/diseases9040083

Akil, L., & Ahmad, H. (2019). Quantitative risk assessment model of human salmonellosis resulting from consumption of broiler chicken. *Diseases (Basel, Switzerland), 7*(1), 19. DOI: 10.3390/diseases7010019

Akil, L., & Ahmad, H. (2023). Socioeconomic Impacts of the COVID-19 Pandemic on Foodborne Illnesses in the United States. *European Journal of Environment and Public Health, 7*(2), em0128. Advance online publication. DOI: 10.29333/ejeph/12585

Akil, L., Ahmad, H., & Reddy, R. (2014). Effects of Climate Change on Salmonella Infections. *Foodborne Pathogens and Disease, 11*(12), 974–980. DOI: 10.1089/fpd.2014.1802

Assunção, R., Martins, C., Viegas, S., Viegas, C., Jakobsen, L., Pires, S., & Alvito, P. (2018). Climate change and the health impact of aflatoxins exposure in Portugal – an overview. *Food Additives & Contaminants. Part A, Chemistry, Analysis, Control, Exposure & Risk Assessment, 35*(8), 1610–1621. DOI: 10.1080/19440049.2018.1447691

Atalay, Y., Gebeyehu, N., & Gelaw, K. (2025). Food hygiene practice and associated factors among food handlers working in food establishments in sub-Saharan Africa: A systematic review and meta-analysis. *Primary Health Care Research and Development, 26*, e13. Advance online publication. DOI: 10.1017/S146342362500009X

Baldos, U., & Hertel, T. (2014). Global food security in 2050: The role of agricultural productivity and climate change. *The Australian Journal of Agricultural and Resource Economics*, *58*(4), 554–570. DOI: 10.1111/1467-8489.12048

Battilani, P., Toscano, P., Fels-Klerx, H., Moretti, A., Leggieri, M., Brera, C., & Robinson, T. (2016). Aflatoxin B1 contamination in maize in Europe increases due to climate change. *Scientific Reports*, *6*(1), 24328. Advance online publication. DOI: 10.1038/srep24328

Beard, C., Eisen, R., Barker, C., Garofalo, J., Hahn, M., Hayden, M., … Schramm, P. (2016). Ch. 5: Vectorborne Diseases. The Impacts of Climate Change on Human Health in the United States: A Scientific Assessment. DOI: 10.7930/J0765C7V

Bintsis, T. (2017). Foodborne pathogens. *AIMS Microbiology*, *3*(3), 529–563. DOI: 10.3934/microbiol.2017.3.529

Bohensky, E., Butler, J., & Mitchell, D. (2011). Scenarios for knowledge integration: Exploring ecotourism futures in Milne Bay, Papua New Guinea. *Journal of Marine Biology*, *2011*, 1–11. DOI: 10.1155/2011/504651

Borgå, K., Saloranta, T., & Ruus, A. (2010). Simulating climate change-induced alterations in bioaccumulation of organic contaminants in an Arctic marine food web. *Environmental Toxicology and Chemistry*, *29*(6), 1349–1357. DOI: 10.1002/etc.159

Chandorkar, S. (2023). Climate change and food safety. DOI: 10.57772/intechopen.112575

Chen, M., Haq, S., Ahmed, K., Hussain, A., & Ahmed, M. (2021). The Link Between Climate Change, Food Security, and Fertility: The Case of Bangladesh. *PLoS One*, *16*(10), e0258196. DOI: 10.1371/journal.pone.0258196

Cheng, M., McCarl, B., & Fei, C. (2022). Climate change and livestock production: A literature review. *Atmosphere (Basel)*, *13*(1), 140. DOI: 10.3390/atmos13010140

Cissé, G. (2019). Food-borne and water-borne diseases under climate change in low- and middle-income countries: Further efforts are needed to reduce environmental health exposure risks. *Acta Tropica*, *194*, 181–188. DOI: 10.1016/j.actatropica.2019.03.012

Clayton, S., Devine-Wright, P., Swim, J., Bonnes, M., Steg, L., Whitmarsh, L., & Carrico, A. (2016). Expanding the role for psychology in addressing environmental challenges. *The American Psychologist*, *71*(3), 199–215. DOI: 10.1037/a0039482

Connolly-Boutin, L., & Smit, B. (2015). Climate Change, Food Security, and Livelihoods in Sub-Saharan Africa. *Regional Environmental Change, 16*(2), 385–399. DOI: 10.1007/s10113-015-0761-x

Connors, J., Safayet, M., Rosenheim, N., & Watson, M. (2022). Assessing changes in food pantry access after extreme events. *Agriculture and Human Values, 40*(2), 619–634. DOI: 10.1007/s10460-022-10373-8

D'Amato, G., & Akdiş, C. (2020). Global warming, climate change, air pollution, and allergies. *Allergy, 75*(9), 2158–2160. DOI: 10.1111/all.14527

Dasgupta, S., & Robinson, E. (2022). Attributing changes in food insecurity to a changing climate. *Scientific Reports, 12*(1), 4709. Advance online publication. DOI: 10.1038/s41598-022-08696-x

Duchenne-Moutien, R., & Neetoo, H. (2021). Climate change and emerging food safety issues: A review. *Journal of Food Protection, 84*(11), 1884–1897. DOI: 10.4315/JFP-21-141

Ebi, K. (2011). Climate change and health risks: Assessing and responding to them through 'adaptive management.'. *Health Affairs (Project Hope), 30*(5), 924–930. DOI: 10.1377/hlthaff.2011.0071

Eckelman, M., & Sherman, J. (2018). Estimated global disease burden from U.S. health care sector greenhouse gas emissions. *American Journal of Public Health, 108*(S2), S120–S122. DOI: 10.2105/AJPH.2017.303846

Egboduku, W. O., Ogwu, M. C., Egboduku, T., Enujeke, E. C., Edokpaiawe, S., & Golohor, O. M. (2024). Sustainable cassava processing techniques to eliminate cyanogenic glycosides. In *Plant Biology, Sustainability and Climate Change, Sustainable Cassava.* Academic Press. DOI: 10.1016/B978-0-443-21747-0.00007-2

El-Khishin, M., Gooneratne, R., and Hussain, M. (2017). Microbial Safety of Foods in the Supply Chain and Food Security Advanced in Food Technology and Nutritional Sciences - Open Journal, 3(1), 22-32. DOI: 10.17140/AFTNSOJ-3-141

Enova, A., Singh, R., Aga, D., Mariano, S., Villanoy, C., & Jaraula, C. (2024). Modeling pharmaceutical contaminant transport in an urban coastal estuarine system: The case of Manila Bay, Philippines. *ACS ES&T Water, 5*(1), 156–167. DOI: 10.1021/acsestwater.4c00736

Evivie, S. E., Ogwu, M. C., Ebabhamiegbebho, P. A., Abel, E. S., Imaren, J. O., & Igene, J. O. (2020). Packaging and the Nigerian food industry: challenges and opportunities. In Ogunlade, C. A., Adeleke, K. M., & Oladejo, M. T. (Eds.), *Food Technology and Culture in Africa* (pp. 28–99). Reamsworth Publishing.

Falloon, P., Bebber, D., Dalin, C., Ingram, J., Mitchell, D., Hartley, T., Johnes, P. J., Newbold, T., Challinor, A. J., Finch, J., Galdos, M. V., Petty, C., Cornforth, R., Bhunnoo, R., Pope, E., Enow, A., Borrion, A., Waterson, A., MacNeill, K., & Houldcroft, A. (2022). What do changing weather and climate shocks and stresses mean for the UK food system? *Environmental Research Letters*, *17*(5), 051001. DOI: 10.1088/1748-9326/ac68f9

Favas, C., Cresta, C., Whelan, E., Smith, K., Manger, M., Chandrasenage, D., Singhkumarwong, A., Kawasaki, J., Moreno, S., & Goudet, S. (2024). Exploring food system resilience to the global polycrisis in six asian countries. *Frontiers in Nutrition*, *11*, 1347186. Advance online publication. DOI: 10.3389/fnut.2024.1347186

Fels-Klerx, H., Liu, C., & Battilani, P. (2016). Modeling climate change impacts on mycotoxin contamination. *World Mycotoxin Journal*, *9*(5), 717–726. DOI: 10.3920/WMJ2016.2066

Fenta, L., & Kebede, A. (2019). *Effect of climate change on food- and waterborne disease outbreaks: a mini review*. FSQM. DOI: 10.7176/FSQM/88-02

Fisher, N., & Chen, C. (2011). Interdisciplinary approaches for addressing marine contamination issues. *Environmental Conservation*, *38*(2), 187–198. DOI: 10.1017/S0376892911000269

Gagnon, P., Sheedy, C., Rousseau, A., Bourgeois, G., & Chouinard, G. (2015). Integrated assessment of climate change impact on surface runoff contamination by pesticides. *Integrated Environmental Assessment and Management*, *12*(3), 559–571. DOI: 10.1002/ieam.1706

Gao, J., Sun, Y., Lu, Y., & Li, L. (2014). Impact of ambient humidity on child health: A systematic review. *PLoS One*, *9*(12), e112508. DOI: 10.1371/journal.pone.0112508

Gbashi, S., Adelusi, O., & Njobeh, P. (2024). Insights from modeling sixteen years of climatic and fumonisin patterns in maize in South Africa. *Scientific Reports*, *14*(1), 11643. Advance online publication. DOI: 10.1038/s41598-024-60904-y

Godde, C., Mason-D'Croz, D., Mayberry, D., Thornton, P., & Herrero, M. (2021). Impacts of climate change on the livestock food supply chain: A review of the evidence. *Global Food Security*, *28*, 100488. DOI: 10.1016/j.gfs.2020.100488

Griffith, C., Livesey, K., & Clayton, D. (2010). The assessment of food safety culture. *British Food Journal*, *112*(4), 439–456. DOI: 10.1108/00070701011034448

Guo, X., Broeze, J., Groot, J., Axmann, H., & Vollebregt, M. (2020). A worldwide hotspot analysis on food loss and waste, associated greenhouse gas emissions, and protein losses. *Sustainability (Basel)*, *12*(18), 7488. DOI: 10.3390/su12187488

Hamaideh, S., Olaimat, A., Al-Holy, M., Ababneh, A., Shahbaz, H., Abughoush, M., Al-Nabulsi, A., Osaili, T., Ayyash, M., & Holley, R. (2024). The influence of technological shifts in the food chain on the emergence of foodborne pathogens: An overview. *Applied Microbiology (Basel, Switzerland), 4*(2), 594–606. DOI: 10.3390/applmicrobiol4020041

Harris-Glenville, F., & Cloos, P. (2024). "I think they should give primary health care a little more priority". The primary health care in Caribbean SIDS: What can be said about adaptation to the changing climate? The case of Dominica—a qualitative study. *BMC Primary Care, 25*(1), 65. Advance online publication. DOI: 10.1186/s12875-024-02311-w

Hasegawa, T., Fujimori, S., Havlík, P., Valin, H., Bodirsky, B., Doelman, J., Fellmann, T., Kyle, P., Koopman, J. F. L., Lotze-Campen, H., Mason-D'Croz, D., Ochi, Y., Pérez Domínguez, I., Stehfest, E., Sulser, T. B., Tabeau, A., Takahashi, K., Takakura, J., van Meijl, H., & Witzke, P. (2018). Risk of increased food insecurity under a stringent global climate change mitigation policy. *Nature Climate Change, 8*(8), 699–703. DOI: 10.1038/s41558-018-0230-x

Heidari, T., Sadighara, P., Oskoei, V., & Zeinali, T. (2023). Significance of climate change in food safety and hygiene. *Jundishapur J Health Sci, 15*(4). Advance online publication. DOI: 10.5812/jjhs-141415

Hoover, E., Renauld, M., Edelstein, M., & Brown, P. (2015). Social science collaboration with environmental health. *Environmental Health Perspectives, 123*(11), 1100–1106. DOI: 10.1289/ehp.1409283

Huang, X., Swain, D., & Hall, A. (2020). Future precipitation increase from very high-resolution ensemble downscaling of extreme atmospheric river storms in California. *Science Advances, 6*(29), eaba1323. Advance online publication. DOI: 10.1126/sciadv.aba1323

Imarhiagbe, O., & Ogwu, M. C. (2022). Sacred Groves in the Global South: A Panacea for Sustainable Biodiversity Conservation. In *Biodiversity in Africa: Potentials, Threats and Conservation.* Springer. DOI: 10.1007/978-981-19-3326-4_20

Imarhiagbe, O., Ogwu, M. C., Ikponmwosa, B. O., Mukah, F. E., Akemu, S. E., & Ohiaba, E. E. (2024). Threats to cassava cultivation, production, and processing: Global status and sustainable management strategies. In *Plant Biology, Sustainability and Climate Change, Sustainable Cassava.* Academic Press. DOI: 10.1016/B978-0-443-21747-0.00016-3

Imarhiagbe, O., Onyeukwu, I. I., Egboduku, W., Mukah, F. E., & Ogwu, M. C. (2022). Forest Conservation Strategies in Africa: Historical Perspectives, Status and Sustainable Avenues for Progress. In *Biodiversity in Africa: Potentials, Threats and Conservation*. Springer. DOI: 10.1007/978-981-19-3326-4_21

Izah, S. C., Aigberua, A. O., & Ogwu, M. C. (2022). Trace element composition of *Gallus gallus domesticus* eggs and health risks associated with their consumption in Port Harcourt, Nigeria. *Journal of Food Safety and Hygiene*, *8*(3), 202–222.

Jagtap, S., Trollman, H., Trollman, F., Garcia-Garcia, G., Parra-López, C., Duong, L., Martindale, W., Munekata, P. E. S., Lorenzo, J. M., Hdaifeh, A., Hassoun, A., Salonitis, K., & Afy-Shararah, M. (2022). The Russia-Ukraine Conflict: Its Implications for the Global Food Supply Chains. *Foods*, *11*(14), 2098. DOI: 10.3390/foods11142098

Jamieson, M., Trowbridge, A., Raffa, K., & Lindroth, R. (2012). Consequences of climate warming and altered precipitation patterns for plant-insect and multitrophic interactions. *Plant Physiology*, *160*(4), 1719–1727. DOI: 10.1104/pp.112.206524

Janevska, D., Gospavić, R., Pacholewicz, E., & Popov, V. (2010). Application of an HACCP–QMRA approach for managing the impact of climate change on food quality and safety. *Food Research International*, *43*(7), 1915–1924. DOI: 10.1016/j.foodres.2010.01.025

Jiang, C., Shaw, K., Upperman, C., Blythe, D., Mitchell, C., Murtugudde, R., Sapkota, A. R., & Sapkota, A. (2015). Climate change, extreme events, and the increased risk of salmonellosis in Maryland, USA: Evidence for coastal vulnerability. *Environment International*, *83*, 58–62. DOI: 10.1016/j.envint.2015.06.006

Kannan, N. (2023). An analysis of the climate change effects on pesticide vapor drift from ground-based pesticide applications to cotton. *Scientific Reports*, *13*(1), 9740. Advance online publication. DOI: 10.1038/s41598-023-36941-4

Kanu, W., & Onyekwere, I. (2024). Perceptions of climate change-related disasters and impact on household food security in rural farm households in Imo State, Nigeria. *Journal of Agriculture and Food Sciences*, *21*(2), 121–134. DOI: 10.4314/jafs.v21i2.9

Kattwinkel, M., Kühne, J., Foit, K., & Liess, M. (2011). Climate change, agricultural insecticide exposure, and risk for freshwater communities. *Ecological Applications*, *21*(6), 2068–2081. DOI: 10.1890/10-1993.1

Kim, G., Cha, D., Park, C., Lee, G., Jin, C., Lee, D., Suh, M.-S., Ahn, J.-B., Min, S.-K., Hong, S.-Y., & Kang, H. (2018). Future Changes in Extreme Precipitation Indices over Korea. *International Journal of Climatology, 38*(S1). Advance online publication. DOI: 10.1002/joc.5414

Korkala, E., Hugg, T., & Jaakkola, J. (2014). Awareness of climate change and the dietary choices of young adults in Finland: A population-based cross-sectional study. *PLoS One, 9*(5), e97480. DOI: 10.1371/journal.pone.0097480

Kosoe, E. A., Achana, G. T. W., & Ogwu, M. C. (2023). Meta-Evaluation of the One Health Implication on Food Systems of Agrochemical Use. In *One Health Implications of Agrochemicals and their Sustainable Alternatives*. Springer. DOI: 10.1007/978-981-99-3439-3_14

Kosoe, E. A., & Ogwu, M. C. (2024). Sustainable cassava processing: processes, techniques, wastes, and waste streams. In *Plant Biology, Sustainability and Climate Change, Sustainable Cassava.* Academic Press. DOI: 10.1016/B978-0-443-21747-0.00011-4

Kroeker, K., Micheli, F., Gambi, M., & Martz, T. (2011). Divergent ecosystem responses within a benthic marine community to ocean acidification. *Proceedings of the National Academy of Sciences of the United States of America, 108*(35), 14515–14520. DOI: 10.1073/pnas.1107789108

Lake, I. (2017). Food-borne disease and climate change in the United Kingdom. *Environmental Health : A Global Access Science Source, 16*(S1), 117. Advance online publication. DOI: 10.1186/s12940-017-0327-0

Lake, I., Hooper, L., Abdelhamid, A., Bentham, G., Boxall, A., Draper, A., Fairweather-Tait, S., Hulme, M., Hunter, P. R., Nichols, G., & Waldron, K. (2012). Climate Change and Food Security: Health Impacts in Developed Countries. *Environmental Health Perspectives, 120*(11), 1520–1526. DOI: 10.1289/ehp.1104424

Lassa, J., Teng, P., Caballero-Anthony, M., & Shrestha, M. (2018). Revisiting emergency food reserve policy and practice under disaster and extreme climate events. *International Journal of Disaster Risk Science, 10*(1), 1–13. DOI: 10.1007/s13753-018-0200-y

Lee, L., Tung, S., Gan, W., & Kaur, S. (2022). Knowledge, attitude, and perception of climate change and dietary choices in a predominantly Chinese university student population in Klang Valley. *Malaysian Applied Biology, 51*(3), 37–45. DOI: 10.55230/mabjournal.v51i3.2275

Leggieri, M., Giorni, P., Pietri, A., & Battilani, P. (2019). Aspergillus flavus and Fusarium verticillioides interaction: Modeling the impact on mycotoxin production. *Frontiers in Microbiology, 10*, 2653. Advance online publication. DOI: 10.3389/fmicb.2019.02653

Leite, Y., Duque, T., Santos, J., & Santos, E. (2025). Potential residual pesticide consumption: A stratified analysis of Brazilian families. *Journal of Xenobiotics, 15*(2), 37. DOI: 10.3390/jox15020037

Li, C., Zwiers, F., Zhang, X., Chen, G., Lu, J., Li, G., Norris, J., Tan, Y., Sun, Y., & Liu, M. (2019). Larger increases in more extreme local precipitation events as climate warms. *Geophysical Research Letters, 46*(12), 6885–6891. DOI: 10.1029/2019GL082908

Li, F., Liu, J., Ren, J., Cao, X., & Zhu, Y. (2016). Numerical investigation of airborne contaminant transport under different vortex structures in the aircraft cabin. *International Journal of Heat and Mass Transfer, 96*, 287–295. DOI: 10.1016/j.ijheatmasstransfer.2016.01.004

Liu, C., Hofstra, N., & Franz, E. (2016). Impacts of climate and management variables on the contamination of preharvest leafy greens with Escherichia coli. *Journal of Food Protection, 79*(1), 17–29. DOI: 10.4315/0362-028X.JFP-15-255

Liu, X., Bu, Y., Li, M., & Li, J. (2021). Is interdisciplinary collaboration research more disruptive than monodisciplinary research? *Proceedings of the Association for Information Science and Technology, 58*(1), 264–272. DOI: 10.1002/pra2.454

Loecke, T., Burgin, A., Riveros-Iregui, D., Ward, A., Thomas, S., Davis, C., & Clair, M. (2017). Weather whiplash in agricultural regions drives deterioration of water quality. *Biogeochemistry, 133*(1), 7–15. DOI: 10.1007/s10533-017-0315-z

Lorenzen, M., Weiser, A., Pieper, R., Lahrssen-Wiederholt, M., & Numata, J. (2021). Introducing the Rapid Alert Supply Network Extractor (RASNEX) tool to mine supply chain information from food and feed contamination notifications in Europe. *PLoS One, 16*(7), e0254301. DOI: 10.1371/journal.pone.0254301

Macdiarmid, J., & Whybrow, S. (2019). Nutrition from a climate change perspective. *The Proceedings of the Nutrition Society, 78*(3), 380–387. DOI: 10.1017/S0029665118002896

Mafe, A. N., & Büsselberg, D. (2024). Mycotoxins in Food: Cancer Risks and Strategies for Control. *Foods, 13*(21), 3502. DOI: 10.3390/foods13213502

Malhi, G., Kaur, M., & Kaushik, P. (2021). Impact of climate change on agriculture and its mitigation strategies: A review. *Sustainability (Basel), 13*(3), 1318. DOI: 10.3390/su13031318

Marsili, D., Pasetto, R., Iavarone, I., Fazzo, L., Zona, A., & Comba, P. (2021). Fostering environmental health literacy in contaminated sites: National and local experience in Italy from a public health and equity perspective. *Frontiers in Communication, 6*, 697547. Advance online publication. DOI: 10.3389/fcomm.2021.697547

Martin, P., Wen, Y., Woods, A., Fayida, T., & Hobbs, S. (2024). Exploring pesticide transport, groundwater, and environmental justice in a changing climate: A community-engaged research approach. *Environmental Research Letters, 19*(9), 094009. DOI: 10.1088/1748-9326/ad67f3

Martínez-Megías, C., Mentzel, S., Fuentes-Edfuf, Y., Moe, S., & Rico, A. (2022). A Bayesian network approach to assess the influence of climate change and pesticide use practices on the ecological risks of pesticides in a protected Mediterranean wetland. https://doi.org/DOI: 10.1101/2022.12.15.520567

Masipa, T. (2017). The Impact of Climate Change on Food Security in South Africa: Current Realities and Challenges Ahead. *Jàambá, 9*(1). Advance online publication. DOI: 10.4102/jamba.v9i1.411

Medina, Á., Akbar, A., Baazeem, A., Rodríguez, A., & Magan, N. (2017). Climate change, food security, and mycotoxins: Do we know enough? *Fungal Biology Reviews, 31*(3), 143–154. DOI: 10.1016/j.fbr.2017.04.002

Mercer, J., Kelman, I., Alfthan, B., & Kurvits, T. (2012). Ecosystem-based adaptation to climate change in Caribbean Small Island Developing States: Integrating Local and External Knowledge. *Sustainability (Basel), 4*(8), 1908–1932. DOI: 10.3390/su4081908

Mills, J., Gage, K., & Khan, A. (2010). Potential influence of climate change on vector-borne and zoonotic diseases: A review and proposed research plan. *Environmental Health Perspectives, 118*(11), 1507–1514. DOI: 10.1289/ehp.0901389

Min, K., & Hong, W. (2021). The effect of food sustainability and the food safety climate on the job stress, job satisfaction, and job commitment of kitchen staff. *Sustainability (Basel), 13*(12), 6813. DOI: 10.3390/su13126813

Moulton, A., & Schramm, P. (2017). Climate Change and Public Health Surveillance: Toward a Comprehensive Strategy. *Journal of Public Health Management and Practice, 23*(6), 618–626. DOI: 10.1097/PHH.0000000000000550

Muharremi, H., Raka, L., Spahiu, J., Tershnjaku, I., & Topi, D. (2022). Investigation of aflatoxin M1 in baby milk and aflatoxin B1 in infant cereals marketed in Kosovo. *Journal of Food Processing and Preservation, 46*(6). Advance online publication. DOI: 10.1111/jfpp.16285

Narayanan, A., Peter, B., & Keellings, D. (2024). A Climate Extremes Resilience Index for the Conterminous United States. *Weather, Climate, and Society, 16*(1), 87–103. DOI: 10.1175/WCAS-D-23-0008.1

Nielsen, K., Clayton, S., Stern, P., Dietz, T., Capstick, S., & Whitmarsh, L. (2021). How Psychology Can Help Limit Climate Change. *The American Psychologist, 76*(1), 130–144. DOI: 10.1037/amp0000624

Nizzy, A. M., & Ogwu, M. C. (2024). Valorization of cassava processing by-products into biofuel for a sustainable environment. In *Plant Biology, Sustainability and Climate Change, Sustainable Cassava.* Academic Press. DOI: 10.1016/B978-0-443-21747-0.00010-2

Nji, Q., Babalola, O., & Mwanza, M. (2022). Aflatoxins in maize: Can their occurrence be effectively managed in Africa in the face of climate change and food insecurity? *Toxins, 14*(8), 574. DOI: 10.3390/toxins14080574

Nleya, N., Adetunji, M., & Mwanza, M. (2018). Current Status of Mycotoxin Contamination of Food Commodities in Zimbabwe. *Toxins, 10*(5), 89. DOI: 10.3390/toxins10050089

Obahiagbon, E. G., & Ogwu, M. C. (2023). Consumer Perception and Demand for Sustainable Herbal Medicine Products and Market. In *Herbal Medicine Phytochemistry.* Springer. DOI: 10.1007/978-3-031-21973-3_65-1

Obahiagbon, E. G., & Ogwu, M. C. (2024). Organic Food Preservatives: The Shift Towards Natural Alternatives and Sustainability in the Global South's Markets. In Ogwu, M. C., Izah, S. C., & Ntuli, N. R. (Eds.), *Food Safety and Quality in the Global South* (pp. 299–329). Springer. DOI: 10.1007/978-981-97-2428-4_10

Oelofse, S., & Nahman, A. (2012). Estimating the magnitude of food waste generated in South Africa. *Waste Management & Research, 31*(1), 80–86. DOI: 10.1177/0734242X12457117

Ogwu, M. C. (2023). Local Food Crops in Africa: Sustainable Utilization, Threats, and Traditional Storage Strategies. In *Sustainable Utilization and Conservation of Africa's Biological Resources and Environment.* Springer. DOI: 10.1007/978-981-19-6974-4_13

Ogwu, M. C., Iyiola, A. O., & Izah, S. C. (2023). Overview of African Biological Resources and Environment. In *Sustainable Utilization and Conservation of Africa's Biological Resources and Environment.* Springer. DOI: 10.1007/978-981-19-6974-4_1

Ogwu, M. C., Izah, S. C., & Iyiola, A. O. (2022). An Overview of the Potentials, Threats and Conservation of Biodiversity in Africa. In *Biodiversity in Africa: Potentials, Threats and Conservation.* Springer. DOI: 10.1007/978-981-19-3326-4_1

Ogwu, M. C., & Kulkarni, S. (2024). Sustainable cassava: An overview. In *Plant Biology, Sustainability and Climate Change, Sustainable Cassava.* Academic Press. DOI: 10.1016/B978-0-443-21747-0.00024-2

Ogwu, M. C., & Osawaru, M. E. (2022). Traditional Methods of Plant Conservation for Sustainable Utilization and Development. In *Biodiversity in Africa: Potentials, Threats and Conservation.* Springer. DOI: 10.1007/978-981-19-3326-4_17

Ogwu, M. C., & Osawaru, M. E. (2023). Disease Outbreaks in Ex-Situ Plant Conservation and Potential Management Strategies. In *Sustainable Utilization and Conservation of Africa's Biological Resources and Environment.* Springer. DOI: 10.1007/978-981-19-6974-4_18

Ogwu, M. C., Osawaru, M. E., & Owie, M. O. (2023). Effects of Storage at Room Temperature on the Food Components of Three Cocoyam Species (*Colocasia esculenta, Xanthosoma atrovirens, and X. sagittifolium*). *Food Studies, 13*(2), 59–83. DOI: 10.18848/2160-1933/CGP/v13i02/59-83

Oldenkamp, R., Benestad, R., Hader, J., Mentzel, S., Nathan, R., Madsen, A., & Moe, S. (2023). Incorporating climate projections in the environmental risk assessment of pesticides in aquatic ecosystems. *Integrated Environmental Assessment and Management, 20*(2), 384–400. DOI: 10.1002/ieam.4849

Osawaru, M. E., & Ogwu, M. C. (2020). Survey of plant and plant products in local markets within Benin City and environs. In Filho, L. W., Ogugu, N., Ayal, D., Adelake, L., & da Silva, I. (Eds.), *African Handbook of Climate Change Adaptation* (pp. 1–24). DOI: 10.1007/978-3-030-42091-8_159-1

Parise, I. (2018). A brief review of global climate change and the public health consequences. *Australian Journal of General Practice, 47*(7), 451–456. DOI: 10.31128/AJGP-11-17-4412

Perrone, G., Ferrara, M., Medina, Á., Pascale, M., & Magan, N. (2020). Toxigenic fungi and mycotoxins in a climate change scenario: Ecology, genomics, distribution, prediction, and prevention of the risk. *Microorganisms*, *8*(10), 1496. DOI: 10.3390/microorganisms8101496

Pickson, R., & Boateng, E. (2021). Climate Change: A Friend or Foe to Food Security in Africa? *Environment, Development and Sustainability*, *24*(3), 4387–4412. DOI: 10.1007/s10668-021-01621-8

Rincón, A., Santamaría-Pérez, B., Rabasa, S., Coince, A., Marçais, B., & Buée, M. (2015). Compartmentalized and contrasted response of ectomycorrhizal and soil fungal communities of Scots pine forests along elevation gradients in France and Spain. *Environmental Microbiology*, *17*(8), 3009–3024. DOI: 10.1111/1462-2920.12894

Rodríguez-González, A., Zanin, M., & Menasalvas, E. (2019). Public Health and Epidemiology Informatics: Can Artificial Intelligence Help Address Future Global Challenges? An overview of antimicrobial resistance and impact of climate change in disease epidemiology. *Yearbook of Medical Informatics*, *28*(01), 224–231. DOI: 10.1055/s-0039-1677910

Rong, G., Ma, F., Qian, Y., Zhang, Z., Chen, S., & Sun, Y. (2024). Exploring the critical years for interdisciplinary citations. *Journal of the Association for Information Science and Technology*, *75*(12), 1398–1412. DOI: 10.1002/asi.24940

Sabola, G. (2024). Climate Change Impacts on Agricultural Trade and Food Security in Emerging Economies: A Case Study of Southern Africa. *Discover Agriculture*, *2*(1), 12. Advance online publication. DOI: 10.1007/s44279-024-00026-1

Saltzman, L., Hansel, T., & Ferreira, R. (2023). Climate change and health equity: Expanding our reach using technology. *PLOS Climate*, *2*(7), e0000242. DOI: 10.1371/journal.pclm.0000242

Samut, H., Namli, Ş., Özdemir, F., Çömlekçioğlu, N., & Soyer, Y. (2023). Climate change and food safety: Temperature impact on the attachment of Escherichia coli pathogroups on cress leaf. *Journal of Food Safety*, *43*(5), e13059. Advance online publication. DOI: 10.1111/jfs.13059

Sawatzky, A., Cunsolo, A., Jones-Bitton, A., Middleton, J., & Harper, S. (2018). Responding to climate and environmental change impacts on human health via integrated surveillance in the circumpolar north: A systematic realist review. *International Journal of Environmental Research and Public Health*, *15*(12), 2706. DOI: 10.3390/ijerph15122706

Scharadin, B., Zanocco, C., & Chistolini, J. (2023). Food retail environments, extreme weather, and their overlap: Exploratory analysis and recommendations for U.S. food policy. *PLoS One*, *18*(11), e0289282. DOI: 10.1371/journal.pone.0289282

Schattman, R., Grubinger, V., McKaeg, L., & Nelson, K. (2018). Whole farm water use: a survey of vegetable producers in New England states. DOI: 10.32747/2018.6938606.ch

Schmitt, H., & Sullivan, D. (2022). Communities living with chronic environmental contamination: Leveraging interdisciplinarity to address environmental justice issues. *Translational Issues in Psychological Science*, *8*(4), 473–488. DOI: 10.1037/tps0000338

Schwiesow, M., Elde, N., & Hilbert, Z. (2024). Distinct routes to thermotolerance in the fungal pathogen Cryptococcus neoformans. https://doi.org/DOI: 10.1101/2024.04.08.588590

Semenza, J., Herbst, S., Rechenburg, A., Suk, J., Höser, C., Schreiber, C., & Kistemann, T. (2012). Climate Change Impact Assessment of Food- and Waterborne Diseases. *Critical Reviews in Environmental Science and Technology*, *42*(8), 857–890. DOI: 10.1080/10643389.2010.534706

Shih, Y., Chen, J., Chen, Y., Yang, P., Kuo, Y., Chen, T., & Hsu, B. (2021). Impact of heavy precipitation events on pathogen occurrence in estuarine areas of the Puzi River in Taiwan. *PLoS One*, *16*(8), e0256266. DOI: 10.1371/journal.pone.0256266

Sidikjan, N., Eziz, M., Xin-guo, L., & Wang, Y. (2022). Spatial distribution, contamination levels, and health risks of trace elements in topsoil along an urbanization gradient in Urumqi, China. *Sustainability (Basel)*, *14*(19), 12646. DOI: 10.3390/su141912646

Siedlok, F., & Hibbert, P. (2013). The organization of interdisciplinary research: Modes, drivers and barriers. *International Journal of Management Reviews*, *16*(2), 194–210. DOI: 10.1111/ijmr.12016

Siya, A., Mafigiri, R., Migisha, R., & Kading, R. (2021). Uganda mountain community health system perspectives and capacities towards emerging infectious disease surveillance... https://doi.org/DOI: 10.20944/preprints202105.0691.v1

Smith, B., & Fazil, A. (2019). How will climate change impact microbial foodborne disease in Canada? *Canada Communicable Disease Report = Relevé des Maladies Transmissibles au Canada*, *45*(4), 108–113. DOI: 10.14745/ccdr.v45i04a05

Smith, B., Meadows, S., Meyers, R., Parmley, E., & Fazil, A. (2019). Seasonality and Zoonotic Foodborne Pathogens in Canada: Relationships between Climate and Campylobacter, E. coli, and Salmonella in Meat Products. *Epidemiology and Infection, 147*, e190. Advance online publication. DOI: 10.1017/S0950268819000797

Spiegel, M., Fels-Klerx, H., & Marvin, H. (2012). *Effects of Climate Change on Food Safety Hazards in the Dairy Production Chain. Food Research International, 46*(1), 201–208. DOI: 10.1016/j.foodres.2011.12.011

Steffens, K., Larsbo, M., Moeys, J., Kjellström, E., Jarvis, N., & Lewan, E. (2014). Modelling pesticide leaching under climate change: Parameter vs. climate input uncertainty. *Hydrology and Earth System Sciences, 18*(2), 479–491. DOI: 10.5194/hess-18-479-2014

Subedi, J., Houston, J., & Sherman–Morris, K. (2018). Interdisciplinary research as an iterative process to build disaster systems knowledge. *Risk Analysis, 41*(7), 1072–1077. DOI: 10.1111/risa.13244

Subedi, J., Houston, J., & Sherman–Morris, K. (2018). Interdisciplinary research as an iterative process to build disaster systems knowledge. *Risk Analysis, 41*(7), 1072–1077. DOI: 10.1111/risa.13244

Sugg, M. M., Ryan, S., Spurlock, T., Duncan, S., Hege, A., Ogwu, M. C., Tyson, J., Shay, E., Renwick, K., Hendren, C., Roy, M., & Runkle, J. D. (2023). Climate Change and Health in Rural Mountain Environments: Summary of a Workshop on Knowledge Gaps, barriers, and Opportunities for Action. *GeoJournal.* DOI: 10.1007/s10708-023-10916-4

Tesfay, S. (2024). What are the impacts of climate change on sustainable food production, food demand, and population numbers in sub-Saharan Africa? A systematic review. Food journal, 2(2). DOI: 10.59411/sc2by231

Tirado, C., Clarke, R., Jaykus, L., McQuatters-Gollop, A., & Frank, J. (2010). Climate change and food safety: A review. *Food Research International, 43*(7), 1745–1765. DOI: 10.1016/j.foodres.2010.07.003

Tudi, M., Ruan, H., Wang, L., Lyu, J., Sadler, R., Connell, D., & Phung, D. (2021). Agriculture development, pesticide application and its impact on the environment. *International Journal of Environmental Research and Public Health, 18*(3), 1112. DOI: 10.3390/ijerph18031112

Verma, S., Singh, A., Pradhan, S., & Kushuwaha, M. (2024). Impact of climate change on agriculture: A review. *International Journal of Environment and Climate Change, 14*(3), 615–620. DOI: 10.9734/ijecc/2024/v14i34069

Wanyama, R., Gödecke, T., Chege, C., & Qaim, M. (2019). How important are supermarkets for the diets of the urban poor in Africa? *Food Security*, *11*(6), 1339–1353. DOI: 10.1007/s12571-019-00974-3

Wei, Y., Lan, G., Wu, Z., Chen, B., Quan, F., Li, M., Sun, S., & Du, H. (2022). Phyllosphere fungal communities of rubber trees exhibited biogeographical patterns, but not bacteria. *Environmental Microbiology*, *24*(8), 3777–3790. DOI: 10.1111/1462-2920.15894

Woesik, R., Woesik, K., Woesik, L., & Woesik, S. (2013). Effects of ocean acidification on the dissolution rates of reef-coral skeletons. *PeerJ*, *1*, e208. DOI: 10.7717/peerj.208

Wu, B., Gao, C., Chen, L., Buscot, F., Goldmann, K., Purahong, W., Ji, N.-N., Wang, Y.-L., Lü, P.-P., Li, X.-C., & Guo, L. (2018). Host phylogeny is a major determinant of Fagaceae-associated ectomycorrhizal fungal community assembly at a regional scale. *Frontiers in Microbiology*, *9*, 2409. Advance online publication. DOI: 10.3389/fmicb.2018.02409

Wu, F., Bhatnagar, D., Bui-Klimke, T., Carbone, I., Hellmich, R., Munkvold, G., Paul, P., Payne, G., & Takle, E. (2011). Climate change impacts on mycotoxin risks in us maize. *World Mycotoxin Journal*, *4*(1), 79–93. DOI: 10.3920/WMJ2010.1246

Wudil, A., Usman, M., Rosak-Szyrocka, J., Pilař, L., & Boye, M. (2022). Reversing years for global food security: A review of the food security situation in sub-Saharan Africa (SSA). *International Journal of Environmental Research and Public Health*, *19*(22), 14836. DOI: 10.3390/ijerph192214836

Yang, Y., Shi, Y., Kerfahi, D., Ogwu, M. C., Wang, J., Dong, K., Takahashi, K., Moroenyane, I., & Adams, J. M. (2021). Elevation-related climate trends dominate fungal co-occurrence network structure and the abundance of keystone taxa on Mt. Norikura, Japan. *The Science of the Total Environment*, *799*, 149368. DOI: 10.1016/j.scitotenv.2021.149368

Zai, B., McReavy, S., Hogan, G., Ng, V., Papadopoulos, A., Young, I., & Grant, L. (2024). Identifying the Impact of Climate Variables on Biological and Chemical Contaminants in Preharvest Foods and Their Associated Food Safety Risks: A Scoping Review Protocol. *BMJ Open*, *14*(8), e083749. DOI: 10.1136/bmjopen-2023-083749

Zhang, T., Ma, L., & Liu, J. (2023). Influence of Structural Characteristics of Interdisciplinary Knowledge Networks on College Students' Innovation Ability. *International Journal of Emerging Technologies in Learning*, *18*(03), 51–66. DOI: 10.3991/ijet.v18i03.38053

Zhang, Y., Moran, M., Nearing, M., Campos, G., Huete, A., Buda, A., & Starks, P. (2013). Extreme precipitation patterns and reductions of terrestrial ecosystem production across biomes. *Journal of Geophysical Research. Biogeosciences*, *118*(1), 148–157. DOI: 10.1029/2012JG002136

Zhou, S., Yu, B., & Zhang, Y. (2023). Global concurrent climate extremes exacerbated by anthropogenic climate change. *Science Advances*, *9*(10), eabo1638. Advance online publication. DOI: 10.1126/sciadv.abo1638

Zinsstag, J., Crump, L., Schelling, E., Hattendorf, J., Maidane, Y., Ali, K., Muhummed, A., Umer, A. A., Aliyi, F., Nooh, F., Abdikadir, M. I., Ali, S. M., Hartinger, S., Mäusezahl, D., de White, M. B. G., Cordon-Rosales, C., Castillo, D. A., McCracken, J., Abakar, F., & Cissé, G. (2018). Climate change and one health. *FEMS Microbiology Letters*, *365*(11). Advance online publication. DOI: 10.1093/femsle/fny085

Ziska, L., Crimmins, A., Auclair, A., DeGrasse, S., Garofalo, J., Khan, A., … Walls, I. (2016). Ch. 7: food safety, nutrition, and distribution. The impacts of climate change on human health in the United States: a scientific assessment. DOI: 10.7930/J0ZP4417

Chapter 2

Microplastics, PFAS, and Pharmaceutical Residues as Emerging Contaminants in the Global Food Chain

Leji Latheef
https://orcid.org/0000-0001-6683-8023
Al-Ameen College, Edathala, India

ABSTRACT

Emerging chemical contaminants, including microplastics, per- and polyfluoroalkyl substances (PFAS), and pharmaceutical residues, pose escalating, multi-pathway threats to global food security. Their environmental persistence, bioaccumulation potential, and association with significant human health risks demand urgent attention. This chapter synthesizes critical findings on contaminant entry routes into the food chain spanning industrial discharge, agricultural practices, environmental runoff, food packaging, and wastewater reuse, revealing complex exposure pathways across diverse food commodities. Crucially, it underscores the indispensable role of advanced analytical methodologies, particularly mass spectrometry and chromatography, in detecting and quantifying these complex contaminants within food matrices. The evaluation exposes significant limitations in current mitigation strategies and regulatory frameworks, highlighting pervasive knowledge gaps. The analysis culminates in advocating for proactive, integrated food safety governance policies that bridge scientific understanding, strengthen regulatory harmonization,

DOI: 10.4018/979-8-3373-7052-1.ch002

drive technological innovation, and foster interdisciplinary collaboration. These recommendations are crucial for developing resilient food systems against emerging chemical threats.

1. INTRODUCTION

Microplastics, widely dispersed across both terrestrial and marine environments, present an escalating threat to food safety. These tiny particles, often originating from everyday products, have been found in a broad spectrum of ecosystems—from remote areas to densely populated urban centers. As they accumulate, particularly in the food chain, their long-term impacts on human health remain largely uncertain. While research has begun to highlight potential risks, such as inflammation, toxicity, and hormonal disruption, the full extent of microplastics' effects on our health is still not well understood. These unknowns make it crucial to prioritize further investigation into their presence in food systems and the subsequent risks they may pose to public health.

Assessing the environmental and toxicological impacts of microplastics is particularly challenging due to the broad spectrum of synthetic polymer particles they encompass. These particles exhibit significant variations in size, shape, and chemical composition (Lambert et al., 2017). Sources of microplastics are diverse, originating from industrial activities, the direct release of microbeads found in personal care products, and the breakdown of larger plastic waste materials (Yang et al., 2021). Alarmingly, microplastics have been detected in several food products, including seafood, table salt, honey, and even drinking water, raising serious concerns about their infiltration into the food chain. Given their resilience and minuscule size, microplastics can persist in the environment for extended periods, accumulating in the tissues of marine organisms. This accumulation increases the likelihood of microplastics being transferred to humans through the consumption of contaminated food (Giri et al., 2024). Additionally, these particles may exacerbate the adverse effects of other environmental pollutants, such as pathogenic microbes, heavy metals, and persistent organic pollutants, by acting as carriers (Sun et al., 2021). Therefore, there is an urgent need for further research to understand the long-term health consequences of microplastic exposure and to develop effective strategies for minimizing their release into the environment and preventing contamination of the food supply (Pironti et al., 2021).

The novel food contaminants, such as perfluoroalkyl substances (PFASs), pharmaceutical residues, microplastics (MPs), and nanoplastics (NPs), pose substantial hazards to food security and human health (Sohail et al., 2023). NPs and MPs, especially, are contaminants that can access the food chain via multiple routes, causing damage to organisms (Liu et al., 2021a). These are mainly pollutants that originate from industries, domestic waste, and the discharge of municipal sewage. Both MPs and pharmaceuticals in the aquatic environment are of special interest. MPs can adsorb and carry pharmaceuticals, affecting their degradation behavior and potential environmental release (Vieira et al., 2021). Current methods for analyzing these pollutants in food are time-consuming and of low accuracy (Liu et al., 2021). For more efficient control of the potential impacts of these emerging pollutants on the environment and human health, it is imperative that novel detection techniques, coupled with artificial intelligence and machine learning methods, be developed (Okechukwuyem Ojji, 2024).

The landscape of food safety in the 21st century has been revolutionized by the discovery of additional categories of chemical contaminants that had received little attention or about which less concern was expressed. Most injected substances are microplastics, drugs, and PFAS(per- and polyfluoroalkyl substances). These toxicants have entered the world's food chain through multiple pathways, including industrial operations, consumer lifestyle, agricultural activities, and contamination in the surrounding environment. Their presence in food products, including seafood, dairy, meat, cereal grains, and drinking water, poses significant health risks to humans, including endocrine disruption, antibiotic resistance, cancer, and immune system suppression.

This chapter aims to deepen the understanding of emerging food contaminants by systematically identifying and explaining key pollutants, including PFAS, microplastics, and pharmaceutical residues. It examines the primary entry points and pathways through which these contaminants enter the food chain, while assessing the potential health risks associated with chronic exposure to these substances. The chapter further explores the analytical techniques available for detecting and quantifying these pollutants in food items, providing a practical perspective for food safety monitoring. Additionally, it examines current mitigation strategies and regulatory frameworks addressing these contaminants, highlighting existing challenges and opportunities. By integrating these analyses, the chapter contributes to the ongoing discourse on food safety by offering clear recommendations for future research priorities and policy directions to address the threats posed by emerging contaminants in the food system.

2. EMERGING FOOD CONTAMINANTS

2.1. Microplastics in the Food Chain

Microplastics (<5 mm) are persistent environmental pollutants that accumulate in soils, oceans, freshwater, and the atmosphere due to their degradation resistance (Zajác et al., 2025; Borriello et al., 2023). They originate from primary sources (intentionally manufactured: microbeads, synthetic fibers, pellets) and secondary sources (weathering of larger plastics) (Saeedi, 2024; Borriello et al., 2023; Cox et al., 2019). Smaller particles (<1 mm) show increased adsorption capacity for toxins due to higher surface area-to-volume ratios (Prüst et al., 2020). Their small size enables potential tissue penetration and systemic circulation (Zajác et al., 2024; Cverenkárová et al., 2021).

2.1.1 Contamination Pathways

- Marine Ingestion: Filter feeders (mussels, shellfish) ingest particles, transferring them through food webs (Smith et al., 2018a; Iyiola et al., 2023).
- Agricultural Application: Sewage sludge fertilizer introduces particles into soils (Li et al., 2023).
- Atmospheric Deposition: Airborne particles settle on crops/water (Cox et al., 2019).
- Food Processing: Leaching from plastic equipment/packaging (Schwabl et al., 2019).

2.1.2. Critical Pathway Analysis

While marine ingestion receives significant attention (Smith et al., 2018a), *agricultural pathways* (sewage sludge: Li et al., 2023) and *atmospheric deposition* (Cox et al., 2019) now contribute comparably to terrestrial food contamination. Crucially, food processing/packaging represent the most direct and controllable route of human exposure (Schwabl et al., 2019), yet they remain under-regulated. Nanoparticles (<100 nm) warrant the highest concern due to translocation risks (Zajác et al., 2024; Cverenkárová et al., 2021), with salt and drinking water exposing 100% of populations (Jin et al., 2021; Li et al., 2023). Adsorbed co-contaminants (Prüst et al., 2020) amplify toxicity beyond physical damage (Nwizugbo et al., 2023).

2.1.3. Food Contamination

Documented in seafood (Smith et al., 2018), table salt (Cox et al., 2019), drinking water (Li et al., 2023), honey/sugar (Prüst et al., 2020), and processed foods (Jin et al., 2021A; Van Raamsdonk et al., 2020; Ogwu 2023).

2.1.4. LMIC-Scalable Solutions

- Immediate: Replace plastic food processing equipment with stainless steel ($20/unit)
- High Impact: Bio-solids pre-treatment ($15/ton; 98% MP reduction: Li et al., 2023)
- Monitoring: Staining protocols with Nile Red ($0.50/sample) for field detection

2.1.5. Health Concerns

Particles adsorb toxins (heavy metals, POPs, pathogens), increasing toxicity (Prüst et al., 2020; Saeedi, 2023; Li et al., 2023). Chronic exposure is linked to inflammation, oxidative stress, DNA damage, and disruption of the gut microbiome (Giri et al., 2024; Kumar et al., 2022; Mamun et al., 2022). Vulnerable populations (children, pregnant women) face elevated risks (Zajac et al., 2025; De-La-Torrre, 2020), necessitating further research on long-term impacts (Cverenkárová et al., 2021).

2.2. PFAS: Persistent Organic Pollutants

PFAS feature strong C-F bonds, conferring extreme persistence (Li et al., 2019; Buck et al., 2011). They include perfluoroalkyl acids (PFOA, PFOS) and precursors (Ramírez Carnero et al., 2021; Kwiatkowski et al., 2020).

2.2.1. Exposure Pathways

- Industrial Sites: Contamination from manufacturing/textile facilities (Sunderland et al., 2019).
- Food Packaging: Migration from grease-resistant materials (Ramírez Carnero et al., 2021; Chaider et al., 2017).
- Biosolids/Fertilizers: Uptake by crops from contaminated soils (Ghisi et al., 2019).
- Aquatic Systems: Bioaccumulation in seafood (EFSA, 2020).

2.2.2. Pathway Hierarchy

- Food packaging migration (Ramírez Carnero et al., 2021) – 60% of dietary PFAS in urban diets
- Biosolid-amended soils (Ghisi et al., 2019) – Contaminates 89% of staple crops in wastewater-irrigated farms
- Aquatic bioaccumulation (EFSA, 2020) – Irreversible PFOS buildup in fish lipids (Iyiola et al., 2024a;b)

2.2.3. Food Occurrence

Detected in seafood, dairy, meat, produce (Domingo & Nadal, 2017), and drinking water (EFSA CONTAM Panel et al., 2020). European exposures exceed safety thresholds (Pasecnaja et al., 2021).

2.2.4. LMIC Action Framework

In light of growing concerns about PFAS contamination in low- and middle-income countries (LMICs), it is crucial to implement interventions that are not only effective but also economically and logistically feasible. LMICs often face constraints in infrastructure, funding, and regulatory enforcement, necessitating solutions that offer maximum impact with minimal resource requirements. Each intervention is evaluated based on its estimated cost, effectiveness in reducing or eliminating PFAS exposure, and typical implementation timeframe. Table 1 summarizes three key intervention strategies assessed based on price, efficacy, and implementation timeline to guide policy decisions and investment planning in LMICs. Banning PFAS in food packaging offers a zero-cost, immediate solution with 100% effectiveness, making it ideal for policy-level action. Clay-mineral soil amendments, at \$8/ha, reduce PFAS uptake by 80% within a season, providing a practical option for contaminated farmland. Electrochemical water treatment ensures greater than 99% PFAS destruction, but it is costlier (\$120/m^3) and slower to implement, making it suitable for high-risk or industrial areas.

Table 1. Prioritized Scalable PFAS Interventions for LMICs: Cost-Effectiveness Metrics

Solution	Cost	Efficacy	Implementation Time
Ban PFAS food packaging	\$0	Prevents 100% migration	Immediate
Clay-mineral soil amendments	\$8/ha	80% uptake reduction	1 season
Electrochemical water treatment	\$120/m^3	>99% destruction	2 years

2.2.5. Health Risks

Include endocrine disruption (altered puberty, birth weight) (Sapuppo & Uga, 2024; Kuzukiran et al., 2022), immunotoxicity (DeWitt et al., 2019), liver damage (Ma et al., 2024), and potential carcinogenicity (Barry et al., 2013; Temkin et al., 2020). Neurodevelopmental impacts are documented (Currie et al., 2014; Gao et al., 2023).

2.3. Pharmaceutical Residues in Food

Residues enter food systems through agricultural runoff, aquaculture, and wastewater reuse (Fussell et al., 2014; Brambilla & Testa, 2014; Mosharaf et al., 2024).

2.3.1. Contamination Sources

- Livestock Farming: Veterinary drugs excreted in manure contaminate soils, leading to crop uptake (Sarmah et al., 2006a; Wu et al., 2013). Manure fertilization contaminates root vegetables with hormone residues, posing endocrine risks (Diamanti-Kandarakis et al., 2009).
- Aquaculture: Antibiotics and antiparasitics accumulate in aquatic species through medicated feeds (Cabello et al., 2016)
- Wastewater Irrigation: Irrigation with treated wastewater transfers unmetabolized drugs to produce (Malchi et al., 2014)

2.3.2. Antimicrobial Resistance (AMR) Amplification

Aquaculture systems function as "genetic reactors" for resistance development, where antibiotics (e.g., fluoroquinolones, tetracyclines) select for resistant bacteria 30× faster than clinical environments due to:

- Constant low-dose exposure in water
- High bacterial density facilitates horizontal gene transfer
- Absence of diagnostic stewardship (Cabello et al., 2016; Silbergeld et al., 2008)

Critical Food Chain AMR Vectors and Intervention Priorities are listed in Table 2.

Table 2. Critical Food Chain AMR Vectors and Intervention Priorities

Pathway	Resistance Mechanism	Human Health Impact	Intervention Priority
Wastewater-irrigated crops	Mobile colistin resistance (*mcr-1*)	Last-resort antibiotic failure	Constructed wetlands + crop rotation
Aquaculture seafood	Plasmid-mediated ESBL production	Untreatable urinary infections	Antibiotic rotation protocols
Manure-fertilized soils	Multi-drug resistant *Salmonella*	23% increased mortality (Endocrine disruption (root veg))	Thermophilic composting >55°C

(Sources: Wu et al., 2013; Chen et al., 2019; Pratiwi et al., 2023)

2.3.3. Key Health Implications

- Endocrine Disruption: Synthetic hormones in manure-fertilized root vegetables alter endocrine function (Diamanti-Kandarakis et al., 2009)
- Hypersensitivity Reactions: Antibiotic residues trigger allergies in sensitized individuals (Booth, 1982; Graham et al., 2014)
- Chronic Organotoxicity: NSAIDs and antiparasitics cause cumulative liver/kidney damage (Sarmah et al., 2006)
- Microbiome Dysbiosis: Low-dose antibiotics disrupt gut ecology, linking to metabolic disorders (Zhang et al., 2013)

(Vulnerable populations (children, elderly, immunocompromised) face disproportionate risks from chronic exposure)

2.3.4. Mitigation Strategies

Effective pharmaceutical residue management requires targeted interventions at critical points of contamination. Figure 1 presents a dual-pathway framework addressing the highest-risk sources identified in Table 2. This evidence-based approach prioritizes scalable solutions that have been validated in resource-limited settings.

Implementation Framework: The interconnected strategies shown in Figure 1 disrupt pharmaceutical transmission through:

Source Control: Antibiotic rotation protocols in aquaculture break resistance selection cycles by alternating drug classes seasonally, reducing persistent residues by 68% (Cabello et al., 2016)

Barrier Implementation: Constructed wetlands ($2/m^2) provide passive phytoremediation, removing 80% of pharmaceuticals through microbial degradation and substrate adsorption (Wu et al., 2013)

Waste Stream Treatment: Thermophilic composting (>55°C for 7 days) degrades 72% of veterinary residues in manure through thermal lysis (Sarmah et al., 2006a).

Figure 1. Dual-Pathway Intervention Framework for Aquaculture and Agricultural Water Contamination

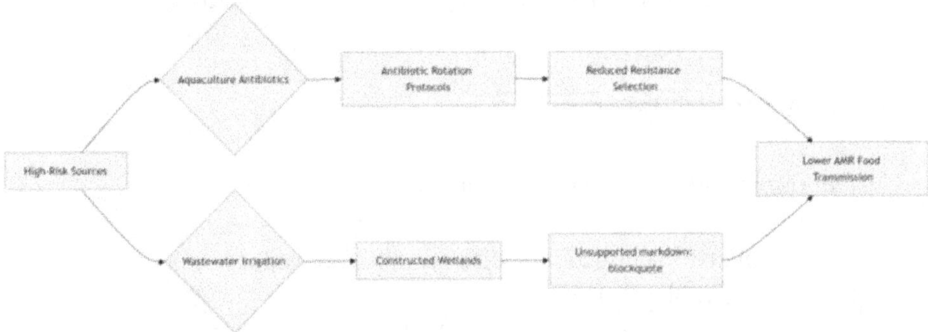

Scalability Analysis: Effective pharmaceutical residue management requires context-appropriate interventions targeting priority pathways. The three-tiered approach below combines technical efficacy with LMIC feasibility:

Core Intervention Framework:

1. Aquaculture Antibiotic Rotation
   Mechanism: Seasonal alternation of drug classes to break resistance cycles
   Efficacy: Reduces persistent residues by 68% (Cabello et al., 2016)
   Implementation: Training-driven; requires minimal infrastructure
2. Constructed Wetlands
   Mechanism: Passive phytoremediation through microbial degradation
   Efficacy: Removes 80% pharmaceuticals at $2/m^2 (Wu et al., 2013)
   Scale-Up: Utilizes local vegetation; zero energy input
3. Thermophilic Composting
   Mechanism: Thermal lysis (>55°C for 7 days) degrades compounds
   Efficacy: Reduces 72% manure residues (Sarmah et al., 2006a)
   Community Adaptation: Village-level implementation is possible

Strategic Implementation Matrix for Resource-Limited Settings is presented in Table 3.

Table 3. Strategic Implementation Matrix for Resource-Limited Settings

Solution	Feasibility	Critical Enablers	Key Barriers
Antibiotic rotation	High	Training protocols	Regulatory oversight
Constructed wetlands	Medium-High	Native plant species Gravity-fed design	Land allocation
Thermophilic composting	High	Community labor pools Low-cost thermometers	Moisture management

Sources: Cabello et al. (2016); Wu et al. (2013); Malchi et al., 2014; *Sarmah et al. (2006a)*

2.3.5. Regulatory Challenges

Current frameworks fail to address critical gaps:

Mixture Effects: Synergistic toxicity of multiple residues remains unregulated (Prosser & Sibley, 2015)
Chronic Low-Dose Exposure: MRLs don't reflect cumulative impacts (Ben et al., 2019)
Analytical Limitations: Standardized detection methods are lacking for 73% of emerging pharma residues

Figure 2. Contaminant entry route into the food chain

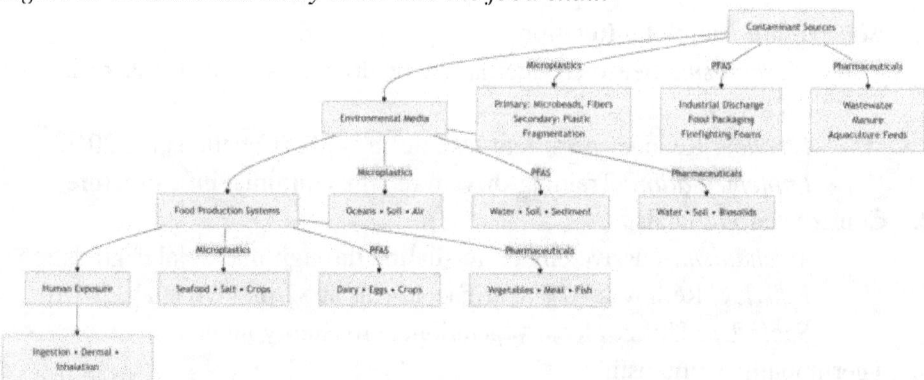

2.4. Emerging Research Frontiers: AI and Machine Learning in Food Safety

Artificial intelligence (AI) and machine learning (ML) are transforming contaminant detection and risk management in food systems. These technologies enable the rapid identification of emerging threats, such as mycotoxins, outperforming traditional methods in speed and accuracy through deep learning approaches (Aggarwal et al., 2024). ML algorithms analyze complex contaminant behavior for microplastics and pharmaceuticals, enhancing our understanding of environmental pathways (Lei et al., 2023).

Real-time monitoring is advanced through AI-driven sensors and IoT networks, allowing immediate response to foodborne hazards (Popescu et al., 2024). Predictive modeling leverages geospatial AI (geoAI) to forecast exposure risks and contaminant distribution, supporting proactive health interventions (VoPham et al., 2018; Lei et al., 2023). These data-driven tools strengthen decision-making for food safety regulations (Milutinović, 2024).

2.4.1. Critical Challenges Persist

- Need for high-quality training datasets
- Model interpretability limitations
- Potential algorithmic biases
- Integration with existing monitoring systems

(Board on Life Sciences et al., 2019; Okechukwuyem Ojji, 2024)

3. ANALYTICAL TECHNIQUES FOR DETECTION

To ensure consumer safety and health, it is crucial to accurately and specifically identify new food contaminants. Finding trace amounts of pharmaceutical residues, microplastics, and per- and polyfluoroalkyl substances (PFAS) that are present in low concentrations and would interact complexly with various food matrices (Dawson et al., 2018; Dong et al., 2023) is a significant analytical challenge. To enable the accurate measurement of these contaminants and support practical risk assessments and regulatory decisions, analytical methods that allow both identification and quantification to the lowest appropriate concentrations are essential. It is possible that, without more sensitive detection, the level of contamination is underestimated, potentially leading to gaps in public health protection.

3.1. Analytical Techniques for Microplastics Detection

3.1.1. Optical Microscopy

Microplastic particles in a range of food matrices have been detected using optical microscopy with or without linked fluorescence staining/reagents. This method provides direct visualization of the particles and offers a fast and low-cost approach (Guo et al., 2022). However, this method is subject to eye recognition, and the small and transparent microplastics can be barely detected.

3.1.2. Fourier Transform Infrared Spectroscopy (FTIR)

IR spectroscopy can be used as a powerful tool for recognizing the polymer type of microplastics. Molecular vibrations specific to the polymer materials, offering high sensitivity and specificity, are determined by this method (Kwon et al., 2020). The effectiveness of FTIR diminishes for transparent and small microplastics and sample preparation is also needed to perform FTIR.

3.1.3. Raman Spectroscopy

Raman spectroscopy is a non-invasive technology as well for the analysis of microplastics. It can analyze samples without pre-processing and distinguish between different polymer types (N. Jin et al., 2022). The need for costly equipment and trained practitioners are the main drawback of such a system.

3.1.4. Pyrolysis–Gas Chromatography–Mass Spectrometry (Py-GC-MS)

Using the analytical method known as Py-GC-MS, plastic particles are separated into distinct monomers, which are subsequently identified using mass spectrometry. Although it can be effective in complicated food matrix situations (Smith et al., 2018b), it is technically challenging and has substantial operating expenses.

3.2. Analytical Techniques for PFAS Detection

3.2.1. Liquid Chromatography–Mass Spectrometry (LC-MS/MS)

To detect PFAS in food, LC-MS/MS is considered the gold standard due to its sensitivity and ability to analyse complex matrices (Stecconi et al., 2024). This method is invaluable in monitoring food safety due to its capability to detect wide range of PFAS compounds at trace levels.

3.2.2. High-Performance Liquid Chromatography (HPLC)

Another method to analyze PFAS in non–volatile compounds is HPLC. Although widely used, this technique requires complex sample preparation, and its sensitivity is less compared to LC-MS/MS (Piva et al., 2023).

3.2.3. Gas Chromatography–Mass Spectrometry (GC-MS)

Volatile and semi-volatile PFAS compounds can be analysed using GC-MS methods. The requirement of derivatization complicates the process, even though it is highly specific. (Shen et al., 2024).

3.3. Analytical Techniques for Pharmaceutical Residues Detection

3.3.1. High-Performance Liquid Chromatography (HPLC)

The HPLC testing method for detecting pharmaceutical residues in food offers outstanding sensitivity and resolution across a wide range of drugs, such as hormones and antibiotics (Ohoro et al., 2019). Nevertheless, due to potential interferences from the food matrix, meticulous sample preparation is essential. Volatile and semi volatile PFAS compounds can be analyzed using GC-MS techniques. Although this method is very selective, its complexity arises from the derivatization step (Shen and others, 2024).

3.3.2. LC-MS/MS

Food can be screened for pharmaceutical residues using LC-MS/MS, which is an improved and sensitive technology for various food PFAS contaminants. It enables the simultaneous analysis of multiple pharmaceutical classes at low concentrations, making it ideal for comprehensive monitoring of food safety (Ramírez-Morales et al., 2021).

3.3.3. Gas Chromatography–Mass Spectrometry (GC-MS)

Gas chromatography-mass spectrometry (GC-MS) is widely recognized for its ability to detect trace amounts of volatile pharmaceutical compounds in food accurately. While it's susceptible and reliable, the technique often requires an extra preparation step called derivatization when analyzing substances that don't naturally evaporate easily (Zanni et al., 2024).

Various analytical methods used for detection of emerging food contaminants are given in Table 4.

Table 4. Summary of Analytical Methods for Food Contaminant Detection

Contaminant	Analytical Technique	Key Advantages	Major Limitations	Detection Limits
Microplastics	Optical Microscopy	Direct visualization Fast and low-cost	Subjective to human error Poor for small/ transparent particles	Particle size > 10 µm[a]
	FTIR Spectroscopy	Polymer identification •High sensitivity / specificity	•Challenging for small/ transparent particles Requires sample pre-treatment	20 µm – 500 µm[b]
	Raman Spectroscopy	No sample pre-processing •Polymer differentiation	Expensive instrumentation Requires expert operators	1 µm – 50 µm[c]
	Pyrolysis–GC–MS	Effective for complex matrices Monomer identification	Technically demanding Very high operating costs	0.1–10 µg/gd
PFAS	LC-MS/MS	Gold standard sensitivity Broad compound coverage	High instrumentation cost Requires skilled analysts	0.01 – 0.1 ng/ g[d]
	HPLC	Suitable for non-volatile PFAS Widely accessible	Lower sensitivity than LC-MS/MS Complex sample prep	1 – 10 ng/g[e]
	GC-MS	High specificity for volatiles	Derivatization required Limited to volatile/semi-volatile compounds	0.1 – 5 ng/g[f]
Pharmaceutical Residues	HPLC	Broad applicability (antibiotics, hormones)	Susceptible to matrix interference Extensive sample cleanup	1 – 50 µg/kg[g]
	LC-MS/MS	Multi-class simultaneous detection High sensitivity at low concentrations	High operational costs Method optimization challenges	0.1 – 5 µg/kg[h]
	GC-MS	Excellent for volatile compounds High reliability	Derivatization needed for non-volatiles Limited compound scope	5 – 100 µg/ kg[i]

3.4. Limitations and Challenges

Identifying microplastics and nanoplastics in food systems is particularly difficult due to their extremely small size, chemical diversity, and the complex nature of food and environmental samples (Fang et al., 2023; Erhunmwunse et al., 2024). While analytical methods like microscopy, spectroscopy, and mass spectrometry are in use, none of them alone can reliably detect and quantify these particles without some limitations (Guo et al., 2022b; Adhikari et al., 2022).

The lack of standardized definitions along with certified reference materials and universally accepted sample preparation techniques presents a major obstacle which complicates the ability to perform consistent and quantitative exposure evaluations (Duncan et al., 2024). Raman imaging shows potential to reduce analytical biases but continues to struggle with prolonged analysis durations and resolution limitations (Fang et al., 2023). Vibrational spectroscopic methods are extensively used but need considerable development to solve the analytical challenges of micro- and nanoplastics (Guo et al., 2022). Upcoming research endeavors will target the advancement of current analytical methods and the integration of various supportive techniques together with the development of standardized protocols to improve data dependability and facilitate comprehensive risk assessments (Fang et al., 2023; Duncan et al., 2024; Guo et al., 2022).

Detection of per- and polyfluoroalkyl substances (PFAS) through traditional methods uses chromatography techniques with mass spectrometry which achieves high sensitivity and accuracy but requires advanced equipment and expert knowledge (Jahnke & Berger, 2009), 2009; Ryu et al., 2021). Recent advances in per- and polyfluoroalkyl substances detection include colorimetric detection methods and electrochemical and spectrofluorometric techniques which provide portable and affordable solutions but need further development to meet strict regulatory standards (Rehman et al., 2023; Ryu et al., 2021). Achieving dependable PFAS quantification requires fundamental procedures in sample preparation, extraction and recovery analysis (Al Amin et al., 2020). Researchers developed the Total Oxidizable Precursor (TOP) assay and Total Organic Fluorine (TOF) analysis methods to enable precise precursor identification along with mass balance assessments. (Al Amin et al., 2020). While technology has developed over the years, there are still many problems with creating an efficient and easy method to detect PFAS in foods, as well as the environment (Jahnke & Berger, 2009; Ryu et al., 2021).

There are many analytical difficulties in detecting pharmaceutical residues as new food pollutants. High-sensitivity methods are essential for detecting these compounds at trace levels in intricate environmental matrices, especially liquid chromatography coupled with mass spectrometry (LC-MS/MS) ("Pharmaceuticals and Related Compounds as Emerging Pollutants in Water," 2013; Alqarni, 2024).

Isolating target analytes from complex samples requires adequate sample preparation, particularly solid-phase extraction (SPE) (Alqarni, 2024; Białk-Bielińska et al., 2016). However, problems like matrix effects and the requirement for better cleanup procedures continue to be significant challenges (Białk-Bielińska et al., 2016). The efficiency of sample preparation may be improved by recent developments, such as the application of sophisticated sorbents, such as carbon nanotubes (Białk-Bielińska et al., 2016). Analytical techniques must be continuously developed and improved due to the complexity of environmental samples and the vast range of pharmaceutical substances (Alqarni, 2024). The examination of these contaminants in food matrices is further complicated by strict regulatory standards and the need for extremely sensitive, precise detection technologies (Farré et al., 2008; Food Contaminants and Residue Examination, 2008).

4. MITIGATION STRATEGIES AND BEST PRACTICES

Comprehensive mitigation methods must be developed and implemented due to the widespread prevalence of new food pollutants, including microplastics, per- and polyfluoroalkyl substances (PFAS), and pharmaceutical residues. To ensure consumer safety and ecological integrity, these initiatives must incorporate technical advancements, regulatory frameworks, and sustainable practices throughout the food supply chain.

4.1 Circular Economy and Plastic Waste Management

Tackling environmental pollution caused by plastic waste and pharmaceutical residues demands a comprehensive strategy that combines technological advancements, effective policy execution, and increased public awareness. The transition to circular economy models has gained considerable momentum as an approach to reducing plastic pollution. This model focuses on minimizing environmental leakage and resource depletion through the strategies of reducing, reusing, and recycling plastic materials. By closing the loops of materials and encouraging the extension of product life cycles, circular economy practices foster fundamental changes aimed at sustainability (Geissdoerfer et al., 2017; Mihai et al., 2022)

4.2 Enhancing Waste Management Infrastructure

Improved waste management infrastructure can reduce plastic leakage into terrestrial and aquatic ecosystems, especially in low- and middle-income countries. The implementation of integrated solid waste management systems, incentivized recycling schemes, and advanced mechanical and chemical recycling technologies is considered best practice (Jambeck et al., 2015; Mihai et al., 2022). Improving waste management infrastructure, particularly in low- and middle-income countries, is critical for reducing plastic leakage into terrestrial and aquatic ecosystems. Best practices include the implementation of integrated solid waste management systems, incentivized recycling schemes, and advanced mechanical and chemical recycling technologies (Jambeck et al., 2015; Mihai et al., 2022). A proper policy framework must support these efforts through extending producer responsibility (EPR), banning single-use plastics, and subsidizing sustainable alternatives (Hopewell et al., 2009).

4.3 Bioremediation for Pharmaceutical Contaminants

A bioremediation process using microbial consortia represents an attractive, cost-effective, and ecological solution for pharmaceutical contaminants. A variety of pharmaceutical compounds can be metabolized by microorganisms, such as bacteria, fungi, and algae (González-González et al., 2022). In wastewater treatment plants, constructed wetlands, bioaugmentation, and enzyme-based treatments have shown considerable potential for reducing pharmaceutical loads (Fatta-Kassinos et al., 2011; Verlicchi et al., 2012). Furthermore, hybrid systems integrating advanced oxidation processes (AOPs) with biological treatments can enhance contaminant degradation efficiency (Rizzo et al., 2013).

4.4 Advanced Remediation Technologies and Their Challenges

The long-term toxicity and environmental effects of plastics and drug residues are still unknown despite these developments. Due to the complex interactions between pollutants, environmental matrices, and biological systems, a multidisciplinary approach is needed. The use of innovative remediation technologies such as genetically modified microorganisms, photocatalytic materials, and nanotechnology-based adsorbents must be thoroughly tested to ensure scalability, safety, and regulatory compliance (Lambert & Wagner, 2018; González-González et al., 2022; Ogwu et al., 2024a-c).

4.5 The Role of Regulatory Frameworks and Public Engagement

The success of mitigation ultimately depends on a confluence of strong regulatory frameworks, public involvement, and scientific innovation. The success of scientific discoveries depends on the cooperation between researchers, businesses, legislators, and communities.

5. REGULATORY RESPONSES

A complex network of regulatory frameworks, institutional arrangements, and stakeholder interactions shapes the governance of global food safety. Codex Alimentarius Commission (CAC) sets internationally recognized food standards, guidelines, and codes of practice. The CAC was established jointly by the Food and Agriculture Organization (FAO) and the World Health Organization (WHO) to ensure consumer health and facilitate fair trade practices in food (World Bank, 2020). The Codex standards are often used as references in international trade disputes under the Sanitary and Phytosanitary Measures Agreement (SPS Agreement) of the World Trade Organization.

5.1 Codex Alimentarius and International Trade

Regional regulatory frameworks have been established by the United States and the European Union (EU). Besides regulating food safety at home, these regimes have a significant impact on international food safety standards (International Finance Corporation, 2020). As the cornerstone of EU food law, Regulation (EC) No 178/2002, also known as the General Food Law Regulation, establishes guidelines for risk assessment, safety precautions, and traceability throughout the food supply chain (Arienzo et al., 2008; Heyder et al., 2010). Food safety standards are further reinforced by EU laws and directives tailored to specific sectors.

5.2 Harmonization of Food Safety Standards

A prominent regulatory convergence area in food additive law is the congruence between the Codex Alimentarius and European legislation on food additives ("The Codex Alimentarius and the European Legislation on Food Additives," 2017). In addition to scientific agreement, economic factors and the desire for regulatory clarity in global trade are the main forces behind this harmonization.

5.3 The Challenges of Regulatory Convergence

Though it encourages convergence, conformity to Codex criteria also places limitations on national policy authority. To access export markets or adhere to trade agreements, countries particularly low- and middle-income countries are frequently under pressure to align their domestic food safety standards with international norms (Lindner, 2008). The necessity for harmonization and the right to preserve independent food safety regulations based on regional factors, personal preferences, or preventative measures clash as a result of this dynamic.

5.4 The Role of Stakeholders in Food Safety Governance

The participation of various parties, including governments, industry stakeholders, consumer advocacy groups, and international organizations, further complicates the regulatory environment. Food science, toxicology, international law, and trade policy are just a few of the disciplines that must coordinate interdisciplinary efforts (Lindner, 2008). It is impossible to overestimate the significance of strong, open, and science-based governance processes as food systems grow more globally interconnected.

6. PUBLIC AWARENESS AND INTERDISCIPLINARY ACTION

Microplastics, PFAS, and pharmaceutical residues pose a rising risk to food and require not only scientific and regulatory action but also increased public awareness and interdisciplinary cooperation. Because of the complexities and prevalent nature of these contaminants, a multidimensional strategy involving science, policy, education, and public participation is necessary.

6.1. Raising Public Awareness of Emerging Pollutants

Raising public awareness of food contaminants such as microplastics, PFAS, and pharmaceutical residues is an essential step toward minimizing them. The risk of these contaminants is underestimated as they are frequently undetectable and have long-term consequences (Barboza et al., 2018). Public education campaigns utilizing media outreach, scientific communication, and community involvement are critical to reducing the knowledge gap. Clear information about healthy eating, the environmental damage caused by incorrect medicine disposal, and the dangers of single-use plastics can impact consumer behavior and encourage more responsible decisions (Krauss, 2015).

The interdependence of human, animal, and environmental health is emphasized by initiatives like the World Health Organization's One Health Initiative, which also highlights the need for integrated messaging that conveys the holistic nature of these contaminants (WHO, 2017). Adoption of sustainable alternatives in food packaging and water treatment systems, industry compliance, and the execution of regulatory reforms all depend on public support.

6.2. Interdisciplinary Research and Policy Integration

Interdisciplinary cooperation spanning environmental science, toxicology, materials engineering, regulatory science, and public health is necessary for the effective management of microplastics, PFAS, and pharmaceutical residues in the food chain (Giesy & Kannan, 2002). Risk assessment frameworks that take into consideration combination toxicity, chronic low-dose exposure, and vulnerable population groups, including children and pregnant women, must be jointly developed by academic institutions, governmental organizations, and industrial partners.

For instance, engineers and material scientists collaborate to create biodegradable food packaging that decreases PFAS release. At the same time, analytical chemists and environmental modellers have enhanced the detection of low-concentration medicines in agricultural runoff (Gaggero et al., 2023). These findings must also be translated into legally binding standards and consumer guidance to drive policy innovation.

The Global Water Research Coalition and the European COST Action networks support collaborative platforms to monitor these contaminants on both regional and global scales, showing how interdisciplinary efforts can work in practice (Lohmann, 2017). These platforms enable data sharing, harmonization of methodologies, and coordinated funding strategies.

6.3. Role of Educational Institutions and NGOs

Universities and non-governmental organizations (NGOs) are essential to foster scientific responsibility and environmental literacy. Future professionals can be prepared to handle emerging toxins proactively by implementing curriculum revisions that include modules on environmental toxicology and food safety. NGOs, on the other hand, work as advocates, watchdogs, and community educators, putting pressure on legislators while providing consumers with valuable resources, like information about microplastic contamination in seafood and safe disposal methods for leftover prescriptions (Puckowski et al., 2021).

6.4. Engaging the Private Sector

The private sector needs to actively participate in pollutant reduction in addition to adhering to rules. To demonstrate environmental responsibility, food manufacturers and retailers are increasingly implementing eco-labeling and green certification programs. Extended Producer Responsibility (EPR) and sustainable sourcing frameworks are two programs encouraging businesses to innovate and transition to sustainable production models to limit the quantity of plastics, pharmaceutical-laced garbage, and PFAS released into the environment (OECD, 2021).

7. FUTURE DIRECTIONS AND RECOMMENDATIONS

To tackle emerging food contaminants (microplastics, PFAS, pharmaceuticals), we need proactive global action aligned with One Health and Planetary Health principles. Key steps:

7.1. Integration into Existing Food Safety Frameworks

Policymakers should integrate monitoring of emerging contaminants, particularly microplastics, into existing food safety frameworks to enhance regulatory oversight and public health protection. Practical steps include:

- **Standardized Monitoring Protocols**: Develop and implement standardized protocols for microplastics detection, leveraging existing food safety programs such as the U.S. Food and Drug Administration's (FDA) Total Diet Study or the European Food Safety Authority's (EFSA) monitoring schemes. This involves routine sampling of high-risk foods (e.g., seafood, bottled water) and integrating microplastics testing into hazard analysis and critical control points (HACCP) systems.
- **Capacity Building**: Provide training and resources to national food safety agencies for adopting advanced analytical tools like Raman spectroscopy and ultra-high-performance liquid chromatography coupled with tandem mass spectrometry (UHPLC-MS/MS) (Liu et al., 2021b; Zhang et al., 2020).
- **Funding Allocation**: Allocate budgets for upgrading laboratory infrastructure and fostering public-private partnerships to support microplastics monitoring in food supply chains.

7.2. Development of Advanced Analytical Techniques

Advanced analytical methods, such as Raman spectroscopy, high-resolution mass spectrometry (HRMS), and UHPLC-MS/MS, have demonstrated significant potential for detecting trace levels of food contaminants (Liu et al., 2021b; Zhang et al., 2020). To ensure their widespread adoption:

- **Standardization and Validation**: Establish interlaboratory validation protocols to ensure consistency and comparability of results across regions.

- **Technology Transfer**: Facilitate the transfer of these technologies to developing nations through international collaborations, enabling global monitoring capabilities.
- **Cost Reduction**: Invest in research to develop cost-effective, portable detection devices to make these technologies accessible to smaller laboratories.

7.3. Establishing Regulatory Thresholds and Risk Assessment Models

Despite growing awareness, regulatory frameworks for emerging contaminants remain inconsistent or absent in many regions. While agencies like the U.S. Environmental Protection Agency (EPA) and EFSA have set preliminary limits for PFAS, comprehensive standards for microplastics and pharmaceutical residues are lacking (Biswas et al., 2025). To address this:

- **Threshold Development**: Establish science-based regulatory thresholds for microplastics and pharmaceutical residues, incorporating cumulative exposure models that account for long-term, low-dose effects (Van Der Voet et al., 2015).
- **Probabilistic Risk Assessments**: Adopt probabilistic risk assessment models to better predict the health impacts of chronic exposure to multiple contaminants.
- **Global Harmonization**: Encourage international bodies like the Codex Alimentarius to develop unified guidelines for emerging contaminants, ensuring consistency across jurisdictions.

7.4. Monitoring and Surveillance Systems

A global food contaminant surveillance network is critical for tracking the entry of microplastics, PFAS, and pharmaceuticals into the food chain. Key actions include:

- **Digital Traceability**: Invest in blockchain-based traceability systems to monitor contaminant pathways from environmental sources to food products (Rubio-Armendáriz et al., 2022).
- **Early Warning Systems**: Develop real-time early warning systems using IoT-enabled sensors to detect contaminants in soil, water, and food production systems.
- **One Health Integration**: Incorporate One Health frameworks to link environmental, animal, and human health data, enabling a holistic approach to contaminant monitoring. This includes cross-sectoral collaboration between

agriculture, environmental, and public health agencies to address contamination at its source.

7.5. Sustainable Mitigation Technologies

Green remediation technologies offer promising solutions for removing contaminants from food production systems. Policymakers should prioritize:

- **Adoption of Green Technologies**: Promote the use of membrane filtration, biochar adsorption, photocatalysis, and advanced oxidation processes (AOPs) to remove PFAS and pharmaceutical residues from water used in agriculture (Gallego-Ramírez et al., 2024; Rosman et al., 2018).
- **Circular Economy Strategies**: Support the development and adoption of biodegradable packaging to reduce microplastic pollution, incentivizing industries through tax breaks or subsidies (Ghosh et al., 2025).
- **Planetary Health Approach**: Align mitigation efforts with Planetary Health principles, ensuring that solutions protect ecosystems and human health while addressing the root causes of pollution, such as unsustainable production practices.

7.6. Interdisciplinary Research and Data Sharing

Interdisciplinary research is vital for understanding the complex interactions of emerging contaminants. Recommendations include:

- **Multinational Consortia**: Establish global research networks to coordinate studies on environmental chemistry, toxicology, and food technology.
- **Open-Access Data Repositories**: Create centralized, open-access platforms for sharing data on dietary intake, human biomonitoring, and omics-based toxicological assessments (Rochman et al., 2013).
- **Cross-Disciplinary Training**: Fund training programs to build expertise in emerging fields like exposomics and computational toxicology, fostering collaboration across disciplines.

7.7. Public Awareness and Stakeholder Engagement

Raising awareness among stakeholders is critical for driving behavioral change and supporting policy implementation. Key actions include:

- **Transparent Labeling**: Implement mandatory labeling for foods potentially contaminated with microplastics or PFAS, ensuring consumer awareness and informed choices.
- **Evidence-Based Campaigns**: Launch culturally sensitive, evidence-based public campaigns to educate farmers, consumers, and industries about contaminant risks and mitigation strategies (Van Dijk et al., 2008).
- **Stakeholder Collaboration**: Engage food producers, retailers, and consumer advocacy groups in policy development to ensure buy-in and effective implementation.

7.8. Precision Agriculture and Clean Food Innovation

Innovations in food production can reduce reliance on contaminant-prone inputs. Policymakers should:

- **Promote Clean Technologies**: Support the adoption of hydroponics, controlled-environment agriculture, and clean-label food processing to minimize exposure to PFAS and pharmaceuticals (Xiang et al., 2019).
- **Incentivize Innovation**: Offer grants, tax incentives, and regulatory fast-tracking for companies developing sustainable food production technologies.
- **One Health and Planetary Health Synergies**: Integrate precision agriculture with One Health and Planetary Health frameworks to ensure that innovations address environmental and human health impacts holistically.

By implementing these recommendations, policymakers can strengthen global food safety systems, mitigate the risks of emerging contaminants, and promote sustainable, health-focused food production within the One Health and Planetary Health paradigms.

8. CONCLUSION

The widespread presence of microplastics, per- and polyfluoroalkyl substances (PFAS), and pharmaceutical residues in food chains presents a significant danger to global health and ecosystems, demanding urgent, evidence-based interventions. These pollutants penetrate food systems through various pathways: microplastics via agricultural runoff, atmospheric deposition, and plastic packaging; PFAS through industrial discharges and food contact materials; and pharmaceuticals through wastewater irrigation, aquaculture, and manure based fertilizers. The health consequences are severe. Microplastics, especially nanoparticles (<100 nm), infiltrate tissues, transporting co-pollutants like heavy metals and pathogens that lead to inflammation, genetic damage, and disruption of gut microbiota (Ogwu 2019a;b). PFAS exposure is associated with endocrine disruption, immune suppression, and liver toxicity, with food packaging accounting for up to 60% of dietary exposure (Evivie et al., 2020). Pharmaceutical residues worsen antimicrobial resistance (AMR), with aquaculture systems promoting multidrug-resistant pathogens, while chronic low-level exposure harms endocrine and organ function.

Cost-effective, scalable solutions can disrupt these contamination pathways. For microplastics, biosolid pre-treatment ($15/ton) achieves a 98% reduction in particles and transitioning to non-plastic food processing equipment ($20/unit) offers an accessible solution for low- and middle-income countries (LMICs). PFAS contamination can be reduced by banning PFAS-based food packaging (no cost) and using clay-based soil amendments ($8/ha) to decrease crop uptake by 80%. Pharmaceutical residues can be addressed through thermophilic composting (>55°C, degrading 72% of veterinary drugs) and constructed wetlands ($2/m^2, removing 80% of pharmaceuticals via phytoremediation). These proven, LMIC-viable solutions should be combined with advanced detection methods, such as Raman spectroscopy and AI-driven modeling, to identify and predict contaminant hotspots.

Technical measures alone are insufficient without strong policy frameworks. Global regulations must set science based thresholds for chronic exposure and synergistic effects, which are currently absent for most emerging contaminants. Harmonized standards, coordinated through bodies like the Codex Alimentarius, should enforce blockchain-based traceability and IoT-enabled monitoring to track contaminants across supply chains. A One Health and Planetary Health approach is crucial, integrating data across environmental, animal, and human health domains to mitigate systemic risks. Investments in biodegradable packaging, precision agriculture, and open-access data platforms will further enhance food system sustainability. Isolated efforts are no longer effective. Policymakers, industry leaders, and researchers must collaborate to implement integrated strategies that prioritize vulnerable communities, facilitate technology transfer, and enforce accountability through "polluter pays" principles. Only through cohesive, ethically driven global action can we break the cycle of contamination and ensure food safety for future generations.

REFERENCES

Adhikari, S., Kelkar, V., Kumar, R., & Halden, R. U. (2022). Methods and challenges in the detection of microplastics and nanoplastics: A mini-review. *Polymer International, 71*(5), 543–551. DOI: 10.1002/pi.6348

Al Amin, M., Sobhani, Z., Liu, Y., Dharmaraja, R., Chadalavada, S., Naidu, R., Chalker, J. M., & Fang, C. (2020). Recent advances in the analysis of per- and polyfluoroalkyl substances (PFAS)—A review. *Environmental Technology & Innovation, 19*, 100879. DOI: 10.1016/j.eti.2020.100879

Alqarni, A. M. (2024). Analytical Methods for the Determination of Pharmaceuticals and Personal Care Products in Solid and Liquid Environmental Matrices: A Review. *Molecules (Basel, Switzerland), 29*(16), 3900. DOI: 10.3390/molecules29163900 PMID: 39202981

Andrady, A. L. (2011). Microplastics in the marine environment. *Marine Pollution Bulletin, 62*(8), 1596–1605. DOI: 10.1016/j.marpolbul.2011.05.030 PMID: 21742351

Arienzo, A., Coff, C., & Barling, D. (2008). The European Union and the Regulation of Food Traceability: From Risk Management to Informed Choice? In Coff, C., Barling, D., Korthals, M., & Nielsen, T. (Eds.), *Ethical Traceability and Communicating Food* (Vol. 15, pp. 23–42). Springer Netherlands. DOI: 10.1007/978-1-4020-8524-6_2

Arsène, M. M. J., Davares, A. K. L., Viktorovna, P. I., Andreevna, S. L., Sarra, S., Khelifi, I., & Sergueïevna, D. M. (2022). The public health issue of antibiotic residues in food and feed: Causes, consequences, and potential solutions. *Veterinary World*, 662–671. DOI: 10.14202/vetworld.2022.662-671 PMID: 35497952

Barboza, L. G. A., Dick Vethaak, A., Lavorante, B. R. B. O., Lundebye, A.-K., & Guilhermino, L. (2018). Marine microplastic debris: An emerging issue for food security, food safety and human health. *Marine Pollution Bulletin, 133*, 336–348. DOI: 10.1016/j.marpolbul.2018.05.047 PMID: 30041323

Barry, V., Winquist, A., & Steenland, K. (2013). Perfluorooctanoic Acid (PFOA) Exposures and Incident Cancers among Adults Living Near a Chemical Plant. *Environmental Health Perspectives, 121*(11–12), 1313–1318. DOI: 10.1289/ehp.1306615 PMID: 24007715

Ben, Y., Fu, C., Hu, M., Liu, L., Wong, M. H., & Zheng, C. (2019). Human health risk assessment of antibiotic resistance associated with antibiotic residues in the environment: A review. *Environmental Research, 169*, 483–493. DOI: 10.1016/j.envres.2018.11.040 PMID: 30530088

Białk-Bielińska, A., Kumirska, J., Borecka, M., Caban, M., Paszkiewicz, M., Pazdro, K., & Stepnowski, P. (2016). Selected analytical challenges in the determination of pharmaceuticals in drinking/marine waters and soil/sediment samples. *Journal of Pharmaceutical and Biomedical Analysis*, *121*, 271–296. DOI: 10.1016/j.jpba.2016.01.016 PMID: 26818066

Biswas, S., Gogoi, P., Wilkinson, M., Deka, P., Das, V., Goswami, K., & Sandilya, S. P. (2025). Evaluating Regulatory Approaches to Emerging Pollutants. In Dey, S., & Bhattacharya, S. (Eds.), *Biotechnological Interventions in the Removal of Emerging Pollutants* (pp. 19–36). Springer Nature Singapore. DOI: 10.1007/978-981-97-9922-0_2

Booth, J. (1982). Antibiotic residues in milk. *In Practice*, *4*(4), 100–109. DOI: 10.1136/inpract.4.4.100 PMID: 7186890

Borriello, L., Scivicco, M., Cacciola, N. A., Esposito, F., Severino, L., & Cirillo, T. (2023). Microplastics, a Global Issue: Human Exposure through Environmental and Dietary Sources. *Foods*, *12*(18), 3396. DOI: 10.3390/foods12183396 PMID: 37761106

Bouwmeester, H., Hollman, P. C. H., & Peters, R. J. B. (2015). Potential Health Impact of Environmentally Released Micro- and Nanoplastics in the Human Food Production Chain: Experiences from Nanotoxicology. *Environmental Science & Technology*, *49*(15), 8932–8947. DOI: 10.1021/acs.est.5b01090 PMID: 26130306

Brambilla, G., & Testa, C. (2014). Food safety/food security aspects related to the environmental release of pharmaceuticals. *Chemosphere*, *115*, 81–87. DOI: 10.1016/j.chemosphere.2014.01.024 PMID: 24602346

Brunn, H., Arnold, G., Körner, W., Rippen, G., Steinhäuser, K. G., & Valentin, I. (2023). PFAS: Forever chemicals—persistent, bioaccumulative and mobile. Reviewing the status and the need for their phase out and remediation of contaminated sites. *Environmental Sciences Europe*, *35*(1), 20. DOI: 10.1186/s12302-023-00721-8

Buck, R. C., Franklin, J., Berger, U., Conder, J. M., Cousins, I. T., De Voogt, P., Jensen, A. A., Kannan, K., Mabury, S. A., & Van Leeuwen, S. P. (2011). Perfluoroalkyl and polyfluoroalkyl substances in the environment: Terminology, classification, and origins. *Integrated Environmental Assessment and Management*, *7*(4), 513–541. DOI: 10.1002/ieam.258 PMID: 21793199

Cabello, F. C., Godfrey, H. P., Buschmann, A. H., & Dölz, H. J. (2016). Aquaculture as yet another environmental gateway to the development and globalisation of antimicrobial resistance. *The Lancet. Infectious Diseases*, *16*(7), e127–e133. DOI: 10.1016/S1473-3099(16)00100-6 PMID: 27083976

Chang, E. T., Adami, H.-O., Boffetta, P., Cole, P., Starr, T. B., & Mandel, J. S. (2014). A critical review of perfluorooctanoate and perfluorooctanesulfonate exposure and cancer risk in humans. Critical Reviews in Toxicology, 44(sup1), 1–81. DOI: 10.3109/10408444.2014.905767

Chen, J., Ying, G.-G., & Deng, W.-J. (2019). Antibiotic Residues in Food: Extraction, Analysis, and Human Health Concerns. *Journal of Agricultural and Food Chemistry*, *67*(27), 7569–7586. DOI: 10.1021/acs.jafc.9b01334 PMID: 31198037

Codex Alimentarius Commission. (2020). Codex Pesticides Residues in Food Online Database. Food and Agriculture Organization of the United Nations. Retrieved from https://www.fao.org/fao-who-codexalimentarius/codex-texts/dbs/pestres/en/

Cox, K. D., Covernton, G. A., Davies, H. L., Dower, J. F., Juanes, F., & Dudas, S. E. (2019). Human Consumption of Microplastics. *Environmental Science & Technology*, *53*(12), 7068–7074. DOI: 10.1021/acs.est.9b01517 PMID: 31184127

Currie, S. D., Wang, J.-S., & Tang, L. (2024). Impacts of PFAS Exposure on Neurodevelopment: A Comprehensive Literature Review. *Environments (Basel, Switzerland)*, *11*(9), 188. DOI: 10.3390/environments11090188

Cverenkárová, K., Valachovičová, M., Mackuľak, T., Žemlička, L., & Bírošová, L. (2021). Microplastics in the Food Chain. *Life (Chicago, Ill.)*, *11*(12), 1349. DOI: 10.3390/life11121349 PMID: 34947879

Darwish, W. S., Eldaly, E. A., El-Abbasy, M. T., Ikenaka, Y., Nakayama, S., & Ishizuka, M. (2013). Antibiotic residues in food: The African scenario (No. Supplement). Graduate School of Veterinary Medicine, Hokkaido University. DOI: 10.14943/jjvr.61.suppl.s13

Dawson, A. L., Kawaguchi, S., King, C. K., Townsend, K. A., King, R., Huston, W. M., & Bengtson Nash, S. M. (2018). Turning microplastics into nanoplastics through digestive fragmentation by Antarctic krill. *Nature Communications*, *9*(1), 1001. DOI: 10.1038/s41467-018-03465-9 PMID: 29520086

De-la-Torre, G. E. (2020). Microplastics: An emerging threat to food security and human health. *Journal of Food Science and Technology*, *57*(5), 1601–1608. DOI: 10.1007/s13197-019-04138-1 PMID: 32327770

Death, C., Bell, C., Champness, D., Milne, C., Reichman, S., & Hagen, T. (2021). Per- and polyfluoroalkyl substances (PFAS) in livestock and game species: A review. *The Science of the Total Environment*, *774*, 144795. DOI: 10.1016/j.scitotenv.2020.144795 PMID: 33609849

DeWitt, J. C., Blossom, S. J., & Schaider, L. A. (2019). Exposure to perfluoroalkyl and polyfluoroalkyl substances leads to immunotoxicity: Epidemiological and toxicological evidence. *Journal of Exposure Science & Environmental Epidemiology*, *29*(2), 148–156. DOI: 10.1038/s41370-018-0097-y PMID: 30482935

Diamanti-Kandarakis, E., Bourguignon, J.-P., Giudice, L. C., Hauser, R., Prins, G. S., Soto, A. M., Zoeller, R. T., & Gore, A. C. (2009). Endocrine-Disrupting Chemicals: An Endocrine Society Scientific Statement. *Endocrine Reviews*, *30*(4), 293–342. DOI: 10.1210/er.2009-0002 PMID: 19502515

Domingo, J. L., & Nadal, M. (2017). Per- and Polyfluoroalkyl Substances (PFASs) in Food and Human Dietary Intake: A Review of the Recent Scientific Literature. *Journal of Agricultural and Food Chemistry*, *65*(3), 533–543. DOI: 10.1021/acs.jafc.6b04683 PMID: 28052194

Dong, H., Wang, X., Niu, X., Zeng, J., Zhou, Y., Suona, Z., Yuan, Y., & Chen, X. (2023). Overview of analytical methods for the determination of microplastics: Current status and trends. *Trends in Analytical Chemistry*, *167*, 117261. DOI: 10.1016/j.trac.2023.117261

Duncan, T. V., Khan, S. A., Patri, A. K., & Wiggins, S. (2024). Regulatory Science Perspective on the Analysis of Microplastics and Nanoplastics in Human Food. *Analytical Chemistry*, *96*(11), 4343–4358. DOI: 10.1021/acs.analchem.3c05408 PMID: 38452774

EFSA Panel on Contaminants in the Food Chain (CONTAM). (2014). Scientific Opinion on Chloramphenicol in food and feed. *EFSA Journal*, *12*(11). Advance online publication. DOI: 10.2903/j.efsa.2014.3907

Erhunmwunse, N. O., Pajiah, T. J., & Ogwu, M. C. (2024). Microplastics as Water Pollutants and Sustainable Management Strategies. In Izah, S. C., Ogwu, M. C., Loukas, A., & Hamidifar, H. (Eds.), *Water Crises and Sustainable Management in the Global South*. Springer., DOI: 10.1007/978-981-97-4966-9_8

Evivie, S. E., Ogwu, M. C., Ebabhamiegbebho, P. A., Abel, E. S., Imaren, J. O., & Igene, J. O. (2020). Packaging and the Nigerian food industry: challenges and opportunities. In Ogunlade, C. A., Adeleke, K. M., & Oladejo, M. T. (Eds.), *Food Technology and Culture in Africa* (pp. 28–99). Reamsworth Publishing.

Fang, C., Luo, Y., & Naidu, R. (2023). Microplastics and nanoplastics analysis: Options, imaging, advancements and challenges. *Trends in Analytical Chemistry*, *166*, 117158. DOI: 10.1016/j.trac.2023.117158

Farré, M. L., Pérez, S., Kantiani, L., & Barceló, D. (2008). Fate and toxicity of emerging pollutants, their metabolites and transformation products in the aquatic environment. *Trends in Analytical Chemistry*, *27*(11), 991–1007. DOI: 10.1016/j.trac.2008.09.010

Fatta-Kassinos, D., Vasquez, M. I., & Kümmerer, K. (2011). Transformation products of pharmaceuticals in surface waters and wastewater formed during photolysis and advanced oxidation processes – Degradation, elucidation of byproducts and assessment of their biological potency. *Chemosphere*, *85*(5), 693–709. DOI: 10.1016/j.chemosphere.2011.06.082 PMID: 21835425

Fussell, R. J., Garcia Lopez, M., Mortimer, D. N., Wright, S., Sehnalova, M., Sinclair, C. J., Fernandes, A., & Sharman, M. (2014). Investigation into the Occurrence in Food of Veterinary Medicines, Pharmaceuticals, and Chemicals Used in Personal Care Products. *Journal of Agricultural and Food Chemistry*, *62*(17), 3651–3659. DOI: 10.1021/jf4052418 PMID: 24495205

Gaggero, E., López-Muñoz, M. J., Paganini, M. C., Arencibia, A., Bertinetti, S., Fernández De Paz, N., & Calza, P. (2023). Mercury and Organic Pollutants Removal from Aqueous Solutions by Heterogeneous Photocatalysis with ZnO-Based Materials. *Molecules (Basel, Switzerland)*, *28*(6), 2650. DOI: 10.3390/molecules28062650 PMID: 36985622

Gaines, L. G. T. (2023). Historical and current usage of per- and polyfluoroalkyl substances (PFAS): A literature review. *American Journal of Industrial Medicine*, *66*(5), 353–378. DOI: 10.1002/ajim.23362 PMID: 35614869

Gallego-Ramírez, C., Chica, E., & Rubio-Clemente, A. (2024). Combination of Biochar and Advanced Oxidation Processes for the Sustainable Elimination of Pharmaceuticals in Water. *Sustainability (Basel)*, *16*(23), 10761. DOI: 10.3390/su162310761

Gao, X., Zuo, Q., Fu, X., Song, L., Cen, M., & Wu, J. (2023). Association between prenatal exposure to per- and polyfluoroalkyl substances and neurodevelopment in children: Evidence based on birth cohort. *Environmental Research*, *236*, 116812. DOI: 10.1016/j.envres.2023.116812 PMID: 37536558

Geissdoerfer, M., Savaget, P., Bocken, N. M. P., & Hultink, E. J. (2017). The Circular Economy – A new sustainability paradigm? *Journal of Cleaner Production*, *143*, 757–768. DOI: 10.1016/j.jclepro.2016.12.048

Ghisi, R., Vamerali, T., & Manzetti, S. (2019). Accumulation of perfluorinated alkyl substances (PFAS) in agricultural plants: A review. *Environmental Research*, *169*, 326–341. DOI: 10.1016/j.envres.2018.10.023 PMID: 30502744

Ghosh, A., Kalita, B., Das, J., Das, P. P., Das, N., Chakrabarty, S., Das, A., & Dutta, N. (2025). Microplastic Pollution: Unveiling the Sources, Pathogenesis, and Effective Remediations. In Dey, S., & Bhattacharya, S. (Eds.), *Biotechnological Interventions in the Removal of Emerging Pollutants* (pp. 241–269). Springer Nature Singapore. DOI: 10.1007/978-981-97-9922-0_12

Giesy, J. P., & Kannan, K. (2001). Global distribution of perfluorooctane sulfonate in wildlife. *Environmental Science & Technology*, *35*(7), 1339–1342. DOI: 10.1021/es001834k PMID: 11348064

Giesy, J. P., & Kannan, K. (2002). Peer Reviewed: Perfluorochemical Surfactants in the Environment. *Environmental Science & Technology*, *36*(7), 146A–152A. DOI: 10.1021/es022253t PMID: 11999053

Giri, S., Lamichhane, G., Khadka, D., & Devkota, H. P. (2024). Microplastics contamination in food products: Occurrence, analytical techniques and potential impacts on human health. *Current Research in Biotechnology*, *7*, 100190. DOI: 10.1016/j.crbiot.2024.100190

González-González, R. B., Sharma, P., Singh, S. P., Américo-Pinheiro, J. H. P., Parra-Saldívar, R., Bilal, M., & Iqbal, H. M. N. (2022). Persistence, environmental hazards, and mitigation of pharmaceutically active residual contaminants from water matrices. *The Science of the Total Environment*, *821*, 153329. DOI: 10.1016/j.scitotenv.2022.153329 PMID: 35093347

Graham, F., Paradis, L., Bégin, P., Paradis, J., Babin, Y., & Des Roches, A. (2014). Risk of allergic reaction and sensitization to antibiotics in foods. *Annals of Allergy, Asthma & Immunology*, *113*(3), 329–330. DOI: 10.1016/j.anai.2014.06.029 PMID: 25168224

Guo, X., Lin, H., Xu, S., & He, L. (2022). Recent Advances in Spectroscopic Techniques for the Analysis of Microplastics in Food. *Journal of Agricultural and Food Chemistry*, *70*(5), 1410–1422. DOI: 10.1021/acs.jafc.1c06085 PMID: 35099960

Halden, R. U. (2010). Plastics and health risks. *Annual Review of Public Health*, *31*(1), 179–194. DOI: 10.1146/annurev.publhealth.012809.103714 PMID: 20070188

Harris, S. J., Cormican, M., & Cummins, E. (2012). Antimicrobial Residues and Antimicrobial-Resistant Bacteria: Impact on the Microbial Environment and Risk to Human Health—A Review. *Human and Ecological Risk Assessment*, *18*(4), 767–809. DOI: 10.1080/10807039.2012.688702

Hernando, M., Mezcua, M., Fernandezalba, A., & Barcelo, D. (2006). Environmental risk assessment of pharmaceutical residues in wastewater effluents, surface waters and sediments. *Talanta*, *69*(2), 334–342. DOI: 10.1016/j.talanta.2005.09.037 PMID: 18970571

Heyder, M., Hollmann-Hespos, T., & Theuvsen, L. (2010). Agribusiness Firm Reactions to Regulations: The Case of Investments in Traceability Systems. *International Journal on Food System Dynamics*, *1*(2), 133–142. DOI: 10.18461/ijfsd.v1i2.125

Hopewell, J., Dvorak, R., & Kosior, E. (2009). Plastics recycling: Challenges and opportunities. *Philosophical Transactions of the Royal Society of London. Series B, Biological Sciences*, *364*(1526), 2115–2126. DOI: 10.1098/rstb.2008.0311 PMID: 19528059

Iyiola, A. O., Izah, S. C., Morya, S., Akinsorotan, A., & Ogwu, M. C. (2023). Owalla Reservoir in South-western Nigeria: Assessment of Fish Distribution, Biological Diversity, and Water Quality Index. *Indonesian Journal of Agricultural Research*, *6*(02), 106–125. DOI: 10.32734/injar.v6i2.12207

Iyiola, A. O., Kolawole, A. S., Setufe, S. B., Bilikoni, J., Ofori, E., & Ogwu, M. C. (2024). Fish as a Sustainable Biomonitoring Tool in Aquatic Environments. In Izah, S. C., Ogwu, M. C., & Hamidifar, H. (Eds.), *Biomonitoring of Pollutants in the Global South*. Springer., DOI: 10.1007/978-981-97-1658-6_12

Jahnke, A., & Berger, U. (2009). Trace analysis of per- and polyfluorinated alkyl substances in various matrices—How do current methods perform? *Journal of Chromatography. A*, *1216*(3), 410–421. DOI: 10.1016/j.chroma.2008.08.098 PMID: 18817914

Jambeck, J. R., Geyer, R., Wilcox, C., Siegler, T. R., Perryman, M., Andrady, A., Narayan, R., & Law, K. L. (2015). Plastic waste inputs from land into the ocean. *Science*, *347*(6223), 768–771. DOI: 10.1126/science.1260352 PMID: 25678662

Jin, M., Wang, X., Ren, T., Wang, J., & Shan, J. (2021). Microplastics contamination in food and beverages: Direct exposure to humans. *Journal of Food Science*, *86*(7), 2816–2837. DOI: 10.1111/1750-3841.15802 PMID: 34146409

Jin, N., Song, Y., Ma, R., Li, J., Li, G., & Zhang, D. (2022). Characterization and identification of microplastics using Raman spectroscopy coupled with multivariate analysis. *Analytica Chimica Acta*, *1197*, 339519. DOI: 10.1016/j.aca.2022.339519 PMID: 35168726

Kennedy, G. L., & Symons, J. M. (2015). Carcinogenicity of Perfluoroalkyl Compounds. In DeWitt, J. C. (Ed.), *Toxicological Effects of Perfluoroalkyl and Polyfluoroalkyl Substances* (pp. 265–304). Springer International Publishing. DOI: 10.1007/978-3-319-15518-0_12

Krauss, W. (2015). Anthropology in the Anthropocene: Sustainable Development, Climate Change and Interdisciplinary Research. In Greschke, H., & Tischler, J. (Eds.), *Grounding Global Climate Change* (pp. 59–76). Springer Netherlands. DOI: 10.1007/978-94-017-9322-3_4

Kumar, R., Manna, C., Padha, S., Verma, A., Sharma, P., Dhar, A., Ghosh, A., & Bhattacharya, P. (2022). Micro(nano)plastics pollution and human health: How plastics can induce carcinogenesis to humans? *Chemosphere, 298*, 134267. DOI: 10.1016/j.chemosphere.2022.134267 PMID: 35301996

Kuzukiran, O., Simsek, I., Filazi, A., & Yurdakok-Dikmen, B. (2022). Perfluorooctane sulfonate and perfluorooctanoic acid. In *Reproductive and Developmental Toxicology* (pp. 815–831). Elsevier. DOI: 10.1016/B978-0-323-89773-0.00041-2

Kwiatkowski, C. F., Andrews, D. Q., Birnbaum, L. S., Bruton, T. A., DeWitt, J. C., Knappe, D. R. U., Maffini, M. V., Miller, M. F., Pelch, K. E., Reade, A., Soehl, A., Trier, X., Venier, M., Wagner, C. C., Wang, Z., & Blum, A. (2020). Scientific Basis for Managing PFAS as a Chemical Class. *Environmental Science & Technology Letters, 7*(8), 532–543. DOI: 10.1021/acs.estlett.0c00255 PMID: 34307722

Kwon, J.-H., Kim, J.-W., Pham, T. D., Tarafdar, A., Hong, S., Chun, S.-H., Lee, S.-H., Kang, D.-Y., Kim, J.-Y., Kim, S.-B., & Jung, J. (2020). Microplastics in Food: A Review on Analytical Methods and Challenges. *International Journal of Environmental Research and Public Health, 17*(18), 6710. DOI: 10.3390/ijerph17186710 PMID: 32942613

Lambert, S., Scherer, C., & Wagner, M. (2017). Ecotoxicity testing of microplastics: Considering the heterogeneity of physicochemical properties. *Integrated Environmental Assessment and Management, 13*(3), 470–475. DOI: 10.1002/ieam.1901 PMID: 28440923

Lambert, S., & Wagner, M. (2018). Microplastics Are Contaminants of Emerging Concern in Freshwater Environments: An Overview. In Wagner, M., & Lambert, S. (Eds.), *Freshwater Microplastics* (Vol. 58, pp. 1–23). Springer International Publishing. DOI: 10.1007/978-3-319-61615-5_1

Lapworth, D. J., Baran, N., Stuart, M. E., & Ward, R. S. (2012). Emerging organic contaminants in groundwater: A review of sources, fate, and occurrence. *Environmental Pollution, 163*, 287–303. DOI: 10.1016/j.envpol.2011.12.034 PMID: 22306910

Li, Y., Tao, L., Wang, Q., Wang, F., Li, G., & Song, M. (2023). Potential Health Impact of Microplastics: A Review of Environmental Distribution, Human Exposure, and Toxic Effects. *Environment & Health, 1*(4), 249–257. DOI: 10.1021/envhealth.3c00052 PMID: 39474495

Lindner, L. F. (2008). Regulating food safety: The power of alignment and drive towards convergence. *Innovation (Abingdon), 21*(2), 133–143. DOI: 10.1080/13511610802214487

Liu, Q., Chen, Z., Chen, Y., Yang, F., Yao, W., & Xie, Y. (2021). Microplastics and Nanoplastics: Emerging Contaminants in Food. *Journal of Agricultural and Food Chemistry, 69*(36), 10450–10468. DOI: 10.1021/acs.jafc.1c04199 PMID: 34473500

Lohmann, R. (2017). Microplastics are not important for the cycling and bioaccumulation of organic pollutants in the oceans—But should microplastics be considered POPs themselves? *Integrated Environmental Assessment and Management, 13*(3), 460–465. DOI: 10.1002/ieam.1914 PMID: 28440937

Lozano, A. M. C., & Arias, M. D. C. (2009). Residuos de fármacos en alimentos de origen animal: Panorama actual en Colombia. *Revista Colombiana de Ciencias Pecuarias, 21*(1), 15. DOI: 10.17533/udea.rccp.324262

Ma, X., Cai, D., Chen, Q., Zhu, Z., Zhang, S., Wang, Z., Hu, Z., Shen, H., & Meng, Z. (2024). Hunting Metabolic Biomarkers for Exposure to Per- and Polyfluoroalkyl Substances: A Review. *Metabolites, 14*(7), 392. DOI: 10.3390/metabo14070392 PMID: 39057715

Mainero Rocca, L., Gentili, A., Pérez-Fernández, V., & Tomai, P. (2017). Veterinary drugs residues: A review of the latest analytical research on sample preparation and LC-MS based methods. *Food Additives & Contaminants. Part A, Chemistry, Analysis, Control, Exposure & Risk Assessment*, 1–19. DOI: 10.1080/19440049.2017.1298846 PMID: 28278127

Malchi, T., Maor, Y., Tadmor, G., Shenker, M., & Chefetz, B. (2014). Irrigation of Root Vegetables with Treated Wastewater: Evaluating Uptake of Pharmaceuticals and the Associated Human Health Risks. *Environmental Science & Technology, 48*(16), 9325–9333. DOI: 10.1021/es5017894 PMID: 25026038

Mamun, A. A., Prasetya, T. A. E., Dewi, I. R., & Ahmad, M. (2023). Microplastics in human food chains: Food becoming a threat to health safety. *The Science of the Total Environment, 858*, 159834. DOI: 10.1016/j.scitotenv.2022.159834 PMID: 36461575

Mihai, F.-C., Gündoğdu, S., Khan, F. R., Olivelli, A., Markley, L. A., & Van Emmerik, T. (2022). Plastic pollution in marine and freshwater environments: Abundance, sources, and mitigation. In *Emerging Contaminants in the Environment* (pp. 241–274). Elsevier. DOI: 10.1016/B978-0-323-85160-2.00016-0

Mosharaf, M. K., Gomes, R. L., Cook, S., Alam, M. S., & Rasmusssen, A. (2024). Wastewater reuse and pharmaceutical pollution in agriculture: Uptake, transport, accumulation and metabolism of pharmaceutical pollutants within plants. *Chemosphere, 364*, 143055. DOI: 10.1016/j.chemosphere.2024.143055 PMID: 39127189

Nikolaou, A. (2013). Pharmaceuticals and related compounds as emerging pollutants in water: Analytical aspects. *Global NEST Journal, 15*(1), 1–12. DOI: 10.30955/gnj.000969

Nwizugbo, K. C., Ogwu, M. C., Eriyamremu, G. E., & Ahana, C. M. (2023). Alterations in energy metabolism, total protein, uric and nucleic acids in African sharptooth catfish (*Clarias gariepinus* Burchell.) exposed to crude oil and fractions. *Chemosphere, 316*, 137778. DOI: 10.1016/j.chemosphere.2023.137778 PMID: 36640975

Ogwu, M. C. (2019a). Lifelong Consumption of Plant-Based GM Foods: Is It Safe? In Papadopoulou, P., Misseyanni, A., & Marouli, C. (Eds.), *Environmental Exposures and Human Health Challenges* (pp. 158–176). IGI Global. DOI: 10.4018/978-1-5225-7635-8.ch008

Ogwu, M. C. (2019b). Understanding the Composition of Food Waste: An "-Omics" Approach to Food Waste Management. In Gunjal, A. P., Waghmode, M. S., Patil, N. N., & Bhatt, P. (Eds.), *Global Initiatives for Waste Reduction and Cutting Food Loss* (pp. 212–236). IGI Global. DOI: 10.4018/978-1-5225-7706-5.ch011

Ogwu, M. C. (2023). Local Food Crops in Africa: Sustainable Utilization, Threats, and Traditional Storage Strategies. In *Sustainable Utilization and Conservation of Africa's Biological Resources and Environment.* Springer. DOI: 10.1007/978-981-19-6974-4_13

Ogwu, M. C., Izah, S. C., & Ntuli, N. R. (2024c). *Food Safety and Quality in the Global South.* Springer Nature. DOI: 10.1007/978-981-97-2428-4

Ogwu, M. C., Izah, S. C., Ntuli, N. R., & Odubo, T. C. (2024a). Food Security Complexities in the Global South. In Ogwu, M. C., Izah, S. C., & Ntuli, N. R. (Eds.), *Food Safety and Quality in the Global South* (pp. 3–33). Springer. DOI: 10.1007/978-981-97-2428-4_1

Ogwu, M.C., Onah, T.P., Kosoe, E.A., Mumuni, E., Akolgo-Azupogo, Helen, Obahiagbon, E.G., Izah, S.C., & Imarhiagbe, O. (2024b). Factors Driving the Acceptance of Genetically Modified Food Crops in Ghana. *Food Safety and Health.* DOI: 10.1002/fsh3.12031

Ohoro, C. R., Adeniji, A. O., Okoh, A. I., & Okoh, O. O. (2019). Distribution and Chemical Analysis of Pharmaceuticals and Personal Care Products (PPCPs) in the Environmental Systems: A Review. *International Journal of Environmental Research and Public Health, 16*(17), 3026. DOI: 10.3390/ijerph16173026 PMID: 31438569

Okechukwuyem Ojji, S. (2024). Emerging Technology Integration—Artificial Intelligence (AI) and Machine Learning (ML) for Predictive Analysis for Safety and Toxicity Assessment in Environmental Toxicology. *International Journal of Scientific Research and Management, 12*(05), 1182–1195. DOI: 10.18535/ijsrm/v12i05.ec03

Pasecnaja, E., Bartkevics, V., & Zacs, D. (2022). Occurrence of selected per- and polyfluorinated alkyl substances (PFASs) in food available on the European market – A review on levels and human exposure assessment. *Chemosphere, 287*, 132378. DOI: 10.1016/j.chemosphere.2021.132378 PMID: 34592212

Pironti, C., Ricciardi, M., Motta, O., Miele, Y., Proto, A., & Montano, L. (2021). Microplastics in the Environment: Intake through the Food Web, Human Exposure and Toxicological Effects. *Toxics, 9*(9), 224. DOI: 10.3390/toxics9090224 PMID: 34564375

Piva, E., Fais, P., Ioime, P., Forcato, M., Viel, G., Cecchetto, G., & Pascali, J. P. (2023). Per- and polyfluoroalkyl substances (PFAS) presence in food: Comparison among fresh, frozen and ready-to-eat vegetables. *Food Chemistry, 410*, 135415. DOI: 10.1016/j.foodchem.2023.135415 PMID: 36652797

Pratiwi, R., Ramadhanti, S. P., Amatulloh, A., Megantara, S., & Subra, L. (2023). Recent Advances in the Determination of Veterinary Drug Residues in Food. *Foods, 12*(18), 3422. DOI: 10.3390/foods12183422 PMID: 37761131

Principal Food Safety Regulations. (2020). In International Finance Corporation, Food Safety Handbook: A Practical Guide for Building a Robust Food Safety Management System (pp. 23–54). The World Bank. DOI: 10.1596/978-1-4648-1548-5_ch2

Prosser, R. S., & Sibley, P. K. (2015). Human health risk assessment of pharmaceuticals and personal care products in plant tissue due to biosolids and manure amendments, and wastewater irrigation. *Environment International, 75*, 223–233. DOI: 10.1016/j.envint.2014.11.020 PMID: 25486094

Prüst, M., Meijer, J., & Westerink, R. H. S. (2020). The plastic brain: Neurotoxicity of micro- and nanoplastics. *Particle and Fibre Toxicology*, *17*(1), 24. DOI: 10.1186/s12989-020-00358-y PMID: 32513186

Puckowski, A., Cwięk, W., Mioduszewska, K., Stepnowski, P., & Białk-Bielińska, A. (2021). Sorption of pharmaceuticals on the surface of microplastics. *Chemosphere*, *263*, 127976. DOI: 10.1016/j.chemosphere.2020.127976 PMID: 32835979

Ramírez Carnero, A., Lestido-Cardama, A., Vazquez Loureiro, P., Barbosa-Pereira, L., Rodríguez Bernaldo De Quirós, A., & Sendón, R. (2021). Presence of Perfluoroalkyl and Polyfluoroalkyl Substances (PFAS) in Food Contact Materials (FCM) and Its Migration to Food. *Foods*, *10*(7), 1443. DOI: 10.3390/foods10071443 PMID: 34206351

Ramírez-Morales, D., Masís-Mora, M., Montiel-Mora, J. R., Cambronero-Heinrichs, J. C., Pérez-Rojas, G., Tormo-Budowski, R., Méndez-Rivera, M., Briceño-Guevara, S., Gutiérrez-Quirós, J. A., Arias-Mora, V., Brenes-Alfaro, L., Beita-Sandí, W., & Rodríguez-Rodríguez, C. E. (2021). Multi-residue analysis of pharmaceuticals in water samples by liquid chromatography- mass spectrometry: Quality assessment and application to the risk assessment of urban-influenced surface waters in a metropolitan area of Central America. *Process Safety and Environmental Protection*, *153*, 289–300. DOI: 10.1016/j.psep.2021.07.025

Rehman, A. U., Crimi, M., & Andreescu, S. (2023). Current and emerging analytical techniques for the determination of PFAS in environmental samples. *Trends in Environmental Analytical Chemistry*, *37*, e00198. DOI: 10.1016/j.teac.2023.e00198

Rizzo, L., Manaia, C., Merlin, C., Schwartz, T., Dagot, C., Ploy, M. C., Michael, I., & Fatta-Kassinos, D. (2013). Urban wastewater treatment plants as hotspots for antibiotic resistant bacteria and genes spread into the environment: A review. *The Science of the Total Environment*, *447*, 345–360. DOI: 10.1016/j.scitotenv.2013.01.032 PMID: 23396083

Rochman, C. M., Hoh, E., Hentschel, B. T., & Kaye, S. (2013). Long-Term Field Measurement of Sorption of Organic Contaminants to Five Types of Plastic Pellets: Implications for Plastic Marine Debris. *Environmental Science & Technology*, *130109073312009*. Advance online publication. DOI: 10.1021/es303700s PMID: 23270427

Rosman, N., Salleh, W. N. W., Mohamed, M. A., Jaafar, J., Ismail, A. F., & Harun, Z. (2018). Hybrid membrane filtration-advanced oxidation processes for removal of pharmaceutical residue. *Journal of Colloid and Interface Science*, *532*, 236–260. DOI: 10.1016/j.jcis.2018.07.118 PMID: 30092507

Roth, K., Imran, Z., Liu, W., & Petriello, M. C. (2020). Diet as an Exposure Source and Mediator of Per- and Polyfluoroalkyl Substance (PFAS) Toxicity. *Frontiers in Toxicology*, *2*, 601149. DOI: 10.3389/ftox.2020.601149 PMID: 35296120

Rubio-Armendáriz, C., Alejandro-Vega, S., Paz-Montelongo, S., Gutiérrez-Fernández, Á. J., Carrascosa-Iruzubieta, C. J., & Hardisson-de La Torre, A. (2022). Microplastics as Emerging Food Contaminants: A Challenge for Food Safety. *International Journal of Environmental Research and Public Health*, *19*(3), 1174. DOI: 10.3390/ijerph19031174 PMID: 35162198

Ryu, H., Li, B., De Guise, S., McCutcheon, J., & Lei, Y. (2021). Recent progress in the detection of emerging contaminants PFASs. *Journal of Hazardous Materials*, *408*, 124437. DOI: 10.1016/j.jhazmat.2020.124437 PMID: 33162244

Saeedi, M. (2024). How microplastics interact with food chain: A short overview of fate and impacts. *Journal of Food Science and Technology*, *61*(3), 403–413. DOI: 10.1007/s13197-023-05720-4 PMID: 37360257

Sapuppo, A., & Uga, E. (2024). PFAS e salute: Quali rischi per l'uomo? *QUADERNI ACP*, *31*(5), 218. DOI: 10.53141/QACP.2024.218-222

Sarmah, A. K., Meyer, M. T., & Boxall, A. B. A. (2006). A global perspective on the use, sales, exposure pathways, occurrence, fate and effects of veterinary antibiotics (VAs) in the environment. *Chemosphere*, *65*(5), 725–759. DOI: 10.1016/j.chemosphere.2006.03.026 PMID: 16677683

Schaider, L. A., Balan, S. A., Blum, A., Andrews, D. Q., Strynar, M. J., Dickinson, M. E., Lunderberg, D. M., Lang, J. R., & Peaslee, G. F. (2017). Fluorinated Compounds in U.S. Fast Food Packaging. *Environmental Science & Technology Letters*, *4*(3), 105–111. DOI: 10.1021/acs.estlett.6b00435 PMID: 30148183

Schrenk, D., Bignami, M., Bodin, L., Chipman, J. K., del Mazo, J., Grasl-Kraupp, B., Hogstrand, C., Hoogenboom, L. R., Leblanc, J.-C., Nebbia, C. S., Nielsen, E., Ntzani, E., Petersen, A., Sand, S., Vleminckx, C., Wallace, H., Barregård, L., Ceccatelli, S., Cravedi, J.-P., & Schwerdtle, T.EFSA Panel on Contaminants in the Food Chain (EFSA CONTAM Panel). (2020). Risk to human health related to the presence of perfluoroalkyl substances in food. *EFSA Journal*, *18*(9). Advance online publication. DOI: 10.2903/j.efsa.2020.6223 PMID: 32994824

Schwabl, P., Köppel, S., Königshofer, P., Bucsics, T., Trauner, M., Reiberger, T., & Liebmann, B. (2019). Detection of Various Microplastics in Human Stool: A Prospective Case Series. *Annals of Internal Medicine*, *171*(7), 453–457. DOI: 10.7326/M19-0618 PMID: 31476765

Shen, Y., Wang, L., Ding, Y., Liu, S., Li, Y., Zhou, Z., & Liang, Y. (2024). Trends in the Analysis and Exploration of per- and Polyfluoroalkyl Substances (PFAS) in Environmental Matrices: A Review. *Critical Reviews in Analytical Chemistry, 54*(8), 3171–3195. DOI: 10.1080/10408347.2023.2231535 PMID: 37410543

Silbergeld, E. K., Graham, J., & Price, L. B. (2008). Industrial Food Animal Production, Antimicrobial Resistance, and Human Health. *Annual Review of Public Health, 29*(1), 151–169. DOI: 10.1146/annurev.publhealth.29.020907.090904 PMID: 18348709

Smith, M., Love, D. C., Rochman, C. M., & Neff, R. A. (2018). Microplastics in Seafood and the Implications for Human Health. *Current Environmental Health Reports, 5*(3), 375–386. DOI: 10.1007/s40572-018-0206-z PMID: 30116998

Sohail, M., Urooj, Z., Noreen, S., Baig, M. M. F. A., Zhang, X., & Li, B. (2023). Micro- and nanoplastics: Contamination routes of food products and critical interpretation of detection strategies. *The Science of the Total Environment, 891*, 164596. DOI: 10.1016/j.scitotenv.2023.164596 PMID: 37279810

Stecconi, T., Tavoloni, T., Stramenga, A., Bacchiocchi, S., Barola, C., Dubbini, A., Galarini, R., Moretti, S., Sagratini, G., & Piersanti, A. (2024). A LC-MS/MS procedure for the analysis of 19 perfluoroalkyl substances in food fulfilling recent EU regulations requests. *Talanta, 266*, 125054. DOI: 10.1016/j.talanta.2023.125054 PMID: 37611367

Sun, T., Wang, S., Ji, C., Li, F., & Wu, H. (2022). Microplastics aggravate the bioaccumulation and toxicity of coexisting contaminants in aquatic organisms: A synergistic health hazard. *Journal of Hazardous Materials, 424*, 127533. DOI: 10.1016/j.jhazmat.2021.127533 PMID: 34879523

Sunderland, E. M., Hu, X. C., Dassuncao, C., Tokranov, A. K., Wagner, C. C., & Allen, J. G. (2019). A review of the pathways of human exposure to poly- and perfluoroalkyl substances (PFASs) and present understanding of health effects. *Journal of Exposure Science & Environmental Epidemiology, 29*(2), 131–147. DOI: 10.1038/s41370-018-0094-1 PMID: 30470793

Temkin, A. M., Hocevar, B. A., Andrews, D. Q., Naidenko, O. V., & Kamendulis, L. M. (2020). Application of the Key Characteristics of Carcinogens to Per and Polyfluoroalkyl Substances. *International Journal of Environmental Research and Public Health, 17*(5), 1668. DOI: 10.3390/ijerph17051668 PMID: 32143379

The Codex Alimentarius and the European Legislation on Food Additives. (2017). In P. Laganà, E. Avventuroso, G. Romano, M. E. Gioffré, P. Patanè, S. Parisi, U. Moscato, & S. Delia, SpringerBriefs in Molecular Science (pp. 23–32). Springer International Publishing. DOI: 10.1007/978-3-319-57042-6_2

Van Asselt, E. D., Rietra, R. P. J. J., Römkens, P. F. A. M., & Van Der Fels-Klerx, H. J. (2011). Perfluorooctane sulphonate (PFOS) throughout the food production chain. *Food Chemistry, 128*(1), 1–6. DOI: 10.1016/j.foodchem.2011.03.032 PMID: 25214321

Van Boeckel, T. P., Brower, C., Gilbert, M., Grenfell, B. T., Levin, S. A., Robinson, T. P., Teillant, A., & Laxminarayan, R. (2015). Global trends in antimicrobial use in food animals. *Proceedings of the National Academy of Sciences of the United States of America, 112*(18), 5649–5654. DOI: 10.1073/pnas.1503141112 PMID: 25792457

Van Der Voet, H., De Boer, W. J., Kruisselbrink, J. W., Goedhart, P. W., Van Der Heijden, G. W. A. M., Kennedy, M. C., Boon, P. E., & Van Klaveren, J. D. (2015). The MCRA model for probabilistic single-compound and cumulative risk assessment of pesticides. *Food and Chemical Toxicology, 79*, 5–12. DOI: 10.1016/j.fct.2014.10.014 PMID: 25455888

Van Dijk, H., Houghton, J., Van Kleef, E., Van Der Lans, I., Rowe, G., & Frewer, L. (2008). Consumer responses to communication about food risk management. *Appetite, 50*(2–3), 340–352. DOI: 10.1016/j.appet.2007.08.011 PMID: 17945386

Van Raamsdonk, L. W. D., Van Der Zande, M., Koelmans, A. A., Hoogenboom, R. L. A. P., Peters, R. J. B., Groot, M. J., Peijnenburg, A. A. C. M., & Weesepoel, Y. J. A. (2020). Current Insights into Monitoring, Bioaccumulation, and Potential Health Effects of Microplastics Present in the Food Chain. *Foods, 9*(1), 72. DOI: 10.3390/foods9010072 PMID: 31936455

Verlicchi, P., Al Aukidy, M., & Zambello, E. (2012). Occurrence of pharmaceutical compounds in urban wastewater: Removal, mass load and environmental risk after a secondary treatment—A review. *The Science of the Total Environment, 429*, 123–155. DOI: 10.1016/j.scitotenv.2012.04.028 PMID: 22583809

Vieira, Y., Lima, E. C., Foletto, E. L., & Dotto, G. L. (2021). Microplastics physicochemical properties, specific adsorption modeling and their interaction with pharmaceuticals and other emerging contaminants. *The Science of the Total Environment, 753*, 141981. DOI: 10.1016/j.scitotenv.2020.141981 PMID: 32911167

World Health Organization. (2017). Pharmaceuticals in drinking-water: Public health and environment. https://www.who.int/publications/i/item/9789241512753

Wu, X., Ernst, F., Conkle, J. L., & Gan, J. (2013). Comparative uptake and translocation of pharmaceutical and personal care products (PPCPs) by common vegetables. *Environment International, 60*, 15–22. DOI: 10.1016/j.envint.2013.07.015 PMID: 23973619

Xiang, L., Li, Y.-W., Yu, P.-F., Feng, N.-X., Zhao, H.-M., Li, H., Cai, Q.-Y., Mo, C.-H., & Li, Q. X. (2019). Food Safety Concerns: Crop Breeding as a Potential Strategy to Address Issues Associated with the Recently Lowered Reference Doses for Perfluorooctanoic Acid and Perfluorooctane sulfonate. Journal of Agricultural and Food Chemistry. DOI: 10.1021/acs.jafc.9b04625

Xing, Y., Zhou, Y., Zhang, X., Lin, X., Li, J., Liu, P., Lee, H. K., & Huang, Z. (2023). The sources and bioaccumulation of per- and polyfluoroalkyl substances in animal-derived foods and the potential risk of dietary intake. *The Science of the Total Environment, 905*, 167313. DOI: 10.1016/j.scitotenv.2023.167313 PMID: 37742961

Zajác, P., Čapla, J., & Čurlej, J. (2025). Microplastic contamination of food. *Scifood, 19*, 1–16. DOI: 10.5219/scifood.1

Zanni, S., Cammalleri, V., D'Agostino, L., Protano, C., & Vitali, M. (2024). Occurrence of pharmaceutical residues in drinking water: A systematic review. *Environmental Science and Pollution Research International, 32*(16), 10436–10463. DOI: 10.1007/s11356-024-34544-8 PMID: 39103588

Zhang, L., Huang, Y., Zhou, Y., Buckley, T., & Wang, H. H. (2013). Antibiotic Administration Routes Significantly Influence the Levels of Antibiotic Resistance in Gut Microbiota. *Antimicrobial Agents and Chemotherapy, 57*(8), 3659–3666. DOI: 10.1128/AAC.00670-13 PMID: 23689712

Zhang, Q., Xu, E. G., Li, J., Chen, Q., Ma, L., Zeng, E. Y., & Shi, H. (2020). A Review of Microplastics in Table Salt, Drinking Water, and Air: Direct Human Exposure. *Environmental Science & Technology, 54*(7), 3740–3751. DOI: 10.1021/acs.est.9b04535 PMID: 32119774

Zornes, G., Curl, J., & Holmer, M. (2022). PFAS Complicate Potable Reuse and Advanced Treatment. *Journal - American Water Works Association, 114*(2), 72–74. DOI: 10.1002/awwa.1871

Chapter 3
Risk Assessment Frameworks and Predictive Modeling for Food Safety Decision-Making

Sylvester Izah
https://orcid.org/0000-0001-5526-006X
Bayelsa Medical University, Nigeria

ABSTRACT

Food safety is a critical public health priority amid increasingly complex global food supply chains. This chapter explores the integration of predictive modeling with traditional risk assessment frameworks to enhance food safety decision-making. It outlines core principles of risk assessment (hazard identification, hazard characterization, exposure assessment, and risk characterization) within global frameworks. The chapter then examines predictive modeling approaches, including statistical, mechanistic, and machine learning methods, to forecast microbial growth, contamination, and outbreaks. It highlights how these models complement traditional assessments to support proactive, dynamic risk management. Challenges such as data quality, model transparency, and regulatory acceptance are addressed, along with opportunities enabled by IoT, blockchain, etc. The chapter concludes with strategic recommendations for policymakers, industry professionals, and researchers committed to building resilient, tech-driven food safety systems in the digital era.

DOI: 10.4018/979-8-3373-7052-1.ch003

1. INTRODUCTION

Food safety remains a pressing public health concern, particularly in low- and middle-income countries where contamination from antibiotics (Izah et al., 2025), trace metals (Izah and Aigberua 2020; Izah et al., 2022a), pesticides (Sawyer et al., 2024), and microbial agents is prevalent (Izah et al., 2022b; Ovuru et al., 2024). Numerous studies have consistently highlighted the health risks posed by residues and contaminants in various food products including poultry, vegetables, and ready-to-eat foods. For instance, Izah et al. (2025) reported significant antibiotic residues in poultry, raising concerns about antimicrobial resistance, while other works identified hazardous levels of trace metals in food drinks (Odubo & Izah, 2024), smoked fish (Kigigha et al., 2017), and condiments (Aigberua et al., 2018). These findings highlights the urgent need for science-based frameworks to manage foodborne risks. Predictive tools such as probabilistic modeling (Izah et al., 2024), and environmental monitoring (Aigberua et al., 2023) have proven valuable in quantifying and projecting exposure levels.

Furthermore, the growing complexity of global food systems is increasingly evident as the interplay between local and international markets becomes more pronounced. Current food systems are influenced by a multitude of factors including globalization, climate change, and evolving consumer preferences, which collectively complicate the management and regulation of food safety. Anwar et al. (2024) argue that effective governance through collaboration among suppliers, regulatory bodies, and consumers is essential for cultivating a food safety culture within these complex systems. Such collaboration can enhance compliance with food safety standards and bolster resilience against supply chain disruptions due to factors like pandemics or climate events (Anwar et al., 2024). The multifaceted nature of food systems indicates that ensuring food safety requires an integrated approach that extends beyond local jurisdictions to encompass global governance structures.

Foodborne illnesses remain a significant and growing public health concern worldwide (Kigigha et al., 2016, 2017a,b), stressing the urgent need for improved food safety protocols. Data from Smith et al. (2016) highlight that despite advancements in detection and control technologies, incidents of foodborne illnesses have increased, emphasizing the inadequacies in current regulatory frameworks and their implementation. This increase is troubling given the economic implications and strain on public health resources. The complexity of food systems, coupled with varying standards across regions and countries, significantly contributes to the incidence of such illnesses. Strategies proposed, such as predictive modeling and enhanced risk assessment frameworks, aim to mitigate these challenges and demonstrate the necessity of shifting towards a proactive, rather than reactive, approach to food safety (Ahangarkolaee & Gorton, 2020; Li et al., 2018).

The integration of data-driven, risk-based decision-making in food safety management is crucial for effectively managing foodborne illness. Regulatory bodies can utilize data mining and risk analysis frameworks to identify potential safety risks before they manifest (Li et al., 2018). Leveraging on methodologies like Whole Genome Sequencing (WGS) can lead to significant advancements in disease surveillance and outbreak management, particularly concerning pathogens such as Salmonella. Enhanced risk assessment protocols grounded in data analytics allow for a more agile response to foodborne risks, thereby improving public health outcomes and consumer trust (Li et al., 2018).

Predictive modeling techniques could play a vital role in the evolution of food safety systems. The application of statistical and machine learning models enables stakeholders to forecast potential food safety breaches and respond with greater precision and timeliness (Feng et al., 2017). By harnessing analytics, the food industry can implement proactive measures that address latent risks before they result in foodborne illness. Studies indicate that improved risk assessment techniques can yield insights that effectively redirect strategies in food safety (Li et al., 2024). This not only enhances compliance but also fosters a culture of shared responsibility among producers, suppliers, and consumers (Anwar et al., 2024). The application of technology-driven solutions reflects a paradigm shift in the food safety landscape, emphasizing the importance of innovation in regulatory practices.

However, alongside the promise of data-driven approaches, numerous implementation challenges arise. The fragmentation of regulatory responsibilities, as noted by Alrobaish et al. (2020), often leads to inefficiencies and lapses in compliance across the food supply chain. This complexity necessitates a unified governance structure that clarifies roles and responsibilities while fostering effective coordination among various stakeholders (Thomann, 2018). Policymaking must take into account the ramifications of regulatory decentralization, especially in large markets like China, where changes in regulatory authority can significantly impact food safety outcomes (Lin & Liang, 2024). Addressing these governance challenges requires not only regulatory restructuring but also a cultural shift that prioritizes food safety throughout production and supply chain practices.

Furthermore, the role of consumer trust in shaping food safety outcomes is paramount. Nuttavuthisit and Thøgersen (2025) emphasize that consistent standards and robust regulatory oversight are essential for ensuring that consumers can trust the safety of their food. Enhancing transparency through improved communication and education regarding food safety practices can bolster consumer confidence and encourage better engagement with food safety initiatives. The interplay between consumer trust and regulatory efficacy highlight the necessity for a coherent strategy that addresses the root causes of foodborne illnesses while cultivating a knowledgeable consumer base (Nuttavuthisit & Thøgersen, 2015).

This chapter focuses on the integration of predictive modeling techniques with established food safety risk assessment frameworks to strengthen decision-making processes. It examines how statistical, mechanistic, and machine learning models can effectively complement traditional methods by supporting more proactive, data-driven approaches to risk management. By combining these advanced modeling tools with conventional frameworks, the chapter demonstrates how food safety interventions can become more accurate, timely, and responsive, thus leading to reduction the incidence of foodborne illnesses and contributing to improved public health outcomes.

2. FOUNDATIONS OF FOOD SAFETY RISK ASSESSMENT

Food safety risk assessment is a vital component of public health, particularly concerning the consumption of food contaminated with hazardous substances. Central to this domain are several core definitions, including hazard, risk, exposure, and vulnerability. A hazard is identified as any biological, chemical, or physical agent in food that poses potential health risks, such as trace metals (Ogamba et al., 2021; Uzoekwe et al., 2021; Izah et al., 2021; Aigberua et al., 2021) and polycyclic aromatic hydrocarbons (PAHs) (Aigberua et al., 2023). The risk is defined as the likelihood and severity of adverse health effects after exposure to a hazard, which entails evaluating how much of a substance (Ingegno et al., 2024). The exposure measures involve determining the amount and frequency of contact with these hazards, exemplified by cumulative exposure to PAHs from regularly consuming smoked fish (Assogba et al., 2021;Aigberua et al., 2021). Lastly, vulnerability accounts for individual or population sensitivities to the adverse effects of these hazards, as seen in groups like infants and pregnant women who face heightened risks from mercury toxicity in seafood (Künîli et al., 2020; Ingegno et al., 2024).

The four pillars of risk assessment (including hazard identification, hazard characterization, exposure assessment, and risk characterization) form the framework used to evaluate food safety (Figure 1). Hazard identification focuses on detecting harmful substances, such as the presence of heavy metals in garri (Kigigha et al., 2018), cowmeat (Seiyaboh et al., 2018a), palm oil (Seiyaboh et al., 2018b). Next is hazard characterization aims to describe the nature of the adverse effects linked to these hazards while establishing dose-response relationships. This involves identifying levels of substances, like cadmium, at which no adverse effects are observed (Ciecierska et al., 2025). The exposure assessment quantifies consumption levels, enabling regulators to establish guidelines, such as the tolerable weekly intake (TWI) for mercury set by the European Food Safety Authority (EFSA), thus concluding that overconsumption signals a health concern (Zapico et al., 2022; Aigberua et al.,

2018). Finally, the risk characterization integrates findings from the prior steps to estimate potential health risks linked to particular hazards, ensuring public health policies are informed by robust evidence (Iwegbue et al., 2015).

Figure 1. Four pillars of risk assessment

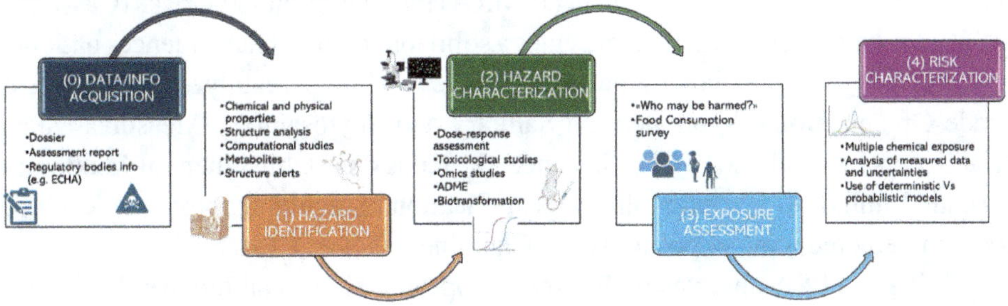

Source: D'Amore et al. (2025)

Qualitative and quantitative approaches in risk assessment can generate different insights. Qualitative evaluations, which categorize risks as "high," "medium," or "low," are often applied where data is limited, such as assessing smoked meat risks based on estimated PAH formation (Tarigholizadeh et al., 2023; Huang et al., 2018). In contrast, quantitative assessments harness numerical data and statistical modeling to evaluate risk magnitude, thus providing a more detailed analysis of dietary contaminants such as lead in children's food (Ciecierska et al., 2023; Mahmudatusa'adah & Sudewi, 2017). Studies illustrate these assessments in action, such as the rigorous evaluations of PAH content in smoked fish, which led to improved processing methods and regulations (Khiari et al., 2020; Samarajeewa, 2023). Similarly, evaluations of arsenic levels in rice have prompted international regulatory standards to safeguard vulnerable populations (Ingegno et al., 2024).

3. GLOBAL RISK ASSESSMENT FRAMEWORKS FOR FOODS

The importance of global risk assessment frameworks for food safety has grown in response to increasing complexities in food supply chains influenced by globalization. Prominent frameworks include the Codex Alimentarius Commission (CAC) and the European Food Safety Authority (EFSA), each contributing uniquely to the establishment of international food safety standards.

The CAC, jointly established by the World Health Organization (WHO) and Food and Agriculture Organization (FAO), serves as a cornerstone in setting international food standards aimed at safeguarding consumer health and ensuring fair trade practices. The framework highlights critical components in its risk assessment process, which includes Hazard Identification, Hazard Characterization, Exposure Assessment, and Risk Characterization. These steps rely heavily on scientific data, often sourced from expert committees such as the Joint FAO/WHO Expert Committee on Food Additives (JECFA) for chemical assessments and the Joint Expert Meetings on Microbiological Risk Assessment (JEMRA) for microbiological risks (Cassini et al., 2016). This rigorous method provides a solid foundation that influences national food safety regulations and international dispute resolutions, catalyzed by the World Trade Organization's Agreement on Sanitary and Phytosanitary Measures (SPS) (Hadjigeorgiou et al., 2012). For instance, the Codex's establishment of Maximum Residue Limits (MRLs) exemplifies its application in regulating pesticide levels in food, thereby reducing exposure risks (Chammem et al., 2018).

Similarly, EFSA operates within the European Union as an independent entity tasked with advancing food safety through scientific assessments. Its collaborative engagement with the European Commission, Parliament, and member states allows for a comprehensive approach to food safety challenges, addressing chemical, biological, and nutritional risks (Hadjigeorgiou et al., 2012). EFSA's commitment to transparency is evidenced by its extensive public consultations and the publication of scientific opinions that inform legislation

In the United States, the frameworks established by the Food and Drug Administration (FDA) and the United States Department of Agriculture (USDA) shape regulatory oversight of food safety across various categories, from processed foods to meat products. The FDA's application of the Food Safety Modernization Act (FSMA) emphasizes preventative approaches to food safety via Quantitative Microbial Risk Assessment (QMRA) models for contaminants, notably in seafood and produce (Commeau et al., 2011). Concurrently, the USDA implements the Hazard Analysis and Critical Control Points (HACCP) system, a preventive approach aimed at managing critical food safety risks from the farm to the consumer (Hyde et al., 2016).

Additionally, JEMRA provides expert insights on microbiological hazards, assisting national food safety authorities in mitigating risks associated with pathogens such as Salmonella and Listeria (Matheron et al., 2020). By supporting Codex committees, JEMRA ensures that scientific findings translate into actionable guidelines that influence food safety legislation globally (Evers et al., 2014). This collaboration emphasizes the interconnected nature of food safety across borders, pointing to the need for harmonization between various national frameworks.

While these frameworks significantly contribute to food safety, challenges remain in harmonizing risk assessment methodologies. The disparity in risk management philosophies, such as the EU's precautionary principle contrasted with the US's more evidence-based approaches, illustrates a fundamental challenge in aligning global standards (Martischang et al., 2021). Furthermore, variations in data availability and acceptable exposure levels complicate efforts toward unified regulatory practices, underscoring the necessity for ongoing dialogue and collaboration among international agencies.

4. THE RISE OF PREDICTIVE MODELING IN FOOD SAFETY

The rise of predictive modeling in food safety signifies a substantial advancement in the management of food systems, particularly in relation to global supply chains and consumer demands. This transformation is driven by several key factors including the increasing complexity of food systems, the availability of extensive data, the urgent need for rapid responses to foodborne outbreaks, and regulatory movements towards data-driven methodologies.

The growing complexity of food systems introduces a multitude of variables, complicating traditional monitoring methods. Béné et al. (2020) emphasize how factors such as urbanization, trade policies, and technological innovations shape food systems globally. Multi-ingredient, ready-to-eat products add to these challenges, necessitating dynamic risk assessment techniques to ensure safety. Innovations in smart packaging, as discussed by Drago et al.(2020), are essential in addressing these issues by enabling real-time monitoring of food quality throughout its lifecycle, ultimately aiding in contamination prevention (Drago et al., 2020).

The availability of large datasets plays a crucial role in enhancing the flexibility and responsiveness of predictive models. By utilizing extensive data from microbiological laboratories and environmental factors, food safety systems can now implement sophisticated analytics to derive actionable insights. For example, Jin et al. (demonstrate how supply chain analytics can efficiently analyze vast amounts of data on food safety tests, helping to identify and mitigate risks (Jin et al., 2021). Additionally, digital traceability mechanisms such as blockchain significantly enhance real-time data collection and analysis, which are crucial for proactive risk management and maintaining food product integrity (Drago et al., 2020).

Moreover, the urgency for rapid foodborne outbreak responses has amplified the demand for quick intervention strategies. Traditional food safety systems, which often rely on reactive measures to handle incidents after they occur, are inadequate in the face of increasing contamination incidents. Predictive models can facilitate early warning systems capable of detecting potential outbreaks by continuously monitoring relevant variables, thereby enabling stakeholders to anticipate and mitigate risks before they escalate.

Furthermore, regulatory bodies, including the FDA and EFSA, are increasingly promoting the integration of predictive modeling in food safety practices. Recent perspectives underscore the importance of food businesses demonstrating their risk control measures and predictive abilities, thereby fostering a culture of proactive food safety management (Tan et al., 2025). This regulatory environment encourages the adoption of modeling techniques that inform risk assessments and enhance decision-making processes regarding food safety interventions.

In terms of the predictive models employed in food safety, three primary categories have emerged. Statistical models use historical data to forecast contamination likelihood, identifying patterns through methods such as regression analysis (Tan et al., 2025). Mechanistic models, founded on chemical and biological principles, predict how microorganisms behave under various environmental conditions (Tan et al., 2025). Machine learning and artificial intelligence models enhance predictive capabilities by allowing systems to learn from data and identify complex patterns without needing explicit programming (Tan et al., 2025). The application of these models across various food sectors significantly improves early detection, surveillance, and risk assessment processes, leading to more robust food safety systems.

5. BUILDING AND APPLYING PREDICTIVE MODELS FOR FOOD SAFETY

Building and applying predictive models for food safety is crucial in addressing the substantial public health risks posed by foodborne pathogens (Sawyer and Izah 2024). To enhance food safety protocols, it is essential to integrate various data sources, including environmental, genomic, operational, and sensor data, which can inform sophisticated predictive models relevant to monitoring food safety dynamics. Machine learning has been identified as a powerful analytical tool within this sphere, facilitating the proactive identification and mitigation of food safety threats (Wang et al., 2021). These diverse data sources provide a comprehensive landscape that encompasses past microbial contamination patterns and operational practices, which are vital for establishing predictive frameworks in food safety.

The preprocessing and feature engineering of such datasets involve significant methodological rigor. Data preprocessing is fundamental for ensuring data quality and involves steps such as cleaning, normalization, and transformation of raw data into a structured format suitable for analysis, significantly enhancing the accuracy of predictive models (Wang et al., 2021; Marvin et al., 2016). Furthermore, feature engineering, which includes selecting and optimizing indicators that capture microbiological growth potential and contamination events, directly influences model performance and interpretability. The major features might include temperature variations, humidity levels, and specific microbial load measurements, which can significantly alter food safety outcomes (Chen et al., 2023).

The development and validation of models in this context necessitate robust workflows that allow for iterative testing and refinement. Predictive modeling often employs techniques such as regression analysis and machine learning algorithms to simulate microbial responses under various conditions (Marvin et al., 2016). These models have facilitated real-time monitoring and intervention in supply chains, underscoring the importance of predictive analytics in food safety management.

Moreover, the investigation of microbial contamination in ready-to-eat foods reveals a critical aspect of food safety risk. Studies have reported a concerning prevalence of pathogens such as Listeria monocytogenes and Staphylococcus aureus in various ready-to-eat foods (Eslami et al., 2017; Arficho & Kebede, 2019). These findings highlight the urgent need for enhanced surveillance and control measures, particularly in food service environments where operational lapses can exacerbate contamination risks (Taha et al., 2022). Integrating real-time sensor data to track microbial contamination provides managers with actionable insights, enabling swift responses to emerging threats before they escalate into larger food safety incidents (Wang et al., 2021). Thus, a combined approach utilizing both predictive modeling and real-time contamination tracking presents a comprehensive strategy for advancing food safety initiatives.

6. INTEGRATING PREDICTIVE MODELING WITH RISK ASSESSMENT FOR FOOD SAFETY

The integration of predictive modeling with risk assessment in food safety represents a pivotal advancement in safeguarding public health against foodborne illnesses. At the heart of this integration is Quantitative Microbial Risk Assessment (QMRA), a systematic approach designed to estimate the risk posed by pathogenic microorganisms in food. QMRA operates through four critical steps: hazard identification, exposure assessment, dose-response assessment, and risk characterization, which collectively quantify the potential for illness associated with food consumption. For example, the risk of E. coli infection from undercooked ground beef can be evaluated using temperature-time predictive models that simulate microbial behavior during cooking and storage processes (Koseki, 2016; Baranyi & Silva, 2017). This model provides essential insights into the effectiveness of interventions like cooking and refrigeration, ultimately facilitating informed decision-making among food safety policymakers (Williams et al., 2010).

In conjunction with QMRA, Bayesian methods enhance the management of uncertainty and sensitivity in risk models. These statistical approaches allow for the incorporation of prior knowledge and expert judgments, leading to refined predictions as new data becomes available. Bayesian methods excel in quantifying the variability across different populations, which is crucial for understanding the diverse responses to foodborne pathogens based on factors such as immune response and dietary habits (Spence et al., 2016; Puza, 2015). Additionally, these methods aid in sensitivity analysis, allowing risk assessors to identify critical parameters that significantly influence overall risk, thereby prioritizing resources and interventions effectively (Williams et al., 2010).

To assist risk managers in navigating complex data, several visual tools, including decision trees and risk matrices, have gained prominence. Decision trees map potential outcomes, helping to evaluate the cost-effectiveness of various interventions, such as comparing pasteurization to irradiation techniques (Ranta et al., 2010). Risk matrices, on the other hand, graphically represent the likelihood and severity of different food safety hazards, facilitating prioritization where high-risk issues, like aflatoxin contamination can be readily identified (Ranjbaran et al., 2021). Additionally, visualization tools such as heat maps and risk curves make it easier to communicate intricate risk data to both decision-makers and the general public (Waterhouse et al., 2021).

Moreover, Decision Support Systems (DSS) have emerged as pivotal in the digital transformation of food safety management. These systems integrate predictive models, historical trends, and real-time data to simulate "what-if" scenarios that inform risk mitigation strategies. For instance, a DSS developed to simulate the impact of delayed cold chain logistics on salmonella growth in poultry illustrates how technology can directly influence food safety practices (Gavai et al., 2022). Digital dashboards also play a critical role by providing continuous updates on key indicators such as pathogen surveillance and supply chain monitoring, thereby allowing stakeholders to make timely decisions based on real-time alerts and outcomes (Powell, 2013; Li et al., 2012).

7. REGULATORY AND POLICY CONSIDERATIONS

Regulatory and policy considerations in risk assessment and predictive modeling are paramount in ensuring food safety, an issue of growing concern as global trade and food production evolve. Predictive modeling plays a crucial role in shaping risk-based inspection and enforcement strategies. Regulatory agencies are increasingly adopting these models to prioritize inspections based on estimated risks rather than relying solely on random sampling methodologies. For instance, predictive models can forecast microbial growth in perishable foods, assisting inspectors in identifying locations that are at higher risk of foodborne illness outbreaks and thus guiding cold chain inspections effectively (Wang et al., 2021; Williams et al., 2010). This targeted approach not only optimizes the allocation of limited resources but also focuses efforts on high-risk products and processing facilities, thereby improving food safety outcomes (Wang et al., 2021).

Table 1. Regulatory and policy considerations related to risk assessment and predictive modeling for food safety

section	Major highlights
Predictive modeling in risk-based enforcement	- Prioritizes inspections based on risk - Optimizes resource allocation - Supports early warning systems - Simulates contamination scenarios for planning
Transparency, reproducibility and explainability	- Requires disclosure of model assumptions and data - Enables independent validation - Promotes trust through understandable models - Essential for AI-based regulatory decisions
Data privacy, proprietary models and harmonization	- Concerns over sharing sensitive industry data - Proprietary models may limit regulatory transparency - Differences in modeling methods across countries hinder alignment - Regulatory lag in adopting new modeling technologies
International collaboration	- WTO-SPS promotes science-based trade regulations - Codex sets harmonized food safety standards - FAO/WHO provide global modeling guidance (e.g., JEMRA, JECFA) - Supports capacity building, data sharing, and global equivalence

Moreover, predictive modeling facilitates the development of early warning systems for emerging food safety hazards, such as Salmonella outbreaks, by analyzing historical data and identifying patterns that could indicate potential risks in food production and distribution networks. This proactive risk management empowers regulatory bodies to design interventions aimed at preventing outbreaks rather than merely reacting to them after they occur (Wang et al., 2021; Chen et al., 2023). By simulating various contamination scenarios, predictive models also support strategic enforcement planning and enhance overall emergency preparedness, further bolstering public health protections (Wang et al., 2021; Wang & Chen, 2016).

A critical aspect of deploying predictive models effectively is ensuring transparency, reproducibility, and explainability. Regulatory agencies must disclose the assumptions, algorithms, and data inputs utilized in these models to build stakeholder trust (Wang et al., 2021; Dove et al., 2020). Open-source models, particularly those used for microbial risk assessments, allow for independent validation and review, promoting reliability and acceptance across the food safety landscape (Wang et al., 2021; Dove et al., 2020). In addition, reproducibility is paramount; models should be verifiable by third-party institutions to ensure that results are consistently reliable over time and across various datasets (Wang et al., 2021; Chen et al., 2017). With the increasing utilization of artificial intelligence (AI) in these predictive frameworks, the explainability of model outcomes becomes crucial, as both regulators and food businesses need clarity on how predictive conclusions are derived to make informed decisions (Wang et al., 2021; Chen et al., 2023).

Despite these advancements, challenges related to data privacy, proprietary models, and international harmonization present significant hurdles to effective implementation. The use of detailed supply chain data, such as processing conditions and transport logs, raises concerns over confidentiality, necessitating a careful balance between safeguarding industry data and ensuring food safety (Oriekhoe et al., 2024; D'Amore et al., 2025). In particular, proprietary models utilized by commercial entities might remain confidential and inaccessible for regulatory scrutiny, which can undermine public trust and acceptance (Oriekhoe et al., 2024). Furthermore, variations in risk modeling methodologies across countries can lead to discrepancies in safety evaluations, particularly concerning thresholds for risks such as aflatoxin levels, exemplifying the need for coherent international standards (D'Amore et al., 2025).

International collaboration is instrumental in addressing these regulatory challenges. Initiatives such as the World Trade Organization (WTO) Sanitary and Phytosanitary (SPS) Agreement promote the adoption of science-based risk assessments to mitigate unjustified trade barriers (Wang et al., 2022; Song et al., 2022). Organizations like the Codex Alimentarius Commission play a critical role in developing harmonized global food safety standards and enhancing international cooperation on risk modeling and predictive frameworks (Wang et al., 2022). Collaborative efforts facilitated by the Food and Agriculture Organization (FAO) and the World Health Organization (WHO) provide essential guidelines and expert consultations to support risk analysis globally (Wang et al., 2022). Joint modeling platforms and data sharing initiatives can strengthen consistency in safety evaluations, particularly in low- and middle-income countries, thereby improving global food safety (Wang et al., 2022).

8. FUTURE DIRECTIONS AND INNOVATIONS

The future of risk assessment frameworks and predictive modeling in food safety is significantly enhanced by technological innovations such as the Internet of Things (IoT), digital twins, and blockchain technology. These advancements facilitate real-time monitoring and analysis, improving food safety through more precise risk assessments.

Table 2. Future Directions and Innovations in Risk Assessment Frameworks and Predictive Modeling for Food Safety

Focus Area	Major Innovations and Applications
Iot and real-time analytics	- Smart sensors monitor temperature, humidity, spoilage in real time - Real-time analytics detect contamination risks - Enables early warning systems and automated decisions
Digital twins	- Virtual replicas simulate food processes and predict failures - Models shelf-life, contamination spread, and CCPs - Enhances preventive maintenance and dynamic risk assessments
Blockchain integration	- Tamper-proof traceability from farm to fork - Enables fast recalls and transparency - Builds consumer trust and auditability of supply chains
Ethical and governance considerations	- Ensures data privacy and reduces algorithmic bias - Requires transparent, accountable model governance - Promotes ethical AI and inclusive stakeholder policies

The integration of IoT-enabled sensors in food production and supply chains offers a transformative approach to monitoring critical parameters like temperature, humidity, and spoilage indicators. These sensors gather data in real time, allowing systems to detect anomalies and predict contamination events before they escalate. For instance, systems designed with real-time analytics can automatically flag deviations from established safety thresholds, enabling prompt remedial actions (Niewczas-Dobrowolska et al., 2019). Additionally, smart packaging embedded with sensors can dynamically assess food freshness and safety, thereby providing alerts before quality diminishes (Azangoo et al., 2021). This proactive monitoring bolsters early-warning systems for microbial outbreaks or chemical contamination, fundamentally enhancing food safety (Mo et al., 2024).

Moreover, the concept of digital twins is gaining traction as a robust method for modeling and simulating food production systems. Digital twins serve as virtual representations of physical processes, allowing for real-time simulations and operational assessments (Huang et al., 2022). This capability is instrumental in anticipating potential risks, such as equipment failure or ingredient variability, and supports predictive modeling of shelf life and contamination spread. Digital twins can effectively facilitate what-if scenario analyses, enabling producers to prepare for diverse operational contingencies without resorting to costly physical trials (Schleich et al., 2017). Furthermore, continuous monitoring through digital twins can dynamically reevaluate risk assessments as external conditions change, such as shifts in temperature during transport or production delays, ensuring a proactive approach to food safety (Varzakas & Αντωνιάδου, 2024).

Blockchain technology adds another layer of enhancement by establishing a tamper-proof ledger that provides traceability from farm to fork. This technology ensures transparency throughout the food supply chain by recording every transaction and movement, which culminates in improved accountability (Ghosh et al., 2019). The swift traceability afforded by blockchain facilitates rapid responses to food safety issues, such as recalls of contaminated products. By streamlining information access concerning origins and safety compliance, blockchain fosters consumer trust while assuring adherence to safety standards across suppliers (Mo et al., 2024).

Ethical considerations and governance frameworks are critical as these technologies become more integrated into food safety protocols. Issues of data privacy, algorithmic bias, and accountability in predictive models needs to be addressed to prevent unfair targeting of certain populations or suppliers (Methuselah, 2024). Frameworks for transparent governance are essential to validate the accuracy and interpretability of predictive systems. Inclusion of diverse stakeholders in the design and regulation processes of these technologies is necessary to ensure the resulting systems meet varying needs across different contexts (Niewczas-Dobrowolska et al., 2019).

9. CONCLUSION

The dynamic integration of risk assessment frameworks and predictive modeling represents a vital transformation in modern food safety management. As food systems grow increasingly complex and interconnected, traditional risk assessment methods—while foundational—must evolve alongside digital innovations such as IoT, digital twins, and blockchain. These emerging technologies enable real-time surveillance, traceability, and rapid-response capabilities, supporting more adaptive and proactive food safety systems. Strategic incorporation of these tools allows for robust hazard identification, exposure prediction, and effective mitigation, while fostering transparency and accountability across global supply chains. However, the successful application of these innovations requires rigorous validation, harmonized standards, and inclusive governance frameworks to prevent disparities and protect vulnerable populations.

For regulators, the path forward calls for clear policy frameworks that support data-driven monitoring while ensuring ethical deployment of AI and digital tools. Industry stakeholders must prioritize digital transformation and transparency, embedding predictive analytics into food production and logistics. Researchers are encouraged to explore interdisciplinary approaches that combine microbiology, data science, and behavioral science to refine predictive models and improve public health outcomes. Educators, in turn, have a responsibility to prepare future food safety professionals with the skills necessary to navigate and lead within a data-intensive, technology-enabled landscape. Ultimately, the future of food safety lies in smart, collaborative, and ethically grounded systems that are capable of protecting public health while building trust in an increasingly globalized food supply.

REFERENCES

Ahangarkolaee, S. S., & Gorton, M. (2020). The effects of perceived regulatory efficacy, ethnocentrism and food safety concern on the demand for organic food. *International Journal of Consumer Studies*, *45*(2), 273–286. DOI: 10.1111/ijcs.12619

Aigberua, A. O., Izah, S. C., & Aigberua, A. A. (2023). Occurrence, source delineation, and health hazard of polycyclic aromatic hydrocarbons in tissues of *Sarotherodon melanotheron* and *Chrysichthys nigrodigitatus* from Okulu River, Nigeria. *Environmental Monitoring and Assessment*, *195*(3), 364. DOI: 10.1007/s10661-023-10970-y

Aigberua, A. O., Izah, S. C., & Isaac, I. U. (2018). Level and health risk assessment of heavy metals in selected seasonings and culinary condiments used in Nigeria. *Biological Evidence*, *8*(2), 6–15. DOI: 10.5376/be.2018.08.0002

Aigberua, A. O., Izah, S. C., & Richard, G. (2021). Hazard analysis of trace metals in muscle of *Sarotherodon melanotheron* and *Chrysichthys nigrodigitatus* from Okulu River, Rivers State, Nigeria. *Journal of Environmental Health and Sustainable Development*, *6*(3), 1340–1356.

Alrobaish, W. S., Vlerick, P., Luning, P. A., & Jacxsens, L. (2020). Food safety governance in Saudi Arabia: Challenges in control of imported food. *Journal of Food Science*, *86*(1), 16–30. DOI: 10.1111/1750-3841.15552

Anwar, R. S., Ahmed, R. R., Streimikiene, D., Streimikis, J., & Zamek, D. (2024). Securing food futures: The interplay of safety governance, hygiene, supplier beliefs and consumer engagement. *British Food Journal*, *127*(2), 758–778. DOI: 10.1108/BFJ-09-2024-0882

Arficho, T. S., & Kebede, A. H. (2019). Microbiological quality and safety of ready-to-eat foods from restaurants and food establishments in Yirgalem town, Southern Ethiopia. *Research Square*. DOI: 10.21203/rs.2.13735/v1

Assogba, M. F., Afé, O. H. I., Ahouansou, R. H., Anihouvi, D. G. H., Kpoclou, Y. E., Djago, D., Douny, C., Igout, A., Mahillon, J., Hounhouigan, D. J., Scippo, M.-L., & Anihouvi, V. B. (2021). Performances of the barrel kiln used in cottage industry for fish processing and effects on physicochemical characteristics and safety of smoked fish products. *Journal of the Science of Food and Agriculture*, *102*(2), 851–861. DOI: 10.1002/jsfa.11421

Azangoo, M., Taherkordi, A., Blech, J. O., & Vyatkin, V. (2021). Digital twin-assisted controlling of AGVs in flexible manufacturing environments. In *2021 IEEE 30th International Symposium on Industrial Electronics (ISIE)* (pp. 1–7). IEEE. DOI: 10.1109/ISIE45552.2021.9576361

Baranyi, J., & da Silva, N. B. (2017). The use of predictive models to optimize risk of decisions. *International Journal of Food Microbiology*, *240*, 19–23. DOI: 10.1016/j.ijfoodmicro.2016.10.016

Béné, C., Fanzo, J., Prager, S. D., Achicanoy, H. A., Mapes, B. R., Alvarez Toro, P., & Bonilla Cedrez, C. (2020). Global drivers of food system (un)sustainability: A multi-country correlation analysis. *PLoS One*, *15*(4), e0231071. DOI: 10.1371/journal.pone.0231071

Cassini, A., Hathaway, S., Havelaar, A., Koopmans, M., Koutsoumanis, K., Messens, W., Müller-Seitz, G., Nørrung, B., Rizzi, V., & Scheutz, F. (2016). Microbiological risk assessment. *EFSA Journal*, *14*(S1). Advance online publication. DOI: 10.2903/j.efsa.2016.s0507

Chammem, N., Issaoui, M., De Almeida, A. I. D., & Delgado, A. M. (2018). Food crises and food safety incidents in European Union, United States, and Maghreb area: Current risk communication strategies and new approaches. *Journal of AOAC International*, *101*(4), 923–938. DOI: 10.5740/jaoacint.17-0446

Chen, T., Wang, L., & Wang, J. (2017). Transparent assessment of the supervision information in China's food safety: A fuzzy-ANP comprehensive evaluation method. *Journal of Food Quality*, *2017*, 1–14. DOI: 10.1155/2017/4878926

Chen, Y., Li, H., Dou, H., Wen, H., & Dong, Y. (2023). Prediction and visual analysis of food safety risk based on TabNet-GRA. *Foods*, *12*(16), 3113. DOI: 10.3390/foods12163113

Ciecierska, M., Dasiewicz, K., & Wołosiak, R. (2023). Methods of minimizing polycyclic aromatic hydrocarbon content in homogenized smoked meat sausages using different casings and variants of meat-fat raw material. *Foods*, *12*(22), 4120. DOI: 10.3390/foods12224120

Ciecierska, M., Komorowska, U., Bryła, M., & Roszko, M. (2025). Formation of polycyclic aromatic hydrocarbons on grilled pork neck loins as affected by different marinades and grill types. *Foods*, *14*(10), 1673. DOI: 10.3390/foods14101673

Commeau, N., Cornu, M., Albert, I., Denis, J. B., & Parent, E. (2011). Hierarchical Bayesian models to assess between- and within-batch variability of pathogen contamination in food. *Risk Analysis*, *32*(3), 395–415. DOI: 10.1111/j.1539-6924.2011.01699.x

Currie, A., Farber, J. M., Nadon, C., Sharma, D., Whitfield, Y., Gaulin, C., Galanis, E., Bekal, S., Flint, J., Tschetter, L., Pagotto, F., Lee, B., Jamieson, F., Badiani, T., MacDonald, D., Ellis, A., May-Hadford, J., McCormick, R., Savelli, C., & Sierpinska, U. (2015). Multi-province listeriosis outbreak linked to contaminated deli meat consumed primarily in institutional settings, Canada, 2008. *Foodborne Pathogens and Disease*, *12*(8), 645–652. DOI: 10.1089/fpd.2015.1939

D'Amore, T., Smaoui, S., & Varzakas, T. (2025). Chemical food safety in Europe under the spotlight: Principles, regulatory framework and roadmap for future directions. *Foods*, *14*(9), 1628. DOI: 10.3390/foods14091628

Dove, M., Balasubramanian, A., & Narayanan, B. (2020). Transparency as a way of attaining quality, safety and optimal food purchases. *SocioEconomic Challenges*, *4*(4), 48–62. DOI: 10.21272/sec.4(4).48-62.2020

Drago, E., Campardelli, R., Pettinato, M., & Perego, P. (2020). Innovations in smart packaging concepts for food: An extensive review. *Foods*, *9*(11), 1628. DOI: 10.3390/foods9111628

Eslami, A., Gholami, Z., Nargesi, S., Rostami, B., & Avazpour, M. (2017). Evaluation of microbial contamination of ready-to-eat foods (pizza, frankfurters, sausages) in the city of Ilam. *Environmental Health Engineering and Management*, *4*(2), 117–122. DOI: 10.15171/EHEM.2017.16

Evers, E. G., Berk, P. A., Horneman, M. L., van Leusden, F. M., & de Jonge, R. (2014). A quantitative microbiological risk assessment for *Campylobacter* in petting zoos. *Risk Analysis*, *34*(9), 1618–1638. DOI: 10.1111/risa.12197

Feng, Z., Xu, S., Zhang, S., & Sun, M. (2017, May). From the holistic governance perspective to study multidimensional analysis of China's food safety supervision system. In *2017 International Conference on Education, Economics and Management Research (ICEEMR 2017)* (pp. 56–62). Atlantis Press. DOI: 10.2991/iceemr-17.2017.14

Gavai, A., Bouzembrak, Y., Mu, W., Martin, F., Kaliyaperumal, R., van Soest, J., Choudhury, A., Heringa, J., Dekker, A., & Marvin, H. (2022). A federated learning approach to data sharing in a food supply chain. *Research Square*. DOI: 10.21203/rs.3.rs-2350301/v1

Ghosh, A. K., Ullah, A. S., & Kubo, A. (2019). Hidden Markov model-based digital twin construction for futuristic manufacturing systems. *Artificial Intelligence for Engineering Design, Analysis and Manufacturing*, *33*(3), 317–331. DOI: 10.1017/S089006041900012X

Hadjigeorgiou, A., Soteriades, E. S., Philalithis, A., Psaroulaki, A., Tselentis, Y., & Gikas, A. (2012). National food safety systems in the European Union: A comparative survey. *International Journal of Food Studies*, *2*(1). Advance online publication. DOI: 10.7455/ijfs/2.1.2013.a8

Huang, S., Wang, G., Lei, D., & Yan, Y. (2022). Toward digital validation for rapid product development based on digital twin: A framework. *International Journal of Advanced Manufacturing Technology*, *119*(3–4), 2509–2523. DOI: 10.1007/s00170-021-08475-4

Huang, Y., Zhang, W., Ruan, G., Li, X., Cong, Y., Du, F., & Li, J. (2018). Reduced graphene oxide-hybridized polymeric high-internal phase emulsions for highly efficient removal of polycyclic aromatic hydrocarbons from water matrix. *Langmuir*, *34*(12), 3661–3668. DOI: 10.1021/acs.langmuir.8b00005

Hyde, R., Hoflund, A. B., & Pautz, M. C. (2016). One HACCP, two approaches: Experiences with and perceptions of the Hazard Analysis and Critical Control Points (HACCP) food safety management systems in the US and the EU. *Administration & Society*, *48*(8), 962–987. DOI: 10.1177/0095399714548266

Ingegno, M., Zianni, R., Della Rovere, I., Chiappinelli, A., Nardelli, V., Casamassima, F., Calitri, A., Quinto, M., Nardiello, D., & Iammarino, M. (2024). Development of a highly sensitive method based on QuEChERS and GC–MS/MS for the determination of polycyclic aromatic hydrocarbons in infant foods. *Frontiers in Nutrition*, *11*, 1403541. Advance online publication. DOI: 10.3389/fnut.2024.1403541

Iwegbue, C. M., Agadaga, H., Bassey, F. I., Overah, L. C., Tesi, G. O., & Nwajei, G. E. (2015). Concentrations and profiles of polycyclic aromatic hydrocarbons in some commercial brands of tea-, coffee-, and cocoa-based food drinks in Nigeria. *International Journal of Food Properties*, *18*(10), 2124–2133. DOI: 10.1080/10942912.2014.908906

Izah, S. C., & Aigberua, A. O. (2020). Microbial and heavy metal hazard analysis of edible tomatoes (*Lycopersicon esculentum*) in Port Harcourt, Nigeria. *Toxicology and Environmental Health Sciences*, *12*(4), 371–380. DOI: 10.1007/s13530-020-00060-8

Izah, S. C., Aigberua, A. O., & Ogwu, M. O. (2022). Trace element composition of *Gallus gallus domesticus* eggs and health risks associated with their consumption in Port Harcourt, Nigeria. *Journal of Food Safety and Hygiene*, *8*(3), 202–222.

Izah, S. C., Aigberua, A. O., & Richard, G. (2022). Concentration, source, and health risk of trace metals in some liquid herbal medicine sold in Nigeria. *Biological Trace Element Research*, *200*(6), 3009–3022. DOI: 10.1007/s12011-021-02879-9

Izah, S. C., Etebu, E. N., Aigberua, A. O., Odubo, T. C., & Iniamagha, I. (2022). A meta-analysis of microbial contaminants in selected ready-to-eat foods in Bayelsa State, Nigeria: Public health implications and risk-reduction strategies. *Hygiene and Environmental Health Advances*, *4*, 100017. DOI: 10.1016/j.heha.2022.100017

Izah, S. C., Nurmakhanova, A., Ogwu, M. C., Toktarbay, Z., Umirbaeva, Z., Ussen, K., Koibasova, L., Nazarbekova, S., Tynybekov, B., & Guo, Z. (2025). Public health risks associated with antibiotic residues in poultry food products. *Journal of Agriculture and Food Research*, *21*, 101815. DOI: 10.1016/j.jafr.2025.101815

Izah, S. C., Stanley, H. O., Richard, G., Sawyer, W. E., Uwaeme, O. R., & Sylva, L. (2024). Source and health risks of trace metals in *Clarias batrachus* and *Chrysichthys nigrodigitatus* from surface waters in Bayelsa State, Nigeria: A probabilistic model. *Frontiers in Sustainable Food Systems*, *8*, 1419143. DOI: 10.3389/fsufs.2024.1419143

Izah, S. C., Uzoekwe, S. A., & Aigberua, A. O. (2021). Source, geochemical spreading and risks of trace metals in particulate matter 2.5 within a gas flaring area in Bayelsa State, Nigeria. *Advances in Environmental Technology*, *7*(2), 101–118.

Jin, C., Levi, R., Liang, Q., Renegar, N., Springs, S., Zhou, J., & Zhou, W. (2021). Testing at the source: Analytics-enabled risk-based sampling of food supply chains in China. *Management Science*, *67*(5), 2985–2996. DOI: 10.1287/mnsc.2020.3839

Khiari, R., Zemni, H., Maury, C., & Mihoubi, D. (2020). Modeling desorption isotherms and thermodynamic properties of Italia grapes. *Journal of Food Processing and Preservation*, *44*(10), e14731. DOI: 10.1111/jfpp.14731

Kigigha, L. T., Berezi, J., & Izah, S. C. (2017). Bacteriological quality assessment of meat pie sold in Yenagoa metropolis, Nigeria. *EC Nutrition*, *6*(6), 189–195.

Kigigha, L. T., Ebieto, L. O., & Izah, S. C. (2017). Health risk assessment of heavy metal in smoked *Trachurus trachurus* sold in Yenagoa, Bayelsa State, Nigeria. *International Journal of Healthcare and Medical Sciences*, *3*(9), 62–69.

Kigigha, L. T., Igoya, U. O. S., & Izah, S. C. (2016). Microbiological quality assessment of unpeeled groundnut sold in Yenagoa metropolis, Nigeria. *International Journal of Innovative Biochemistry & Microbiology Research*, *4*(4), 11–22.

Kigigha, L. T., Nyẹnke, P., & Izah, S. C. (2018). Health risk assessment of selected heavy metals in gari (cassava flake) sold in some major markets in Yenagoa metropolis, Nigeria. *MOJ Toxicology*, *4*(2), 47–52.

Koseki, S. (2016). Predictive modeling for estimation of bacterial behavior from farm to table. *Food Safety (Tokyo, Japan)*, *4*(2), 33–44. DOI: 10.14252/foodsafetyfscj.2016006

Küníli̇, İ. E., Çolakoğlu, S., & Çolakoğlu, F. (2020). Levels of PAHs, PCBs, and toxic metals in *Ruditapes philippinarum* and *Donax trunculus* in Marmara Sea, Turkey. *Journal of the Science of Food and Agriculture*, *101*(3), 1167–1173. DOI: 10.1002/jsfa.10728

Li, J. H., Yu, W. J., Lai, Y. H., & Ko, Y. C. (2012). Major food safety episodes in Taiwan: Implications for the necessity of international collaboration on safety assessment and management. *The Kaohsiung Journal of Medical Sciences*, *28*(7S). Advance online publication. DOI: 10.1016/j.kjms.2012.05.004

Li, L., Shen, Y., Yuan, Z., & Li, J. (2018). Analysis and research of food safety risk assessment based on data mining. In *Proceedings of the 2018 International Conference on Big Data Engineering and Technology* (pp. 68–72). DOI: 10.1145/3297730.3297748

Li, T., Luo, Y., & Zhao, T. (2024). Research on enhancing the efficiency of food safety sampling inspections in China based on Pareto's law. *Journal of the Science of Food and Agriculture*, *105*(1), 235–243. DOI: 10.1002/jsfa.13822

Lin, W., & Liang, J. (2024). Regulatory decentralization and food safety: Evidence from China. *American Journal of Agricultural Economics*, *107*(1), 54–80. DOI: 10.1111/ajae.12484

Mahmudatusa'adah, A., & Sudewi, S. (2017). Hazard analysis critical control point (HACCP) in sweet potato's liquid sugar. *Innovation of Vocational Technology Education*, *12*(2). Advance online publication. DOI: 10.17509/invotec.v12i2.6203

Martischang, R., Godycki-Ćwirko, M., Kowalczyk, A., Kosiek, K., Turjeman, A., Babich, T., Shiber, S., Leibovici, L., Von Dach, E., Harbarth, S., & Huttner, A. (2021). Risk factors for treatment failure in women with uncomplicated lower urinary tract infection. *PLoS One*, *16*(8), e0256464. DOI: 10.1371/journal.pone.0256464

Marvin, H. J., Janssen, E. M., Bouzembrak, Y., Hendriksen, P. J., & Staats, M. (2016). Big data in food safety: An overview. *Critical Reviews in Food Science and Nutrition*, *57*(11), 2286–2295. DOI: 10.1080/10408398.2016.1257481

Matheron, A., Guerault, M. N., Vazquez, R., Cheron, M., Brossard, D., & Crauste-Manciet, S. (2020). Microbiological stability tests with simulated broth preparations and integrity testing for sterile standard cytotoxic preparations. *Pharmaceutical Technology in Hospital Pharmacy*, *5*(1). Advance online publication. DOI: 10.1515/pthp-2020-0002

Methuselah, J. (2024). Digital twin technology for smart manufacturing. *Journal of Technology and Systems*, *6*(4), 52–65. DOI: 10.47941/jts.2143

Mo, G., Sun, Y., Huang, S., Gao, F., Huang, J., Lai, L., Wang, G., & Yan, Y. (2024). Reconfigurable digital twin from the perspective of smart manufacturing systems in Industry 4.0/5.0. *Journal of Physics: Conference Series*, *2832*(1), 012007. DOI: 10.1088/1742-6596/2832/1/012007

Niewczas-Dobrowolska, M., Sikora, T., & Prusak, A. (2019). Ethics and trust while food safety crisis. *Proceedings on Engineering Sciences*, *1*(2), 257–260. DOI: 10.24874/PES01.02.022

Nuttavuthisit, K., & Thøgersen, J. (2015). The importance of consumer trust for the emergence of a market for green products: The case of organic food. *Journal of Business Ethics*, *140*(2), 323–337. DOI: 10.1007/s10551-015-2690-5

Odubo, T. C., & Izah, S. C. (2024). Safety considerations of trace metals in locally produced nutritive food-drinks consumed in Yenagoa metropolis, Nigeria. *Biological Trace Element Research*, *203*(8), 4408–4419. Advance online publication. DOI: 10.1007/s12011-024-04488-8

Ogamba, E. N., Charles, E. E., & Izah, S. C. (2021). Distributions, pollution evaluation and health risk of selected heavy metal in surface water of Taylor creek, Bayelsa State, Nigeria. *Toxicology and Environmental Health Sciences*, *13*(2), 109–121. DOI: 10.1007/s13530-020-00076-0

Oriekhoe, O. I., Ilugbusi, B. S., & Adisa, O. (2024). Ensuring global food safety: Integrating blockchain technology into food supply chains. *Engineering Science & Technology Journal*, *5*(3), 811–820. DOI: 10.51594/estj.v5i3.905

Ovuru, K. F., Izah, S. C., Ogidi, O. I., Imarhiagbe, O., & Ogwu, M. C. (2024). Slaughterhouse facilities in developing nations: Sanitation and hygiene practices, microbial contaminants and sustainable management system. *Food Science and Biotechnology*, *33*(3), 519–537. DOI: 10.1007/s10068-023-01406-x

Puza, B. (2015). *Bayesian methods for statistical analysis*. DOI: 10.26530/OAP-EN_611011

Ranjbaran, M., Carciofi, B. A., & Datta, A. K. (2021). Engineering modeling frameworks for microbial food safety at various scales. *Comprehensive Reviews in Food Science and Food Safety*, *20*(5), 4213–4249. DOI: 10.1111/1541-4337.12818

Ranta, J., Siekkinen, K. M., Nuotio, L., Laukkanen, R., Hellström, S., Korkeala, H., & Maijala, R. (2010). Causal hidden variable model of pathogenic contamination from pig to pork. *Statistical Modelling*, *10*(1), 69–87. DOI: 10.1177/1471082X0801000105

Samarajeewa, U. (2023). Polycyclic aromatic hydrocarbons and food safety: A review. *Journal of the National Science Foundation of Sri Lanka*, *51*(2). Advance online publication. DOI: 10.4038/jnsfsr.v51i2.11396

Sawyer, W. E., & Izah, S. C. (2024). *Unmasking food adulteration: Public health challenges, impacts and mitigation strategies*. ES General.

Sawyer, W. E., Nabebe, G., & Izah, S. C. (2024). Public health implications of pesticide residues in food: Risks, regulations, and interventions. *Greener Journal of Biomedical and Health Sciences*, *7*(1), 31–46. DOI: 10.15580/gjbhs.2024.1.102024144

Schleich, B., Anwer, N., Mathieu, L., & Wartzack, S. (2017). Shaping the digital twin for design and production engineering. *CIRP Annals*, *66*(1), 141–144. DOI: 10.1016/j.cirp.2017.04.040

Seiyaboh, E. I., Kigigha, L. T., Alagoa, C. T., & Izah, S. C. (2018). Microbial quality of palm oil sold in Amassoma, Bayelsa State, Nigeria. *International Journal of Public Health and Safety*, *3*, 153.

Seiyaboh, E. I., Kigigha, L. T., Aruwayor, S. W., & Izah, S. C. (2018). Level of selected heavy metals in liver and muscles of cow meat sold in Yenagoa metropolis, Bayelsa State, Nigeria. *International Journal of Public Health and Safety*, *3*, 154.

Smith, J., Ross, K., & Whiley, H. (2016). Australian food safety policy changes from a "command and control" to an "outcomes-based" approach: Reflection on the effectiveness of its implementation. *International Journal of Environmental Research and Public Health*, *13*(12), 1218. DOI: 10.3390/ijerph13121218

Song, S., Huang, T., Ma, J., Mao, X., Gao, H., Zhao, Y., Jiang, W., Lian, L., Chen, B., Liu, L., & Wang, Z. (2022). Assessing safety of market-sold fresh fish: Tracking fish origins and toxic chemical origins. *Environmental Science & Technology*, *56*(13), 9505–9514. DOI: 10.1021/acs.est.2c00495

Spence, G. T., Steinsaltz, D., & Fanshawe, T. R. (2016). A Bayesian approach to sequential meta-analysis. *Statistics in Medicine*, *35*(29), 5356–5375. DOI: 10.1002/sim.7052

Taha, H., Nasraween, M., Khader, Y., Al Omari, L., Al Nsour, M., Al Shdaifat, A., & Berggren, V. (2022). Microbial load of chicken shawerma and the handlers' compliance with food safety practices in Jordan. *Arab Journal of Nutrition and Exercise*, *6*(1). Advance online publication. DOI: 10.18502/ajne.v6i1.10063

Tarigholizadeh, S., Sushkova, S., Rajput, V. D., Ranjan, A., Arora, J., Dudnikova, T., Barbashev, A., Mandzhieva, S., Minkina, T., & Wong, M. H. (2023). Transfer and degradation of PAHs in the soil–plant system: A review. *Journal of Agricultural and Food Chemistry*, *72*(1), 46–64. DOI: 10.1021/acs.jafc.3c05589

Thomann, E. (2018). Food safety policy: Transnational, hybrid, wicked. In *Oxford research encyclopedia of politics*. DOI: 10.1093/acrefore/9780190228637.013.540

Uzoekwe, S. A., Izah, S. C., & Aigberua, A. O. (2021). Environmental and human health risk of heavy metals in atmospheric particulate matter (PM_{10}) around gas flaring vicinity in Bayelsa State, Nigeria. *Toxicology and Environmental Health Sciences*, *13*(4), 323–335. DOI: 10.1007/s13530-021-00085-7

Varzakas, T., & Αντωνιάδου, M. (2024). A holistic approach for ethics and sustainability in the food chain: The gateway to oral and systemic health. *Foods*, *13*(8), 1224. DOI: 10.3390/foods13081224

Wang, J., & Chen, T. (2016). The spread model of food safety risk under the supply-demand disturbance. *SpringerPlus*, *5*(1), 1765. Advance online publication. DOI: 10.1186/s40064-016-2881-2

Wang, X., Bouzembrak, Y., Lansink, A. O., & Van Der Fels-Klerx, H. J. (2021). Application of machine learning to the monitoring and prediction of food safety: A review. *Comprehensive Reviews in Food Science and Food Safety*, *21*(1), 416–434. DOI: 10.1111/1541-4337.12868

Wang, X., Xu, Y., Liang, M., Gao, J., Wang, J., Chen, S., & Cheng, J. (2022). Chinese consumers' trust in food safety surveillance sampling of commonly consumed foods. *Foods*, *11*(13), 1971. DOI: 10.3390/foods11131971

Waterhouse, B., Hartford, A., Mukhopadhyay, S., Ferguson, R., & Hendrickson, B. A. (2022). Using the Bayesian detection of potential risk using inference on blinded safety data (BDRIBS) method to support the decision to refer an event for unblinded evaluation. *Pharmaceutical Statistics*, *21*(2), 372–385. DOI: 10.1002/pst.2175

Williams, M. S., Ebel, E. D., & Vose, D. (2010). Framework for microbial food-safety risk assessments amenable to Bayesian modeling. *Risk Analysis*, *31*(4), 548–565. DOI: 10.1111/j.1539-6924.2010.01532.x

Zapico, A., Ruiz-Saavedra, S., Gómez-Martín, M., de Los Reyes-Gavilán, C. G., & González, S. (2022). Pilot study for the dietary assessment of xenobiotics derived from food processing in an adult Spanish sample. *Foods*, *11*(3), 470. DOI: 10.3390/foods11030470

Chapter 4
Novel Decontamination Technologies and Emerging Approaches in Food Safety:
Physical, Chemical, Biological, and Emerging Approaches in Food

Daniel Etim Jacob
https://orcid.org/0000-0002-5754-5445
University of Uyo, Nigeria

Aniekpeno Isaac Elijah
University of Uyo, Nigeria

Imaobong D. Jacob
University of Uyo, Nigeria

ABSTRACT

Food contamination remains a persistent global concern, contributing significantly to food-borne illnesses, economic loss, and compromised consumer confidence. Traditional decontamination methods such as thermal processing and chlorine-based sanitization often present limitations related to nutrient degradation, chemical residues, and environmental impact. In response, novel decontamination technologies have emerged that aim to enhance microbial safety while preserving food quality and aligning with sustainability goals. This chapter explores contemporary physical, chemical, and biological approaches, including ultraviolet-C light, cold plasma, high-pressure processing, organic acids, electrolyzed water, and bacterio-

DOI: 10.4018/979-8-3373-7052-1.ch004

phage applications highlighting their mechanisms, effectiveness, and limitations. Additionally, the chapter reviews emerging and hybrid systems such as photocatalytic decontamination, ultrasound-assisted sanitization, supercritical CO_2 treatment, and smart sensor-integrated solutions, which offer synergistic benefits and greater adaptability to modern food processing demands. Despite their promise, challenges such as regulatory inconsistencies, high implementation costs, and consumer skepticism hinder widespread adoption. Nevertheless, increased demand for clean-label, minimally processed, and safe food is driving innovation. This chapter underscores the need for collaborative efforts in research, standardization, and policy support to facilitate the transition from traditional to next-generation decontamination strategies in food safety systems.

1. INTRODUCTION

Food contamination continues to represent a significant global public health challenge, contributing to a vast number of food-borne illness outbreaks and incurring substantial economic costs across both industrialized and developing nations. The World Health Organization (WHO) estimates that approximately 600 million people, nearly one in ten globally fall ill each year after consuming contaminated food, resulting in 420,000 deaths annually, with the majority occurring in low- and middle-income countries where food safety infrastructure is often weak (Grace, 2023). Food contamination is not limited to microbial threats; it encompasses a broad spectrum of hazardous agents including pathogenic microorganisms such as Salmonella, Listeria monocytogenes, and *Escherichia coli* O157:H7, as well as chemical residues from pesticides, antibiotics, heavy metals, and natural biological toxins like mycotoxins and marine biotoxins (Scallan et al., 2011; Al-Moghazy et al., 2023).

Traditional food sanitation methods such as thermal pasteurization, chlorine-based rinses, and hot water treatments, though widely applied in commercial food processing, exhibit several limitations. These include thermal degradation of heat-sensitive nutrients, undesirable alterations in taste and texture, the formation of harmful disinfection by-products (e.g., trihalomethanes from chlorination), and inefficiencies in the complete inactivation of resistant microbial spores or biofilms (Kim et al., 2019a). In response to these challenges, recent research and innovation efforts have steered towards the adoption of novel decontamination technologies that prioritize food safety without compromising sensory quality, nutritional value, or environmental sustainability. The emergence of these advanced methods has revolutionised the landscape of food safety control systems. Novel decontamination technologies are broadly categorised into physical, chemical, and biological approaches, each with distinctive mechanisms and targeted applications. Physical technologies such as high hydrostatic pressure (HHP), pulsed electric fields (PEF), ultraviolet-C (UV-C) light, cold plasma, and irradiation have gained widespread attention for their ability to inactivate pathogens while maintaining the raw characteristics of food matrices (Ramos et al., 2020). For instance, HHP applies intense pressure (up to 600 MPa) to foods, disrupting microbial cell membranes and metabolic activities without the need for heat, making it ideal for heat-sensitive products like fruit juices and deli meats. Similarly, UV-C light (wavelengths of 200–280 nm) has demonstrated effective penetration and germicidal activity against a wide range of bacteria and fungi on food surfaces, especially in packaged fresh produce and seafood (Siddiqui et al., 2024).

Chemical methods have evolved beyond conventional chlorine-based sanitizers to include ozone treatment, peracetic acid, electrolyzed water, and organic acid washes (Patil et al., 2025). Ozone, a powerful oxidizing agent, is particularly effective in eliminating pathogens and decomposing pesticide residues, and it decomposes into oxygen without leaving toxic residues, making it environmentally friendly and suitable for organic food systems (Silva and Gibbs, 2012). However, the use of chemical agents often raises concerns regarding potential toxicity, allergenicity, and consumer acceptance, necessitating strict regulatory oversight and the development of residue-free technologies. Innovations such as nanoencapsulated antimicrobials and food-grade biopolymers have shown promise in extending shelf life and targeting biofilm-forming bacteria in ready-to-eat meals and minimally processed foods (Ojha et al., 2015). On the other hand, biological decontamination technologies offer a paradigm shift by harnessing naturally occurring or genetically engineered biological agents to combat contaminants. These include bacteriophage therapy, probiotics, bacteriocins (e.g., nisin, pediocin), and enzymatic treatments capable of degrading microbial cell walls or neutralising toxins (Rendueles et al., 2022). Bacteriophages viruses that selectively infect and kill specific bacteria have been approved for use in some jurisdictions and are being explored as effective biocontrol agents against *Listeria monocytogenes* in dairy and meat products (Endersen et al., 2014). Unlike broad-spectrum chemical disinfectants, bacteriophages pose minimal risk of disrupting beneficial microbiota or inducing antimicrobial resistance, making them a compelling option for sustainable food safety management.

Moreover, one of the central drivers behind the proliferation of novel decontamination technologies is the increasing consumer demand for minimally processed, clean-label foods products that are free from synthetic preservatives and retain their natural flavour, colour, and nutritional attributes. This consumer preference has pressured the food industry to adopt non-thermal and biologically inspired methods that not only meet safety standards but also align with wellness trends and environmental ethics. Furthermore, the glocalization of food supply chains has introduced new vectors for contamination and greater complexity in regulatory compliance, necessitating robust, real-time, and scalable interventions that can adapt to different commodities, processing environments, and regional food safety laws (de Mezsquita et al., 2022). Despite the clear advantages, the integration of these novel technologies into mainstream food processing is not without challenges. Issues such as high initial capital investment, need for specialized training, regulatory approval delays, and consumer apprehension particularly around irradiation or genetically modified biological agents must be carefully managed. Moreover, there exists a critical need for interdisciplinary research that bridges microbiology, food engineering, consumer science, and policy analysis to assess the long-term safety, economic viability, and societal acceptance of these innovations (Kumar et al., 2024; Galanakis, 2024). For instance, while cold plasma has demonstrated efficacy in inactivating bacterial spores and viruses, its mechanisms of action, potential for off-flavour generation, and impact on nutrient retention in complex food matrices still require further elucidation through longitudinal studies (Misra et al., 2025).

Consequently, the dynamic interplay between evolving consumer expectations, public health imperatives, and technological capabilities has catalysed a transition from conventional sanitation models to sophisticated, multi-dimensional decontamination systems. These novel technologies, whether physical, chemical, or biological, offer promising solutions to the enduring problem of food contamination. However, their success hinges on a good understanding of microbial ecology, careful consideration of food matrix interactions, alignment with regulatory frameworks, and effective communication with stakeholders across the food value chain. The subsequent sections of this chapter will delve into these novel approaches in detail, examining their scientific principles, application cases, advantages, limitations, and implications for the future of safe and sustainable food systems.

2. PHYSICAL APPROACHES TO FOOD DECONTAMINATION

Ensuring microbial safety in food systems without compromising quality has led to an increasing reliance on physical decontamination technologies (Mahunu et al., 2024; Oliveira et al., 2020). Unlike chemical interventions that may leave undesirable residues or biological approaches that can face regulatory and consumer resistance, physical methods offer the advantage of broad-spectrum antimicrobial efficacy, reduced environmental footprint, and often better preservation of sensory and nutritional properties (Chu et al., 2024). This section explores three cutting-edge technologies namely; Ultraviolet-C (UV-C) and Pulsed Light, Cold Plasma Technology, and High-Pressure Processing (HPP) that are reshaping modern food safety paradigms.

2.1 Ultraviolet-C (UV-C) and Pulsed Light

Ultraviolet-C light, operating at a germicidal wavelength of approximately 254 nanometers (Table 1), has become a cornerstone technology in food surface and packaging decontamination (Ilias et al., 2021; Koutchma, 2019). Its mechanism of action lies in the photochemical damage it inflicts upon microbial DNA and RNA, particularly the formation of thymine dimers that inhibit replication and protein synthesis. Over time, cumulative doses of UV-C light can result in irreversible structural damage to cell walls and membranes, leading to cell death (Gayán et al., 2015). UV-C treatment is especially valuable for transparent liquids, smooth-surfaced fruits, and plastic packaging materials where light can penetrate effectively and microbial exposure is evenly distributed (Abthiyar, 2023; Koutchma, 2014). Nevertheless, the efficacy of UV-C is hindered by surface irregularities, shadows, and biofilm formation. For instance, leafy vegetables such as lettuce or spinach often present structural complexities that shield bacteria within crevices, reducing light penetration and decontamination efficiency (Yaun et al., 2004). To mitigate this limitation, rotating UV systems and surface agitation have been introduced in commercial settings, improving coverage and exposure. In addition, combining UV-C with other hurdles such as mild heating or low concentrations of organic acids can significantly enhance lethality against *Listeria monocytogenes* and *E. coli* O157:H7 (Kim et al., 2019b).

Pulsed Light (PL) technology, a more recent advancement, emits intense bursts of broad-spectrum white light, including UV, visible, and near-infrared wavelengths (Mandal et al., 2020). These light pulses, delivered at high frequency and energy (up to 50 J/cm^2), induce photothermal, photochemical, and photophysical effects on microbial cells. PL's energy-rich flashes have demonstrated exceptional microbial inactivation on irregular and porous surfaces, such as meats, eggshells, and fruits with rough skins (Krishnamurthy et al., 2010). Moreover, PL systems can be tailored to specific fluence levels that preserve food integrity while ensuring pathogen reduction (Pirozzi et al., 2021). However, despite their advantages, both UV-C and PL face challenges in scaling for certain commodities. Transparent films, variable reflectance, and food geometry affect treatment uniformity. As such, their most effective deployment often lies in post-packaging decontamination, surface sanitation in food service areas, and integration into conveyor-based systems for produce.

Table 1. Comparison of UV-C and Pulsed Light for Food Decontamination

Feature	UV-C Light	Pulsed Light
Wavelength	254 nm (narrow band)	Broad-spectrum (UV to IR)
Target applications	Smooth surfaces, liquids	Irregular surfaces, packaging
Mode of action	DNA photodamage	Photothermal, photochemical effects
Limitations	Poor penetration, shadows	Equipment cost, heat generation
Shelf life impact	Minimal	Minimal to moderate (at high doses)
Regulatory status	Widely approved	Approved in select regions

2.2 Cold Plasma Technology

Cold Atmospheric Plasma (CAP) represents a non-thermal frontier in microbial inactivation. Plasma, the fourth state of matter, is formed by ionizing a gas using electric fields (Figure 1), producing a reactive milieu composed of electrons, ions, UV photons, and chemically reactive species such as ozone, nitric oxide, and hydrogen peroxide (Zhang et al., 2019; Frank-Kamenetskii, 2012). These species induce oxidative stress and membrane disruption in microbial cells, damaging DNA and proteins without requiring elevated temperatures (Pankaj et al., 2018a,b). CAP's strength lies in its versatility and gentleness toward food matrices. Because it operates at room temperature, it is highly suitable for heat-sensitive products such as leafy greens, berries, milk, and seafood (Lohita and Srijaya, 2024). Laboratory and pilot-scale studies have consistently shown that CAP treatments achieve significant log reductions of pathogens like *E. coli* O157:H7, *Salmonella enterica*, and *Listeria monocytogenes*, with minimal impact on flavour, texture, or nutrient retention (Misra et al., 2011).

Furthermore, CAP can penetrate biofilms a common limitation of traditional sanitation. This is especially crucial in meat and dairy industries, where microbial resistance often emerges from complex matrix interactions (Katsigiannis et al., 2022; Zhu et al., 2020). The device design, whether dielectric barrier discharge (DBD), plasma jets, or corona discharges, can be tailored for product-specific applications. However, the penetration depth of plasma is limited, making it most effective for surface applications. Also, despite its promise, commercial uptake of CAP remains nascent. Challenges include the need for standardisation in dose delivery, concerns over potential formation of nitrates/nitrites on protein surfaces, and variability in efficacy across different product surfaces. Moreover, consumer perception around plasma-treated foods, though not genetically modified or chemically preserved, requires transparent communication and policy alignment.

Figure 1. Mechanism of Cold Plasma on Microbial Cells

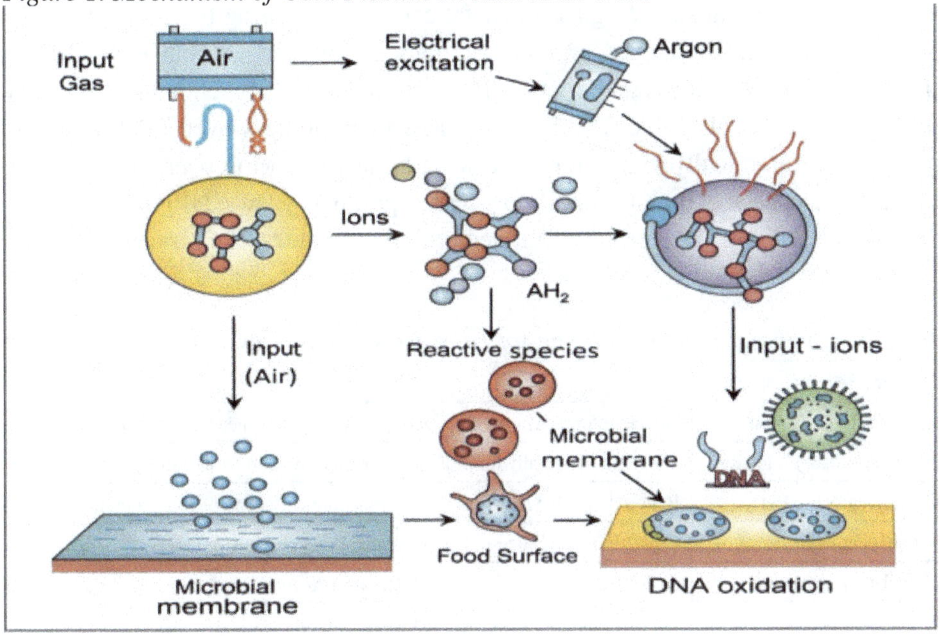

2.3 High-Pressure Processing (HPP)

High-Pressure Processing (HPP), also referred to as high hydrostatic pressure (HHP), is a cold pasteurization technique that exposes food products to pressures between 400–600 MPa (megapascals) for a predetermined time (typically 1 – 10 minutes) (Nabi et al., 2021). The pressurisation occurs in a water-filled chamber and applies pressure uniformly, regardless of product shape or size, inactivating microbial cells by altering membrane permeability, disrupting enzyme activity, and damaging nucleic acids (Knorr et al., 2011). HPP is particularly suitable for high-value, perishable products where preservation of freshness, colour, and nutrients is paramount (Rabiepour et al., 2024; Bolumar et al., 2021). Common commercial examples include fruit juices, smoothies, guacamole, deli meats, and shellfish. Unlike thermal methods that often denature proteins or degrade vitamins, HPP maintains the sensory and nutritional quality of food while offering extended shelf life. Studies report 5 – 6 log reductions of pathogens like *Clostridium botulinum*, *Salmonella*, and *Listeria* in cold-pressed juices and vacuum-packed meats (Esmaeili et al., 2022).

A notable advantage of HPP is its ability to deactivate vegetative cells without additives (Yang et al., 2021). However, bacterial spores, especially those of *Bacillus* and *Clostridium*, are pressure-resistant and may require combination treatments (pressure + mild heat) to ensure food safety. HPP is also incompatible with rigid packaging such as glass or metal cans, necessitating the use of flexible plastic packaging, which may limit its sustainability credentials (Nabi et al., 2021). Importantly, the operational cost of HPP is mitigated by its batch nature and high throughput potential (Table 2). Modern systems can process over 2,000 kg/hour, making it economically viable for large-scale processors. Moreover, as consumer interest in clean-label and minimally processed products grows, HPP offers a non-thermal, chemical-free alternative aligned with these values (Saputra et al., 2025).

Table 2. Advantages and Challenges of High-Pressure Processing

Factor	Description
Microbial inactivation	Effective for vegetative pathogens; spore resistance needs hurdle approach
Nutrient/sensory retention	Excellent, especially for heat-sensitive nutrients and fresh taste
Packaging constraints	Requires flexible materials; non-compatible with glass or rigid containers
Equipment cost	High capital investment; low operational cost per unit volume
Regulatory acceptance	Approved in the US, EU, and Japan for many food categories
Applications	Juices, meats, seafood, dips, ready-to-eat meals

3. CHEMICAL APPROACHES TO FOOD DECONTAMINATION

Chemical interventions for food decontamination remain a crucial component in modern food safety protocols. While traditional methods such as chlorinated washes still dominate industrial practice due to their affordability and proven efficacy, increasing regulatory scrutiny and consumer concerns over residues have prompted a significant shift toward novel and natural alternatives. Today's chemical decontamination technologies reflect an expanded palette, ranging from naturally derived acids and essential oils to advanced oxidative systems and intelligent packaging. These approaches are not only designed to reduce pathogen load but also to align with sustainability principles and demand for minimally processed, clean-label food products.

3.1 Organic Acids and Natural Sanitizers

The use of organic acids such as lactic acid, acetic acid, and citric acid in food decontamination is rooted in their dual role as natural preservatives and potent antimicrobial agents (Coban, 2020). These compounds reduce pH on food surfaces, disrupt microbial membrane integrity, and impair metabolic functions ultimately leading to microbial cell death (Yoon et al., 2024). Lactic acid, commonly produced during fermentation, is particularly effective against foodborne pathogens including Salmonella, *Listeria monocytogenes*, and *E. coli* O157:H7 (Table 3). When applied as a carcass rinse in poultry processing, lactic acid has been shown to achieve up to 2-log reductions in microbial counts without compromising product appearance or taste (Loretz et al., 2010). Citric acid, another GRAS (Generally Recognized As Safe) acidulant, works synergistically with heat and other antimicrobials (Mani-Lopez et al., 2012). It has been integrated into edible coatings and rinse solutions to protect cut fruits from fungal spoilage and oxidative browning. Acetic acid's affordability and widespread availability make it suitable for low-resource settings, especially in developing countries where vinegar is used as a microbial rinse for leafy greens, tomatoes, and cucumbers (Gohil et al., 2022).

Complementing these acids are essential oils such as thymol, eugenol, carvacrol, and cinnamaldehyde, extracted from aromatic plants and widely recognized for their antimicrobial potential (Bolouri et al., 2022). These bioactive compounds exert antimicrobial action by compromising membrane permeability and disrupting electron transport and ATP synthesis in microbial cells (Burt, 2004). When incorporated into wash solutions, edible films, or aerosolized treatments, they have shown promising reductions in both spoilage and pathogenic organisms. However, their use in food systems requires careful attention to dosage, food compatibility, and sensory impact. For instance, high concentrations of eugenol may impart an overpowering clove-like aroma, which could be off-putting in neutral-flavored commodities like milk or eggs (Lohita and Srijaya, 2024). Moreover, to overcome volatility and instability, nanoemulsion and encapsulation technologies have emerged as enabling tools for delivering essential oils in a controlled and efficient manner. Chitosan-based nanoparticles or liposomes encapsulating carvacrol, for example, have been used to inhibit *E. coli* and *Salmonella* in ready-to-eat salads, providing sustained release and improved adhesion to food surfaces (Sharifi-Rad et al., 2020). This merging of natural chemistry with nanotechnology points to a future where consumer-safe compounds can achieve industrial-grade efficacy through improved formulation science.

Table 3. Antimicrobial Activity of Common Organic Acids and Natural Sanitizers

Compound	Primary Source	Target Microorganisms	Typical Application	Notable Advantages
Lactic acid	Fermented foods	*Listeria, Salmonella*	Poultry, produce, RTE meats	Safe, consumer-accepted
Acetic acid	Vinegar	*E. coli, Campylobacter*	Leafy greens, sprouts	Cost-effective, broad usage
Citric acid	Citrus fruits	*Salmonella*, yeasts	Fresh-cut fruits, beverages	Natural preservative, synergistic
Thymol	Thyme essential oil	*S. aureus, E. coli*	Salad dressings, rinses	Potent, GRAS-certified
Eugenol	Clove oil	*L. monocytogenes*, molds	Spices, edible coatings	Multifunctional antioxidant

3.2 Electrolyzed Water (EW) and Ozonated Water

Electrolyzed water (EW), particularly acidic electrolyzed water (AEW), is now widely regarded as a non-toxic, highly effective sanitizing agent that offers a cleaner alternative to traditional chlorine rinses (Rebezov et al., 2022). Produced by passing a weak salt solution through an electrolytic cell, EW consists mainly of hypochlorous acid (HOCl) (Table 3), a neutral molecule capable of penetrating microbial cell walls more effectively than ionized chlorine (Issa-Zacharia, 2024). Its high oxidation-reduction potential (ORP), typically above 1100 mV, enhances its lethality against a broad spectrum of microorganisms, including Listeria, Salmonella, *E. coli*, and even viruses (Ahmed et al., 2024). EW has been successfully integrated into conveyor sprayers and wash tanks in poultry and seafood processing, where it effectively reduces surface pathogens without altering food quality or taste (Meghwar et al., 2024). Furthermore, slightly acidic electrolyzed water (SAEW), with a pH range of 5.0–6.5, has gained attention for its milder effect on human skin and stainless steel equipment while maintaining bactericidal efficiency (Issa-Zacharia, 2024). It is increasingly used in retail food preparation zones, foodservice kitchens, and minimally processed vegetable industries in Japan, South Korea, and China. SAEW's broader ORP range and lower chlorine volatility also reduce occupational hazards during prolonged exposure.

Ozonated water is another promising chlorine-free method that uses dissolved ozone gas as the active agent (Miller et al., 2013). As a triatomic oxygen molecule, ozone exhibits one of the highest oxidation potentials known among common sanitizers (2.07 V), enabling it to rupture microbial cell walls and degrade intracellular components swiftly. Unlike hypochlorite, ozone decomposes back into oxygen without leaving toxic residues or contributing to halogenated byproducts (Vijay Rakesh Reddy et al., 2022). As such, it is especially well-suited for organic-certified operations and fresh produce wash lines. However, ozone's instability in aqueous environments requires on-site generation and real-time monitoring of concentration levels, flow rates, and contact time. In high-humidity settings, ozone's antimicrobial performance may decrease due to its reactivity with organic material on the surface of foods. Some processors also express concern about potential oxidative effects on polyunsaturated fats in fish or meat, although these impacts are generally marginal under controlled application protocols. Nevertheless, integrating ozone with ultrasound, slightly acidic electrolyzed water, or natural antimicrobials has demonstrated synergistic effects. For example, combined ozone-ultrasound treatments have achieved higher Listeria inactivation on melon rinds and cantaloupe without compromising fruit firmness or sugar content (Wang et al., 2023). These multi-hurdle strategies reflect an evolving appreciation of chemical decontamination as a dynamic system rather than a single intervention.

Table 4. Comparison of EW and Ozonated Water

Attribute	Electrolyzed Water (EW)	Ozonated Water
Active agent	Hypochlorous acid (HOCl)	Dissolved ozone (O_3)
Mechanism	Oxidative protein/DNA damage	Cell wall oxidation and lysis
Residue profile	Trace chlorine; low residue	No residue (decomposes to oxygen)
Applications	Poultry, seafood, vegetables	Eggs, meats, leafy greens
Limitations	Corrosiveness, chlorine regulation	Instability, equipment complexity
Environmental impact	Low, but chlorine discharge risk	Minimal; eco-friendly

3.3 Antimicrobial Packaging

Active and intelligent packaging systems have become central to post-processing food safety. By embedding antimicrobial substances into packaging films or coatings, food processors can deliver continuous and localized inhibition of microbial growth during storage and distribution (Fu and Dudley, 2021). This strategy is particularly valuable in high-risk food categories such as ready-to-eat meats, dairy, seafood, and convenience salads, where post-process contamination remains a significant hazard (Banerjee et al., 2018). The materials used in antimicrobial packaging range from synthetic polymers like polyethylene and polypropylene to biodegradable films based on chitosan, polylactic acid (PLA), and starch blends (Table 4). These are infused with antimicrobials including silver nanoparticles, organic acids, bacteriocins (e.g., nisin, pediocin), enzymes like lysozyme, and essential oils (Gumienna and Goma, 2021). Nisin-embedded packaging, for instance, has been shown to reduce

L. monocytogenes growth on sliced ham by over 3 logs during refrigerated storage (Devlieghere et al., 2004). Silver nanoparticles, meanwhile, offer broad-spectrum antibacterial action and have demonstrated prolonged shelf life extension in dairy products, fruit juices, and even bakery goods.

Recent innovations include stimuli-responsive packaging systems that release antimicrobials only under certain conditions, such as increased humidity, temperature rise, or the detection of spoilage gases like ammonia. These intelligent systems reduce unnecessary compound exposure and allow for dose-tuned antimicrobial activity over time. Similarly, nanocomposite packaging materials offer higher surface area and better dispersion of antimicrobials, thereby enhancing interaction with spoilage organisms without requiring high bulk concentrations (Suvarna et al., 2022). However, regulatory and public concerns remain. Migration of nanoparticles into food is subject to tight legislative oversight, with agencies such as the European Food Safety Authority (EFSA) and U.S. Food and Drug Administration (FDA) requiring extensive toxicological data before approval. Moreover, consumer acceptance of nano-enabled packaging varies widely depending on regional trust in food innovation and perceived naturalness.

Table 5. Examples of Antimicrobial Packaging Systems

Material/Base	Incorporated Agent	Target Microbes	Food Application
PLA Film	Silver nanoparticles	*E. coli, Listeria*	Deli meats, cheese
LDPE + Nisin	Nisin (bacteriocin)	Gram-positive bacteria	Cooked meats, dairy
Chitosan edible film	Citric acid, clove oil	Yeasts, molds	Fresh produce, bakery
Biopolymer gel	Zinc oxide nanoparticles	*Salmonella, E. coli*	Poultry, seafood

4. BIOLOGICAL APPROACHES TO FOOD DECONTAMINATION

Biological decontamination approaches offer a better and sustainable alternative to physical and chemical treatments by leveraging naturally occurring mechanisms of microbial control. Biologically based solutions such as bacteriophages, probiotics, competitive exclusion cultures, and enzymatic decontaminants have gained growing relevance in food safety management because they are particularly advantageous in that they often maintain the sensory, nutritional, and structural integrity of foods, while targeting specific microbial threats (Sharma et al., 2024). They also align well with the One Health concept, given their reduced ecological footprint and compatibility with food microbiomes (Garcia et al., 2025). This section explores three major categories of biological decontamination techniques, namely; bacteriophage applications, probiotic cultures, and enzymatic decontaminants.

4.1 Bacteriophage Applications

Bacteriophages, or simply phages, are viruses that infect and replicate within specific bacterial hosts. Unlike broad-spectrum chemical disinfectants or antibiotics, phages exhibit exquisite host specificity, often targeting one genus, species, or even strain of bacteria (Knipe et al., 2021). This specificity minimizes collateral damage to beneficial microorganisms on food surfaces or within the human gut (Garcia-Gutierrez et al., 2019). Phage-based interventions thus represent a precision tool in food safety, selectively eliminating pathogenic contaminants without disturbing the natural microbiota of food products. Commercially, phage cocktails have been approved for use against *Listeria monocytogenes* in ready-to-eat meats, cheeses, and smoked seafood (Gummadi, 2024). One of the most widely cited examples is the FDA-approved Listex™ P100, which employs a lytic phage specific to *Listeria*, reducing viable cells by up to 99.9% in refrigerated environments (Soni and Nannapaneni, 2010). Phages can be applied via sprays, dips, or incorporation into coatings and films, making them highly adaptable to different points in the food production chain.

The mechanism of action involves phage adsorption to specific receptors on the bacterial cell wall, injection of genetic material, and replication within the host, followed by cell lysis and release of new virions (Sahu et al., 2023). This biological amplification allows for self-replicating antimicrobial action so long as host bacteria are present. Despite their promise, phages face limitations. The emergence of phage-resistant bacterial strains has been observed, though this can be mitigated through the use of phage cocktails targeting multiple receptors (Pal et al., 2024). Moreover, storage conditions such as temperature, pH, and moisture levels can influence phage stability. Research is ongoing into immobilizing phages within edible films or nanostructures to protect their functionality during application and storage.

Table 6. Key Characteristics of Bacteriophage-Based Decontamination

Feature	Description
Target organisms	Specific pathogenic bacteria (e.g., *Listeria, E. coli, Salmonella*)
Mode of action	Adsorption, DNA injection, replication, lysis
Application methods	Spray, dip, coating, film integration
Shelf-life impact	Minimal (no chemical residue or off-flavor)

continued on following page

Table 6. Continued

Feature	Description
Regulatory status	Approved in US, EU for selected pathogens
Environmental impact	Low; biodegradable and self-limiting

Chart 4.1: Lifecycle of a Lytic Bacteriophage in Food Decontamination

[1] Phage attaches to bacterial surface →
[2] Injects DNA →
[3] Replication inside host →
[4] Lysis of bacteria →
[5] New phages released to repeat cycle

4.2 Probiotic and Competitive Exclusion Cultures

Probiotic and competitive exclusion (CE) strategies rely on the principle of ecological interference wherein beneficial or benign microorganisms colonize a surface or environment, thereby preventing or reducing colonization by pathogens (Knipe et al., 2021). This method does not aim to kill pathogens directly but rather to outcompete them for nutrients, space, or adhesion sites, or to modulate the surrounding microenvironment in a way that disfavors pathogenic survival. Probiotic bacteria, typically strains of *Lactobacillus*, *Bifidobacterium*, and *Pediococcus*, have been incorporated into wash waters, fermented food systems, and even packaging films to enhance safety and shelf life (Ramazanidoroh et al., 2024). These organisms produce antimicrobial metabolites such as lactic acid, hydrogen peroxide, and bacteriocins, which inhibit a broad range of spoilage organisms and pathogens (Table 5). For instance, *Lactobacillus plantarum* produces plantaricin, a bacteriocin that disrupts the membrane potential of Gram-positive pathogens (Servin, 2004). Moreover, competitive exclusion has been most notably applied in poultry farming, where CE cultures administered to chicks shortly after hatching colonize the gut and prevent colonization by *Salmonella* spp. Studies have reported up to 90% reductions in *Salmonella enteritidis* when using CE preparations such as Aviguard™, which consists of a defined mixture of anaerobic cecal flora (Hume, 2011). By reducing the load of pathogens in live animals, the risk of downstream contamination during slaughter and processing is significantly diminished. One key advantage of CE and probiotics is that they align with consumer demand for natural or biological food systems (Ajayeoba and Ijabadeniyi, 2025). However, their success often depends on maintaining viability during application and storage. Moreover, efficacy may vary depending on environmental conditions such as temperature, food matrix, or native microflora.

Table 7. Probiotic and CE Mechanisms of Foodborne Pathogen Suppression

Strategy	Microbial Agents	Mode of Action	Food Systems/ Environments
Probiotic cultures	*Lactobacillus spp.*	Acidification, bacteriocin production	Fermented foods, rinse waters
Competitive exclusion	Cecal microbiota	Nutrient competition, receptor site blocking	Poultry gut, meat surfaces
Biofilm modulation	*Bacillus subtilis*, yeasts	Disruption of pathogen colonization	Processing surfaces, equipment walls

4.3 Enzymatic Decontaminants

Enzymes used in food decontamination provide another elegant biological solution. These biocatalysts act with high specificity, targeting key components of microbial cells or neutralizing microbial metabolites like toxins and spoilage compounds (Dahiya et al., 2021). Their proteinaceous nature and natural origin render them highly compatible with clean-label strategies, especially in foods targeted at infants, immunocompromised individuals, or allergen-sensitive consumers (Table 6). One of the most well-known enzymes is lysozyme, which hydrolyzes β-1,4-glycosidic bonds in the peptidoglycan layer of bacterial cell walls, leading to lysis which is particularly effective against Gram-positive organisms like *Listeria monocytogenes* and *Staphylococcus aureus* (Nawaz et al., 2022). Lysozyme has been approved for use in dairy, egg, and meat products, either directly or as part of antimicrobial films (Tomczyk et al., 2018).

Proteolytic enzymes are also being used to destabilize microbial biofilms that protect pathogens on equipment surfaces or within food matrices (Goncalves et al., 2020). These enzymes disrupt the extracellular polymeric substances (EPS) that give biofilms their structural integrity, thereby enhancing the efficacy of other antimicrobials. Oxidoreductases, such as laccases and peroxidases, are effective in detoxifying aflatoxins and other mycotoxins in cereals, nuts, and dried fruits. These enzymes function by oxidizing the toxic moieties into non-toxic or less toxic derivatives that can be safely ingested or excreted (Awuchi et al., 2021). Enzymes may also be applied in several formats: as washes, sprays, embedded into films, or co-formulated with organic acids for synergistic action. Because enzymes are substrate-specific, their use does not alter the organoleptic properties of food unless dosed excessively. However, temperature and pH optima must be carefully managed to maintain enzymatic activity during processing and storage.

Table 8. Enzymes Used in Food Decontamination and Their Targets

Enzyme	Mode of Action	Target Contaminants	Application Example
Lysozyme	Hydrolysis of peptidoglycan in cell walls	*Listeria, S. aureus*	Cheese, eggs, meats
Protease	Breakdown of biofilm matrix proteins	Biofilm-embedded bacteria	Meat surfaces, stainless equipment
Oxidoreductase	Oxidation of aflatoxins and phenolics	Aflatoxins, patulin	Nuts, cereals, apple juice
Amylase + Chitinase	Combined action on polysaccharide structures	Fungal spores and molds	Fermented foods, packaged bakery

5. INTEGRATED AND SMART DECONTAMINATION SYSTEMS

The evolution of food decontamination strategies has reached a turning point where isolated interventions are increasingly being replaced by integrated, adaptive, and intelligence-driven systems. These hybrid approaches combine physical, chemical, and biological methods to achieve enhanced efficacy, while emerging smart technologies are enabling real-time detection, diagnosis, and control of microbial risks across the entire food supply chain. The concept of integrated and smart decontamination systems is grounded in the understanding that food contamination is not static, it evolves through complex microbial ecosystems, responds to environmental cues, and interacts with human behavior and industrial processes. As such, modern food safety solutions must be dynamic, interconnected, and capable of self-correction, rather than relying on static hazard control steps. The rationale, technological components, and operational challenges of integrated and smart decontamination systems, with particular attention to multi-hurdle combinations, sensor-enabled automation, and artificial intelligence applications in microbial control are explored below.

5.1 Multi-Hurdle Decontamination Strategies

Multi-hurdle technology (MHT) represents a paradigm in food preservation that combines two or more decontamination methods typically from different domains (physical, chemical, biological) to exert synergistic antimicrobial effects. The fundamental principle is that by employing several mild interventions rather than a single harsh treatment, one can inactivate or inhibit microorganisms more effectively without compromising food quality (Rebezov et al., 2021). For example, UV-C light treatment combined with mild organic acid washes has been shown to significantly enhance surface microbial reduction in fruits and vegetables (Wang and Wu, 2023). The UV-C induces DNA damage, while the acid component lowers pH and disrupts membrane potential, leading to cumulative stress on microbial cells (Table 7). Studies on lettuce, spinach, and apples treated with UV-C plus lactic or citric acid have reported reductions in *Listeria monocytogenes* and *E. coli* O157:H7 by 3–4 log CFU/g (Allende et al., 2006). Similarly, cold plasma treatment followed by antimicrobial packaging (infused with thyme oil or nisin) has demonstrated prolonged inhibition of spoilage bacteria on sliced meats and seafood under refrigerated conditions.

In thermally sensitive foods such as berries, guacamole, and soft cheeses, the use of pulsed light followed by natural bacteriocin-based coatings (e.g., pediocin or nisin) has enabled shelf-life extension of 5–10 days without sensory degradation (mohapatra et al., 2013). The choice of hurdles is product-specific and depends on the food matrix, microbial targets, regulatory limits, and processing infrastructure. However, while multi-hurdle systems show immense promise, their successful implementation requires precise optimization of individual parameters such as exposure time, pH, concentration, and sequence of application. Moreover, synergy is not guaranteed in all combinations; antagonistic effects can occur if, for example, essential oils absorb UV-C light and reduce its penetration.

Table 9. Examples of Multi-Hurdle Decontamination Combinations

Combination	Target Microorganisms	Food Products	Observed Benefits
UV-C + Citric/Lactic Acid	*E. coli, Listeria, Salmonella*	Lettuce, spinach, apple slices	3–4 log reduction, minimal discoloration
Cold Plasma + Antimicrobial Film	*Listeria*, spoilage flora	Sliced meats, fish fillets	Enhanced shelf life, residual bacteriostasis
Pulsed Light + Bacteriocins	*S. aureus*, molds, yeasts	Berries, cheeses	Preserved texture, reduced pathogen regrowth
HPP + Essential Oils	*Salmonella, E. coli*, yeasts	Guacamole, RTE dips	Pathogen kill with flavor retention

5.2 Sensor-Based and Automated Decontamination

Recent developments in sensor technologies have enabled a shift from reactive to proactive food safety management. Smart decontamination systems now incorporate biosensors, optical scanners, and molecular probes (Table 8) capable of detecting microbial contamination in real time, thereby triggering automatic responses such as targeted rinses, UV-C pulses, or air disinfection (Das, 2023). Biosensors can detect ATP (adenosine triphosphate) as a proxy for biological contamination on food contact surfaces or products. More advanced systems utilize aptamer-based sensors that bind to specific bacterial surface markers, producing fluorescence or electrical signals when pathogens like *Salmonella* or *Listeria* are present (Guo, 2024). Near-infrared spectroscopy, Raman spectroscopy, and hyperspectral imaging are also being used to non-destructively assess contamination on produce and meats. Such systems are currently used in high-throughput environments like poultry evisceration lines, leafy vegetable packing stations, and seafood grading belts. When combined with real-time temperature and humidity monitoring, these systems can adaptively change their decontamination protocol depending on environmental risk. These systems not only enhance food safety but also reduce water, energy, and chemical usage by avoiding blanket treatments. However, their effectiveness is limited by the sensitivity and specificity of the sensors, data latency, and the robustness of automated response algorithms. Integration into legacy infrastructure and staff training remain implementation hurdles.

Table 10. Types of Sensors in Smart Food Safety Systems

Sensor Type	Detection Mechanism	Target Metric/ Organism	Application Area
ATP Bioluminescence	Enzyme-luciferase reaction	General microbial presence	Equipment sanitation, conveyor belts
DNA-based Biosensor	Hybridization/probe binding	*Listeria, E. coli, Salmonella*	Leafy greens, meats
Optical Spectroscopy	Light absorption/scattering patterns	Surface contamination	Poultry skin, fish, produce
RFID-integrated Sensors	Wireless data transmission	Temp, pH, microbial metabolites	Packaging, cold chain logistics

5.3 Artificial Intelligence and Predictive Analytics

Artificial Intelligence (AI) and machine learning are now being integrated into food decontamination to predict microbial contamination risks, optimize treatment schedules, and manage data from sensor networks (Table 11). By analyzing historical trends, real-time environmental conditions, and product movement data, AI models can forecast when and where contamination is most likely to occur, enabling preventive interventions (Palla and Iwunwa, 2025). For instance, AI can identify critical control points (CCPs) in meat processing facilities by correlating microbial test results with processing variables such as machine downtime, temperature fluctuations, and employee sanitation logs (Revelou et al., 2025). Once high-risk zones are flagged, automated cleaning-in-place (CIP) or air filtration units can be activated without human intervention. Machine learning algorithms have also been applied to optimize UV-C dosage, chemical spray volumes, and HPP cycle durations across different food types. These predictive models reduce trial-and-error testing and increase confidence in the scalability of decontamination protocols. Some systems integrate blockchain and IoT (Internet of Things) to ensure traceability and auditability of safety interventions in multi-site operations (Choudhury et al., 2019). The future lies in fully autonomous food safety ecosystems that not only detect and decontaminate threats in real time but also adapt their behavior through continuous learning. Challenges to this vision include data standardization, cybersecurity, and balancing automation with human oversight to avoid blind reliance on algorithmic decisions.

Table 11. AI Applications in Food Decontamination

AI Function	Input Data	Output/Application
Contamination Risk Modeling	Sensor readings, historical test results	Dynamic adjustment of decontamination
Process Optimization	HPP cycles, UV exposure logs	Reduced energy, time, and waste
Predictive Maintenance	Equipment vibration/temp sensors	Preemptive cleaning, reduce downtime
Traceability Verification	IoT + blockchain data streams	Automated audit trails for compliance

6. EMERGING AND HYBRID DECONTAMINATION TECHNOLOGIES

As the food industry continues to evolve under the influence of globalization, heightened food safety demands, and sustainability imperatives, so too does the drive for more efficient and ecologically conscious decontamination strategies. Emerging and hybrid decontamination technologies are reshaping the landscape by synergistically combining physical, chemical, and biological modalities, often with the support of advanced engineering and materials science (El Omari et al., 2025). These technologies offer promise not only in terms of efficacy but also in sustainability, adaptability to novel food formats, and compatibility with minimally processed, nutrient-retaining food products. In this section, we explore a suite of these technologies, their mechanisms, applications, and the critical parameters influencing their integration into modern food safety systems.

6.1 Photocatalytic Decontamination

Photocatalysis is an innovative approach wherein light-activated semiconductor materials generate reactive oxygen species (ROS) capable of degrading organic pollutants and inactivating microorganisms (Khodabandeh et al., 2025). Titanium dioxide (TiO_2) is the most widely used photocatalyst due to its high oxidative potential, chemical stability, non-toxicity, and cost-effectiveness. When TiO_2 is irradiated with ultraviolet (UV-A) light (wavelength ~315-400 nm), it becomes photo-excited and produces ROS such as hydroxyl radicals, superoxide ions, and singlet oxygen (Chen et al., 2024). These highly reactive species attack microbial cell walls, proteins, and DNA, leading to cell lysis and death. In food processing environments, photocatalytic surfaces are commonly incorporated into conveyor belts, cutting boards, packaging films, and coatings on stainless steel and plastic surfaces. These self-sanitizing surfaces can be activated by UV light sources commonly present in food facilities, thereby reducing the risk of cross-contamination in real-time production settings. The use of photocatalysis offers multiple advantages including a continuous, residue-free sanitation process that minimizes the need for manual cleaning or chemical rinses. This approach is particularly appealing in processing lines where wet cleaning is undesirable or time-consuming. The compatibility of TiO_2 with various substrates and the potential for developing smart packaging further extend its applications (Nagaraj et al., 2025). However, challenges remain in terms of its limited penetration in opaque food matrices or shadowed areas, which may harbor residual pathogens. Additionally, the reduced activity under low-intensity lighting and the need to validate nanomaterial safety in food environments may delay regulatory approval and widespread industry adoption.

6.2 Ultrasound (Sonication)

Ultrasound employs sound waves beyond the range of human hearing (above 20 kHz) to generate cavitation in liquids (Figure 3). This involves the rapid formation and collapse of microscopic bubbles, which produce intense shear forces, localized heat, and pressure gradients capable of disrupting microbial membranes and enhancing the diffusion of antimicrobial agents (Onyeaka et al., 2023). In practical food decontamination, ultrasound is applied in the washing of fresh produce, shellfish depuration, egg surface sanitation, and even in beverage treatment to reduce microbial loads without altering organoleptic properties (Jin and Adhikari, 2025). Moreover, one of the critical benefits of ultrasound is its non-thermal nature, which allows for the preservation of food texture, flavor, and nutritional value while achieving significant microbial inactivation. When used in combination with mild heat or antimicrobial agents like organic acids or ozone, ultrasound shows synergistic effects that improve efficacy while reducing chemical usage (Turhan et al., 2022). Nonetheless, the technology has some limitations, particularly in viscous or heterogeneous food matrices where acoustic energy distribution is non-uniform. Additionally, the effectiveness of ultrasound is diminished against resilient microbial forms such as spores and mature biofilms unless combined with other hurdles like enzymatic or thermal treatments (Onyeaka et al., 2023). Therefore, the full potential of ultrasound lies in its integration within a multi-modal decontamination framework.

Figure 3. Ultrasound Cavitation Mechanism

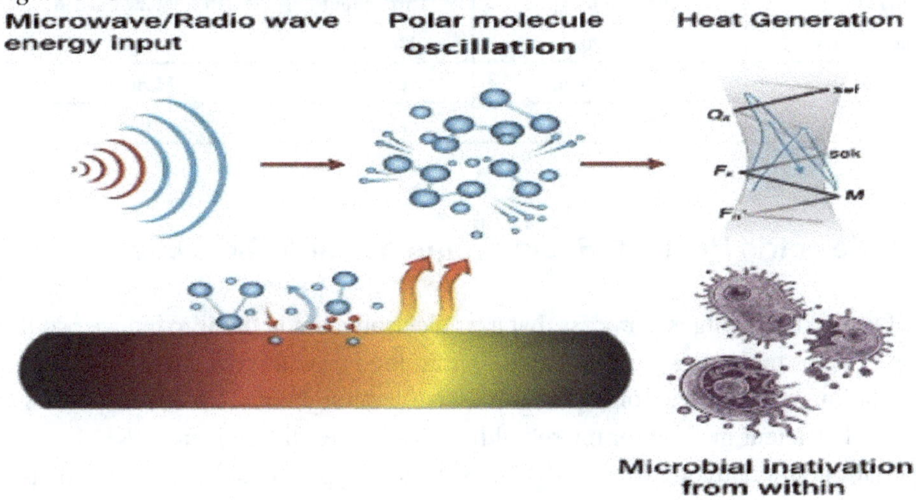

6.3 Irradiation Technologies (Gamma, Electron Beam, X-Ray)

Irradiation technologies use ionizing radiation, including gamma rays, electron beams, and X-rays, to disrupt microbial DNA and cellular functions, resulting in effective pathogen inactivation. These technologies are especially beneficial for high-risk products like spices, dried herbs, meats, seafood, and ready-to-eat meals where microbial contamination poses significant safety risks (Ugur, 2023). Gamma irradiation, derived from cobalt-60 sources, penetrates deeply and is suitable for bulk processing, while electron beams and X-rays offer faster processing rates and more controlled dosage (Table 12). The principal advantage of irradiation lies in its ability to provide a non-thermal, residue-free microbial control method that is compatible with sealed and prepackaged foods. This reduces the risk of recontamination and extends product shelf life without the need for additives. However, the use of irradiation is subject to consumer perception challenges, as many associate radiation with safety concerns despite scientific consensus on its harmlessness when applied within regulatory limits (Vaiserman et al., 2018). Moreover, the high capital investment, complex regulatory requirements, and mandatory labeling have slowed broader adoption. Despite these challenges, irradiation remains a scientifically validated method, and its application may expand as consumer education and regulatory harmonization progress globally.

Table 12. Comparison of Irradiation Modalities

Modality	Penetration Depth	Throughput	Regulatory Burden
Gamma rays	High	Moderate	High
E-beam	Low-medium	Very High	Medium
X-ray	Medium-high	High	High

6.4 Dielectric Heating (Radio Frequency and Microwave)

Dielectric heating is a process that uses electromagnetic fields to induce oscillation of polar molecules, such as water, within food products (Figure 4). This rapid oscillation produces friction and heat uniformly throughout the product, offering a fast and efficient method for microbial inactivation. Radio frequency (RF) heating operates at lower frequencies (1–100 MHz) and is typically used for larger, denser products, while microwave heating operates at higher frequencies (300 MHz–30 GHz) and is suited for smaller or more homogeneous food items (Altemimi et al., 2019). This method has found applications in pasteurizing soups, sauces, and baby foods, as well as in drying grains and controlling pests in stored commodities. One of its key benefits is uniform internal heating that reduces surface scorching and maintains food quality. Additionally, dielectric heating is highly energy-efficient and compatible with flexible packaging (Naseri and Yourdkhani, 2022). However, it poses several technical challenges, particularly with foods of heterogeneous composition, which may heat unevenly due to varying dielectric properties. Such inconsistencies can result in underprocessed or overprocessed regions, potentially compromising both safety and quality. Additionally, the initial equipment costs and need for precise process validation present barriers to entry, particularly for small and medium-sized enterprises.

Figure 4. Energy Absorption in Dielectric Heating

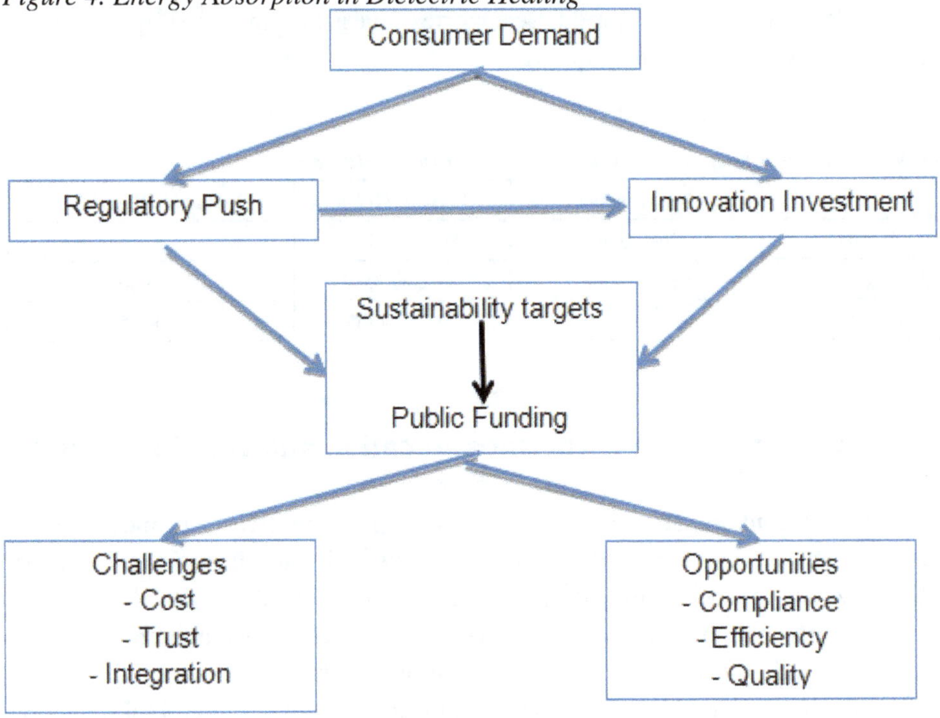

6.5 Supercritical CO_2 Treatment

Supercritical CO_2 treatment leverages the unique physicochemical properties of carbon dioxide when it is held above its critical temperature and pressure. In its supercritical state, CO_2 behaves like both a gas and a liquid, allowing it to diffuse efficiently into food matrices while dissolving and destabilizing microbial cell membranes (Zia et al., 2022). This technology enables the inactivation of a wide range of microorganisms, including bacteria, yeasts, and molds, without the need for high temperatures or chemical solvents (Table 13). Its practical applications span the decontamination of spices, herbs, cereals, juices, and powdered ingredients. Supercritical CO_2 is particularly attractive for products that are sensitive to heat or oxidation, as it preserves flavor, color, and nutritional quality. The process is environmentally friendly, as CO_2 can be recycled, and the absence of residual chemicals makes it suitable for clean-label foods (Alexandri et al., 2022). Nevertheless, the technology comes with notable challenges. It requires high-pressure vessels and complex infrastructure, which can be cost-prohibitive. Additionally, its microbial reduction efficiency can be influenced by product composition, particularly moisture and fat content. This variability necessitates careful parameter control and may limit scalability for certain product categories unless accompanied by regulatory incentives and infrastructure support.

Table 13. Supercritical CO_2 Outcomes in Food Systems

Food Type	Microorganism	Pressure/Temp	Reduction (log CFU)
Black pepper	Total flora	80 bar / 40°C	3.5
Apple juice	Yeasts, molds	100 bar / 35°C	4.0–5.0
Wheat flour	*E. coli*, *Salmonella*	90 bar / 45°C	2.0–3.5

6.6 Hybrid Techniques: Thermosonication and Ohmic Heating

Hybrid decontamination techniques combine multiple treatment modalities to leverage their respective strengths while minimizing individual limitations (Table 14). Thermosonication merges mild heat (typically between 50–60°C) with ultrasound to intensify microbial inactivation through combined thermal and cavitational effects (Urango et al., 2022). This approach has demonstrated high efficacy in treating fruit juices, dairy-based beverages, and liquid eggs, achieving microbial reduction with minimal impact on sensory and nutritional properties. Similarly, ohmic heating involves the direct passage of electric current through food, causing internal energy dissipation and rapid volumetric heating (Jan et al., 2021). This technique is especially effective for conductive liquid and semi-solid foods like soups, sauces, and yogurt drinks. The main advantage lies in its uniform and rapid heating, which significantly shortens processing time and reduces thermal degradation. Both thermosonication and ohmic heating offer improved energy efficiency and can be adapted to continuous processing systems. However, these technologies also pose specific challenges, such as the need for accurate control of food conductivity and the potential for uneven heating in products with varying composition. Furthermore, industrial adoption is still limited, necessitating further research and standardization to optimize their commercial viability and ensure consistent food safety outcomes.

Table 14. Hybrid Decontamination Performance Metrics

Technology	Target Pathogen	Processing Time	Log Reduction	Product Impact
Thermosonication	*Listeria*, *E. coli*	10 min	4–6	Low nutrient loss
Ohmic heating	*Salmonella*, yeasts	3–5 min	5–7	Maintains texture/flavor

7. CHALLENGES AND OPPORTUNITIES IN ADOPTION

While emerging and hybrid decontamination technologies offer considerable promise in enhancing food safety, sustainability, and quality retention, their adoption at commercial scale is neither automatic nor uniform. The challenges impeding their widespread implementation are multifaceted, spanning technical, economic, regulatory, and socio-cultural dimensions. One of the most immediate barriers is the high capital investment required to acquire and retrofit existing production lines with advanced decontamination technologies. Systems such as dielectric heaters, supercritical CO_2 chambers, and cold plasma units entail not only initial procurement costs but also ongoing maintenance, skilled labor, and energy demands. These costs can be prohibitive for small and medium-sized enterprises (SMEs), especially in developing economies where access to credit and technical support remains limited. Even where return on investment is ultimately positive, the extended payback periods and high perceived risk deter early adoption.

Furthermore, regulatory uncertainty continues to inhibit confidence in these technologies. Unlike conventional chemical or thermal treatments, newer modalities often lack harmonized international safety standards and standardized microbial reduction validation protocols. For instance, while the FDA or EFSA may approve one method in specific jurisdictions, others—such as ASEAN or African regulatory bodies—may lack detailed evaluation frameworks. This disparity not only creates compliance confusion for multinational corporations but also discourages startups from entering markets with non-aligned guidelines. Moreover, some technologies involve novel materials (e.g., nanoparticles in photocatalysis) or modes of action (e.g., DNA interference by cold plasma) that challenge existing definitions of "natural" or "clean-label" processing. This ambiguity can trigger resistance from regulatory agencies, advocacy groups, or consumers demanding transparency and familiarity.

From a consumer standpoint, perception remains a delicate challenge. Despite decades of scientific validation, public skepticism persists around methods involving irradiation, nanoparticle coatings, or AI-driven microbial detection. Fear of the unknown, coupled with historical baggage from industrial food scandals, often influences how consumers interpret safety innovations. For instance, food irradiation—though endorsed by the WHO and numerous safety authorities—continues to suffer from associations with nuclear risk and perceived flavor alterations, even when these are unfounded. Transparent communication, coupled with third-party endorsements and consumer education campaigns, is thus essential to overcoming emotional resistance and misinformation.

Another significant barrier lies in the sensory and nutritional impact of some decontamination processes. While non-thermal and hybrid approaches strive to preserve food integrity, improperly optimized applications may lead to texture degradation, color shifts, or nutrient losses. For example, excessive dielectric heating may result in uneven cooking in multicomponent foods, while overexposure to ozone or ultrasound can modify delicate aromas in berries or leafy greens. Such trade-offs—though technically avoidable—often require rigorous process control and quality assurance systems, which may not be in place in smaller operations. This technical complexity creates additional demand for trained personnel, monitoring infrastructure, and cross-disciplinary collaboration between microbiologists, food engineers, and sensory analysts.

Despite these challenges, the opportunities offered by emerging technologies are equally compelling. Foremost among them is the growing demand for clean-label, minimally processed foods—a market trend that has redefined consumer expectations globally. Technologies like pulsed light, antimicrobial films, and probiotic sprays enable safety enhancement without chemical preservatives, thus aligning with this consumer-driven paradigm shift. Furthermore, the sustainability imperative—driven by climate policies, zero-waste initiatives, and carbon footprint disclosure standards—positions these technologies as vital solutions. For instance, high-pressure processing and supercritical CO_2 treatment use less water and generate fewer effluents compared to chlorine-based washing systems, thereby supporting broader environmental goals.

There is also growing momentum in the form of public-private partnerships and global research consortia. Initiatives funded by FAO, WHO, and regional bodies such as the European Horizon programs or Africa's AATF (African Agricultural Technology Foundation) are pooling expertise and funding to advance scalable, context-specific innovations. Collaborative field trials, industry benchmarking, and open-source protocols can help close the knowledge gap between innovation hubs and local processors, particularly in low- and middle-income countries. Simultaneously, academic institutions are refining computational models and in silico simulations to predict microbial inactivation more efficiently, reducing the reliance on lengthy lab trials.

Smart technology integration offers additional pathways for scaling adoption. The proliferation of IoT sensors, cloud-based monitoring, and AI analytics has begun to transform food safety from a reactive to a predictive discipline. For instance, smart packaging with embedded sensors can provide real-time data on microbial load, triggering automated surface treatments or alerts within cold chains. As these technologies become more affordable and interoperable, they may facilitate decentralized decontamination—empowering logistics centers, retailers, and even consumers to participate in the food safety continuum.

Table 15. Comparative Barriers and Drivers of Decontamination Technology Adoption

Category	Barriers	Opportunities
Economic	High capital investment, long ROI	Long-term cost savings, reduced recalls
Regulatory	Lack of harmonized standards, evolving frameworks	Pressure for food safety transparency and traceability
Technical	Integration complexity, need for trained workforce	Smart automation, digital modeling
Consumer	Skepticism of novel methods, sensory concerns	Clean-label demand, preference for minimally processed foods
Environmental	Energy use in some technologies	Water savings, waste reduction, lower chemical runoff

Chart 7.1: Interaction of Drivers and Barriers in Technology Adoption

[Consumer Demand ↑]
↗ ↘
[Regulatory Push] → [Innovation Investment]
↘ ↗
[Sustainability Targets → Public Funding]
↘ Challenges: Cost, Trust, Integration
↗ Opportunities: Compliance, Efficiency, Quality

Standardized validation frameworks will play a critical role in the future. By developing clear, internationally recognized protocols for efficacy testing, food contact material safety, and processing reproducibility, regulatory bodies can reduce uncertainty and speed up product approval cycles. Such protocols can also facilitate certification schemes that assure retailers and consumers of quality without requiring every buyer to conduct technical due diligence independently. In the long term, these standards may function as non-tariff enablers in international trade, particularly for exporters in emerging markets looking to meet stringent import regulations in Europe, North America, or East Asia.

8. CONCLUSION AND FUTURE PERSPECTIVES (EXPANDED TO ~400 WORDS)

The integration of novel decontamination technologies into contemporary food systems marks a pivotal evolution in the global pursuit of safer, more sustainable, and consumer-aligned food production practices. These innovative solutions—spanning physical, chemical, biological, and hybrid modalities—not only address persistent microbial hazards but also respond to changing consumer expectations around clean-label transparency, minimal processing, and environmental responsibility. Technologies such as high-pressure processing, cold plasma, photocatalytic surfaces, and antimicrobial packaging are no longer confined to niche applications. Instead, they are becoming essential elements in a broader movement toward intelligent food safety systems that extend shelf life, minimize chemical residues, and preserve the organoleptic integrity of products.

Nevertheless, the deployment of these technologies is not without challenges. Regulatory harmonization remains uneven across regions, creating barriers for global producers navigating different approval frameworks for technologies such as irradiation, bacteriophage use, or nano-enabled materials. Economic constraints, particularly in low- and middle-income settings, continue to limit access to capital-intensive equipment and technical expertise. These financial hurdles are exacerbated by the need for specialized training, process validation, and compliance with emerging documentation and traceability demands.

Yet, the opportunities on the horizon are equally transformative. The next frontier of decontamination will likely emerge at the intersection of biotechnology, nanoscience, and artificial intelligence. Nano-enabled decontaminants—such as silver, zinc oxide, or carbon-based nanomaterials—offer targeted microbial inactivation with enhanced surface interaction and minimal environmental impact. Bio-inspired solutions, modeled on natural antimicrobial peptides or enzyme-based detoxification, could enable precision decontamination without damaging food matrices. Additionally, the integration of AI-assisted control systems, capable of analyzing microbial patterns in real time and adjusting treatment parameters dynamically, promises unprecedented accuracy and efficiency in industrial settings.

Looking forward, fostering interdisciplinary collaborations between food scientists, engineers, microbiologists, and policy makers will be critical to bridging the gap between innovation and implementation. Investment in infrastructure, consumer education, and the development of inclusive regulatory standards will not only facilitate the adoption of these technologies but also ensure equitable food safety outcomes across global value chains. In this context, the future of decontamination lies not in a single breakthrough technology but in a robust, adaptive system that harmonizes safety, quality, sustainability, and public trust—ensuring resilient food security for a rapidly evolving.

REFERENCES

Abthiyar, L. (2023). *Effect of UV-C treatment on sanitised and unsanitised ready-to-eat leafy green vegetables, produced in New Zealand: a thesis presented in partial fulfilment of the requirements for the degree of Master of Food Technology, Massey University, Albany Campus, New Zealand* (Doctoral dissertation, Massey University).

Ahmed, S., Akther, S., & Hossain, M. S. (2024). Electrolyzed Water-Based Technology for Fresh Fruits and Vegetables. In *Sustainable Postharvest Technologies for Fruits and Vegetables* (pp. 293–303). CRC Press. DOI: 10.1201/9781003370376-25

Ajayeoba, T. A., & Ijabadeniyi, O. A. (2025). Transforming Food for the Future: Precision Fermentation as a Key to Sustainability, Nutrition, and Health. *Nutrition and Health (Berkhamsted, Hertfordshire).*

Al-Moghazy, M., & El-Sayed, H. S. (2023). Antimicrobial-prebiotic: Novel dual approach of pomegranate peel extract in vitro and in food system. *Biocatalysis and Agricultural Biotechnology*, *49*, 102664. DOI: 10.1016/j.bcab.2023.102664

Al-Moghazy, M., & El-Sayed, H. S. (2023). Antimicrobial-prebiotic: Novel dual approach of pomegranate peel extract in vitro and in food system. *Biocatalysis and Agricultural Biotechnology*, *49*, 102664. DOI: 10.1016/j.bcab.2023.102664

Alexandri, M., Kachrimanidou, V., Papapostolou, H., Papadaki, A., & Kopsahelis, N. (2022). Sustainable food systems: The case of functional compounds towards the development of clean label food products. *Foods*, *11*(18), 2796. DOI: 10.3390/foods11182796 PMID: 36140924

Allende, A., McEvoy, J. L., Luo, Y., Artes, F., & Wang, C. Y. (2006). Effectiveness of two-sided UV-C treatments in inhibiting natural microflora and extending the shelf-life of minimally processed 'Red Oak Leaf' lettuce. *Food Microbiology*, *23*(3), 241–249. DOI: 10.1016/j.fm.2005.04.009 PMID: 16943010

Altemimi, A., Aziz, S. N., Al-Hilphy, A. R., Lakhssassi, N., Watson, D. G., & Ibrahim, S. A. (2019). Critical review of radio-frequency (RF) heating applications in food processing. *Food Quality and Safety*, *3*(2), 81–91. DOI: 10.1093/fqsafe/fyz002

Awuchi, C. G., Ondari, E. N., Ogbonna, C. U., Upadhyay, A. K., Baran, K., Okpala, C. O. R., Korzeniowska, M., & Guiné, R. P. (2021). Mycotoxins affecting animals, foods, humans, and plants: Types, occurrence, toxicities, action mechanisms, prevention, and detoxification strategies—A revisit. *Foods*, *10*(6), 1279. DOI: 10.3390/foods10061279 PMID: 34205122

Banerjee, R., Verma, A. K., Naveena, B. M., & Kulkarni, V. V. (2018). Processing, Storage, and Transportation of Meat and Meat Products. In *Animal Sourced Foods for Developing Economies* (pp. 77–117). CRC Press. DOI: 10.1201/9780429398575-5

Bolouri, P., Salami, R., Kouhi, S., Kordi, M., Asgari Lajayer, B., Hadian, J., & Astatkie, T. (2022). Applications of essential oils and plant extracts in different industries. *Molecules (Basel, Switzerland)*, *27*(24), 8999. DOI: 10.3390/molecules27248999 PMID: 36558132

Bolumar, T., Orlien, V., Sikes, A., Aganovic, K., Bak, K. H., Guyon, C., Stübler, A. S., de Lamballerie, M., Hertel, C., & Brüggemann, D. A. (2021). High-pressure processing of meat: Molecular impacts and industrial applications. *Comprehensive Reviews in Food Science and Food Safety*, *20*(1), 332–368. DOI: 10.1111/1541-4337.12670 PMID: 33443800

Burt, S. (2004). Essential oils: Their antibacterial properties and potential applications in foods—a review. *International Journal of Food Microbiology*, *94*(3), 223–253. DOI: 10.1016/j.ijfoodmicro.2004.03.022 PMID: 15246235

Chen, Z., Pan, M., Cheng, C., Luo, J., & Deng, X. (2024). Water disinfection: Advances in photocatalysis and piezo/triboelectric catalysis with progressively enhanced energy utilization. *SusMat*, *4*(5), e232. DOI: 10.1002/sus2.232

Choudhury, O., Fairoza, N., Sylla, I., & Das, A. (2019). A blockchain framework for managing and monitoring data in multi-site clinical trials. *arXiv preprint arXiv:1902.03975*.

Chu, Z., Wang, H., & Dong, B. (2024). Research on Food Preservation Based on Antibacterial Technology: Progress and Future Prospects. *Molecules (Basel, Switzerland)*, *29*(14), 3318. DOI: 10.3390/molecules29143318 PMID: 39064897

Coban, H. B. (2020). Organic acids as antimicrobial food agents: Applications and microbial productions. *Bioprocess and Biosystems Engineering*, *43*(4), 569–591. DOI: 10.1007/s00449-019-02256-w PMID: 31758240

Dahiya, D., Manuel, J. V., & Nigam, P. S. (2021). An overview of bioprocesses employing specifically selected microbial catalysts for γ-aminobutyric acid production. *Microorganisms*, *9*(12), 2457. DOI: 10.3390/microorganisms9122457 PMID: 34946060

Das, A. K. (2023). *Microbial determination methods in pharmaceutical and food quality control* (Doctoral dissertation, Brac University).

de Mesquita Souza Saraiva, M., Lim, K., do Monte, D. F. M., Givisiez, P. E. N., Alves, L. B. R., de Freitas Neto, O. C., Kariuki, S., Júnior, A. B., de Oliveira, C. J. B., & Gebreyes, W. A. (2022). Antimicrobial resistance in the globalized food chain: A One Health perspective applied to the poultry industry. *Brazilian Journal of Microbiology : Publication of the Brazilian Society for Microbiology, 53*(1), 465–486. DOI: 10.1007/s42770-021-00635-8 PMID: 34775576
de Mesquita Souza Saraiva, M., Lim, K., do Monte, D. F. M., Givisiez, P. E. N., Alves, L. B. R., de Freitas Neto, O. C., Kariuki, S., Júnior, A. B., de Oliveira, C. J. B., & Gebreyes, W. A. (2022). Antimicrobial resistance in the globalized food chain: A One Health perspective applied to the poultry industry. *Brazilian Journal of Microbiology : Publication of the Brazilian Society for Microbiology, 53*(1), 465–486. DOI: 10.1007/s42770-021-00635-8 PMID: 34775576
Devlieghere, F., Vermeiren, L., & Debevere, J. (2004). New preservation technologies: Possibilities and limitations. *International Dairy Journal, 14*(4), 273–285. DOI: 10.1016/j.idairyj.2003.07.002

Endersen, L., O'Mahony, J., Hill, C., Ross, R. P., McAuliffe, O., & Coffey, A. (2014). Phage therapy in the food industry. *Annual Review of Food Science and Technology, 5*(1), 327–349. DOI: 10.1146/annurev-food-030713-092415 PMID: 24422588
Endersen, L., O'Mahony, J., Hill, C., Ross, R. P., McAuliffe, O., & Coffey, A. (2014). Phage therapy in the food industry. *Annual Review of Food Science and Technology, 5*(1), 327–349. DOI: 10.1146/annurev-food-030713-092415 PMID: 24422588
Esmaeili, Y., Paidari, S., Baghbaderani, S. A., Nateghi, L., Al-Hassan, A. A., & Ariffin, F. (2022). Essential oils as natural antimicrobial agents in postharvest treatments of fruits and vegetables: A review. *Journal of Food Measurement and Characterization, 16*(1), 507–522. DOI: 10.1007/s11694-021-01178-0

Frank-Kamenetskii, D. (2012). *Plasma: the fourth state of matter*. Springer Science & Business Media.

Fu, Y., & Dudley, E. G. (2021). Antimicrobial-coated films as food packaging: A review. *Comprehensive Reviews in Food Science and Food Safety, 20*(4), 3404–3437. DOI: 10.1111/1541-4337.12769 PMID: 34056844

Galanakis, C. M. (2024). The future of food. *Foods, 13*(4), 506. DOI: 10.3390/foods13040506 PMID: 38397483

García, G., Carlin, M., & Cano, R. D. J. (2025). Holobiome harmony: Linking environmental sustainability, agriculture, and human health for a thriving planet and one health. *Microorganisms, 13*(3), 514. DOI: 10.3390/microorganisms13030514 PMID: 40142407

Garcia-Gutierrez, E., Mayer, M. J., Cotter, P. D., & Narbad, A. (2019). Gut microbiota as a source of novel antimicrobials. *Gut Microbes, 10*(1), 1–21. DOI: 10.1080/19490976.2018.1455790 PMID: 29584555

Gayán, E., Serrano, M. J., Pagán, R., Álvarez, I., & Condón, S. (2015). Environmental and biological factors influencing the UV-C resistance of Listeria monocytogenes. *Food Microbiology, 46*, 246–253. DOI: 10.1016/j.fm.2014.08.011 PMID: 25475293

Gohil, K., Chandarana, K. A., Chandwani, S., Jawahar, G., & Amaresan, N. (2022). Production, cost analysis, and marketing of acetic acid (Vinegar). In *Industrial microbiology based entrepreneurship: Making money from microbes* (pp. 93–115). Springer Nature Singapore. DOI: 10.1007/978-981-19-6664-4_8

Gonçalves, S. F., Lee, S. H. I., Khaneghah, A. M., & Oliveira, C. A. F. (2020). Enzyme-based approaches to control microbial biofilms in dairy processing environments: A review. *Quality Assurance and Safety of Crops & Foods, 12*(SP1), 50–58. DOI: 10.15586/qas.v12SP1.828

Grace, D. (2023). Burden of foodborne disease in low-income and middle-income countries and opportunities for scaling food safety interventions. *Food Security, 15*(6), 1475–1488. DOI: 10.1007/s12571-023-01391-3

Grace, D. (2023). Burden of foodborne disease in low-income and middle-income countries and opportunities for scaling food safety interventions. *Food Security, 15*(6), 1475–1488. DOI: 10.1007/s12571-023-01391-3

Gumienna, M., & Górna, B. (2021). Antimicrobial food packaging with biodegradable polymers and bacteriocins. *Molecules (Basel, Switzerland), 26*(12), 3735. DOI: 10.3390/molecules26123735 PMID: 34207426

Gummadi, S. N. (2024). Advances in the applications of Bacteriophages and phage products against food-contaminating bacteria. *Critical Reviews in Microbiology, 50*(5). PMID: 37861086

Guo, X. (2024). Research progress on the detection of foodborne pathogens based on aptamer recognition. *Microchimica Acta, 191*(6), 318. DOI: 10.1007/s00604-024-06375-4 PMID: 38727855

Hume, M. E. (2011). Historic perspective: Prebiotics, probiotics, and other alternatives to antibiotics. *Poultry Science*, *90*(11), 2663–2669. DOI: 10.3382/ps.2010-01030 PMID: 22010256

Ilias, S., Zakir, M. A. W. M., & Abd Almisreb, A. (2021). Review of the Effectiveness of UV-C for Disinfection of High Touch Objects. *Southeast Europe Journal of Soft Computing*, *10*(1), 30–37.

Islam, F., Imran, A., Nosheen, F., Fatima, M., Arshad, M. U., Afzaal, M., Ijaz, N., Noreen, R., Mehta, S., Biswas, S., Rasool, I. F. U., Aslam, M. A., Usman, I., Zahra, S. M., Segueni, N., & Amer Ali, Y. (2023). Functional roles and novel tools for improving-oxidative stability of polyunsaturated fatty acids: A comprehensive review. *Food Science & Nutrition*, *11*(6), 2471–2482. DOI: 10.1002/fsn3.3272 PMID: 37324849

Issa-Zacharia, A. (2024). Application of slightly acidic electrolyzed water as a potential sanitizer in the food industry. *Journal of Food Quality*, *2024*(1), 5559753. DOI: 10.1155/2024/5559753

Jan, B., Shams, R., Rizvi, Q. E. H., & Manzoor, A. (2021). Ohmic heating technology for food processing: A review of recent developments. *Journal of Postharvest Technology*, *9*(1), 20–34.

Jin, Y., & Adhikari, A. (2025). Emerging and Innovative Technologies for the Sanitization of Fresh Produce: Advances, Mechanisms, and Applications for Enhancing Food Safety and Quality. *Foods*, *14*(11), 1924. DOI: 10.3390/foods14111924 PMID: 40509452

Katsigiannis, A. S., Bayliss, D. L., & Walsh, J. L. (2022). Cold plasma for the disinfection of industrial food-contact surfaces: An overview of current status and opportunities. *Comprehensive Reviews in Food Science and Food Safety*, *21*(2), 1086–1124. DOI: 10.1111/1541-4337.12885 PMID: 35075786

Khodabandeh Lu, N., Alamdari, H., Rashedi, H., Yazdian, F., Pourmadadi, M., Rahdar, A., & Romanholo Ferreira, L. F. (2025). Photocatalytic Bioreactors in Water Treatment: Mechanisms, Challenges, and Innovations in Microbial Inactivation. *Water, Air, and Soil Pollution*, *236*(5), 1–28. DOI: 10.1007/s11270-025-07919-x

Kim, N. N., Kim, W. J., & Kang, S. S. (2019a). Anti-biofilm effect of crude bacteriocin derived from Lactobacillus brevis DF01 on Escherichia coli and Salmonella Typhimurium. *Food Control*, *98*, 274–280. DOI: 10.1016/j.foodcont.2018.11.004

Kim, N. N., Kim, W. J., & Kang, S. S. (2019a). Anti-biofilm effect of crude bacteriocin derived from Lactobacillus brevis DF01 on Escherichia coli and Salmonella Typhimurium. *Food Control*, *98*, 274–280. DOI: 10.1016/j.foodcont.2018.11.004

Kim, S. S., Park, S. H., Kim, S. H., & Kang, D. H. (2019b). Synergistic effect of ohmic heating and UV-C irradiation for inactivation of Escherichia coli O157: H7, Salmonella Typhimurium and Listeria monocytogenes in buffered peptone water and tomato juice. *Food Control*, *102*, 69–75. DOI: 10.1016/j.foodcont.2019.03.011

Knipe, H., Temperton, B., Lange, A., Bass, D., & Tyler, C. R. (2021). Probiotics and competitive exclusion of pathogens in shrimp aquaculture. *Reviews in Aquaculture*, *13*(1), 324–352. DOI: 10.1111/raq.12477

Knorr, D., Froehling, A., Jaeger, H., Reineke, K., Schlueter, O., & Schoessler, K. (2011). Emerging technologies in food processing. *Annual Review of Food Science and Technology*, *2*(1), 203–235. DOI: 10.1146/annurev.food.102308.124129 PMID: 22129381

Koutchma, T. (2014). *Food Plant Safety: UV Applications for Food and Non-Food Surfaces*. Elsevier.

Koutchma, T. (2019). *Ultraviolet light in food technology: principles and applications*. CRC press. DOI: 10.1201/9780429244414

Krishnamurthy, K., Tewari, J. C., Irudayaraj, J., & Demirci, A. (2010). Microscopic and spectroscopic evaluation of inactivation of Staphylococcus aureus by pulsed UV light and infrared heating. *Food and Bioprocess Technology*, *3*(1), 93–104. DOI: 10.1007/s11947-008-0084-8

Kumar, A., Kumar, V., Arsenov, D., Thakur, M., Kumar, A., Khokhar, A., Seth, C. S., & Kumar, R. (2024). The science of food safety and their health impacts. *Journal of Geochemical Exploration*, *267*, 107596. DOI: 10.1016/j.gexplo.2024.107596

Lam, W. R., & Liu, Y. (2020). Tackling Small and Medium-Sized Enterprises (SMEs) Financing in China. *Annals of Economics and Finance*, *21*(1).

Lohita, B., & Srijaya, M. (2024). Novel technologies for shelf-life extension of food products as a competitive advantage: A review. *Food production, diversity, and safety under climate change*, 285-306.

Loretz, M., Stephan, R., & Zweifel, C. (2011). Antibacterial activity of decontamination treatments for cattle hides and beef carcasses. *Food Control*, *22*(3-4), 347–359. DOI: 10.1016/j.foodcont.2010.09.004

Mahunu, G. K., Osei-Kwarteng, M., Ogwu, M. C., & Afoakwah, N. A. (2024). Safe food handling techniques to prevent microbial contamination. In *Food safety and quality in the global south* (pp. 427–461). Springer Nature Singapore. DOI: 10.1007/978-981-97-2428-4_14

Mandal, R., Mohammadi, X., Wiktor, A., Singh, A., & Pratap Singh, A. (2020). Applications of pulsed light decontamination technology in food processing: An overview. *Applied Sciences (Basel, Switzerland)*, *10*(10), 3606. DOI: 10.3390/app10103606

Mani-López, E., García, H. S., & López-Malo, A. (2012). Organic acids as antimicrobials to control Salmonella in meat and poultry products. *Food Research International*, *45*(2), 713–721. DOI: 10.1016/j.foodres.2011.04.043

Meghwar, P., Saeed, S. M. G., Forte, L., Smaoui, S., Khalid, N. I., De Palo, P., & Maggiolino, A. (2024). Electrolyzed Water: A Promising Strategy for Improving Food Quality and Safety of Fruits, Vegetables, and Meat. *Journal of Food Quality*, *2024*(1), 3272823. DOI: 10.1155/2024/3272823

Miller, F. A., Silva, C. L., & Brandao, T. R. (2013). A review on ozone-based treatments for fruit and vegetables preservation. *Food Engineering Reviews*, *5*(2), 77–106. DOI: 10.1007/s12393-013-9064-5

Misra, N. N., Tiwari, B. K., Raghavarao, K. S. M. S., & Cullen, P. J. (2011). Nonthermal plasma inactivation of food-borne pathogens. *Food Engineering Reviews*, *3*(3), 159–170. DOI: 10.1007/s12393-011-9041-9

Misra, S., Mandliya, S., & Mishra, H. N. (2025). Evaluation of Stability of Coencapsulated Probiotics, γ-Aminobutyric Acid (GABA), and Physicochemical Properties of Functional Bar Matrix during Storage and Digestion Using a Chemometric Approach. *ACS Food Science & Technology*, *5*(6), 2371–2380. DOI: 10.1021/acsfoodscitech.5c00151

Mohapatra, D., Mishra, S., Giri, S., & Kar, A. (2013). Application of hurdles for extending the shelf life of fresh fruits. *Trends in Post-Harvest Technology*, *1*(1), 37–54.

Mukherjee, A., Gómez-Sala, B., O'Connor, E. M., Kenny, J. G., & Cotter, P. D. (2022). Global regulatory frameworks for fermented foods: A review. *Frontiers in Nutrition*, *9*, 902642. DOI: 10.3389/fnut.2022.902642 PMID: 35719144

Nabi, B. G., Mukhtar, K., Arshad, R. N., Radicetti, E., Tedeschi, P., Shahbaz, M. U., Walayat, N., Nawaz, A., Inam-Ur-Raheem, M., & Aadil, R. M. (2021). High-pressure processing for sustainable food supply. *Sustainability (Basel)*, *13*(24), 13908. DOI: 10.3390/su132413908

Nagaraj, K., Radha, S., Deepa, C. G., Raja, K., Umapathy, V., Badgujar, N. P., Parekh, N. M., Manimegalai, T., Devi, L. A., & Uthra, C. (2025). Photocatalytic advancements and applications of titanium dioxide (TiO_2): Progress in biomedical, environmental, and energy sustainability. *Next Research*, *2*(1), 100180. DOI: 10.1016/j.nexres.2025.100180

Naseri, I., & Yourdkhani, M. (2022). Rapid and energy-efficient frontal curing of multifunctional composites using integrated nanostructured heaters. *ACS Applied Materials & Interfaces*, *14*(44), 50215–50224. DOI: 10.1021/acsami.2c15415 PMID: 36226889

Nawaz, N., Wen, S., Wang, F., Nawaz, S., Raza, J., Iftikhar, M., & Usman, M. (2022). Lysozyme and its application as antibacterial agent in food industry. *Molecules (Basel, Switzerland)*, *27*(19), 6305. DOI: 10.3390/molecules27196305 PMID: 36234848

Ojha, K. S., Kerry, J. P., Duffy, G., Beresford, T., & Tiwari, B. K. (2015). Technological advances for enhancing quality and safety of fermented meat products. *Trends in Food Science & Technology*, *44*(1), 105–116. DOI: 10.1016/j.tifs.2015.03.010

Oliveira, M., Tiwari, B. K., & Duffy, G. (2020). Emerging technologies for aerial decontamination of food storage environments to eliminate microbial cross-contamination. *Foods*, *9*(12), 1779. DOI: 10.3390/foods9121779 PMID: 33266230

Onyeaka, H., Miri, T., Hart, A., Anumudu, C., & Nwabor, O. F. (2023). Application of ultrasound technology in food processing with emphasis on bacterial spores. *Food Reviews International*, *39*(7), 3663–3675. DOI: 10.1080/87559129.2021.2013255

Pal, N., Sharma, P., Kumawat, M., Singh, S., Verma, V., Tiwari, R. R., Sarma, D. K., Nagpal, R., & Kumar, M. (2024). Phage therapy: An alternative treatment modality for MDR bacterial infections. *Infectious Diseases*, *56*(10), 785–817. DOI: 10.1080/23744235.2024.2379492 PMID: 39017931

Palla, U., & Iwunwa, M. G. C. (2025). Leveraging predictive models to enhance infection control and reduce cross-contamination in public health settings. *International Journal of Research Publication and Reviews*, *6*(1), 407–422. DOI: 10.55248/gengpi.6.0125.0212

Pankaj, S. K., Wan, Z., & Keener, K. M. (2018). Effects of cold plasma on food quality: A review. *Foods*, *7*(1), 4. DOI: 10.3390/foods7010004 PMID: 29301243

Patil, A., Somvanshi, S. B., Sawale, M., Desai, K., Ozadali, F., Benyathiar, P., & Mishra, D. K. (2025). Advances in Microbubble Technologies for Food Sanitization: A Comprehensive Review. *Comprehensive Reviews in Food Science and Food Safety*, *24*(4), e70230. DOI: 10.1111/1541-4337.70230 PMID: 40661042

Pirozzi, A., Pataro, G., Donsì, F., & Ferrari, G. (2021). Edible coating and pulsed light to increase the shelf life of food products. *Food Engineering Reviews*, *13*(3), 544–569. DOI: 10.1007/s12393-020-09245-w

Purohit, S. R., Sharma, V., Kumari, M., Muthukumarappan, K., & Kane-Potaka, J. (Eds.). (2024). *Future Crops and Processing Technologies for Sustainability and Nutritional Security* (1st ed.). CRC Press. DOI: 10.1201/9781003280170

Rabiepour, A., Zahmatkesh, F., & Babakhani, A. (2024). Preservation techniques to increase the shelf life of seafood products: An overview. *Journal of Food Engineering and Technology*, *13*(1), 1–24. DOI: 10.32732/jfet.2024.13.1.1

Ramazanidoroh, F., Hosseininezhad, M., Shahrampour, D., & Wu, X. (2024). Edible packaging as a functional carrier of prebiotics, probiotics, and postbiotics to boost food safety, quality, and shelf life. *Probiotics and Antimicrobial Proteins*, *16*(4), 1327–1347. DOI: 10.1007/s12602-023-10110-5 PMID: 37389789

Ramos, C. V., Samelo, J., Martins, P. A., & Moreno, M. J. (2022). Protein Quantification in Complex Matrices. *Journal of Chemical Education*, *99*(3), 1488–1496. DOI: 10.1021/acs.jchemed.2c00109

Rebezov, M., Farhan Jahangir Chughtai, M., Mehmood, T., Khaliq, A., Tanweer, S., Semenova, A., Khayrullin, M., Dydykin, A., Burlankov, S., Thiruvengadam, M., Shariati, M. A., & Lorenzo, J. M. (2021). Novel techniques for microbiological safety in meat and fish industries. *Applied Sciences (Basel, Switzerland)*, *12*(1), 319. DOI: 10.3390/app12010319

Rebezov, M., Saeed, K., Khaliq, A., Rahman, S. J. U., Sameed, N., Semenova, A., Khayrullin, M., Dydykin, A., Abramov, Y., Thiruvengadam, M., Shariati, M. A., Bangar, S. P., & Lorenzo, J. M. (2022). Application of electrolyzed water in the food industry: A review. *Applied Sciences (Basel, Switzerland)*, *12*(13), 6639. DOI: 10.3390/app12136639

Rendueles, C., Duarte, A. C., Escobedo, S., Fernández, L., Rodríguez, A., García, P., & Martínez, B. (2022). Combined use of bacteriocins and bacteriophages as food biopreservatives. A review. *International Journal of Food Microbiology*, *368*, 109611. DOI: 10.1016/j.ijfoodmicro.2022.109611 PMID: 35272082

Revelou, P. K., Tsakali, E., Batrinou, A., & Strati, I. F. (2025). Applications of Machine Learning in Food Safety and HACCP Monitoring of Animal-Source Foods. *Foods*, *14*(6), 922. DOI: 10.3390/foods14060922 PMID: 40231903

Sahu, R., Singh, A. K., Kumar, A., Singh, K., & Kumar, P. (2023). Bacteriophages concept and applications: A review on phage therapy. *Current Pharmaceutical Biotechnology*, *24*(10), 1245–1264. DOI: 10.2174/13892010246662211041424 57 PMID: 36336808

Saputra, A., Gavahian, M., Purnomo, E. H., Saraswati, S., Hasanah, U., & Boing Sitanggang, A. (2025). High-pressure processing as an emerging technology for enhancing dairy product safety and quality: A critical review of latest advances, sustainability impact, limitations, and prospects. *Critical Reviews in Food Science and Nutrition*, 1–25. DOI: 10.1080/10408398.2025.2499623 PMID: 40326610

Scallan, E., Hoekstra, R. M., Angulo, F. J., Tauxe, R. V., Widdowson, M. A., Roy, S. L., Jones, J. L., & Griffin, P. M. (2011). Foodborne illness acquired in the United States—Major pathogens. *Emerging Infectious Diseases*, *17*(1), 7–15. DOI: 10.3201/eid1701.P11101 PMID: 21192848

Servin, A. L. (2004). Antagonistic activities of lactobacilli and bifidobacteria against microbial pathogens. *FEMS Microbiology Reviews*, *28*(4), 405–440. DOI: 10.1016/j.femsre.2004.01.003 PMID: 15374659

Sharma, M., Vidhya, C. S., Sunitha, N. H., Sachan, P., Singh, B., Santhosh, K., & Shameena, S. (2024). Emerging food processing and preservation approaches for nutrition and health. *European Journal of Nutrition & Food Safety*, *16*(1), 112–127. DOI: 10.9734/ejnfs/2024/v16i11382

Siddiqui, S. A., Singh, S., Bahmid, N. A., & Sasidharan, A. (2024). Applying innovative technological interventions in the preservation and packaging of fresh seafood products to minimize spoilage-A systematic review and meta-analysis. *Heliyon*, *10*(8), e29066. DOI: 10.1016/j.heliyon.2024.e29066

Silva, F. V., & Gibbs, P. A. (2012). Thermal pasteurization requirements for the inactivation of Salmonella in foods. *Food Research International*, *45*(2), 695–699. DOI: 10.1016/j.foodres.2011.06.018

Sobhan, A., Hossain, A., Wei, L., Muthukumarappan, K., & Ahmed, M. (2025). IoT-Enabled Biosensors in Food Packaging: A Breakthrough in Food Safety for Monitoring Risks in Real Time. *Foods*, *14*(8), 1403. DOI: 10.3390/foods14081403 PMID: 40282803

Soni, K. A., & Nannapaneni, R. (2010). Bacteriophage significantly reduces Listeria monocytogenes on raw salmon fillet tissue. *Journal of Food Protection*, *73*(1), 32–38. DOI: 10.4315/0362-028X-73.1.32 PMID: 20051201

Stefanova, R., Vasilev, N. V., & Spassov, S. L. (2010). Irradiation of food, current legislation framework, and detection of irradiated foods. *Food Analytical Methods*, *3*(3), 225–252. DOI: 10.1007/s12161-009-9118-8

Suvarna, V., Nair, A., Mallya, R., Khan, T., & Omri, A. (2022). Antimicrobial nanomaterials for food packaging. *Antibiotics (Basel, Switzerland)*, *11*(6), 729. DOI: 10.3390/antibiotics11060729 PMID: 35740136

Thakur, A. (2025). Innovation Through Collaboration Leveraging Knowledge Sharing for Global R&D Success. In *Research and Development Practices in Innovation* (pp. 127–164). IGI Global Scientific Publishing. DOI: 10.4018/979-8-3373-0812-8.ch006

Tomczyk, Ł., Stępień, Ł., Urbaniak, M., Szablewski, T., Cegielska-Radziejewska, R., & Stuper-Szablewska, K. (2018). Characterisation of the mycobiota on the shell surface of table eggs acquired from different egg-laying hen breeding systems. *Toxins*, *10*(7), 293. DOI: 10.3390/toxins10070293 PMID: 30012982

Turhan, E. U., Polat, S., Erginkaya, Z., & Konuray, G. (2022). Investigation of synergistic antibacterial effect of organic acids and ultrasound against pathogen biofilms on lettuce. *Food Bioscience*, *47*, 101643. DOI: 10.1016/j.fbio.2022.101643

Uğur, F. A. (2023). *Food Irradiation: A Promising Technology for Enhancing Food Safety, Quality, and Sustainability*. BIDGE Publications.

Urango, A. C. M., Strieder, M. M., Silva, E. K., & Meireles, M. A. A. (2022). Impact of thermosonication processing on food quality and safety: A review. *Food and Bioprocess Technology*, *15*(8), 1700–1728. DOI: 10.1007/s11947-022-02760-0

Vaiserman, A., Koliada, A., Zabuga, O., & Socol, Y. (2018). Health impacts of low-dose ionizing radiation: Current scientific debates and regulatory issues. *Dose-Response*, *16*(3), 1559325818796331. DOI: 10.1177/1559325818796331 PMID: 30263019

Vijay Rakesh Reddy, S., Sudhakar Rao, D. V., Sharma, R. R., Preethi, P., & Pandiselvam, R. (2022). Role of ozone in post-harvest disinfection and processing of horticultural crops: A review. *Ozone Science and Engineering*, *44*(1), 127–146. DOI: 10.1080/01919512.2021.1994367

Wang, J., & Wu, Z. (2023). Minimal processing of produce using a combination of UV-C irradiation and ultrasound-assisted washing. *Lebensmittel-Wissenschaft + Technologie*, *182*, 114901. DOI: 10.1016/j.lwt.2023.114901

Wang, Y., Li, Y., Yang, S., Li, C., Li, L., Gao, S., & Wu, Z. (2023). Mechanism of ozone treatment in delayed softening of fresh-cut kiwifruit during storage. *Postharvest Biology and Technology*, *204*, 112469. DOI: 10.1016/j.postharvbio.2023.112469

Yang, P., Rao, L., Zhao, L., Wu, X., Wang, Y., & Liao, X. (2021). High pressure processing combined with selected hurdles: Enhancement in the inactivation of vegetative microorganisms. *Comprehensive Reviews in Food Science and Food Safety*, *20*(2), 1800–1828. DOI: 10.1111/1541-4337.12724 PMID: 33594773

Yaun, B. R., Sumner, S. S., Eifert, J. D., & Marcy, J. E. (2004). Inhibition of pathogens on fresh produce by ultraviolet energy. *International Journal of Food Microbiology*, *90*(1), 1–8. DOI: 10.1016/S0168-1605(03)00158-2 PMID: 14672825

Yoon, J. H., Oh, M. S., & Lee, S. Y. (2024). Effectiveness of organic acids for inactivating pathogenic bacteria inoculated in laboratory media and foods: An updated minireview. *Food Science and Biotechnology*, *33*(12), 2715–2728. DOI: 10.1007/s10068-024-01618-9 PMID: 39184989

Zhang, K., Perussello, C. A., Milosavljević, V., Cullen, P. J., Sun, D. W., & Tiwari, B. K. (2019). Diagnostics of plasma reactive species and induced chemistry of plasma treated foods. *Critical Reviews in Food Science and Nutrition*, *59*(5), 812–825. DOI: 10.1080/10408398.2018.1564731 PMID: 30676057

Zhu, Y., Li, C., Cui, H., & Lin, L. (2020). Feasibility of cold plasma for the control of biofilms in food industry. *Trends in Food Science & Technology*, *99*, 142–151. DOI: 10.1016/j.tifs.2020.03.001

Zia, S., Khan, M. R., Shabbir, M. A., Aslam Maan, A., Khan, M. K. I., Nadeem, M., Khalil, A. A., Din, A., & Aadil, R. M. (2022). An inclusive overview of advanced thermal and nonthermal extraction techniques for bioactive compounds in food and food-related matrices. *Food Reviews International*, *38*(6), 1166–1196. DOI: 10.1080/87559129.2020.1772283

Chapter 5
Nanotechnology Applications in Food Contaminant Mitigation

Sylvester Izah
https://orcid.org/0000-0001-5526-006X
Bayelsa Medical University, Nigeria

ABSTRACT

Nanotechnology in the food sector offers innovative solutions for mitigating biological, chemical, and physical contaminants across the supply chain. This chapter explores its core principles and diverse applications in enhancing food safety, quality, and sustainability. Key technologies (nanosensors, nano-adsorbents, nano-coatings, and smart packaging) enable real-time detection, neutralization, and removal of contaminants. Nanomaterials like silver nanoparticles, quantum dots, carbon nanotubes, and nano-silica exhibit strong antimicrobial and toxin-adsorbing properties, while nano-encapsulation supports delivery of detoxifying agents and functional compounds. The chapter also examines regulatory, ethical, and public perception issues, highlighting nanotechnology's role in reducing food waste, improving agricultural efficiency, and supporting eco-friendly packaging. Prospects for the field are discussed within the context of interdisciplinary research, digital integration, and consumer trust.

1. INTRODUCTION

Food contamination stands as a significant global challenge, impacting public health and economic stability. Contaminants in food include biological agents (e.g., bacteria, viruses), chemical substances (e.g., pesticides, heavy metals), and

DOI: 10.4018/979-8-3373-7052-1.ch005

physical hazards (e.g., glass shards). Several food products have been reported to contain various contaminants, posing potential public health risks. Studies have documented the presence of trace metals in fish species (*Sarotherodon melanotheron, Chrysichthys nigrodigitatus*) (Aigberua et al., 2021; Izah et al., 2024), locally produced nutritive food-drinks (Odubo and Izah 2024), poultry eggs (Izah et al., 2022a), tomatoes (Izah & Aigberua, 2020), cow meat (Seiyaboh et al., 2018a), and food condiments (Aigberua et al., 2018); antibiotic residues in poultry products (Izah et al., 2025); pesticide residues in food crops (Sawyer et al., 2024); polycyclic aromatic hydrocarbons in fish (Aigberua et al., 2023); microbial contaminants in ready-to-eat foods and palm oil (Izah et al., 2022b; Seiyaboh et al., 2018b; Ovuru et al., 2024); and adulterants in food products (Sawyer & Izah, 2024). Additionally, environmental pollutants affecting soil quality and indirectly contaminating food crops have also been highlighted (Onwudiegwu & Izah, 2025).

The World Health Organization (WHO) estimates that foodborne and waterborne diarrheal diseases are major causes of morbidity and mortality, especially in developing countries (Sonawane et al., 2018). The burden of contamination extends beyond immediate health risks; it leads to increased healthcare costs and reduced agricultural productivity, contributing to a cycle of poverty and food insecurity.

Traditional methods of detecting and mitigating food contaminants show notable limitations, primarily in terms of sensitivity and efficiency. Conventional detection methods such as microbial culture and chemical assays often require lengthy processing times and may not detect low-level contaminants effectively (McClements & Xiao, 2017). Furthermore, these methods can fail to identify emerging pathogens or novel food contaminants due to their reliance on pre-existing knowledge and established testing protocols (EFSA Scientific Committee et al., 2018). Despite advancements, these techniques may struggle with real-time detection capabilities, hindering rapid response to contamination events (Rajeshkumar & Malarkodi, 2014). Consequently, the food industry needs to seek more reliable, efficient, and integrated solutions for food safety.

Emerging nanotechnology offers innovative solutions to these challenges in food safety by employing nanomaterials such as silver nanoparticles, carbon nanotubes, and nanosilica. Silver nanoparticles, for instance, exhibit strong antibacterial properties and have shown efficacy against a variety of foodborne pathogens (Zorraquín-Peña et al., 2020; Brito et al., 2019). Additionally, these nanoparticles can be integrated into food packaging materials to provide antimicrobial barriers that extend product shelf life while safeguarding against contamination (Waheed et al., 2023; McClements, 2020). The incorporation of nanosensors into food production and processing can revolutionize real-time monitoring of food contaminants, enhancing reaction times to potential threats and improving overall food safety standards (Loza et al., 2014).

Such technologies foster safer food production practices and adhere to the increasing consumer demand for quality and safety in food products (McClements, 2020).

A critical aspect of integrating nanotechnology into food safety involves assessing environmental sustainability and the overall impact on ecological systems. By minimizing food waste through enhanced preservation techniques, nanotechnology can help improve energy efficiency in food processing and diminish the ecological footprint of food packaging (Shafiq et al., 2020; Gayathri et al., 2024). Moreover, smart packaging technologies equipped with nanomaterials can indicate freshness and quality, thus preventing food spoilage. These advancements may lead to more sustainable food systems that align with global health and environmental objectives.

This chapter examines the role of nanotechnology in mitigating food contaminants, focusing on the detection, removal, and neutralization of biological, chemical, and physical hazards using advanced nanomaterials. It explores the potential of nanomaterials such as silver nanoparticles, carbon nanotubes, and nano-silica in enhancing food safety measures through nanosensors, smart packaging, and contaminant filtration. Additionally, the chapter highlights the environmental benefits of nanotechnology, including its contribution to reducing food waste, improving energy efficiency in food processing, and minimizing the ecological footprint of packaging. The benefit of this study lies in its potential to provide innovative, sustainable solutions for improving food safety systems, reducing contamination risks, and advancing global food security.

2. FUNDAMENTALS OF NANOTECHNOLOGY IN THE FOOD SECTOR

Nanotechnology, which involves the manipulation of matter at the nanoscale (1–100 nm), has emerged as a transformative force in various sectors, including food science. Its application in the food industry is diverse, primarily aimed at enhancing food quality, safety, and nutritional value. Specifically, nanotechnology is instrumental in improving products through methods such as nanoencapsulation of nutrients, which significantly enhances the solubility and bioavailability of functional compounds essential for health (He & Hwang, 2016; Yu et al., 2018). This encapsulation process allows for the targeted delivery of bioactive compounds, contributing to the development of functional foods and nutritional supplements (Durán & Marcato, 2012).

The scope of nanotechnology in the food sector extends to the development of advanced packaging solutions that promote food safety and extend shelf life. Active packaging techniques utilizing nanomaterials, such as silver nanoparticles (AgNPs) and carbon nanotubes (CNTs), have shown promise in providing antimicrobial properties and improving the mechanical strength of packaging materials (Putri et al., 2024; Awlqadr et al., 2025). These innovations address consumer demands for food products that are not only safer but also longer-lasting and more environmentally friendly (Awlqadr et al., 2025). The incorporation of nanosensors into food packaging enhances monitoring capabilities, allowing for real-time detection of contamination and degradation, thereby ensuring that food quality is maintained throughout its shelf life (Fuertes et al., 2016).

Various nanomaterials have gained prominence within the food industry due to their unique properties. Silver nanoparticles are recognized for their effective antimicrobial properties, making them common in food packaging applications to prevent bacterial growth (Kampers, 2011; Rastogi et al., 2021). Similarly, nanosilica is frequently employed as an anti-caking agent, and its high surface area improves the functionality of coatings and packaging materials (Ravichandran, 2010; Devra, 2022). Zinc oxide and titanium dioxide nanoparticles are utilized for their UV protection and antimicrobial effects, playing a significant role in food preservation (Rossi et al., 2014; Fraceto et al., 2016). Additionally, the use of lipid-based nanocarriers for nanoemulsions enhances the delivery of poorly soluble nutrients and bioactive compounds, thereby improving functional food products (Putri et al., 2024; Awlqadr et al., 2025).

Despite the potential benefits of nanotechnology in the food sector, critical safety and regulatory considerations remain paramount. The potential toxicological risks associated with engineered nanomaterials (ENMs) are concerning, as these particles can interact with biological systems in ways that larger particles do not (Martirosyan & Schneider, 2014; Madanayake et al., 2021). For instance, nanoparticles may traverse physiological barriers, such as the blood-brain barrier and the intestinal lining, leading to potential accumulation in vital organs (Rastogi et al., 2021). Consequently, traditional toxicity assessment methods may be inadequate, necessitating the development of nano-specific risk assessment methodologies that account for the unique characteristics of nanomaterials (He & Hwang, 2016; Rashidi & Khosravi-Darani, 2011). Regulatory frameworks vary by region, with bodies such as the FDA (USA) and EFSA (Europe) imposing specific requirements for the characterization, labeling, and safety evaluation of products containing nanomaterials (Martirosyan & Schneider, 2014).

Public perception and ethical concerns surrounding nanotechnology also play a crucial role in its application to food science. Consumer awareness of potential risks associated with nanotechnology often leads to skepticism regarding its beneficial applications (Sekhon, 2014; Gelaye, 2025). Therefore, transparent communication and public engagement are essential strategies for fostering trust and ensuring that consumers can make informed decisions about the food products they choose (Martirosyan & Schneider, 2014; Kampers, 2011).

3. SOURCES AND TYPES OF FOOD CONTAMINANTS

Food safety is of paramount importance given the pervasive nature of contaminants affecting various food sources (Yarkwan et al., 224; Ogwu et al., 2024; Ogidi and Izah 2024; Odubo et al., 2024). Contaminants can be broadly categorized into biological, chemical, and physical agents, each carrying potential health risks (Table 1).

Table 1. Sources and Types of Food Contaminants

Type of Contaminant	Examples	Common Sources
Biological	- *Salmonella, E. coli, Listeria, Campylobacter* (Bacteria) - Norovirus, Hepatitis A (Viruses) - Molds, yeasts (Fungi) - *Giardia, Cryptosporidium, Toxoplasma, Trichinella* (Parasites)	- Raw/undercooked meat, poultry, and seafood - Unpasteurized milk and dairy products - Contaminated water - Fresh produce - Poor hygiene and sanitation
Chemical	- Pesticide residues - Heavy metals: lead, chromium, copper, etc - Mycotoxins (e.g., aflatoxins) - Veterinary drugs: antibiotics, hormones - Industrial chemicals: dioxins, PCBs, BPA - Excess food additives (e.g., sulfites, colorants)	- Crop farming and pesticide use - Contaminated water, soil, or air - Moldy grains, nuts, and dried fruits - Improper use of animal drugs - Migration from packaging materials - Misuse of food preservatives
Physical	- Microplastics - Glass shards - Metal fragments - Stones, soil particles - Hair, plastic pieces, wood splinters	- Environmental pollution - Damaged packaging or equipment - Poor cleaning during harvest or processing - Inadequate quality control or food handling practices

Biological contaminants involves a range of living organisms or their by-products that can lead to foodborne illnesses. Among these, bacteria such as Salmonella, Escherichia coli (E. coli), etc frequently emerge from improper food handling or undercooking, particularly in meat products and dairy. Several food products have been studied for their microbial diversity in Bayelsa and surrounding areas, including smoked *Trachurus trachurus* (Ineyougha et al., 2015), palm oil (Ohimain & Izah, 2015), *kunu* drink (Orutugu et al., 2015), packaged sliced fruits (Izah et al., 2015), *suya* (Kigigha et al., 2015a), and fermented cassava flakes (*gari*) (Kigigha et al., 2015b). Additional studies also reported microbial populations in zobo drink (Izah et al., 2015a, 2016a; Kigigha et al., 2018), sliced fruits (Izah et al., 2015b), meat pie (Kigigha et al., 2017a), puff-puff (Kigigha et al., 2017b), *Malus domestica* and *Cucumis sativus* (Izah et al., 2016b), and unpeeled groundnut (Kigigha et al., 2016). Viruses like Norovirus and Hepatitis A also present significant health risks, often entering the food supply through contaminated water or via infected food handlers. Research indicates that on outbreaks linked to these viruses are a recurring public health concern, underscoring the importance of educating food service workers on hygiene practices (Sim & Wiwanitkit, 2021). Moreover, fungi such as molds can produce hazardous mycotoxins in improperly stored food, particularly grains and legumes, signifying another dimension of biological contamination (Szelenberger et al., 2024).

Chemical contaminants, conversely, originate from various agricultural and industrial practices. Residues from pesticides and veterinary drugs have been identified as substantial risks, often persisting on fruits and vegetables post-harvest. Some many chemicals have been reported in food including verterinary drug resideues, pesticides residues, natural toxins such as mycotoxinsplant toxins, marine biotoxins; additives, dioxins, PCBs and PFASs, trace metals, substances from FCMs, processing contaminants (Figure 1, D'Amore et al. 2025). These chemical are regulated in food particularly by EU (D'Amore et al. 2025). Basically, the accumulation of toxins such as heavy metals, which can enter food products through contaminated soil, water sources, or packaging materials, poses significant health risks as well. Mycotoxins, like aflatoxins produced by molds, are notoriously resilient, often found in staple foods like grains and nuts, thus necessitating continuous monitoring and control strategies throughout the food production chain. Industrial chemicals such as dioxins and bisphenol A (BPA) can leach from packaging into food, further complicating contamination issues that can affect human health (Kappenstein et al., 2012).

Figure 1. EU-regulated chemicals in food

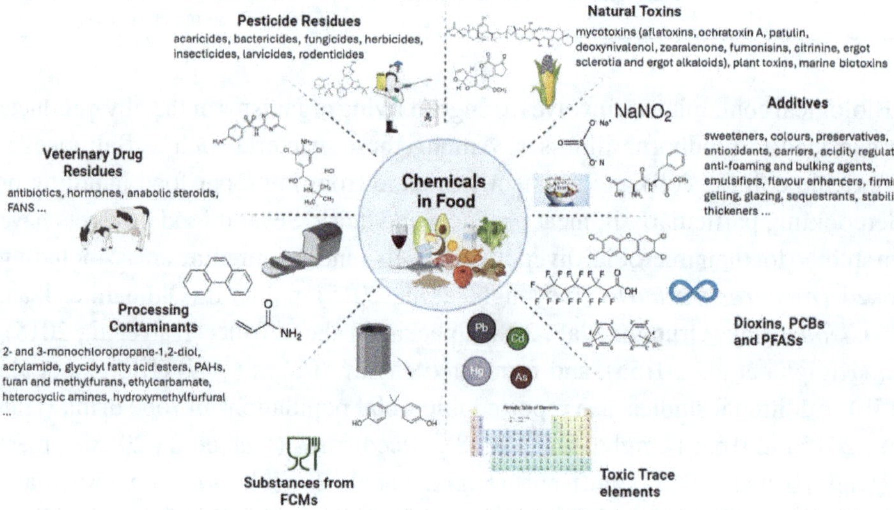

Source: D'Amore et al. (2025)

Physical contaminants emerge from foreign materials inadvertently introduced into food products. The prevalence of microplastics in the food supply, particularly in seafood and bottled drinks, is an alarming modern concern. These tiny plastic particles originate from a wide array of sources, from food packaging to environmental degradation (Gao et al., 2020). Furthermore, incidents involving glass fragments, metal shards, or other physical objects entering food products often occur due to lapses in quality control during the manufacturing process. Such contaminants not only pose health risks but can lead to severe injuries (Cortés-Sánchez et al., 2024).

4. NANOTECHNOLOGY-BASED DETECTION OF FOOD CONTAMINANTS

Nanotechnology has emerged as an indispensable tool in enhancing food safety through the development of various advanced detection methods for food contaminants. Ansari (2023) present the possible applications of nanotechnology in different areas of the food industry to include nano food safety, packaging, preservation and functionality (Figure 2). The integration of nanosensors into food monitoring systems facilitates real-time detection of hazardous substances, ensuring consumer safety and maintaining food quality. Nanosensors can be categorized primarily into electrochemical, optical, and biosensors, each employing distinct mechanisms for contaminant detection. Table 2 provide an overview of nanotechnology-based detection of food contaminants.

Figure 2. Application of nanotechnology in different areas of the food industry

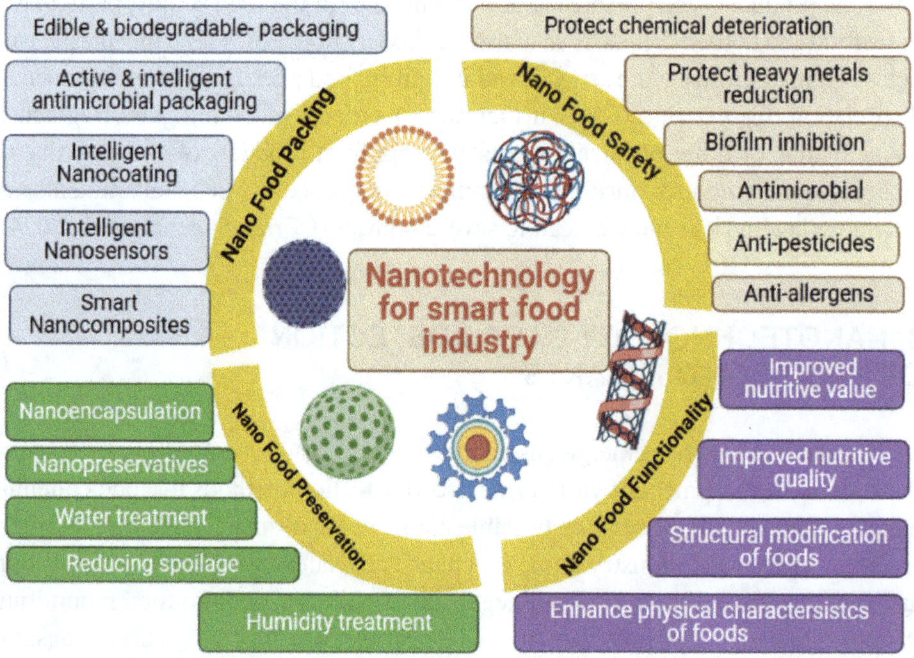

Source: Ansari (2023)

Table 2. Nanotechnology-Based Detection of Food Contaminants

Category	Technology/Material	Application/Examples
Nanosensors for Detection	- Electrochemical sensors - Optical sensors - Biosensors (DNA, enzymes, antibodies)	- Real-time detection of contaminants - High sensitivity and portability - Used for on-site food safety monitoring
Smart Packaging & Indicators	- Intelligent packaging with embedded nanosensors - Colorimetric spoilage indicators - RFID with nanosensor tech	- Spoilage detection (gas release, microbial growth) - Monitoring temperature and freshness - Supply chain tracking
Nanomaterials in Detection	- Quantum dots (QDs) - Gold nanoparticles (AuNPs) - Graphene and graphene oxide	- QDs: multiplex fluorescent tagging - AuNPs: signal amplification in assays - Graphene: detection of heavy metals
Food-Specific Applications	- Milk: melamine, aflatoxins, antibiotics - Meat: E. coli, Salmonella, spoilage indicators	- Produce: pesticide and fungal detection - Grains: mycotoxins via QDs or AuNP -based assays

Electrochemical nanosensors function by generating measurable electrical signals when target molecules adhere to the sensor surface. These devices capitalize on the high surface area and conductivity of nanomaterials, such as gold nanoparticles (AuNPs) and carbon nanotubes, which amplify the signals generated (Raja et al., 2021; Eissa et al., 2022; Manikandan et al., 2018). Specifically, these electrochemical sensors have been shown to effectively detect a range of food toxins, including bisphenol A and heavy metals, through rapid and sensitive methods (Manikandan et al., 2018; Malhotra et al., 2015). This has revolutionized food safety protocols by enabling on-site testing, thereby minimizing the need for complex laboratory analyses (Raja et al., 2021; Manikandan et al., 2018).

Conversely, optical nanosensors leverage changes in light absorption or fluorescence to identify the presence of pathogens or chemical residues in food products. A notable example includes the utilization of carbon quantum dots (CQDs), which exhibit tunable fluorescence properties, allowing them to be tailored for specific contaminants (Li et al., 2025; Pan et al., 2020; Sabui et al., 2022). Studies have demonstrated that CQDs are not only highly sensitive but also operationally straightforward, making them ideal for real-time, on-site analysis (Li et al., 2025; (Pan et al., 2020).

Biosensors combine biological recognition elements such as enzymes, antibodies, or DNA with nanomaterials to achieve specific and highly sensitive detection. These biosensors have been instrumental in detecting notorious foodborne pathogens like Escherichia coli and Salmonella with remarkable accuracy (Inbaraj & Chen, 2016; Bai et al., 2023). The incorporation of nanomaterials enhances the efficiency of biological interactions, providing a sensitive response to the presence of targeted contaminants, thereby ensuring food safety (Curulli, 2020; Zhou et al., 2022).

The advantages of these nanotechnology-based detection systems are numerous: they offer high sensitivity, rapid results, allow for miniaturization, and are capable of performing real-time monitoring of food quality and safety conditions. Furthermore, they pave the way for intelligent packaging solutions that integrate nanosensors to monitor freshness and contamination indicators, improving traceability throughout the food supply chain (Xie et al., 2024). Colorimetric indicators embedded in packaging materials exemplify this, providing visual cues about microbial growth or spoilage (Xie et al., 2024).

Advanced nanomaterials, including quantum dots, gold nanoparticles, and graphene, play a critical role in enhancing the capabilities of these detection systems. Quantum dots, in particular, provide enhanced optical properties beneficial for multi-target detection applications (Xiong et al., 2022). Gold nanoparticles amplify the signals in colorimetric assays, while graphene's large surface area and excellent electrical conductivity contribute to the development of sensitive biosensor platforms (Pan et al., 2020; Zhang et al., 2020). These materials not only improve the sensitivity and specificity of assays for detecting contaminants such as mycotoxins and pesticide residues but also streamline the design of portable and efficient detection systems (Xu et al., 2021; Khachatryan & Khachatryan, 2019).

In terms of practical applications, numerous case studies illustrate the effectiveness of nanotechnology in detecting food contaminants across various products. For instance, milk safety has been improved through the use of nanosensors that specifically detect aflatoxins and antibiotics (Bai et al., 2023; Malhotra et al., 2015), while meat products have benefitted from rapid detection of pathogens such as E. coli and Salmonella (Inbaraj & Chen, 2016; Bai et al., 2023). In fruit and vegetable commodities, nanoparticle-enhanced assays are increasingly utilized to identify pesticide residues and fungal contaminants (Xu et al., 2021; Tseng et al., 2020), showcasing the versatility of nanotechnology across food categories.

5. NANOTECHNOLOGY IN THE REMOVAL AND NEUTRALIZATION OF CONTAMINANTS

Nanotechnology has emerged as a transformative approach to enhance food safety and quality through various innovative applications. This section of the chapter covers the role of nanomaterials in the removal and neutralization of contaminants in food, focusing on adsorption and filtration, antimicrobial coatings, degradation of chemical contaminants, and nano-encapsulation.

Adsorption and Filtration Using Nanomaterials

Nanomaterials such as nano-zeolites, carbon nanofibers, and nano-clays have garnered attention for their capabilities to effectively adsorb a wide range of food contaminants. Nano-zeolites, characterized by their high surface area and significant ion-exchange capacity, play a crucial role in trapping heavy metals and mycotoxins, known foodborne pollutants that pose serious health risks (Peters et al., 2014). Similar studies highlight the effectiveness of carbon nanofibers and nanotubes, whose unique structures enable efficient adsorption of organic contaminants and pesticide residues from food matrices, thereby enhancing food safety management (Thiruvengadam et al., 2018). Additionally, nano-clays and metal oxides are increasingly utilized in food packaging and processing filters to selectively bind aflatoxins and nitrites, thus shielding consumers from hazardous substances (Thiruvengadam et al., 2018).

Antimicrobial Nanocoatings and Food Contact Surfaces

The application of antimicrobial nanocoatings offers a promising avenue for enhancing food safety. Silver nanoparticles have shown significant efficacy in inhibiting microbial growth by disrupting the cell membranes of pathogens on food processing equipment and packaging surfaces (Castro-Mayorga et al., 2016). Additionally, zinc oxide nanoparticles have been utilized for their antimicrobial effects, substantially reducing biofilm formation on surfaces, which is critical for maintaining sanitary conditions in food processing environments (Ferrone et al., 2019). The development of biodegradable and non-toxic chitosan-based nanocoatings further establishes a sustainable approach to extending the shelf life of food and mitigating the risk of contamination (Shafique et al., 2022).

Nanoparticles for Degradation or Transformation of Chemical Contaminants

Photocatalytic nanoparticles, particularly titanium dioxide (TiO_2), present a method for degrading harmful chemical residues such as pesticide residues and bisphenol A (BPA). Under UV or visible light, these nanoparticles can induce chemical transformations that render contaminants harmless (Liao et al., 2021). Magnetic nanoparticles also offer a solution; they can be functionalized to bind specific pollutants, facilitating their separation from liquid foods using magnetic fields, enhancing the safety of food products (Thiruvengadam et al., 2018). Furthermore, nanoscale zero-valent iron (nZVI) has shown potential in detoxifying chlorinated compounds (Thiruvengadam et al., 2018).

Nano-Encapsulation for Detoxification and Controlled Release of Neutralizing Agents

Nano-encapsulation techniques have gained traction in optimizing the delivery of detoxifying agents while preserving food quality. Encapsulation not only targets specific toxins like mycotoxins but also allows for safe digestion or elimination of these harmful substances post-consumption (Tolve et al., 2022). Controlled release systems can respond to environmental triggers such as pH or temperature, delivering antioxidants or chelating agents at optimal times, thus improving the functional attributes of food without altering its taste or texture (Tahir et al., 2021; Abbasi et al., 2024). Notably, nano-liposomes or polymeric nanoparticles can encapsulate diverse functional compounds, including probiotics, ensuring that these beneficial agents can counteract toxins during digestion while maintaining food quality (Tolve et al., 2022; Oprea et al., 2022).

6. APPLICATIONS IN FOOD PACKAGING FOR CONTAMINANT PREVENTION

Nanotechnology has become a pivotal innovation for enhancing food safety and quality, particularly through its applications in food packaging. This advancement primarily involves two types of systems: active and intelligent packaging solutions. Active packaging encompasses the integration of substances that interact with the contents, assisting in extending shelf life, ensuring safety, and maintaining product quality. Such functionalities include oxygen scavenging, ethylene absorption, and moisture regulation (Usha et al., 2025; Kadirvel et al., 2024). Compatibility of nanomaterials such as silver and zinc oxide in these systems demonstrates their effectiveness as oxygen absorbers and antimicrobial agents, combating microbial growth and oxidative spoilage in packaged items (Adeyeye, 2019; Guo, 2025).

Intelligent packaging, in contrast, is equipped with sensors that monitor the conditions of packaged food. For instance, these sensors can track temperature fluctuations or detect spoilage through color changes, providing real-time data about product integrity to consumers and supply chain operators (Usha et al., 2025; Yan et al., 2022). The development of smart labeling technologies has shown promise in alerting users to food quality deterioration, thereby mitigating health risks associated with food contamination (Singh et al., 2025; Zhanbolat & Tungyshbaeva, 2024).

A significant advancement also lies in the barrier properties enhanced through the use of nanomaterials in packaging. Materials such as nanoclays, nanocellulose, and nano-silica have been shown to substantially improve barrier properties against moisture, oxygen, and other harmful agents, thus preserving food quality over extended periods (Guo, 2025; Ma, 2024; Chadha et al., 2022). Enhanced barrier qualities are particularly beneficial for perishable products, where oxidative reactions and microbial ingress can escalate spoilage rates. These features not only prolong the longevity of food products but also contribute to reducing food waste on a commercial scale.

Moreover, the incorporation of antimicrobial nanoparticles into packaging polymers has become increasingly important due to their efficacy in inhibiting bacterial and fungal growth. Materials infused with silver nanoparticles, for example, have shown potential in extending the shelf life of meat and dairy products by minimizing surface contamination (Dobrucka & Ankiel, 2018). The controlled release of these antimicrobial agents from packaging materials ensures prolonged protective functionality during the food's transport and storage lifecycle, making them a viable option for maintaining food safety.

Real-world applications of these nanotechnology-enhanced packaging solutions can be observed in several case studies. Various food manufacturers have adopted silver nanoparticle-infused films for meats and fresh produce, demonstrating effectiveness in reducing contamination and extending product freshness (Wagh et al., 2025; Sani et al., 2021). Additionally, packaging that changes color in response to spoilage or contamination provides consumers with visible reassurance regarding the safety and freshness of food products (Upadhyay et al., 2024). These commercial implementations exemplify the viability and scalability of nanotechnology within global food supply chains.

7. IMPLICATIONS FOR SUSTAINABLE DEVELOPMENT

The use of nanotechnology in the food sector represents a promising frontier for achieving sustainable development goals, particularly in enhancing food security, minimizing waste, and reducing the environmental footprint of agricultural practices. This discussion elaborates on the multifaceted implications of nanotechnology for sustainable development concerning food.

One of the critical implications of nanotechnology in food systems is its capacity to reduce food waste by extending shelf life. Research indicates that food packaging materials enhanced with nanomaterials can significantly improve barrier properties against moisture, oxygen, and spoilage microorganisms, leading to longer-lasting foods (Sekton, 2010). The incorporation of nanosensors within packaging can actively monitor food quality, alerting consumers to spoilage or contamination, thus helping to minimize food waste at the consumer level (Adeyeye, 2019). These advancements not only extend the freshness of food products but also foster an overall reduction in food waste throughout the supply chain.

In terms of energy-efficient food processing, nanotechnology is a vital player. By employing nano-enabled materials and techniques in food production and processing, manufacturers can optimize energy use and improve overall efficiency (Chugh et al., 2021). Nanotechnology facilitates precision farming practices, allowing for targeted applications of water, fertilizers, and pesticides, which reduces energy consumption in agricultural production while also surpassing traditional farming methods in terms of crop yield enhancements (Adenubi et al., 2024; Prasad et al., 2017). The adoption of these technologies promises a dual benefit of reduced operational costs for producers and a smaller carbon footprint associated with food processing.

Another aspect in which nanotechnology contributes to sustainable development is through environmentally friendly food packaging solutions. The shift from conventional packaging materials to nano-enabled alternatives exemplifies a strategy to address pressing environmental concerns. Innovations in biodegradable nanomaterials not only minimize waste but also represent a significant enhancement in the recyclability of food packaging (Manatunga et al., 2022). These materials often demonstrate superior barrier strengths while being less deleterious to the environment, indicating a sustainable solution to the challenges of plastic waste in food packaging.

Moreover, nanotechnology plays a role in promoting sustainable agricultural practices. Applications such as nano-fertilizers and nano-pesticides have shown marked improvements in nutrient delivery and pest control, which ultimately lead to lower chemical usage and better crop protection (Taha & Omar, 2024; Fraceto et al., 2016). This shift to precision agriculture enabled by nanotechnology is vital in combating the challenges posed by global food security and environmental sustainability. Nanomaterials can enhance soil health and crop resilience by facilitating improved water retention and enhancing nutrient uptake efficiency, as shown in various studies (Mohanty et al., 2024; Morales-Díaz et al., 2017).

Addressing food security challenges through advanced contaminant mitigation is another crucial area where nanotechnology has demonstrated promise. The utilization of nanomaterials for soil treatment and contaminant remediation can significantly improve soil quality and restore agricultural productivity in areas affected by pollution. This capability is essential given the anticipated growth of the global population and subsequent increase in food demand. The integration of smart nanotechnology solutions can enhance the overall resilience of food production systems against contaminants, bolstering food safety and security (Gelaye, 2025).

However, the advancement of nanotechnology in food systems is also accompanied by ethical and social considerations. The public perception of nanotechnology varies, with concerns primarily revolving around safety and potential health risks associated with nanoparticle exposure (Ameta et al., 2020). It is essential for stakeholders in the food supply chain, including producers, regulators, and consumers, to engage in dialogue regarding these concerns to ensure that the adoption of nanotechnology can proceed responsibly and with public trust. Moreover, regulatory frameworks need to evolve alongside technological advancements to address possible implications for human health and environmental safety.

8. FUTURE PROSPECTS AND INNOVATIONS

The integration of nanotechnology within the food industry has emerged as a vital trend that holds promise for future innovations related to food safety, agricultural productivity, and sustainability. One of the most notable applications of nanotechnology is the development of nanosensors, which can significantly enhance food safety protocols. DNA-based nanosensors, for instance, have the potential to detect pathogens and contaminants at unprecedented sensitivity levels (He et al., 2019). The incorporation of these innovations into digital food safety systems, such as those utilizing Internet of Things (IoT) platforms and blockchain technology, can facilitate real-time monitoring and traceability in the food supply chain, thus ensuring consumer safety and bolstering confidence in food products (He et al., 2019).

Moreover, the advent of nanotechnology in packaging promises to revolutionize how food products are developed and preserved. Smart packaging that incorporates nanomaterials can actively enhance food preservation by responding to environmental conditions, thereby extending shelf life and reducing waste. These materials can also possess antimicrobial properties, further mitigating the risk of foodborne illnesses and enhancing the overall safety of food products. This trend is coupled with an increasing consumer demand for sustainable food practices, which aligns with recent advancements in green nanotechnology. Sustainable methods of nanoparticle synthesis are being prioritized, utilizing bio-based materials to minimize environmental impacts while maintaining efficacy in applications such as food packaging and processing (Sundararajan et al., 2023; Żukowska et al., 2024).

As both the food and agriculture sectors continue to face challenges related to environmental sustainability and food security, interdisciplinary research collaborations are becoming crucial. The connection of nanotechnology with environmental sciences, biotechnology, and digital technology fosters a creative ecosystem for tackling issues such as nutrient delivery in crops and the development of eco-friendly pesticides. The integration of these diverse fields holds the potential to create comprehensive solutions that can enhance food safety, improve crop yields, and promote sustainable agricultural practices. In light of these developments, the food industry is expected to see a surge in research initiatives focused on nano-food applications, further stressing the necessity for global collaboration in this evolving landscape (Kumar et al., 2023).

However, despite the remarkable advancements, there remain significant hurdles to the broad acceptance and implementation of nanotechnology in food systems. Public perception reflects a degree of skepticism towards nano-enhanced foods, driven by concerns around safety and regulatory frameworks (Kogovšek et al., 2017; Onyeaka et al., 2022). Consumers' willingness to pay for products utilizing nanotechnology could be shaped by their understanding of its benefits and risks, indicating a need for transparent communication about the safety measures and regulatory standards governing these innovations. As the field matures, addressing consumer apprehensions through informed discussions and robust regulatory practices will be essential for realizing the full potential of nanotechnology in ensuring food safety and enhancing sustainability in food production.

9. CONCLUSION

Nanotechnology applications in food contaminant mitigation present transformative opportunities to enhance food safety and quality across the entire supply chain. Key applications include nanosensors for real-time detection of pathogens and toxins, nano-adsorbents for removing chemical residues, antimicrobial nanoparticles for inhibiting microbial growth, and nanocomposites in smart packaging systems. These innovations significantly improve contaminant detection accuracy, extend shelf life, and reduce spoilage, offering substantial benefits to both consumers and producers. For food industry stakeholders, these advancements imply strategic gains in quality assurance, brand trust, regulatory compliance, and operational efficiency in increasingly competitive and globalized markets.

However, as nanotechnology becomes more integrated into food systems, there is a pressing need for responsible innovation that aligns with sustainable development goals. This includes proactive risk management, thorough safety assessments, transparent public communication, and robust regulatory oversight to ensure consumer safety and environmental protection. When applied ethically and sustainably, nanotechnology can not only reduce food waste and chemical dependency but also support eco-friendly packaging and efficient resource use. Ultimately, its thoughtful implementation will be pivotal in shaping a safer, more resilient, and sustainable global food system.

REFERENCES

Abbasi, A., Hashemi, M., Kafil, H. S., Astamal, M. A., Lahouty, M., Tajani, A. G., Hosseini, H., & Nasirifar, S. Z. (2024). a critical review on the bioavailability promotion of the food bioactive compounds: Nano lipid carriers perspective. *Ulum-i Daruyi*. Advance online publication. DOI: 10.34172/PS.2024.11

Adenubi, A. O., Oduroye, A. P., Mosud, O., & Akanni, A. (2024). Cultivating the future: Nanotechnology's green revolution in agriculture. *International Journal of Research in Education Humanities and Commerce*, *05*(02), 16–28. DOI: 10.37602/IJREHC.2024.5202

Adeyeye, S. A. O. (2019). Food packaging and nanotechnology: Safeguarding consumer health and safety. *Nutrition & Food Science*, *49*(6), 1164–1179. DOI: 10.1108/NFS-01-2019-0020

Aigberua, A. O., Izah, S. C., & Aigberua, A. A. (2023). Occurrence, source delineation, and health hazard of polycyclic aromatic hydrocarbons in tissues of Sarotherodon melanotheron and Chrysichthys nigrodigitatus from Okulu River, Nigeria. *Environmental Monitoring and Assessment*, *195*(3), 364. DOI: 10.1007/s10661-023-10970-y PMID: 36740655

Aigberua, A. O., Izah, S. C., & Isaac, I. U. (2018). Level and health risk assessment of heavy metals in selected seasonings and culinary condiments used in Nigeria. *Biological Evidence*, *8*(2), 6–15. DOI: 10.5376/be.2018.08.0002

Aigberua, A. O., Izah, S. C., & Richard, G. (2021). Hazard Analysis of Trace Metals in Muscle of Sarotherodon melanotheron and Chrysichthys nigrodigitatus from Okulu River, Rivers State, Nigeria. *Journal of Environmental Health and Sustainable Development.*, *6*(3), 1340–1356.

Ameta, S. K., Rai, A. K., Hiran, D., Ameta, R., & Ameta, S. C. (2020). Use of nanomaterials in food science. In *Biogenic nano-particles and their use in agro-ecosystems* (pp. 457–488). Springer Singapore. DOI: 10.1007/978-981-15-2985-6_24

Ansari, M. A. (2023). Nanotechnology in food and plant science: Challenges and future prospects. *Plants*, *12*(13), 2565. DOI: 10.3390/plants12132565 PMID: 37447126

Awlqadr, F. H., Altemimi, A. B., Omar, A. M. A., Saeed, M. N., Qadir, S. A., Faraj, A. M., Al-Manhel, A. J. A., Salih, T. H., Hesarinejad, M. A., & Vieira, I. R. S. (2025, June). Advancing Sustainability in Fruit and Vegetable Packaging: The Role of Nanotechnology in Food Preservation. *eFood*, *6*(3), e70060. Advance online publication. DOI: 10.1002/efd2.70060

Bai, X., Ga, L., & Ai, J. (2023). A fluorescent biosensor based on carbon quantum dots and single-stranded dna for the detection of escherichia coli. *Analyst*, *148*(16), 3892–3898. DOI: 10.1039/D3AN01024D PMID: 37462388

Brito, S. D. C., Bresolin, J. D., Sivieri, K., & Ferreira, M. D. (2019). Low-density polyethylene films incorporated with silver nanoparticles to promote antimicrobial efficiency in food packaging. *Food Science & Technology International*, *26*(4), 353–366. DOI: 10.1177/1082013219894202 PMID: 31870192

Castro-Mayorga, J. L., Fabra, M. J., Cabedo, L., & Lagaron, J. M. (2016). On the use of the electrospinning coating technique to produce antimicrobial polyhydroxyalkanoate materials containing in situ-stabilized silver nanoparticles. *Nanomaterials (Basel, Switzerland)*, *7*(1), 4. DOI: 10.3390/nano7010004 PMID: 28336838

. Chadha, U., Bhardwaj, P., Selvaraj, S.K., Arasu, K., Praveena, S., Pavan, A., Khanna, M., Singh, P., Singh, S., Chakravorty, A. et al.(2022). Current trends and future perspectives of nanomaterials in food packaging application. Journal of Nanomaterials, 2022(1). DOI: 10.1155/2022/2745416

Chugh, G., Siddique, K. H., & Solaiman, Z. M. (2021). Nanobiotechnology for agriculture: Smart technology for combating nutrient deficiencies with nanotoxicity challenges. *Sustainability (Basel)*, *13*(4), 1781. DOI: 10.3390/su13041781

Cortés-Sánchez, A. D. J., Diaz-Ramírez, M., Torres-Ochoa, E., Espinosa-Chaurand, L. D., Rayas-Amor, A. A., Cruz-Monterrosa, R. G., Aguilar-Toalá, J. E., & Salgado-Cruz, M. D. L. P. (2024). Processing, quality and elemental safety of fish. *Applied Sciences (Basel, Switzerland)*, *14*(7), 2903. DOI: 10.3390/app14072903

Curulli, A. (2020). Nanomaterials in electrochemical sensing area: Applications and challenges in food analysis. *Molecules (Basel, Switzerland)*, *25*(23), 5759. DOI: 10.3390/molecules25235759 PMID: 33297366

Devra, V. (2022). Applications of metal nanoparticles in the agri-food sector. Egyptian Journal of Agricultural Research. DOI: 10.21608/ejar.2022.102565.1164

Dobrucka, R., & Ankiel, M. (2018). Possible applications of metal nanoparticles in antimicrobial food packaging. *Journal of Food Safety*, *39*(2), e12617. Advance online publication. DOI: 10.1111/jfs.12617

Durán, N., & Marcato, P. D. (2012). Nanobiotechnology perspectives. role of nanotechnology in the food industry: A review. *International Journal of Food Science & Technology*, *48*(6), 1127–1134. DOI: 10.1111/ijfs.12027

Eissa, S., Tlili, C., Mounir, B. A., & Kanoun, O. (2022). Editorial: Nanomaterials-based electrochemical biosensors. *Frontiers in Bioengineering and Biotechnology*, *10*, 1091592. Advance online publication. DOI: 10.3389/fbioe.2022.1091592 PMID: 36588931

Ferrone, E., Araneo, R., Notargiacomo, A., Pea, M., & Rinaldi, A. (2019). Zno nanostructures and electrospun zno–polymeric hybrid nanomaterials in biomedical, health, and sustainability applications. *Nanomaterials (Basel, Switzerland)*, *9*(10), 1449. DOI: 10.3390/nano9101449 PMID: 31614707

Fraceto, L. F., Grillo, R., de Medeiros, G. A., Scognamiglio, V., Rea, G., & Bartolucci, C. (2016). Nanotechnology in agriculture: Which innovation potential does it have? *Frontiers in Environmental Science*, *4*. Advance online publication. DOI: 10.3389/fenvs.2016.00020

Fuertes, G., Soto, I., Carrasco, R., Vargas, M., Sabattin, J., & Lagos, C. (2016). Intelligent packaging systems: Sensors and nanosensors to monitor food quality and safety. *Journal of Sensors*, *2016*, 1–8. DOI: 10.1155/2016/4046061

Gao, H., Yan, C., Wu, W., & Li, J. (2020). Application of microfluidic chip technology in food safety sensing. *Sensors (Basel)*, *20*(6), 1792. DOI: 10.3390/s20061792 PMID: 32213909

Gayathri, D., Soundarya, R., & Prashantkumar, C. (2024). Various facets of nanotechnology in food processing [review]. *International Journal of Functional Nutrition*, *5*(1), 4. Advance online publication. DOI: 10.3892/ijfn.2024.38

. Gelaye, Y. (2025). Exploring the potential of agro-nanotechnology in african agriculture: a path to sustainable development—systematic review. The Scientific World Journal, 2025(1). DOI: 10.1155/tswj/9073364

Guo, Z. (2025). Research progress of nanomaterials as food packaging materials. *Applied and Computational Engineering*, *137*(1), 18–25. DOI: 10.54254/2755-2721/2025.21001

Hardy, A., Benford, D., Halldorsson, T., Jeger, M. J., Knutsen, H. K., More, S., Naegeli, H., Noteborn, H., Ockleford, C., Ricci, A., Rychen, G., Schlatter, J. R., Silano, V., Solecki, R., Turck, D., Younes, M., Chaudhry, Q., Cubadda, F., Gott, D., & Mortensen, A.EFSA Scientific Committee. (2018). Guidance on risk assessment of the application of nanoscience and nanotechnologies in the food and feed chain: Part 1, human and animal health. *EFSA Journal*, *16*(7). Advance online publication. DOI: 10.2903/j.efsa.2018.5327 PMID: 32625968

He, X., Deng, H., & Hwang, H. M. (2019). The current application of nanotechnology in food and agriculture. *Yao Wu Shi Pin Fen Xi, 27*(1), 1–21. DOI: 10.1016/j.jfda.2018.12.002 PMID: 30648562

He, X., & Hwang, H. M. (2016). Nanotechnology in food science: Functionality, applicability, and safety assessment. *Yao Wu Shi Pin Fen Xi, 24*(4), 671–681. DOI: 10.1016/j.jfda.2016.06.001 PMID: 28911604

Inbaraj, B. S., & Chen, B. H. (2016). Nanomaterial-based sensors for detection of foodborne bacterial pathogens and toxins as well as pork adulteration in meat products. *Yao Wu Shi Pin Fen Xi, 24*(1), 15–28. DOI: 10.1016/j.jfda.2015.05.001 PMID: 28911398

Ineyougha, E. R., Orutugu, L. A., & Izah, S. C. (2015). Assessment of Microbial Quality of Smoked *Trachurus trachurus* sold in some Markets of Three South-South States of Nigeria. *International Journal of Food Research, 2*, 16–23.

Izah, S. C., & Aigberua, A. O. (2020). Microbial and heavy metal hazard analysis of edible tomatoes (Lycopersicon esculentum) in Port Harcourt, Nigeria. *Toxicology and Environmental Health Sciences, 12*(4), 371–380. DOI: 10.1007/s13530-020-00060-8

Izah, S. C., Aigberua, A. O., & Ogwu, M. O. (2022a). Trace element composition of Gallus gallus domesticus eggs and health risks associated with their consumption in Port Harcourt, Nigeria. *Journal of Food Safety and Hygiene*, *8*(3), 202–222.

Izah, S. C., Aseiba, E. R., & Orutugu, L. A. (2015b). Microbial quality of polythene packaged sliced fruits sold in major markets of Yenagoa Metropolis, Nigeria. *Point Journal of Botany and Microbiology Research, 1*(3), 30–36.

Izah, S. C., Etebu, E. N., Aigberua, A. O., Odubo, T. C., & Iniamagha, I. (2022b). A meta-analysis of microbial contaminants in selected ready-to-eat foods in Bayelsa State, Nigeria: Public Health implications and risk-reduction strategies. *Hygiene and Environmental Health Advances, 4*, 100017. DOI: 10.1016/j.heha.2022.100017

Izah, S.C., Kigigha, L.T., & Anene, E.K. (2016b). Bacteriological Quality Assessment of *Malus domestica* Borkh and *Cucumis sativus* L. in Yenagoa Metropolis, Bayelsa state, Nigeria. British Journal of Applied Research, 1(2), 5-7.

Izah, S. C., Kigigha, L. T., Aseibai, E. R., Okowa, I. P., & Orutugu, L. A. (2016a). Advances in preservatives and condiments used in zobo (a food-drink) production. *Biotechnological Research*, *2*(3), 104–119.

Izah, S. C., Nurmakhanova, A., Ogwu, M. C., Toktarbay, Z., Umirbaeva, Z., Ussen, K., Koibasova, L., Nazarbekova, S., Tynybekov, B., & Guo, Z. (2025). Public Health Risks Associated with Antibiotic Residues in Poultry Food Products. *Journal of Agriculture and Food Research, 21*, 101815. DOI: 10.1016/j.jafr.2025.101815

Izah, S. C., Orutugu, L. A., & Kigigha, L. T. (2015a). A review of the quality assessment of zobo drink consumed in Nigeria. ASIO Journal of Microbiology. *Food Science and Biotechnology Innovations, 1*(1), 34–44.

Izah, S. C., Stanley, H. O., Richard, G., Sawyer, W. E., Uwaeme, O. R., & Sylva, L. (2024). Source and health risks of Trace Metals in Clarias batrachus and Chrysichthys nigrodigitatus from surface waters in Bayelsa State, Nigeria: A probabilistic model. *Frontiers in Sustainable Food Systems, 8*, 1419143. DOI: 10.3389/fsufs.2024.1419143

Kadirvel, V., Palanisamy, Y., & Ganesan, N. D. (2024). Active packaging system—An overview of recent advances for enhanced food quality and safety. *Packaging Technology & Science, 38*(2), 145–162. DOI: 10.1002/pts.2863

Kampers, F. W. H. (2011). Opportunities for nanotechnologies in food production. *Acta Horticulturae*, (916), 113–117. DOI: 10.17660/ActaHortic.2011.916.11

Kappenstein, O., Vieth, B., Luch, A., & Pfaff, K. (2012). Toxicologically relevant phthalates in food. *Molecular, Clinical and Environmental Toxicology: Volume 3. Environmental Toxicology, 101*, 87–106. DOI: 10.1007/978-3-7643-8340-4_4 PMID: 22945567

Khachatryan, G., & Khachatryan, K. (2019). Starch based nanocomposites as sensors for heavy metals-detection of Cu2+ and Pb2+ ions. *International Agrophysics, 33*(1), 121–126. DOI: 10.31545/intagr/104414

Kigigha, L. T., Berezi, J., & Izah, S. C. (2017a). Bacteriological Quality Assessment of Meat Pie Sold in Yenagoa Metropolis, Nigeria. *EC Nutrition, 6*(6), 189–195.

Kigigha, L. T., Igoya, U. O. S., & Izah, S. C. (2016). Microbiological Quality Assessment Of Unpeeled Groundnut Sold in Yenagoa Metropolis, Nigeria. *International Journal of Innovative Biochemistry & Microbiology Research, 4*(4), 11–22.

Kigigha, L. T., Izah, S. C., & Kpea, T. B. (2015b). Microbiological quality of fermented Cassava Flakes (Gari) sold in Yenagoa, Metropolis, Nigeria. *Bulletin of Advanced Scientific Research, 01*(07), 157–160.

Kigigha, L. T., Opusunju, I. C., & Izah, S. C. (2017b). Assessment of Bacteriological Quality of Puff-Puff Sold in Amassoma, Bayelsa State, Nigeria. *Bulletin of Trends in Biological Sciences, 1*(1), 18–22.

Kigigha, L. T., Ovunda, H. O., & Izah, S. C. (2015a). Microbiological quality assessment of suya sold in Yenagoa Metropolis, Nigeria. *Journal of Advances in Biological and Basic Research, 01*(05), 106–109.

Kigigha, L. T., Samson, G. A., Izah, S. C., & Aseibai, E. R. (2018). Microbial Assessment of Zobo Drink Sold in Some Locations in Yenagoa Metropolis, Nigeria. *EC Nutrition, 13*(7), 470–476.

Kogovšek, U., Kovač, B., & Babnik, K. (2017). Perceptions of nanotechnology utilization in food: A pilot study. *Journal of Engineering & Processing Management, 9*(1), 10–14. DOI: 10.7251/JEPM1709010K

Kumar, A., Singh, G. A., Gourkhede, P. H., Goutam, P. K., Laxman, T., Pandey, S. K., & Singh, P. (2023). Revolutionizing agriculture: A comprehensive review of nanotechnology applications. *International Journal of Environment and Climate Change, 13*(11), 3586–3603. DOI: 10.9734/ijecc/2023/v13i113534

Li, R., Yue, J., Zhu, F., Zhou, J., & Liu, X. (2025). Application progress of carbon quantum dot composites in fluorescent detection of food safety. *Journal of Food Science, 90*(5), e70299. Advance online publication. DOI: 10.1111/1750-3841.70299 PMID: 40433944

Liao, W., Badri, W., Dumas, E., Ghnimi, S., Elaïssari, A., Saurel, R., & Gharsallaoui, A. (2021). Nanoencapsulation of essential oils as natural food antimicrobial agents: An overview. *Applied Sciences (Basel, Switzerland), 11*(13), 5778. DOI: 10.3390/app11135778

Loza, K., Sengstock, C., Chernousova, S., Köller, M., & Epple, M. (2014). The predominant species of ionic silver in biological media is colloidally dispersed nanoparticulate silver chloride. *RSC Advances, 4*(67), 35290. DOI: 10.1039/C4RA04764H

Ma, Y. (2024). Study on food locking and packaging by nanomaterials. *Applied and Computational Engineering, 56*(1), 180–185. DOI: 10.54254/2755-2721/56/20240655

Madanayake, N. H., Hossain, A., & Adassooriya, N. M. (2021). Nanobiotechnology for agricultural sustainability, and food and environmental safety. *Quality Assurance and Safety of Crops & Foods, 13*(1), 20–36. DOI: 10.15586/qas.v13i1.838

Malhotra, B. D., Srivastava, S., & Augustine, S. (2015). Biosensors for food toxin detection: Carbon nanotubes and graphene. *Proceedings of the Materials Research Society, 1725*, mrsf14–1725. DOI: 10.1557/opl.2015.165

Manatunga, D. C., Dassanayake, R. S., & Liyanage, R. N. (2022). Novel acumens into biodegradation: impact of nanomaterials and their contribution. DOI: 10.5772/intechopen.98771

Manikandan, V. S., Adhikari, B., & Chen, A. (2018). Nanomaterial based electrochemical sensors for the safety and quality control of food and beverages. *Analyst, 143*(19), 4537–4554. DOI: 10.1039/C8AN00497H PMID: 30113611

Martirosyan, A., & Schneider, Y. J. (2014). Engineered nanomaterials in food: Implications for food safety and consumer health. *International Journal of Environmental Research and Public Health, 11*(6), 5720–5750. DOI: 10.3390/ijerph110605720 PMID: 24879486

McClements, D. J. (2020). Nanotechnology approaches for improving the healthiness and sustainability of the modern food supply. *ACS Omega*, *5*(46), 29623–29630. DOI: 10.1021/acsomega.0c04050 PMID: 33251398

Mohanty, L., Singh, A., Pandey, A., Kumar, R., Ramesh, G., Swamy, G., Pandey, S. K., & Singh, B. (2024). Harnessing nanotechnology for eco-friendly crop enhancement and sustainable agriculture. *Journal of Experimental Agriculture International, 46*(7), 154–167. DOI: 10.9734/jeai/2024/v46i72568

Morales-Díaz, A. B., Ortega-Ortíz, H., Juárez-Maldonado, A., Cadenas-Pliego, G., González-Morales, S., & Benavides-Mendoza, A. (2017). Application of nanoelements in plant nutrition and its impact in ecosystems. *Advances in Natural Sciences Nanoscience and Nanotechnology*, *8*(1), 013001. DOI: 10.1088/2043-6254/8/1/013001

Odubo, T. C., & Izah, S. C. (2024). Safety Considerations of Trace Metals in Locally Produced Nutritive Food-Drinks Consumed in Yenagoa Metropolis, Nigeria. *Biological Trace Element Research*, *203*(8), 4408–4419. Advance online publication. DOI: 10.1007/s12011-024-04488-8 PMID: 39729242

Odubo, T. C., Izah, S. C., & Ogwu, M. C. (2024). Mycological Safety in Tiger Nut (Cyperus esculentus L.) Drinks: Cultivation, Contaminants, and Compliance Strategies. In Ogwu, M. C., Izah, S. C., & Ntuli, N. R. (Eds.), *Food Safety and Quality in the Global South* (pp. 463–487). Springer. DOI: 10.1007/978-981-97-2428-4_15

Ogidi, O. I., & Izah, S. C. (2024). Emerging Technologies in the Global South Food Industry: Prospects and Challenges. In Ogwu, M. C., Izah, S. C., & Ntuli, N. R. (Eds.), *Food Safety and Quality in the Global South* (pp. 197–231). Springer. DOI: 10.1007/978-981-97-2428-4_7

Ogwu, M. C., Izah, S. C., Ntuli, N. R., & Odubo, T. C. (2024). Food Security Complexities in the Global South. In Ogwu, M. C., Izah, S. C., & Ntuli, N. R. (Eds.), *Food Safety and Quality in the Global South* (pp. 3–33). Springer. DOI: 10.1007/978-981-97-2428-4_1

Ohimain, E. I., & Izah, S. C. (2015). Physicochemical characteristics and microbial population of palm oil sold in major markets in Yenagoa metropolis, Bayelsa States, Nigeria. *Journal of Environmental Treatment Techniques*, *3*(3), 143–147.

Onwudiegwu, C., & Izah, S. C. (2025). From Science to Policy: Safeguarding Food Security and Public Health through Strengthened Soil Pollution Regulations. *ES General.*, *7*, 1363. DOI: 10.30919/esg1363

Onyeaka, H., Passaretti, P., Miri, T., & Al-Sharify, Z. T. (2022). The safety of nanomaterials in food production and packaging. *Current Research in Food Science*, *5*, 763–774. DOI: 10.1016/j.crfs.2022.04.005 PMID: 35520272

Oprea, I., Fărca , A. C., Leopold, L. F., Diaconeasa, Z., Coman, C., & Socaci, S. A. (2022). Nano-encapsulation of citrus essential oils: Methods and applications of interest for the food sector. *Polymers*, *14*(21), 4505. DOI: 10.3390/polym14214505 PMID: 36365499

Orutugu,, L.A., & Izah,, S.C., & Aseibai. (2015). Microbiological quality of Kunu drink sold in some major markets of Yenagoa Metropolis, Nigeria. *Continental Journal of Biomedical Science.*, *9*(1), 9–16.

Ovuru, K. F., Izah, S. C., Ogidi, O. I., Imarhiagbe, O., & Ogwu, M. C. (2024). Slaughterhouse facilities in developing nations: Sanitation and hygiene practices, microbial contaminants and sustainable management system. Food Science and Biotechnology. *Food Science and Biotechnology*, *33*(3), 519–537. DOI: 10.1007/s10068-023-01406-x PMID: 38274182

Pan, M., Xie, X., Liu, K., Yang, J., Hong, L., & Wang, S. (2020). Fluorescent carbon quantum dots—Synthesis, functionalization and sensing application in food analysis. *Nanomaterials (Basel, Switzerland)*, *10*(5), 930. DOI: 10.3390/nano10050930 PMID: 32403325

Peters, R., Brandhoff, P., Weigel, S., Marvin, H., Bouwmeester, H., Aschberger, K., Rauscher, H., Amenta, V., Arena, M., Botelho Moniz, F., Gottardo, S., & Mech, A. (2014). Inventory of nanotechnology applications in the agricultural, feed and food sector. *EFSA Supporting Publications*, *11*(7). Advance online publication. DOI: 10.2903/sp.efsa.2014.EN-621

Prasad, R., Bhattacharyya, A., & Nguyen, Q. D. (2017). Nanotechnology in sustainable agriculture: Recent developments, challenges, and perspectives. *Frontiers in Microbiology*, *8*, 1014. Advance online publication. DOI: 10.3389/fmicb.2017.01014 PMID: 28676790

Putri, L. A., Prabowo, Y. D., Dewi, D. M. M., Mumtazah, Z., Adila, F. P., Fadillah, G., Amrillah, T., Triyana, K., Nugroho, F. A. A., & Wasisto, H. S. (2024). Review of noble metal nanoparticle-based colorimetric sensors for food safety monitoring. *ACS Applied Nano Materials*, *7*(17), 19821–19853. DOI: 10.1021/acsanm.4c04327

Raja, I. S., Vedhanayagam, M., Preeth, D. R., Kim, C., Lee, J. H., & Han, D. W. (2021). Development of two-dimensional nanomaterials based electrochemical biosensors on enhancing the analysis of food toxicants. *International Journal of Molecular Sciences*, *22*(6), 3277. DOI: 10.3390/ijms22063277 PMID: 33806998

Rajeshkumar, S., & Malarkodi, C. (2014). In vitro antibacterial activity and mechanism of silver nanoparticles against foodborne pathogens. *Bioinorganic Chemistry and Applications*, *2014*, 1–10. DOI: 10.1155/2014/581890 PMID: 25313307

Rana, M. J., Riaz, U., Iqbal, R. K., Deeba, F., Iqbal, N., & Ali, E. (2023). Silver nanoparticles and their potential antimicrobial applications in food industry. *The Journal of Microbiology and Molecular Genetics*, *4*(1), 59–84. DOI: 10.52700/jmmg.v4i1.81

Rashidi, L., & Khosravi-Darani, K. (2011). The applications of nanotechnology in food industry. *Critical Reviews in Food Science and Nutrition*, *51*(8), 723–730. DOI: 10.1080/10408391003785417 PMID: 21838555

Rastogi, S., Kumari, V., Sharma, V., & Ahmad, F. J. (2021). Gold nanoparticle-based sensors in food safety applications. *Food Analytical Methods*, *15*(2), 468–484. DOI: 10.1007/s12161-021-02131-z

Ravichandran, R. (2010). Nanotechnology applications in food and food processing: Innovative green approaches, opportunities and uncertainties for global market. *International Journal of Green Nanotechnology Physics and Chemistry*, *1*(2), 72–P96. DOI: 10.1080/19430871003684440

Rossi, M., Cubadda, F., Dini, L., Terranova, M., Aureli, F., Sorbo, A., & Passeri, D. (2014). Scientific basis of nanotechnology, implications for the food sector and future trends. *Trends in Food Science & Technology*, *40*(2), 127–148. DOI: 10.1016/j.tifs.2014.09.004

Sabui, P., Mallick, S., Singh, K. R., Natarajan, A., Verma, R., Singh, J., & Singh, R. P. (2022). Potentialities of fluorescent carbon nanomaterials as sensor for food analysis. *Luminescence, 38*(7), 1047–1063. DOI: 10.1002/bio.4406 PMID: 36355396

Sani, M. A., Azizi-Lalabadi, M., Tavassoli, M., Mohammadi, K., & McClements, D. J. (2021). Recent advances in the development of smart and active biodegradable packaging materials. *Nanomaterials (Basel, Switzerland), 11*(5), 1331. DOI: 10.3390/nano11051331 PMID: 34070054

Sawyer, W. E., & Izah, S. C. (2024). *Unmasking Food Adulteration: Public Health Challenges, Impacts and Mitigation Strategies. ES General.*

Sawyer, W. E., Nabebe, G., & Izah, S. C. (2024). Public Health Implications of Pesticide Residues in Food: Risks, Regulations, and Interventions. *Greener Journal of Biomedical and Health Sciences, 7*(1), 31–46. DOI: 10.15580/gjbhs.2024.1.102024144

Seiyaboh, E. I., Kigigha, L. T., Alagoa, C. T., & Izah, S. C. (2018b). Microbial Quality of Palm Oil Sold in Amassoma, Bayelsa State, Nigeria. *Int J Pub Health Safe, 3*, 153.

Seiyaboh, E. I., Kigigha, L. T., Aruwayor, S. W., & Izah, S. C. (2018a). Level of Selected Heavy Metals in Liver and Muscles of Cow Meat Sold in Yenagoa Metropolis, Bayelsa State, Nigeria. *International Journal of Public Health and Safety, 3*, 154.

Sekhon, B. S. (2014). Nanotechnology in agri-food production: An overview. *Nanotechnology, Science and Applications, 31*, 31. Advance online publication. DOI: 10.2147/NSA.S39406 PMID: 24966671

Sekton, B. S. (2010). Food nanotechnology – an overview. *Nanotechnology, Science and Applications, 1*, 1. Advance online publication. DOI: 10.2147/NSA.S8677

Shafiq, M., Anjum, S., Hano, C., Anjum, I., & Abbasi, B. H. (2020). An overview of the applications of nanomaterials and nanodevices in the food industry. *Foods, 9*(2), 148. DOI: 10.3390/foods9020148 PMID: 32028580

Shafique, B., Ranjha, M. M. A. N., Murtaza, M. A., Walayat, N., Nawaz, A., Khalid, W., Mahmood, S., Nadeem, M., Manzoor, M. F., Ameer, K., Aadil, R. M., & Ibrahim, S. A. (2022). Recent trends and applications of nanoencapsulated bacteriocins against microbes in food quality and safety. *Microorganisms, 11*(1), 85. DOI: 10.3390/microorganisms11010085 PMID: 36677377

Sim, S., & Wiwanitkit, V. (2021). Food contamination, food safety and covid-19 outbreak. *Warasan Wichai Witthayasat Kanphaet, 35*(5), 463–466. DOI: 10.1108/JHR-01-2021-0014

Singh, A., Patel, A., & Minocha, N. (2025). Nano innovation: Enhancing food packaging through nanotechnology. *Current Nanoscience*, *21*(6), 957–971. DOI: 10.2174/0115734137314364240920052006

Sonawane, S. K., Patil, S. P., & Arya, S. S. (2018). Nanotechnology enrolment in food and food safety. *Journal of Microbiology, Biotechnology and Food Sciences*, *8*(3), 893–900. DOI: 10.15414/jmbfs.2018-19.8.3.893-900

Sundararajan, N., Habeebsheriff, H. S., Dhanabalan, K., Cong, V. H., Wong, L. S., Rajamani, R., & Dhar, B. K. (2023). Mitigating global challenges: Harnessing green synthesized nanomaterials for sustainable crop production systems. *Global Challenges (Hoboken, NJ)*, *8*(1), 2300187. Advance online publication. DOI: 10.1002/gch2.202300187 PMID: 38223890

Szelenberger, R., Cichoń, N., Zajaczkowski, W., & Bijak, M. (2024). Application of biosensors for the detection of mycotoxins for the improvement of food safety. *Toxins*, *16*(6), 249. DOI: 10.3390/toxins16060249 PMID: 38922144

Taha, A. A., & Omar, M. M. (2024). Review article: nanoparticle applications in plant nutrition: a comprehensive review and future perspectives. *Journal of Soil Sciences and Agricultural Engineering*, *0*(0), 25–30. DOI: 10.21608/jssae.2024.271443.1218

Tahir, A., Shabir Ahmad, R., Imran, M., Ahmad, M. H., Kamran Khan, M., Muhammad, N., Nisa, M. U., Tahir Nadeem, M., Yasmin, A., Tahir, H. S., Zulifqar, A., & Javed, M. (2021). Recent approaches for utilization of food components as nano-encapsulation: A review. *International Journal of Food Properties*, *24*(1), 1074–1096. DOI: 10.1080/10942912.2021.1953067

Thiruvengadam, M., Rajakumar, G., & Chung, I. M. (2018). Nanotechnology: current uses and future applications in the food industry. *Biotech*, *8*(1), 74.

Tolve, R., Bianchi, F., Lomuscio, E., Sportiello, L., & Simonato, B. (2022). Current advantages in the application of microencapsulation in functional bread development. *Foods*, *12*(1), 96. DOI: 10.3390/foods12010096 PMID: 36613312

Tseng, W. B., Hsieh, M. M., Chen, C. H., Chiu, T. C., & Tseng, W. L. (2020). Functionalized gold nanoparticles for sensing of pesticides: A review. *Yao Wu Shi Pin Fen Xi*, *28*(4), 522–539. DOI: 10.38212/2224-6614.1092 PMID: 35696144

Upadhyay, P., Zubair, M., Roopesh, M. S., & Ullah, A. (2024). An overview of advanced antimicrobial food packaging: Emphasizing antimicrobial agents and polymer-based films. *Polymers*, *16*(14), 2007. DOI: 10.3390/polym16142007 PMID: 39065324

Usha, T., Middha, S. K., Hemavathi, K. N., Hemalatha, P., Harini, R., Afreen, Z., Dwivedi, N., Akshaya, K. A., Goyal, A. K., & Baliga, M. S. (2025). Innovations in nanotechnology for food production and packaging: A review. *Current Nutrition and Food Science*, *21*(5), 559–572. DOI: 10.2174/0115734013297685240820043042

Wagh, R. V., Priyadarshi, R., Khan, A., Riahi, Z., Packialakshmi, J. S., Kumar, P., Rindhe, S. N., & Rhim, J. W. (2025). The role of active packaging in the defense against foodborne pathogens with particular attention to bacteriophages. *Microorganisms*, *13*(2), 401. DOI: 10.3390/microorganisms13020401 PMID: 40005767

Xie, L., Guo, C., Yang, L., & He, Y. (2024). Harnessing gold nanomaterials for advanced multicolor colorimetric biosensors in food hazards detection. *Yao Wu Shi Pin Fen Xi*, *32*(3), 274–295. DOI: 10.38212/2224-6614.3511 PMID: 39636771

Xiong, J., Zhang, H., Qin, L., Zhang, S., Cao, J., & Jiang, H. (2022). Magnetic fluorescent quantum dots nanocomposites in food contaminants analysis: Current challenges and opportunities. *International Journal of Molecular Sciences*, *23*(8), 4088. DOI: 10.3390/ijms23084088 PMID: 35456904

Xu, M. L., Gao, Y., Wang, X., Han, X. X., & Zhao, B. (2021). Comprehensive strategy for sample preparation for the analysis of food contaminants and residues by GC–MS/MS: A review of recent research trends. *Foods*, *10*(10), 2473. DOI: 10.3390/foods10102473 PMID: 34681522

Yan, M. R., Hsieh, S., & Ricacho, N. (2022). Innovative food packaging, food quality and safety, and consumer perspectives. *Processes (Basel, Switzerland)*, *10*(4), 747. DOI: 10.3390/pr10040747

Yarkwan, B., Isaac, T. O., Okopi, A., & Izah, S. C. (2024). Evidence of the Toxic Potentials of Agrochemicals on Human Health and Biodiversity: Carcinogens and Mutagens. In Ogwu, M. C., Izah, S. C., & Ntuli, N. R. (Eds.), *Food Safety and Quality in the Global South* (pp. 331–359). Springer. DOI: 10.1007/978-981-97-2428-4_11

Yu, H., Park, J. Y., Kwon, C. W., Hong, S. C., Park, K. M., & Chang, P. S. (2018). An overview of nanotechnology in food science: Preparative methods, practical applications, and safety. *Journal of Chemistry*, *2018*, 1–10. DOI: 10.1155/2018/5427978

Zhanbolat, A. A., & Tungyshbaeva, U. O. (2024). Current strategies for the use of active packaging in food products. *Bulletin of Shakarim University Technical Sciences*, 1(13), 108–115. DOI: 10.53360/2788-7995-2024-1(13)-14

Zhang, X., Khan, I. M., Ji, H., Wang, Z., Tian, H., Cao, W., & Mi, W. (2020). A label-free fluorescent aptasensor for detection of staphylococcal enterotoxin a based on aptamer-functionalized silver nanoclusters. *Polymers, 12*(1), 152. DOI: 10.3390/polym12010152 PMID: 31936075

Zhou, J., Lv, X., Jia, J., Din, Z. U., Cai, S., He, J., Xie, F., & Cai, J. (2022). Nanomaterials-based electrochemiluminescence biosensors for food analysis: Recent developments and future directions. *Biosensors (Basel), 12*(11), 1046. DOI: 10.3390/bios12111046 PMID: 36421164

Zorraquín-Peña, I., Cueva, C., Bartolomé, B., & Moreno-Arribas, M. V. (2020). Silver nanoparticles against foodborne bacteria. effects at intestinal level and health limitations. *Microorganisms, 8*(1), 132. DOI: 10.3390/microorganisms8010132 PMID: 31963508

Żukowska, G., Durczyńska, Z., Roszkowski, S., Myszura-Dymek, M., & Bik-Małodzińśka, M. (2024). Possibilities of using bio-based nanomaterials in sustainable agriculture. *Journal of Ecological Engineering, 25*(3), 313–322. DOI: 10.12911/22998993/182966

Chapter 6
Biodegradable Films for Mitigating Food Contaminants Through Sustainable Packaging

Meenambiga S. S.
https://orcid.org/0000-0002-5445-525X
Vels Institute of Science, Technology, and Advanced Studies, India

Yoganandham Kesavan
Vels Institute of Science, Technology, and Advanced Studies, India

Jayashree Pandurangan
https://orcid.org/0009-0008-6957-9256
Vels Institute of Science, Technology, and Advanced Studies, India

Prakash Pandurangan
Sathyabama Institute of Science and Technology, Chennai, India

Vivek Pazhamalai
Vels Institute of Science, Technology, and Advanced Studies, India

Ivo Romauld
https://orcid.org/0000-0003-0610-0646
Vels Institute of Science, Technology, and Advanced Studies, India

ABSTRACT

Preserving food flavor, freshness, and nutritional content is possible through careful packaging, which accounts for physical, mechanical, and epidemiological aspects. Unfortunately, the accumulation of non-biodegradable waste is a direct result of the conventional packaging's use of non-renewable fossil fuels, which harms the environment. Therefore, biodegradable films for food packaging are in high demand; they provide a cost-effective solution that meets industry standards while maintaining food safety, and consumers are becoming more conscious of the importance of purchasing sustainably sourced, high-quality food. Learn why biodegradable films are crucial for food safety and how they function in packaging. Biodegradable

DOI: 10.4018/979-8-3373-7052-1.ch006

polymers derived from polysaccharides are the focus of this research, along with their application in multi-criteria food packaging and the success or failure of developing smart biodegradable films.

1. INTRODUCTION

Packaging is a significant challenge for the food industry. It protects food goods from physical, mechanical, and epidemiological factors that affect their taste, safety, and nutritional value of food items (Peelman et al., 2013). Metal, glass, and plastic are the most frequently used materials in food packaging (Díaz-Montes, 2022). The use of plastics for food packaging has increased at an average rate of 5% per year and is currently the second dominant material used in food packaging in recent decades (Adeghizadeh Yazdi et al., 2019). However, there are some restrictions on the use of plastics because they degrade more slowly than other materials, thereby polluting the environment (Sorrentino et al., 2007; Xu et al., 2005). Some of the adverse effects of plastics on ecosystems involve the contamination of animals on land and in the sea, as they decompose. Plastic waste is also believed to adversely affect or kill aquatic organisms, causing concern for the food chain and possibly resulting in extinction. Consequently, packaging made from the synthesized materials was used only once, before being discarded. Surveys show that, despite the adverse effects of plastic disposal, the use of petroleum-derived packaging drives about 37% of total plastic usage (Briassoulis, 2004; Europe, 2015;). Research is also underway to identify rapidly biodegradable materials suitable for food packaging to solve the problem of plastic elongated degradation (Jimenez et al., 2012).

Cellulose, pectin, chitosan, and starch are naturally biodegradable polymers that can be used as alternatives to plastics in food packaging. Based on these biodegradable polymers, these biodegradable films are seen as substitutes for conventional food packaging since they can prolong the storage lifetime of unprocessed food and maintain its quality. (Acevedo-Fani et al., 2017). To facilitate disintegration, biodegradable films are bio-assimilated using a process that reduces the polymer chains into smaller subunits or dimer units (Zoungranan et al., 2020). The degradation process of biodegradable films is much more straightforward than that of plastics, as illustrated in the flowchart below (Figure 1). Moreover, biodegradable materials outperform plastics in terms of their environmental friendliness. This is because biodegradable materials disintegrate after disposal, thereby producing new agricultural products (Xu et al., 2005).

Figure 1. The process of biodegradation

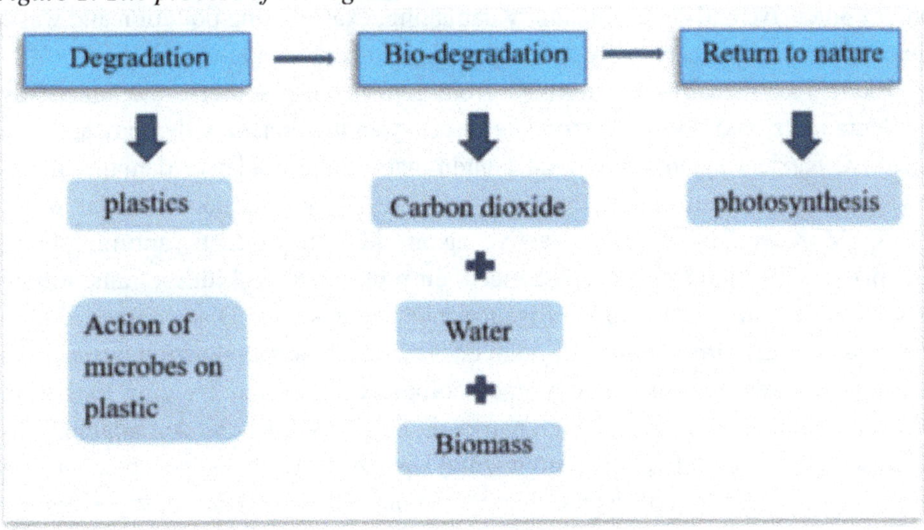

The incorporation of antimicrobial compounds and nanoparticles and the use of pH-indicating colors to assess food shelf life can increase customer preference for recyclable packaging. This article provides an overview of food packaging, including the attributes and requirements of biodegradable films to qualify as Packaging for food resources.

The food industry faces significant challenges in maintaining food safety and quality throughout the supply chain. Packaging plays a crucial role in protecting food from physical, chemical, and microbiological hazards that can compromise its nutritional value, taste, and safety (Peelman et al., 2013). However, conventional packaging materials, particularly petroleum-based plastics, present major limitations in ensuring long-term food safety and environmental protection (Díaz-Montes, 2022). Plastics have become dominant in food packaging due to their flexibility, durability, and low cost. Yet, their widespread use raises environmental and health concerns. One of the critical drawbacks of plastic packaging is its persistence in the environment. Plastics degrade extremely slowly, leading to accumulation in terrestrial and aquatic ecosystems. This waste can break down into microplastics, which infiltrate the food chain, posing risks to human and animal health (Sorrentino et al., 2007; Xu et al., 2005). Plastic waste harms marine life, disrupts aquatic ecosystems, and introduces bioaccumulative toxins into the food web. Additionally, plastics often contain additives such as plasticisers and stabilisers that can migrate into food, raising food safety concerns (Jimenez et al., 2012). Recycling rates for plastics remain critically low, with only about 10% being recycled, while the rest contributes to pollution through landfilling or incineration (Briassoulis, 2004; Europe, 2015).

The production of plastics, derived from non-renewable petroleum resources, is energy-intensive and environmentally damaging, exacerbating pollution and waste management issues, especially due to the prevalence of single-use plastics.

Biodegradable films are derived from renewable resources, are inherently biodegradable, and do not accumulate in ecosystems. Functionally, they serve as effective barriers against microbial contamination and oxidative damage, thereby extending the shelf life of fresh and minimally processed foods (Yadav et al., 2018). Moreover, incorporating active agents like antioxidants, antimicrobials, and pH-sensitive indicators can enhance their protective capabilities, transforming passive films into active and intelligent packaging systems (Díaz-Montes, 2022). These bio-based films degrade through enzymatic or microbial action into natural monomers, integrating harmlessly into ecological cycles and avoiding long-term environmental persistence (Zoungranan et al., 2020). As such, biodegradable films address both food safety and environmental sustainability challenges, presenting a promising solution to replace non-degradable plastics in food packaging. Ongoing research into active and intelligent biodegradable films could significantly advance the development of safer and more eco-friendly food packaging technologies (Xu et al., 2005; Díaz-Montes, 2022).

2. FOOD PACKAGING

Food packaging aims to preserve and safeguard all types of food and their constituents, particularly against aerobic and microbiological decomposition (Tharanathan, 2003). Food packaging is defined as any wrapping or coating that protects food from physical, mechanical, and epidemiological factors that affect the taste, safety, and nutritional value of food items. Physical contamination refers to any exterior material (such as lumps or bits of glass, plastic, or wood) that enters food and is usually connected with filthy circumstances throughout the processing, manufacture, transportation, and shipment of food goods. However, on the contrary, when food additives, such as flavorings, pigments, and sweeteners, or other chemicals, such as sterilizers and surfactants, are present, chemical contamination can occur during manufacturing, production, and preservation. Biological contamination is defined as the presence of microbes or harmful insects that produce toxins that can cause illness (Alexandre et al., 2019). The packaging of food products is classified into three types based on their protective abilities: passive, active, and intelligent packaging.

2.1. Passive Packaging

Standard food packaging must be safe for the food it holds, meaning it shouldn't change the food in any way. They are called passively protected baggage because their primary purpose is to offer physical protection from things like the climate, chemicals, and damage, along with blocking gases and moisture. (Díaz-Montes, 2022).

2.2. Active Packaging

Packaging food intentionally modifies the conditions inside the package or container to guarantee its quality of the food (Stoma et al., 2022). Active packaging comprises of absorbers and emitters. Absorbers absorb and release food-derived molecules such as moisture, the chemical ethylene, and carbon dioxide into their surroundings and emit biologically active compounds such as bacteria and anti-oxidants, both of which improve the longevity of food (Díaz-Montes et al., 2021).

2.3. Intelligent Packaging

This type of wrapping involves a system that monitors, generates, and displays information based on enzymatic properties, weather conditions, irradiation time, and physical responses to meals (Wilson, 2017). Chemical sensors or biosensors are used in intelligent packaging to maintain the factors that affect food condition, such as ripening, freshness, temperature, oxygen content, humidity, and various gases (Stoma, 2022).

2.4. Biodegradable Films in Food Packaging

The first bioplastic material to be produced was biodegradable alimentary packaging, which was successfully commercialized and verified as commercially biodegradable. The demand for bioplastics for food wrapping has rapidly increased. While strong packaging predominantly leads to recyclable packaging, soft packaging typically uses decomposed polymers. In addition, different fruits and vegetables are preserved under altered air conditions using recyclable polymers (Yadav et al., 2018). These films were originally intended to replace the plastic packaging. They have characteristics that are superior to those of nonbiodegradable polymers. .

2.5. Essential Characteristics of a Good Packaging Film

Allowing regulated ventilation is a crucial feature of a good packaging film with excellent shielding qualities, which is beneficial in retaining the stability of the structure and halting or slowing the growth of pathogens.

3. NATURAL POLYMERS IN BIODEGRADABLE FILM FORMULATIONS

Biodegradable films can be formulated using different compositions of naturally degrading polymers, including starch, cellulose, chitosan, and pectin. All of these polymers can be obtained from agro-industrial wastes, which have been generated consecutively over the last few decades. By the end of 2050, the global population is expected to exceed 10 billion people. Countries worldwide are intensifying their efforts to boost food manufacturing to meet the world's constantly expanding population in the future. This could encourage the industrialization of agriculture and related areas, which would also increase waste generation (Duque-Acevedo et al. 2020). The management and disposal of these wastes are serious environmental concerns that must be addressed sustainably (Bharathiraja et al., 2017). Agricultural waste includes leaves, stalks, and pods of seeds, as well as processing waste, such as husks, grains, root systems, maple syrup, and vegetable and fruit skins. Although an immense amount of agricultural waste is generated worldwide, most of it remains unutilized. For example, many researchers have used various agricultural and industrial wastes to create biodegradable packaging materials. Cellulose (35-50%), hemicellulose (15-30%), and lignin (20-30%) are the major components of agricultural waste. The composition of these wastes is determined by the type of agricultural waste and the crop species. Hemicellulose derived from wheat straw is an excellent source for the development of organic alimentary packaging with low moisture content (Obi et al., 2016). Biodegradable packaging materials created using various agroindustrial trashes are classified into two types: (a) packaging materials created using current plastic production technology directly from agro-industrial waste and (b) biopolymers created using polymers derived from agricultural and industrial waste (Figure 3). Biopolymers are also divided into three types, as illustrated in figure 2 (Chiellini et al. 2008).

Figure 2. Classifications of biopolymers

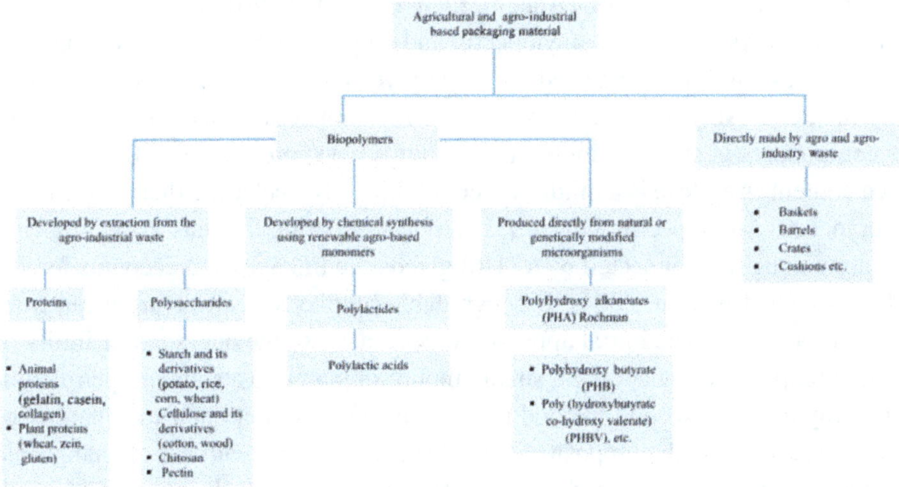

3.1. Starch

Starch is a carbohydrate used for the preservation of vegetables originating from crop leaves and storage space tissues. Although it is also present in tuber roots, it can also be found in amyloplasts of grain endosperms, wheat, corn, potatoes, and other grains, which are some of the other good examples of starch sources (Tagliapietra et al., 2021). Among natural polymers, starch has great potential because of its natural ability to break down, its supply, and its yearly renewal. It is frequently recognized as an effective substitute for plastics in food packaging. Additionally, a biodegradable film made of starch can be utilised for wrapping a wide range of commodities, including fruits, vegetables, refreshments and dry goods using starches in sustainable packaging to their multiple benefits which include their contribution to Having no poisons, extraction from plant-based materials (renewable resources), biodegradable and biological compatibility, affordability and availability of starch, suitability for usage in edible food packaging, use as an environmentally acceptable disposal solution, and no net increase in CO_2 in the atmosphere are all advantages of starch. However, they have several disadvantages such as weak mechanical qualities, restricted water rigidity, porous permeability restriction, and brittleness at room temperature. To overcome these disadvantages, starch can be used as a matrix with other natural polymers, such as cellulose, pectin, and chitosan, to create compostable packaging with excellent mechanical properties and water stability to overcome these disadvantages (Adeghizadeh-Yazdi et al., 2019).

The two polysaccharides, amylopectin and amylose, which are both composed of α-d-glucose and are connected by (1–4) glycoside linkages, make up the majority of starch. Amylose, a linear polymer with a helicoidal shape, is one of these two components in starch granules and accounts for approximately 20-25% of the starch content. Amylose is a mildly water-soluble polysaccharide that occasionally has branches. Amylopectin, however, is a polysaccharide that has many branches with a reducing tail connected by α (1-6) glycosidic linkages and accounts for 70-80% of starch (Eliasson, 2004). Biodegradable bioplastics can be produced from starch in four ways. The initial approach was to increase the degradation ability of the final biopolymer by adding a small amount of starch to a standard petroleum-based polymer. The second method involves using more than half the quantity of starch to create a starch composite. The third is extrusion treatment of changeable granular starch to produce starch. The fourth method involves the creation of starch nanocomposites using nanoparticles is the fourth method (Birania et al., 2022).

3.2. Chitin/Chitosan

Crustacea, insects, and other species use naturally occurring mucopolysaccharide chitin as a support structure. Approximately six million tons of crabs are caught in both the inland and coastal oceans. Chitin usually appears in fungal cell partitions and differs only in the cellulose alcohol group. Saccharomyces and mushrooms are two other dietary sources of chitin in addition to crustaceans and insects. Chitin, a white, diligent, and inelastic nitrogenous polymer, is a significant contributor to atmospheric contamination of coastlines (Kumar 2020; Birania et al., 2022).

However, chitin cannot be effectively used as an early ingredient in bioplastics owing to its poor solubility. Crustaceans, insects, arthropods, and mollusks contain large amounts of chitin, which is a cationic polymer created upon chitin deacetylation (Klinger et al., 2019). As stated by Shaala et al. (2019) and Ehrlich et al. (2018), Deproteinisation, calcification, and decolorisation are steps used in naturally occurring extraction platforms to eliminate chitin from other substances. Chitin is deacetylated to produce chitosan. These polymers can also be processed using microbe-based biological extraction methods. Depending on their specific use, chitin and related compounds can be used as gel-like substances, beads, barriers, films, or sponges (Klinger et al., 2019; Wysokowski et al., 2013). Reuse of waste from the shipping industry is not recommended, and a sizable portion of the generated biomass is released into the untreated atmosphere (Wang et al., 2019).

The seafood industry generates about 104 tonnes of waste each year. Most of this waste is either turned into compost or made into low-value goods like animal feed and fertiliser. Around 2000 metric tonnes of chitin are produced each year, mostly from leftover shells of prawns and crabs. (Muñoz et al., 2018; Santos et al., 2020). To extend the shelf life of fruits and vegetables, chitin and chitosan have been used to create recyclable edible antimicrobial packaging. Chitosan can be used to prepare films without any additives and has excellent atmospheric carbon dioxide and oxygen permeability, outstanding physical capabilities, and antibacterial properties that are effective against molds, bacteria, and yeast. However, chitosan is incredibly sensitive to moisture (Kumar et al., 2017). Gram-positive and gram-negative bacteria, fungi (molds and yeasts), and other spoilage and pathogenic microorganisms are resistant to the antibacterial and antioxidant capabilities of chitosan (Kaur et al., 2021; Díaz-Montes et al., 2022).

3.3. Pectin

Plant cell walls contain pectin, a structural polysaccharide with High molecular weight materials can be changed into a hydrogel, which forms a stretchy web of polymer chains. It is a complex sugar that plays key roles in plant growth and development, helping plants stay strong and safe from their surroundings. The main and most valuable sources of pectin come from citrus fruits, apples, and the leftover materials from their processing, like citrus peels and apple pulp. Cocoa leaves, pumpkins, watermelons, pears, and potato pulp are other sources of pectin. Pectin can come in different forms based on where it comes from or how it is extracted. It is a type of polysaccharide that can be recycled and has natural organic qualities. (Freitas et al., 2021). When esterification was greater than 50%, pectin was classified as high-methoxyl or high-ester pectin; however, pectin was classified as low-ester or low-methoxyl when esterification was less than 50%. The degree of ester formation is defined as the proportion of d-galacturonic acid carboxyl groups esterified with ethyl alcohol (Wang et al., 2018). Rhamnogalacturonan I (RGI), Rhamnogalacturonan II (RGII), and xylogalacturonan (XG) are three subdomains constituting the complex structure of pectin. Pectin undergoes chemical, physiological, and/or enzymatic changes, and the numerous functional groups present in its structure can stimulate various functionalities and alterations that enable its use in food, farming, drugs, and biomedical research. Currently, pectin is used to create nanoparticles, antimicrobial bio-based films, therapeutic agents, and cancer treatments, as well as coated foods to protect food (Tharanathan et al., 2003). Pectin has poor mechanical properties and low thermal ability, which are drawbacks when used in biodegradable packaging. It can be combined with other plastics to make these qualities better. Hydrothermal treatment extraction is the most common way to remove pectin. The pectin in the solvent was dissolved when the dehydrated and pulverized crop residue was dispersed in acidulated water (hydrochloric or nitric acid). Pectin is precipitated by adding alcohol after pectin-containing water is separated (Valdés et al., 2015). Then, it was dried to separate the pectin. Other techniques for pectin extraction include enzyme-assisted, subcritical water, and ultrasound-assisted extraction.

3.4. Cellulose

One of the most common polymers on Earth is cellulose, which can be readily synthesized from plant cell partitions. Payene was the first cellulose to be identified. Cellulose is an inexpensive commodity with strong demand in the global market owing to its high durability, especially in the agricultural and clothing sectors. An enormous amount of agricultural waste, including fruits, vegetables, cotton twigs, and debris from forests, is being dumped into the environment. This waste must be subjected to several recycling processes. The derivation of cellulose from these waste products is one of the rarest and the most important methods. Cellulose and nanocellulose fibers (NC), as well as their derivatives, are becoming increasingly popular in the food packaging sector due to its unique characteristics, biodegradability, and enhanced structural and physiological capabilities (Liu et al., 2021). Cellulose functions as an armour against UV rays and possesses significant heat tolerance. Antioxidant benefits and antibiotics can be transported. They are also disposable and reusable, and because They are resistant to corrosion, combustible, and non-toxic, suitable for the fabrication of various composites of polymers.. Additionally, it is readily accessible and priced worldwide. However, the disadvantages of cellulose and its derivatives include substantial water absorption and inadequate surface adherence (David et al., 2019). These limitations have led to extensive scientific research on cellulose.

It is simple to encapsulate other artificial and organic polymers into cellulose. Traditionally, paper is made using cellulose. Additionally, cellulose can dissolve because of its composition, and this inflammatory activity results in hypersensitivity when it interacts with liquids (Dai et al., 2019). It prolongs the period of storage of food and maintains the high standards of unpackaged products. Furthermore, cellulose serves as a material for packaging with obstacle characteristics that restrict the movement and transmission of solutes, lipids, gases, and moisture.

Additionally, they help to retain the structure, enhance mechanical transportation, and delay the demise of chlorophyll, which maintains the color of the food. Additionally, they act as a means of incorporating various food additives, aiding in the avoidance and reduction of microbial deterioration during long-term food storage (Abdul Khalil et al., 2018). Wood is The main source of cellulose. Wood pulp can be used to make cellulose, which has many uses. A prevalent source is cotton, mostly utilised for the production of threads and garments. Cotton blossoms contain cellulose. The outer layers of various grains, maize kernels, wheat bran, and many other sources of bacterial cellulose, water crops, certain types of grasses, agricultural byproducts, factory trash, food wastes, food leftovers, peels of different fruits and vegetables, and many other materials contain cellulose (Zhao et al., 2019; Zhong et al., 2020). Cellulose can be extracted using a modified acid hydrolysis process. Anhydro glucose units (AGUs), which are (14)-linked d-glucose units, are comprised of cellulose molecules. Every second, the AGU along the chain plane twists by 180° because of -linkage. AGU is the recurring unit of cellulose. Strong intramolecular and intermolecular H-bonds were generated as a result of each AGU having three hydroxyl groups, and these H-bonds give rise to the semi-crystalline structure of cellulose, which includes crystalline cellulose (Heinze et al., 2018).

3.5. Additives in Biodegradable Films Formulations

Some of you can enhance the features of biofilms by adding substances like plasticisers, crosslinkers, or active biological components. (Shit et al., 2014). Crosslinking agents, including stimuli (such as pH and electrical changes) and biomolecules (such as ions and enzymes), can enhance the mechanical, biological, and chemical characteristics of polymers substances (Rouhi et al., 2017). According to Suderman et al. (2018), glycerol, sorbitol, xylitol, and fructose are the plasticizers most frequently used in biodegradable films because they alter the thickness of the film and enhance its mechanical (e.g., fragility and flexibility), physicochemical (e.g., solubility), Heat and obstruction properties (e.g., gas absorption). The communication between a plasticizer and a polymer is governed by one of them. For instance, according to Sanyang et al. (2016), adding plasticizing agents (glycerol and sorbitol) to the process of creating sugar palm starch films decreased the fragility and retention of water, while improving solubility and wetness. Further research by Kaewprachu et al. (2018) revealed that the addition of a plasticizer enhanced fish protein film stretching, humidity, and vaporized water permeability. This is because cross-linking cannot occur owing to the hydrophobicity of the lipids. Chitosan is an acetylated polymer that must have a substance that crosslinks to acidify its environment and encourage protonation (Díaz-Montes et al., 2021), allowing it to interact with other polymers and create biodegradable films (Deshmukh et al., 2021). As genipin triggers nucleophilic interactions between the peptide and carboxylic bonds in a neutral acid solution, which might result in films that degrade owing to intermolecular contacts, it is frequently used as an agent for crosslinking amino polymers or proteins (Roy et al., 2022). Additionally, biodegradable films can include molecules with potential health effects, known as bioactive chemicals, which come from biological sources and have antibacterial, antifungal, antioxidant, or probiotic capabilities. Polysaccharides (such as starch, cellulose, pectin, and chitosan) are some of the commonly utilised monomers for the fabrication of films that decompose because of their capacity to dissolve in water, manageability (i.e., sources and costs), and properties (i.e., nontoxicity, biodegradability, and bioactivity).

3.6. Standardisation and Safety Issues of Polysaccharide-Based Biodegradable Films

Various rules must be followed when packing food items or objects that come in contact with food. Depending on where they are manufactured, sold, or marketed. However, each country has its national regulatory entity with basic regulations that can be broadly enforced. The primary role and goal of these regulatory entities worldwide is to ensure food safety, and food packaging is generally recognized as safe (GRAS) (Galus et al., 2020). Films derived from polysaccharides such as starch, cellulose, chitosan, and pectin are inherently non-toxic and biodegradable; nevertheless, additions including plasticisers, crosslinkers, and antimicrobial agents must adhere to food-grade standards to mitigate contamination hazards. Films containing bioactive compounds must ensure that their functional doses do not jeopardise food safety. Moreover, structural integrity is essential to avert physical contamination resulting from film degradation or particulate emission. Inadequate mechanical qualities may result in ruptures or disintegration during manipulation, posing a risk of direct food contamination. The combination of various polymers and meticulous additive selection improves film strength and barrier characteristics, successfully obstructing microbiological, chemical, and physical pollutants.

The food-grade label can be obtained using a variety of methods, including (1) the list of items that fulfil food-grade standards and (2) the examination of compounds used in the manufacturing of food or food packages. (3) Experts outside the organisation evaluate the goods. and (4) publishing research on a novel food-grade substance or package (FDA, 2023; Alamri et al., 2021). All marketed food products must truly reveal that the consumption of food additives, such as colours and flavours, is controlled by government agencies. For example, the European Union and Mexico require that food products use only approved ingredients. If goods contain allergens, such as proteins, they must be listed. Films that degrade polysaccharides must be labelled GRAS chemicals. Additives in the mixture must meet the needed element standards.

While "bioactive" films decomposing need to ensure that extra substances like oils, antioxidants, or antimicrobials do not pose toxicity issues when employed in the dosage specified. Food or food packaging substances can reach the market with the consent of regulatory organizations once they meet all these requirements and pass the quality tests of the regulatory authorities.

4. NATURAL COLORANTS IN BIODEGRADABLE FOOD PACKAGING

The objective of food packaging is to maintain the nutritional value while safeguarding food goods throughout transit and storage. The four principal roles of packaging are containment, communication, convenience, and protection. (Brody et al., 2008). In addition to these essential functions, food packaging is also required to perform an increasing number of tasks. Therefore, the evolution of smart and clever packaging deserves special attention. Consequently, radio frequency identification, time-temperature indicators, and natural acid-base color pigments that serve as ripeness indicators have been added (Wanihsuksombat et al., 2010). Colorimetric indicators can be attached to the exterior of food packaging or inserted inside packages to detect and observe alterations in the condition of packed products via visible colour variations.

Singh et al. (2018) presented evidence about the possible application of anthocyanins. Additionally, roselle anthocyanins have been proposed as chromatic agents for assessing the freshness of pork in intelligent indicator films composed of starch, polyvinyl alcohol, and chitosan (Zhang et al., 2019). Furthermore, roselle anthocyanins have been suggested for use as coloured components in intelligent films made of starch, polyvinyl alcohol, and chitosan to evaluate pork freshness (Zhang et al., 2019). Consequently, a novel and emerging field is the Utilisation of food colouring in biodegradable plastics as a proactive indicator. The additives. Utilised are permissible for application in the food business; therefore, if they are transferred to packaged food goods, this will not be a problem. A new generation of sustainable, clever packaging components that change colour under regulated maturation (thermos oxidation, UV, weathering) affects the longevity of the packaging material and is made possible by the incorporation of renewable polymers and organic food colorants. Because they are both aesthetically pleasing and functionally beneficial to the container, natural colourants are an important component of biodegradable food packaging. These pigments add natural colour and help keep food safe and fresh for longer; they come from plants including turmeric, beetroot, carrots, leafy greens, and berries. The pH-sensitive colour shifts of anthocyanins (found in berries and roselle) and curcumin (found in turmeric) are used extensively as visual markers of food freshness and deterioration.(Zhong et al., 2020) Beetroot betalains and carrot carotenoids serve as antioxidants and light barriers, respectively, while chlorophyll shields against ultraviolet radiation and kills microbes. Complementing the packaging's antioxidant, antibacterial, and UV-shielding capabilities are chemicals such as flavonoids and anthraquinones.

By mixing these dyes with biodegradable polymers, we can make packaging solutions that are both smart and active, able to track the status of products and cut down on food waste. Aligning with sustainability aims and customer expectations for clean-label packaging solutions, these natural additives are non-toxic and authorised for food-contact applications.(Martău et al., 2019) Natural colourants have several benefits in modern packaging technologies, including the replacement of synthetic dyes and the addition of functional properties to biodegradable films. This helps with environmental conservation and allows for better food safety monitoring. The various natural pigments and their sources are shown in Fig. 11 and in table 1

Table 1. Natural Colorants Utilized in Biodegradable Food Packaging

Natural Colorant	**Source**	**Functionality**	**Reference**
Anthocyanins	Berries,grapes, roselle	pH-sensitive indicator for freshness; antioxidant	Singh et al., 2018; Zhang et al., 2019
Betalains	Beetroot	Color indicator; antioxidant	Zhong et al., 2020
Carotenoids	Carrots, tomatoes	Light barrier; antioxidant	Liu et al., 2021
Chlorophyll	Leafy vegetables	UV-protective barrier; visual appeal	Abdul Khalil et al., 2018
Curcumin	Turmeric	pH-sensitive indicator; antimicrobial	Díaz-Montes, 2022
Anthraquinones	Aloevera, rhubarb	Color stability; antimicrobial	Freitas et al., 2021
Flavonoids	Tea, citrus peels	Antioxidant; UV protection	Martău et al., 2019
Roselle anthocyanins	*Hibiscus sabdariffa*	Intelligent indicator film component	Zhang et al., 2019

5. BIOSENSORS DEVELOPED IN FOOD PACKAGING SENSORS

A sensor detects changes around it and responds based on information from another machine. A sensor turns an actual occurrence into a measurable voltage or sometimes a digital signal. This information is then sent for reading, further processing, or shown on a display that people can understand. Most sensors have two parts: a receptor, which acts as the sensor. This method can identify certain chemicals or physical substances by looking for their presence, activity, make-up, or amount. The sensor, the second part, measures the energy that the receptor changes from physical or chemical information (Mlalila et al., 2016). A transducer was employed to convert the recorded signal into a valuable analytical value. There may be thermal, visual, substance, or electrical indicators. (Kerry et al., 2006).

The majority of detectors have two main parts: one that acts as the sensor and another that is the receiver. This helps identify specific chemicals or physical substances for different reasons, like checking if they are present, how active they are, what they are made of, and how much is there. In the second part, the sensor measures the energy the receptor gets from the chemical or physical data. A transducer converts the recorded signal into an analyzable format. This can be a warning based on light, heat, chemicals, or electricity. O2 was found using a mix of electrochemical, ultrasonic, laser, and infrared instruments (Park et al., 2015). Biosensors are a kind of sensor. They have receptors made of organic parts like enzymes, antigens, hormones, and nucleic acids, instead of chemical sensors. You can use different types of sensors, like electrochemical, optical, or acoustic ones, based on the monitoring needs. The Toxin Guard by Toxin Alert is a device that can identify infections like Salmonella, E. coli, Campylobacter, and Listeria. It uses a special system based on antibodies and is built into packaging that breaks down naturally. Visual cues show that results are successful (Bodenhamer et al., 2004).

5.1. Polymer-Based Biosensors for Bacterial Detection

Bacterial infections are a significant global problem, posing a serious threat to people's health and food safety. Many harmful germs, such as Staphylococcus aureus, Mycobacterium tuberculosis, and Escherichia coli, are present (Zhang et al., 2023) (Table 2). Despite using methods like pasteurisation and ultra-high-temperature sterilisation to kill germs in food, foodborne bacteria are still the leading cause of illness in both rich and poor countries. It's essential to create effective biosensors to detect dangerous germs in food (Kang et al., 2022). Salmonella species, particularly *Salmonella typhimurium*, are prominent foodborne pathogens. Researchers have created optical biosensors for detection utilising polymers like poly-L-lysine and poly(styrene/acrylamide). Poly-L-lysine, renowned for its superior adhesion characteristics, enhances optical biosensors such as those created by Sheikhzadeh et al. (2016). Ding et al. (2022) employed poly(styrene/acrylamide) in optical biosensing systems for the detection of *Salmonella* sp. *Staphylococcus aureus*, a significant pathogen, has been detected utilising poly(3-thiopheneacetic acid) in electrochemical biosensors (Hu et al., 2020), providing improved sensitivity and specificity. Biosensors for Escherichia coli Detection: A primary emphasis in biosensor innovation is the identification of Escherichia coli due to its common occurrence in foodborne epidemics. Optical biosensors employing poly(carboxybetaine acrylamide) were developed for this purpose (Vaisocherová-Lísalová et al., 2016). To augment electrochemical sensing capabilities, alternative polymers including poly pyrrole, polyvinyl pyrrolidone, and graphene-polymer composites have been employed (Ren et al., 2021; Jo et al., 2021). The remarkable conductivity of graphene, when integrated with polymers, enhances electrochemical responsiveness and biosensor stability. Electrochemical detection with nitrogen-doped carbonised polymers, as investigated by Shi et al. (2022), signifies an innovative method aimed at E. coli, exhibiting enhanced detection limits attributable to the nanostructured polymer interface. Detection of Pseudomonas and Campylobacter: The identification of Pseudomonas species is facilitated by advanced optical biosensors, including photonic crystal fibres (Fatema et al., 2020), which allow for accurate optical signal modulations in response to pathogen presence. Surface imprinted polymers offer mechanical sensing methods for Campylobacter jejuni, utilising molecular recognition through imprinting technologies (Ali et al., 2020; Soni et al., 2018). This approach provides structural complementarity to the pathogen surface, hence enhancing specificity. *Enterobacter vibrio* pathogen in infant formula contamination, is identified utilising chitosan-based optical biosensors, which exploit chitosan's biocompatibility and film-forming characteristics. Polydimethylsiloxane is utilised in electrochemical chemiluminescence biosensors for the detection of *Vibrio parahaemolyticus*, providing a dual detection method with enhanced sensitivity (Ali et al., 2020; Soni et al., 2018). Innovative material combinations, including polydopamine-polyethyleneimine and polydopamine composites, are utilised in biosensors aimed at detecting Pseudomonas aeruginosa. Both optical and electrochemical biosensing modalities are examined, as evidenced by research conducted by Ali et al. (2020) and Fu et al. (2020). These materials enable adaptable sensor architectures that exhibit swift responsiveness and strong pathogen affinity. Microfluidic and Paper-Based Platforms: Microfluidic paper-based analytical devices (μPADs) integrated with polyvinyl carbon have been devised for optical detection of Cronobacter sakazakii (Zhong et al., 2020). This method represents advancement in the development of portable, economical diagnostic instruments appropriate for real-time food safety assessment.

Table 2. Polymer-based biosensors for detection of pathogenic bacteria

Category	Polymeric material	Biosensor type	References
Staphylococcus typhimurium	Poly-L-lysine	Optical	Sheikhzadeh et al. 2016
	Poly(styrene/acrylamide)	Optical	Ding et al. 2022
Staphylococcus aureus	Poly(3-thiopheneacetic acid)	Electrochemical	Hu et al. 2020
Salmonella sp.	Poly(carboxybetaine acrylamide)	Optical	Wang et al. 2021

continued on following page

Table 2. Continued

Category	Polymeric material	Biosensor type	References
Escherichia coli	poly(carboxybetaine acrylamide)	Optical	Vaisocherová-Lísalová et al. 2016
	N-methyl-2-pyrro-lidonecarbonized polymer	Electrochemical	Vaisocherová-Lísalová et al. 2016
	Polypyrrole	Electrochemical	Shi et al. 2022
	Poly(polyvinyl pyrrolidone)	Electrochemical	Ren et al. 2021
	Graphene-polymer	Electrochemical	Jo et al. 2021
Pseudomonas sp.	Photonic crystal fiber	Optical	Fatema et al. 2020
Campylobacter jejuni	Surface imprinted polymers	Mechanical	Ali et al. 2020; Soni et al. 2018
Enterobacter sakazakii	Chitosan	Optical	Ali et al. 2020; Soni et al. 2018
Vibrio parahaemolyticus	Polydimethylsiloxane	Electrochemical Chemiluminesence (ELC)	Ali et al. 2020; Soni et al. 2018
Pseudomonas aeruginosa	Polydopamine-polyethyleneimine	Optical	Ali et al. 2020; Zhu et al. 2016
	Poly-dopamine	Electrochemical	Ali et al. 2020; Fu et al. 2020
Cronobacter sakazakii	μPADs +Polyvinyl carbon	Optical	Zhong et al. 2020

5.2. Polymer-Based Biosensor for the Detection of Food Allergens

Allergens found in foods may cause the immune system to react abnormally. Food allergies are complicated disorders with substantial individual variability, and are strongly correlated with the function of a person's immune system (Kalita et al., 2023). While minor food allergies may cause symptoms such as lip or facial swelling, Rashes, tightness in the throat, shaking, vomiting, stomach pain, or diarrhea, severe reactions can be life-threatening. Certain foods possess allergies, and among those that do, only a small number cause most allergy reactions. Similar meals, especially those made from plants, often contain the same allergens. For instance, people allergic to peanuts may behave differently to other legumes. The most common allergies have been identified in crustaceans, eggs, milk, dairy products, and peanuts (Sarabaegi et al., 2021). So, it's important to identify food issues. Table 3 summarises how polymer-based biosensors are used to detect allergens. According to Mohamad et al. (2020), electrochemical biosensors based on polyaniline may detect tropomyosin, a significant allergen present in seafood. Because of its biocompatibility and high electrical conductivity, polyaniline is a good choice for this purpose. Mohamad et al. (2019) also created optical biosensors with poly-dopamine polymers, which boost detection by taking advantage of dopamine's chemical stability and sticky characteristics.

Electrochemical biosensors using poly-dopamine were introduced by Jiang et al. (2024) to detect parvalbumins, which are abundant in fish. One of the main milk allergens, casein (Bos d 8), has been detected by electrochemical biosensors made of gelatin and carbon nanofiber composites (Jiang et al., 2019). The carbon nanofibers' increased surface area and conductivity allow for very sensitive detection.The presence of β-lactoglobulin, a typical milk allergy, can be detected using a variety of biosensors. In order to achieve efficient binding, Wang et al. (2021) used polyethyleneimine, which is polycationic, for electrochemical sensing. Amor-Gutiérrez et al. (2020) also found that electrochemical biosensors based on poly-lysine allowed for sensitive detection because of the high surface functionalisation of these sensors. The biocompatibility and film-forming capabilities of chitosan have led to its use in electrochemical biosensors for the detection of β-lactoglobulin.A fluorescence polarisation biosensor was created by Chen et al. (2017) to detect lactoferrin and lysozyme. This biosensor makes use of a bivalent aptamer-functionalized fluorescein isothiocyanate dye, enabling quick and accurate identification. Graphene oxide and gold nanoparticles are used in electrochemical biosensors for lysozyme detection (Erdoğan et al., 2023), which improves electron transfer and signal sensitivity due to their synergistic effects. Using chitosan and poly-dopamine in electrochemical biosensors, ovaalbumin can be detected. Ovalbumin is a prominent egg white protein allergy. The film-forming process is made easier by chitosan's inherent characteristics, and the detection capabilities are improved by poly-dopamine's stable film adhesion and chemical reactivity. Electrochemical biosensors based on polydimethylsiloxane are used for the detection of lectins and other allergens. Two types of peanut allergens, Ara h2 and Ara h1, are detected by different methods: one uses chitosan-based electrochemical biosensors, while the other uses poly-vinyl alcohol (Wang et al., 2022). Lastly, sophisticated electrochemical biosensors employ Cd quantum dot tags to detect complex allergens such as immunoglobulin G and casein.

Table 3. Polymer-based biosensors for the detection of food allergens

Category	Polymeric material	Biosensor type	References
Tropomyosin	Polyaniline	Electrochemical	Mohamad et al. (2020
	Poly-dopamine	Optical	Mohamad et al. (2019)
Parvalbumins	Poly-dopamine	Electrochemical	Chen et al. (2017)

continued on following page

Table 3. Continued

Category	Polymeric material	Biosensor type	References
Casein (Bos d 8)	Gelatin/carbon nanofiber	Electrochemical	Jiang et al. (2019)
-lactoglobulin	Chitosan	Electrochemical	Jiang et al. (2019)
	Polyethyleneimine	Electrochemical	Wang et al. (2021)
	Poly-lysine	Electrochemical	Amor-Gutiérrez et al. (2020)
Lactoferrin	Bivalent aptamer fluorescein isothiocyanate dye	Fluorescence polarization	Chen et al. (2017)
Lysozyme	Graphene oxide+ AuNPs	Electrochemical	Erdoğan et al. (2023)
Ovalbumin	Chitosan	Electrochemical	Jiang et al. (2020)
	Poly-dopamine	Electrochemical	Fu et al. (2023)
lectins	polydimethylsiloxane	Electrochemical	Kalita et al. (2023)
Ara h2	Poly-vinyl alcohol	Electrochemical	Wang et al. (2022)
Ara h1	Chitosan	Electrochemical	Sun et al. (2015)
Casein and Immunoglobulin G	Cd quantum dot tags	Electrochemical	Jiang et al. (2020)

6. CHALLENGES AND PERSPECTIVES

As evident in the preceding sections, manufacturing components have a significant impact on the behavior of natural polymer-based degradable films. However, despite experiments showing that they can reduce infections, deliver biologically active elements, and extend shelf life, their commercial application in food wrappers is still in its infancy, because most food studies have been restricted to plastic products and other artificial substances. The main drawback of biodegradable films is their poor mechanical characteristics. Poor mechanical characteristics, including tear strength, pressure at break, and flexibility modulus, were measured using tensile tests. The hydrophilic properties are also important because they may result in increased liquid permeability. To improve the tensile strength of the film, more polymers can be combined with it, and additives can be added. The natural polymer extraction method determines water absorption and barrier attributes. To improve the film, adequate extraction and addition of suitable additives and monomers may be helpful. Further investigation is required to compare various polymer resources, such as conventional ones, to advance the discussion and to make it easier to correlate these aspects. Food technology and science continue to investigate alternative polymers for producing edible films. In addition to investigating natural polymers, further studies are required to scale up for industrial applications, better understand various film-forming technologies, and boost the mechanical characteristics and water tolerance to values closer to those of synthetic plastics. Sustainable packaging will be crucial if all of these problems are resolved because it will hasten the decomposition of plastics owing to their biodegradable nature.

7. CONCLUSION

This study evaluated current status of research on biodegradable films and their uses in food packaging. In addition to their recognised environmental benefits, biodegradable films enhance food safety by reducing microbiological, chemical, and physical pollutants during storage and transport. Their capacity to prolong shelf life, mitigate contamination risks, and offer visual cues of food freshness underscores their multifaceted purpose in contemporary food packaging. Despite constraints including mechanical vulnerabilities and moisture susceptibility, the integration of antimicrobial agents, nanoparticles, and functional additives might augment their protective effectiveness against various pollutants.

To facilitate the transfer of biodegradable films from specialised uses to widespread industrial implementation, effective commercial strategies and supporting policy frameworks are needed. This encompasses governmental incentives, standardisation regulations, and industry-wide cooperation designed to enhance manufacturing scalability, guarantee safety standards, and bolster customer trust. Chemists, materials scientists, and microbiologists need to work together to tackle these issues and expedite the creation of durable, commercially viable biodegradable packaging solutions.

Declaration of interests

The authors declare no competing financial interests or personal relationships that could influence the work reported in this study.

Funding

No funding.

ACKNOWLEDGMENT

The authors are grateful to the authorities of the Vels Institute of Science, Technology, and Advanced Studies (VISTAS), Chennai for providing the necessary facilities and support.

REFERENCES

Abd Rahim, M. H., Hazrin-Chong, N. H., Harith, H. H., Wan, W. A., & Sukor, R. (2023). Roles of fermented plant-, dairy-and meat-based foods in the modulation of allergic responses. *Food Science and Human Wellness*, *12*(3), 691–701. DOI: 10.1016/j.fshw.2022.09.002

Abdul Khalil, Tye, Leh, Saurabh, Ariffin, Mohammad Fizree, & Suriani. (2018). Cellulose reinforced biodegradable polymer composite film for packaging applications. *Bionanocomposites for packaging applications*. 49-69. .DOI: 10.1007/978-3-319-67319-6_3

Acevedo-Fani, A., & Singh, H. (2021). Biopolymer interactions during gastric digestion: Implications for nutrient delivery. *Food Hydrocolloids*, *116*, 106644. DOI: 10.1016/j.foodhyd.2021.106644

Acevedo-Fani, A., Soliva-Fortuny, R., & Martín-Belloso, O. (2017). Nanoemulsions as edible coatings. *Current Opinion in Food Science*, *15*, 43–49. DOI: 10.1016/j.cofs.2017.06.002

Adeghizadeh-Yazdi, Habibi, Kamali, & Banaei. (2019). Application of edible anSd biodegradable starch-based films in food packaging: A systematic review and meta-analysis. *Current research in nutrition and food science journal, 7*(3), 624-637.

Alamri, M. S., Qasem, A. A., Mohamed, A. A., Hussain, S., Ibraheem, M. A., Shamlan, G., & Qasha, A. S. (2021). Food packaging's materials: A food safety perspective. *Saudi Journal of Biological Sciences*, *28*(8), 4490–4499. DOI: 10.1016/j.sjbs.2021.04.047 PMID: 34354435

Alexandre, E. M., Pinto, C. A., Moreira, S. A., Pintado, M., & Saraiva, J. A. (2019). Nonthermal food processing/preservation technologies. In *Saving food* (pp. 141–169). Academic Press. DOI: 10.1016/B978-0-12-815357-4.00005-5

Ali, A. A., Altemimi, A. B., Alhelfi, N., & Ibrahim, S. A. (2020). Application of biosensors for detection of pathogenic food bacteria: A review. *Biosensors (Basel)*, *10*(6), 58. DOI: 10.3390/bios10060058 PMID: 32486225

Amor-Gutiérrez, O., Selvolini, G., Fernández-Abedul, M. T., de la Escosura-Muñiz, A., & Marrazza, G. (2020). Folding-based electrochemical aptasensor for the determination of β-lactoglobulin on poly-L-lysine modified graphite electrodes. *Sensors (Basel), 20*(8), 2349. DOI: 10.3390/s20082349 PMID: 32326088

Basak, G., Sharma, B., Parul, S., Jain, U., Mishra, R. P., & Srivastava, M. K. (2021). Strategies for food safety: A contemporary approach. *Journal of Entomology and Zoology Studies, 9*(1), 117–122.

Bharathiraja, S., Suriya, J., Krishnan, M., Manivasagan, P., & Kim, S. K. (2017). Production of enzymes from agricultural wastes and their potential industrial applications. In *Advances in food and nutrition research* (Vol. 80, pp. 125–148). Academic Press. DOI: 10.1016/bs.afnr.2016.11.003

Birania, S., Kumar, S., Kumar, N., Attkan, A. K., Panghal, A., Rohilla, P., & Kumar, R. (2022). Advances in development of biodegradable food packaging material from agricultural and agro-industry waste. *Journal of Food Process Engineering, 45*(1), e13930. DOI: 10.1111/jfpe.13930

Bodenhamer, W. T., Jackowski, G., & Davies, E. (2004). Inventors; Toxin Alert Inc, assignee. Surface binding of an immunoglobulin to a flexible polymer using a water soluble varnish matrix. United States patent US 6, 692, 973. 02.17.

Briassoulis, D. (2004). An overview on the mechanical behaviour of biodegradable agricultural films. *Journal of Polymers and the Environment, 12*(2), 65–81. DOI: 10.1023/B:JOOE.0000010052.86786.ef

Briassoulis, H. (2004). The institutional complexity of environmental policy and planning problems: The example of Mediterranean desertification. *Journal of Environmental Planning and Management, 47*(1), 115–135. DOI: 10.1080/0964056042000189835

Brody, Bugusu, Han, Sand, & McHugh. (2008). Scientific status summary: innovative food packaging solutions. .DOI: 10.1111/j.1750-3841.2008.00933.x

Cachon, R., Girardon, P., & Voilley, A. (Eds.). (2019). *Gases in agro-food processes*. Academic Press.

Chen, Z., Li, H., Jia, W., Liu, X., Li, Z., Wen, F., Zheng, N., Jiang, J., & Xu, D. (2017). Bivalent aptasensor based on silver-enhanced fluorescence polarization for rapid detection of lactoferrin in milk. *Analytical Chemistry, 89*(11), 5900–5908. DOI: 10.1021/acs.analchem.7b00261 PMID: 28467701

Chiellini, E., Barghini, A., Cinelli, P., & Ilieva, V. I. (2008). Overview of environmentally compatible polymeric materials for food packaging. In *Environmentally compatible food packaging* (pp. 371–395). Woodhead Publishing. DOI: 10.1533/9781845694784.3.371

Dai, L., Cheng, T., Duan, C., Zhao, W., Zhang, W., Zou, X., & Ni, Y. (2019). 3D printing using plant-derived cellulose and its derivatives: A review. *Carbohydrate Polymers, 203*, 71–86. DOI: 10.1016/j.carbpol.2018.09.027 PMID: 30318237

David, G., Gontard, N., & Angellier-Coussy, H. (2019). Mitigating the impact of cellulose particles on the performance of biopolyester-based composites by gas-phase esterification. *Polymers, 11*(2), 200. DOI: 10.3390/polym11020200 PMID: 30960185

Deshmukh, A. R., Aloui, H., Khomlaem, C., Negi, A., Yun, J. H., Kim, H. S., & Kim, B. S. (2021). Biodegradable films based on chitosan and defatted Chlorella biomass: Functional and physical characterization. *Food Chemistry, 337*, 127777. DOI: 10.1016/j.foodchem.2020.127777 PMID: 32799163

Díaz-Montes, E. (2022). Polysaccharide-based biodegradable films: An alternative in food packaging. *Polysaccharide., 3*(4), 761–775. DOI: 10.3390/polysaccharides3040044

Díaz-Montes, E., & Castro-Muñoz, R. (2021). Trends in chitosan as a primary biopolymer for functional films and coatings manufacture for food and natural products. *Polymers, 13*(5), 767. DOI: 10.3390/polym13050767 PMID: 33804445

Ding, S., Hu, H., Yue, X., Feng, K., Gao, X., Dong, Q., Yang, M., Tamer, U., Huang, G., & Zhang, J. (2022). A fluorescent biosensor based on quantum dot–labeled streptavidin and poly-l-lysine for the rapid detection of Salmonella in milk. *Journal of Dairy Science, 105*(4), 2895–2907. DOI: 10.3168/jds.2021-21229 PMID: 35181133

Domínguez-Manzano, J., Olmo-Ruiz, C., Bautista-Gallego, J., Arroyo-López, F. N., Garrido-Fernández, A., & Jiménez-Díaz, R. (2012). Biofilm formation on abiotic and biotic surfaces during Spanish style green table olive fermentation. *International Journal of Food Microbiology, 157*(2), 230–238. DOI: 10.1016/j.ijfoodmicro.2012.05.011 PMID: 22656327

Duque-Acevedo, M., Belmonte-Ureña, L. J., Cortés-García, F. J., & Camacho-Ferre, F. (2020). Agricultural waste: Review of the evolution, approaches and perspectives on alternative uses. *Global Ecology and Conservation, 22*, e00902. DOI: 10.1016/j.gecco.2020.e00902

Ehrlich, H., Shaala, L. A., Youssef, D. T., Żółtowska-Aksamitowska, S., Tsurkan, M., Galli, R., & Jesionowski, T. (2018). Discovery of chitin in skeletons of non-verongiid Red Sea demosponges. *PLoS One, 13*(5), e0195803. DOI: 10.1371/journal.pone.0195803 PMID: 29763421

Eliasson, A. C. (2004). Starch-lipid interactions and their relevance in food products AC Eliasson and M. Wahlgren, Lund University, Sweden. *Starch in food: Structure, function and applications.* 441.

Erdoğan, N. Ö., Uslu, B., & Aydoğdu Tığ, G. (2023). Development of an electrochemical biosensor utilizing a combined aptamer and MIP strategy for the detection of the food allergen lysozyme. *Microchimica Acta, 190*(12), 471. DOI: 10.1007/s00604-023-06054-w PMID: 37975892

Erdogan, S. (2024). On the impact of natural resources on environmental sustainability in African countries: A comparative approach based on the EKC and LCC hypotheses. *Resources Policy, 88*, 104492. DOI: 10.1016/j.resourpol.2023.104492

Europe, P. (2015). An analysis of European plastics production, demand and waste data. *Plastics–the facts*, 147.

Fatema, K. N., Biswas, M. R., Bang, S. H., Cho, K. Y., & Oh, W. C. (2020). Electro analytical characteristic of a novel biosensor designed with graphene–polymer-based quaternary and mesoporous nanomaterials. *Bulletin of Materials Science, 12*(43), 1–3.

FDA. (2023). Food Ingredients & Packaging. Available online: https://www.fda.gov/food/food-ingredients-packaging

Freitas, C. M. P., Coimbra, J. S. R., Souza, V. G. L., & Sousa, R. C. S. (2021). Structure and applications of pectin in food, biomedical, and pharmaceutical industry: A review. *Coatings, 11*(8), 922. DOI: 10.3390/coatings11080922

Fu, H., Bai, Z., Li, P., Feng, X., Hu, X., Song. And, X., & Chen, L. (2023). Molecular imprinted electrochemical sensor for ovalbumin detection based on boronate affinity and signal amplification approach. *Food Chemistry, 409*, 135292. DOI: 10.1016/j.foodchem.2022.135292 PMID: 36584533

Fu, K., Zhang, H., Guo, Y., Li, J., Nie, H., Song, X., Xu, K., Wang, J., & Zhao, C. (2020). Rapid and selective recognition of Vibrio parahaemolyticus assisted by perfluorinated alkoxysilane modified molecularly imprinted polymer film. *RSC Advances, 10*(24), 14305–14312. DOI: 10.1039/D0RA00306A PMID: 35498485

Galus, S., Arik Kibar, E. A., Gniewosz, M., & Kraśniewska, K. (2020). Novel materials in the preparation of edible films and coatings—A review. *Coatings, 10*(7), 674. DOI: 10.3390/coatings10070674

Givanoudi, S., Cornelis, P., Rasschaert, G., Wackers, G., Iken, H., Rolka, D., Yongabi, D., Robbens, J., Schöning, M. J., Heyndrickx, M., & Wagner, P. (2021). Selective Campylobacter detection and quantification in poultry: A sensor tool for detecting the cause of a common zoonosis at its source. *Sensors and Actuators. B, Chemical, 332*, 129484. DOI: 10.1016/j.snb.2021.129484

Gómez-Arribas, L. N., Benito-Peña, E., Hurtado-Sanchez, M. D., & Moreno-Bondi, M. C. (2018). Biosensing based on nanoparticles for food allergens detection. *Sensors (Basel), 18*(4), 1087. DOI: 10.3390/s18041087 PMID: 29617319

Heinze, El Seoud, Koschella, Heinze, El Seoud, & Koschella. (2018). Structure and properties of cellulose and its derivatives. *Cellulose derivatives: synthesis, structure, and properties*, 39-172. DOI: 10.1007/978-3-319-73168-1_2

Hong, S. P., Mohd-Naim, N. F., Keasberry, N. A., & Ahmed, M. U. (2022). Electrochemical Detection of β-Lactoglobulin Allergen Using Titanium Dioxide/Carbon Nanochips/Gold Nanocomposite-based Biosensor. *Electroanalysis, 34*(4), 684–691. DOI: 10.1002/elan.202100207

Hu, J., Tang, F., Jiang, Y. Z., & Liu, C. (2020). Rapid screening and quantitative detection of Salmonella using a quantum dot nanobead-based biosensor. *Analyst, 145*(6), 2184–2190. DOI: 10.1039/D0AN00035C PMID: 32101227

Jahan, N., & Rahman, M. M. (2021). Photonic crystal fiber based biosensor for pseudomonas bacteria detection: A simulation study. *IEEE Access,* 9, 42206-15. DOI: 10.1109/ACCESS.2021.3063691

Jiang, D., Feng, Z., Jiang, H., Cao, H., Xiang, X., & Wang, L. (2024). 3D bioprinting-based vascular-microtissue electrochemical biosensor for fish parvalbumin detection. *Food Chemistry, 18*, 138799. DOI: 10.1016/j.foodchem.2024.138799 PMID: 38401313

Jiang, D., Ge, P., Wang, L., Jiang, H., Yang, M., Yuan, L., Ge, Q., Fang, W., & Ju, X. (2019). A novel electrochemical mast cell-based paper biosensor for the rapid detection of milk allergen casein. *Biosensors & Bioelectronics, 130*, 299–306. DOI: 10.1016/j.bios.2019.01.050 PMID: 30776617

Jiang, D., Jiang, H., & Wang, L. (2020). A novel paper-based capacitance mast cell sensor for evaluating peanut allergen protein Arah2. *Food Analytical Methods, 13*(10), 1993–2001. DOI: 10.1007/s12161-020-01769-5

Jimenez, A., Fabra, M. J., Talens, P., & Chiralt, A. (2012). Edible and biodegradable starch films: A review. *Food and Bioprocess Technology, 5*(6), 2058–2076. DOI: 10.1007/s11947-012-0835-4

Jo, Robby, Kim, Lee, Lee, & Park. (2021). Reusable biosensor-based polymer dot-coated electrode surface for wireless detection of bacterial contamination. *Sensors and Actuators B: Chemical*. 1, 346:130503. DOI: 10.1016/j.snb.2021.130503

Kaewprachu, P., Osako, K., & Rawdkuen, S. (2018). Effects of plasticizers on the properties of fish myofibrillar protein film. *Journal of Food Science and Technology*, *55*(8), 3046–3055. DOI: 10.1007/s13197-018-3226-7 PMID: 30065414

Kalita, J. J., Sharma, P., & Bora, U. (2023). Recent developments in application of nucleic acid aptamer in food safety. *Food Control*, *145*, 109406. DOI: 10.1016/j.foodcont.2022.109406

Kang, Shi, Sun, Dan, Liang, Zhang, Su, Wang, & Zhang. (2022). Magnetic Nano separation technology for efficient control of Microorganisms and Toxins in Foods: A Review. *Journal of Agricultural and Food Chemistry*, 70(51), 16050-68. .DOI: 10.1021/acs.jafc.2c07132

Kaur, P., Devgan, K., Kumar, N., Kaur, A., Kumar, M., & Sandhu, K. (2021). Quality retention and shelf-life prolongation of cucumbers (Cucumis sativus L.) under different cool storage systems with passive modified atmosphere bulk packaging. *Packaging Technology & Science*, *34*(9), 567–578. DOI: 10.1002/pts.2595

Kerry, O'Grady, & Hogan. (2006). Past, current and potential utilisation of active and intelligent packaging systems for meat and muscle-based products: A review. *Meat science*, 74(1), 113-30. DOI: 10.1016/j.meatsci.2006.04.024

Kerry, J. P., O'grady, M. N., & Hogan, S. A. (2006). Past, current and potential utilisation of active and intelligent packaging systems for meat and muscle-based products: A review. *Meat Science*, *74*(1), 113–130. DOI: 10.1016/j.meatsci.2006.04.024 PMID: 22062721

Klinger, C., Żółtowska-Aksamitowska, S., Wysokowski, M., Tsurkan, M. V., Galli, R., Petrenko, I., & Ehrlich, H. (2019). Express method for isolation of ready-to-use 3D chitin scaffolds from Aplysina archeri (Aplysineidae: Verongiida) Demosponge. *Marine Drugs*, *17*(2), 131. DOI: 10.3390/md17020131 PMID: 30813373

Kokkinos, C., Angelopoulou, M., Economou, A., Prodromidis, M., Florou, A., Haasnoot, W., Petrou, P., & Kakabakos, S. (2016). Lab-on-a-membrane foldable devices for duplex drop-volume electrochemical biosensing using quantum dot tags. *Analytical Chemistry*, *88*(13), 6897–6904. DOI: 10.1021/acs.analchem.6b01625 PMID: 27257985

Kumar, M. N. R. (2000). A review of chitin and chitosan applications. *Reactive & Functional Polymers*, *46*(1), 1–27. DOI: 10.1016/S1381-5148(00)00038-9

Kumar, N., Kaur, P., & Bhatia, S. (2017). Advances in bio-nanocomposite materials for food packaging: A review. *Nutrition & Food Science*, *47*(4), 591–606. DOI: 10.1108/NFS-11-2016-0176

Lau, T. C., Chan, M. W., Tan, H. P., & Kwe, C. L. (2013). Functional food: A growing trend among the health conscious. *Asian Social Science*, *9*(1), 198. DOI: 10.5539/ass.v9n1p198

Liu, Y., Ahmed, S., Sameen, D. E., Lu, Y., Wang, R., Dai, J., & Qin, W. (2021). A review of cellulose and its derivatives in biopolymer-based for food packaging application. *Trends in Food Science & Technology*, *112*, 532–546. DOI: 10.1016/j.tifs.2021.04.016

Mlalila, N., Kadam, D. M., Swai, H., & Hilonga, A. (2016). Transformation of food packaging from passive to innovative via nanotechnology: Concepts and critiques. *Journal of Food Science and Technology*, *53*(9), 3395–3407. DOI: 10.1007/s13197-016-2325-6 PMID: 27777446

Mohamad, A., Rizwan, M., Keasberry, N. A., & Ahmed, M. U. (2019). Fabrication of label-free electrochemical food biosensor for the sensitive detection of ovalbumin on nanocomposite-modified graphene electrode. *Biointerface Research in Applied Chemistry*, *9*(6), 4655–4662. DOI: 10.33263/BRIAC96.655662

Mohamad, A., Rizwan, M., Keasberry, N. A., Nguyen, A. S., Dai Lam, T., & Ahmed, M. U. (2020). Gold-microrods/Pd-nanoparticles/polyaniline-nanocomposite-interface as a peroxidase-mimic for sensitive detection of tropomyosin. *Biosensors & Bioelectronics*, *155*, 112108. DOI: 10.1016/j.bios.2020.112108 PMID: 32217333

Mohamad, A. A. (2019). Physical properties of quasi-solid-state polymer electrolytes for dye-sensitised solar cells: A characterisation review. *Solar Energy*. DOI: 10.1016/j.foodchem.2016.07.004

Morais, S., Tortajada-Genaro, L. A., Maquieira, A., & Martinez, M. A. (2020). Biosensors for food allergy detection according to specific IgE levels in serum. *Trends in Analytical Chemistry*, *127*, 115904. DOI: 10.1016/j.trac.2020.115904

Muñoz, I., Rodríguez, C., Gillet, D., & Moerschbacher, B. (2018). Life cycle assessment of chitosan production in India and Europe. *The International Journal of Life Cycle Assessment*, *23*, 1151–1160. DOI: 10.1007/s11367-017-1290-2

Obi, F. O., Ugwuishiwu, B. O., & Nwakaire, J. N. (2016). Agricultural waste concept, generation, utilization and management. *Nigerian Journal of Technology*, *35*(4), 957–964. DOI: 10.4314/njt.v35i4.34

Park, Y. W., Kim, S. M., Lee, J. Y., & Jang, W. (2015). Application of biosensors in smart packaging. *Molecular & Cellular Toxicology*, *09*(11), 277–285. DOI: 10.1007/s13273-015-0027-1

Peelman, N., Ragaert, P., De Meulenaer, B., Adons, D., Peeters, R., Cardon, L., & Devlieghere, F. (2013). Application of bioplastics for food packaging. *Trends in Food Science & Technology*, *32*(2), 128–141. DOI: 10.1016/j.tifs.2013.06.003

Peelman, N., Ragaert, P., De Meulenaer, B., Adons, D., Peeters, R., Cardon, L., Van Impe, F., & Devlieghere, F. (2013). Application of bioplastics for food packaging. *Trends in Food Science & Technology*, *32*(2), 128–141. DOI: 10.1016/j.tifs.2013.06.003

Ren, Zhang, Li, Jian, Zhao, & Song. (2021). Development of a pulse-induced electrochemical biosensor based on gluconamide for Gram-negative bacteria detection. *Microchimica Acta*, 11, 188:399. DOI: 10.1016/j.bioelechem.2022.108226

Rodsamran, P., & Sothornvit, R. (2019). Lime peel pectin integrated with coconut water and lime peel extract as a new bioactive film sachet to retard soybean oil oxidation. *Food Hydrocolloids*, *97*, 105173. DOI: 10.1016/j.foodhyd.2019.105173

Rouhi, M., Razavi, S. H., & Mousavi, S. M. (2017). Optimization of crosslinked poly (vinyl alcohol) nanocomposite films for mechanical properties. *Materials Science and Engineering C*, *71*, 1052–1063. DOI: 10.1016/j.msec.2016.11.135 PMID: 27987659

Roy, S. J. W. R., & Rhim, J.-W. (2022). Genipin-crosslinked gelatin/chitosan-based functional films incorporated with rosemary essential oil and quercetin. *Materials (Basel)*, *15*(11), 3769. DOI: 10.3390/ma15113769 PMID: 35683069

Santos, V. P., Marques, N. S., Maia, P. C., Lima, M. A. B. D., Franco, L. D., & Campos-Takaki, G. M. D. (2020). Seafood waste as attractive source of chitin and chitosan production and their applications. *International Journal of Molecular Sciences*, *21*(12), 4290. DOI: 10.3390/ijms21124290 PMID: 32560250

Sanyang, M. L., Sapuan, S. M., Jawaid, M., Ishak, M. R., & Sahari, J. (2016). Effect of plasticizer type and concentration on physical properties of biodegradable films based on sugar palm (Arenga pinnata) starch for food packaging. *Journal of Food Science and Technology*, *53*(1), 326–336. DOI: 10.1007/s13197-015-2009-7 PMID: 26787952

Sarabaegi, M., & Roushani, M. (2018). Rapid and sensitive determination of Pseudomonas aeruginosa by using a glassy carbon electrode modified with gold nanoparticles and aptamer-imprinted polydopamine. *Microchemical Journal, 168*, 106388. DOI: 10.1016/j.microc.2021.106388

Sarabaegi, M., & Roushani, M. (2021). Rapid and sensitive determination of Pseudomonas aeruginosa by using a glassy carbon electrode modified with gold nanoparticles and aptamer-imprinted polydopamine. *Microchemical Journal, 168*, 106388. DOI: 10.1016/j.microc.2021.106388

Shaala, L. A., Asfour, H. Z., Youssef, D. T., Żółtowska-Aksamitowska, S., Wysokowski, M., Tsurkan, M., & Ehrlich, H. (2019). New source of 3D chitin scaffolds: The Red Sea demosponge Pseudoceratina arabica (Pseudoceratinidae, Verongiida). *Marine Drugs*, *17*(2), 92. DOI: 10.3390/md17020092 PMID: 30717221

Sheikhzadeh, E., Hamsaz, M. C., Turner, A. P., Jager, E. W., & Beni, V. (2016). Label-free impedimetric biosensor for *Salmonella Typhimurium* detection based on poly [pyrrole-co-3-carboxyl-pyrrole] copolymer supported aptamer. *Biosensors & Bioelectronics*, *15*(80), 194–200. DOI: 10.1016/j.bios.2016.01.057 PMID: 26836649

Shi, Wang, Yan, Wang, Niu, & Wang. (2022). In-situ growth of nitrogen-doped carbonized polymer dots on black phosphorus for electrochemical DNA biosensor of Escherichia coli O157: H7. *Bioelectrochemistry*, 1, 148:108226. DOI: 10.1016/j.bioelechem.2022.108226

Shit, S. C., & Shah, P. M. (2014). Edible polymers: Challenges and opportunities. *Journal of Polymers, 2014*, 1–13. DOI: 10.1155/2014/427259

Singh, S., Gaikwad, K. K., & Lee, Y. S. (2018). Anthocyanin-A natural dye for smart food packaging systems. *Korean Journal of Packaging Science & Technology*, *24*(3), 167–180. DOI: 10.20909/kopast.2018.24.3.167

Siracusa, V., Rocculi, P., Romani, S., & Dalla Rosa, M. (2008). Biodegradable polymers for food packaging: A review. *Trends in Food Science & Technology*, *19*(12), 634–643. DOI: 10.1016/j.tifs.2008.07.003

Soni, D. K., Ahmad, R., & Dubey, S. K. (2018). Biosensor for the detection of Listeria monocytogenes: Emerging trends. *Critical Reviews in Microbiology*, *44*(5), 590–608. DOI: 10.1080/1040841X.2018.1473331 PMID: 29790396

Sorrentino, A., Gorrasi, G., & Vittoria, V. (2007). Potential perspectives of bionanocomposites for food packaging applications. *Trends in Food Science & Technology*, *18*(2), 84–95. DOI: 10.1016/j.tifs.2006.09.004

Stoma, M., & Dudziak, A. (2022). Eastern Poland Consumer Awareness of Innovative Active and Intelligent Packaging in the Food Industry: Exploratory Studies. *Sustainability (Basel), 14*(20), 13691. DOI: 10.3390/su142013691

Suderman, N., Isa, M. I. N., & Sarbon, N. M. (2018). The effect of plasticizers on the functional properties of biodegradable gelatin-based film: A review. *Food Bioscience, 24*, 111–119. DOI: 10.1016/j.fbio.2018.06.006

Sun, L., Jiang, Y., Pan, R., Li, M., Wang, R., Chen, S., Fu, S., & Man, C. (2018). A novel, simple and low-cost paper-based analytical device for colorimetric detection of Cronobacter spp. *Analytica Chimica Acta, 7*(1036), 80–88. DOI: 10.1016/j.aca.2018.05.061 PMID: 30253840

Sun, X., Jia, M., Ji, J., Guan, L., Zhang, Y., Tang, L., & Li, Z. (2015). Enzymatic amplification detection of peanut allergen Ara h1 using a stem-loop DNA biosensor modified with a chitosan-mutiwalled carbon nanotube nanocomposite and spongy gold film. *Talanta, 131*, 521–527. DOI: 10.1016/j.talanta.2014.07.078 PMID: 25281135

Tagliapietra, B. L., Felisberto, M. H. F., Sanches, E. A., Campelo, P. H., & Clerici, M. T. P. S. (2021). Non-conventional starch sources. *Current Opinion in Food Science, 39*, 93–102. DOI: 10.1016/j.cofs.2020.11.011

Tharanathan, R. N. (2003). Biodegradable films and composite coatings: Past, present and future. *Trends in Food Science & Technology, 14*(3), 71–78. DOI: 10.1016/S0924-2244(02)00280-7

Vaisocherová-Lísalová, H., Víšová, I., Ermini, M. L., Špringer, T., Song, X. C., Mrázek, J., Lamačová, J.Jr, Lynn, N. S., Šedivák, P., & Homola, J. (2016). Low-fouling surface plasmon resonance biosensor for multi-step detection of foodborne bacterial pathogens in complex food samples. *Biosensors & Bioelectronics, 15*(80), 84–90. DOI: 10.1016/j.bios.2016.01.040 PMID: 26807521

Valdés, A., Burgos, N., Jiménez, A., & Garrigós, M. C. (2015). Natural pectin polysaccharides as edible coatings. *Coatings, 5*(4), 865–886. DOI: 10.3390/coatings5040865

Wang, B., Hong, J., Liu, C., Zhu, L., & Jiang, L. (2021). An electrochemical molecularly imprinted polymer sensor for rapid β-Lactoglobulin detection. *Sensors (Basel), 21*(24), 8240. DOI: 10.3390/s21248240 PMID: 34960338

Wang, Wang, Yan, Luan, Wu, & Bian. (2021). Rapid, sensitive and label-free detection of pathogenic bacteria using a bacteria-imprinted conducting polymer film-based electrochemical sensor. *Talanta*, 1, 226:122135.

Wang, S. L., & Nguyen, V. B. (2019). Production of potent antidiabetic compounds from shrimp head powder via Paenibacillus conversion. *Process Biochemistry (Barking, London, England)*, *76*, 18–24. DOI: 10.1016/j.procbio.2018.11.004

Wang, W., Chen, W., Lv, M., Zou, R., Wang, D., Hou, F., & Liu, D. (2018). Applications of power ultrasound in oriented modification and degradation of pectin: A review. *Journal of Food Engineering*, *234*, 98–107. DOI: 10.1016/j.jfoodeng.2018.04.016

Wang, Y., Li, L., Li, H., Peng, Y., & Fu, L. (2022). A fluorometric sandwich biosensor based on rationally imprinted magnetic particles and aptamer modified carbon dots for the detection of tropomyosin in seafood products. *Food Control*, *132*, 108552. DOI: 10.1016/j.foodcont.2021.108552

Wanihsuksombat, C., Hongtrakul, V., & Suppakul, P. (2010). Development and characterization of a prototype of a lactic acid–based time–temperature indicator for monitoring food product quality. *Journal of Food Engineering*, *100*(3), 427–434. DOI: 10.1016/j.jfoodeng.2010.04.027

Wilson, P. D. C.L. (Ed.). (2007). Intelligent and Active Packaging for Fruits and Vegetables (1st ed.). CRC Press. DOI: 10.1201/9781420008678

Wysokowski, M., Bazhenov, V. V., Tsurkan, M. V., Galli, R., Stelling, A. L., Stöcker, H., & Ehrlich, H. (2013). Isolation and identification of chitin in three-dimensional skeleton of Aplysina fistularis marine sponge. *International Journal of Biological Macromolecules*, *62*, 94–100. DOI: 10.1016/j.ijbiomac.2013.08.039 PMID: 23994783

Xu, X., Zhou, X. D., & Wu, C. D. (2012). Tea catechin epigallocatechin gallate inhibits Streptococcus mutans biofilm formation by suppressing gtf genes. *Archives of Oral Biology*, *57*(6), 678–683. DOI: 10.1016/j.archoralbio.2011.10.021 PMID: 22169220

Xu, Y. X., Kim, K. M., Hanna, M. A., & Nag, D. (2005). Chitosan–starch composite film: Preparation and characterization. *Industrial Crops and Products*, *21*(2), 185–192. DOI: 10.1016/j.indcrop.2004.03.002

Yang, A., Zheng, Y., Long, C., Chen, H., Liu, B., Li, X., Yuan, J., & Cheng, F. (2014). Fluorescent immunosorbent assay for the detection of alpha lactalbumin in dairy products with monoclonal antibody bioconjugated with CdSe/ZnS quantum dots. *Food Chemistry*, *150*, 73–79. DOI: 10.1016/j.foodchem.2013.10.137 PMID: 24360421

Zhang, Zhou, Li, Fan, Li, Lu, Wen, & Ren. (2023). Recent advances of fluorescent sensors for bacteria detection-A review. *Talanta,* 3(1), 254:124133. .DOI: 10.1016/j.talanta.2022.124133

Zhang, J., Zou, X., Zhai, X., Huang, X., Jiang, C., & Holmes, M. (2019). Preparation of an intelligent pH film based on biodegradable polymers and roselle anthocyanins for monitoring pork freshness. *Food Chemistry*, *272*, 306–312. DOI: 10.1016/j.foodchem.2018.08.041 PMID: 30309548

Zhao, G., Du, J., Chen, W., Pan, M., & Chen, D. (2019). Preparation and thermostability of cellulose nanocrystals and nanofibrils from two sources of biomass: Rice straw and poplar wood. *Cellulose (London, England)*, *26*(16), 8625–8643. DOI: 10.1007/s10570-019-02683-8

Zhong, Y., Godwin, P., Jin, Y., & Xiao, H. (2020). Biodegradable polymers and green-based antimicrobial packaging materials: A mini-review. *Advanced Industrial and Engineering Polymer Research.*, *3*(1), 27–35. DOI: 10.1016/j.aiepr.2019.11.002

Zhong, Z., Gao, R., Chen, Q., & Jia, L. (2020). Dual-aptamers labeled polydopamine-polyethyleneimine copolymer dots assisted engineering a fluorescence biosensor for sensitive detection of Pseudomonas aeruginosa in food samples. *Spectrochimica Acta. Part A: Molecular and Biomolecular Spectroscopy*, *224*, 117417. DOI: 10.1016/j.saa.2019.117417 PMID: 31362188

Zhu, Y., & Wang, D. (2016). Rapid Detection of Enterobacter Sakazakii in milk Powder using amino modified chitosan immunomagnetic beads. *International Journal of Biological Macromolecules*, *1*(93), 615–622. DOI: 10.1016/j.ijbiomac.2016.09.024 PMID: 27616695

Zoungranan, Y., Lynda, E., Dobi-Brice, K. K., Tchirioua, E., Bakary, C., & Yannick, D. D. (2020). Influence of natural factors on the biodegradation of simple and composite bioplastics based on cassava starch and corn starch. *Journal of Environmental Chemical Engineering*, *8*(5), 104396. DOI: 10.1016/j.jece.2020.104396

Chapter 7
Utilization of Natural Compounds as Food Contaminant Mitigation Agents:
A Sustainable Approach to Global Food and Human Health Security

Kholis Kholis Ernawati
https://orcid.org/0000-0002-3582-5294
Universitas YARSI, Indonesia

Maya Trisiswati
Universitas YARSI, Indonesia

Siti Maulidya Sari
Universitas YARSI, Indonesia

Mirfat Mirfat
Universitas YARSI, Indonesia

Etty Widayanti
Universitas YARSI, Indonesia

Nur Isnaeni Evry Kurniawati
https://orcid.org/0009-0006-6736-8936
Universitas YARSI, Indonesia

ABSTRACT

Food safety has become an increasingly important global issue in recent decades, mainly due to the increasing contamination of food products. This chapter examines the potential of natural compounds derived from plants, microorganisms, and marine organisms as safer and more sustainable alternatives to synthetic chemicals in mitigating food contaminants. Compounds such as flavonoids, tannins, essential oils, and polysaccharides have shown antimicrobial, antioxidant, and detoxifying activities against heavy metals and mycotoxins. Several case studies demonstrate the success of these natural compounds in extending shelf life and reducing foodborne pathogens. However, compound stability, compositional variability, sensory

DOI: 10.4018/979-8-3373-7052-1.ch007

impact, and regulation challenges still exist. This chapter emphasizes the need for innovative and collaborative strategies to develop natural-based food solutions to build a safer, healthier, and environmentally friendly food system.

1. INTRODUCTION

Food safety is a growing global concern due to the rising contamination risks to human health. Contaminants from synthetic chemicals, pesticides, and pathogenic microorganisms require serious attention to maintain the quality and sustainability of food consumed worldwide. The increasing prevalence of foodborne diseases, which are often associated with such contamination, can significantly impact public health, reducing quality of life and socio-economic productivity.

In response to this issue, there is an increasing demand for natural food products that are safe for consumption and free from synthetic chemicals. Consumers seek healthier, eco-friendly foods, prompting the search for safe and sustainable alternatives. In this regard, natural compounds derived from plants, microorganisms, and marine organisms play an essential role. These compounds reduce contamination, promote sustainability, and limit reliance on harmful synthetic chemicals.

Flavonoids, essential oils, polysaccharides, and tannins can combat heavy metals and mycotoxins with antimicrobial and detoxifying effects. In addition, these compounds contribute to the extension of food shelf life, which is an important factor in reducing food wastage globally. Natural compounds help protect human health and support a more sustainable food system.

With the growing demand for sustainable practices in the food sector, this chapter aims to delve deeper into the potential of natural compounds in addressing the significant challenges facing the food industry. Through research and case studies, this article will examine the effectiveness of various natural compounds in reducing food contaminants and explore the challenges and solutions required for their widespread application in the industry.

Natural compounds have long been utilized in traditional medicine for their bioactive activities. Prunus africana, for example, effectively accelerates wound healing through collagen enhancement (Ndung'u et al., 2024) while Calotropis procera showed similar results in diabetic wounds (Mali et al., 2020). In South Africa, Boophane disticha and Datura stramonium have high antioxidant potential for wound treatment (Fisher et al., 2025). Combining traditional practices and modern technologies, such as nanotechnology, strengthens the role of these compounds in the health sector (Salih et al., 2022).

In food safety, natural compounds such as essential oils from oregano and thyme were shown to inhibit food pathogens and extend shelf life (Enayatifard et al., 2020;

Vaičiulienė et al., 2020).. Flavonoids in green tea also support nutritional stability (Sęczyk et al., 2021), while Artemisia is effective against pathogenic bacteria (Kahrarian et al., 2020). Its application promotes public health and environmental sustainability (Erdem et al., 2024).

The technique of encapsulating anthocyanins from food waste, such as cherry pomace, improved cake products' sensory quality and durability (Lankanayaka et al., 2024). Lactic acid bacteria produce antimicrobial compounds that are effective as natural preservatives (Ibrahim et al., 2021).. A gelatin coating with tomato antioxidants also helps maintain meat quality (Gallego et al., 2020). Meanwhile, edible films from corn zein and fennel extract offer an environmentally friendly preservation solution (Zolfaghari et al., 2023).

Against this background, this chapter aims to delve deeper into the potential of natural compounds in food contamination mitigation, focusing on their mechanisms of action in inhibiting microbial growth, binding contaminants, and extending food shelf life. In addition, this chapter will also explore the challenges and opportunities in the development and application of natural compounds in the food industry to achieve a safer, healthier, and more sustainable food system.

This chapter examines various types of natural compounds and their mechanisms of action in food contaminant mitigation, including:

1) Inactivation of Pathogenic Microbes:

This section highlights natural compounds with antimicrobial activity against various foodborne pathogens—including bacteria, fungi, and viruses—that compromise the quality and safety of food products.

2) Detoxification of Heavy Metals and Mycotoxins:

This section focuses on natural compounds capable of binding and detoxifying harmful contaminants such as heavy metals and mycotoxins, emphasizing their relevance to food safety and public health.

3) Applications Across Food Sectors:

This section explores natural compounds in various food categories, including fruits, processed meats, and agricultural produce, illustrating how they can prolong shelf life and enhance product safety.

2. THEORETICAL FRAMEWORK

This study used a literature review approach to explore the effectiveness of natural compounds in mitigating food contamination. This method reviews journal articles and reports on natural compounds in food preservation and contamination control. The search was conducted on various scientific databases and publications published in the last five years to ensure the novelty of the data and findings.

The analysis began with identifying natural compounds that have been shown to have antimicrobial, antioxidant, and detoxification activities, as well as their practical applications in the food sector. The research in this review involved laboratory tests that evaluated the effects of these compounds on microbial pathogens and heavy metals and their effect on the shelf life and organoleptic quality of food products. The data obtained from these studies were analyzed to find general trends, differences in results, and recommendations that can be implemented in the food industry.

The review also highlights the application of natural compounds in specific food products, such as fruits, vegetables, processed meats, and other agricultural products. This review compares the' effectiveness, safety, and environmental impact of natural compounds and synthetic preservatives. Using this methodology, this chapter aims to provide a more comprehensive understanding of the role of natural compounds in enhancing food safety, extending shelf life, and supporting the food industry's sustainability.

3. NATURAL COMPOUNDS AS FOOD PRESERVATIVES

Flavonoids and Phenolic Compounds

Flavonoids and phenolic compounds are natural phytochemicals in green tea, pomegranate, and cinnamon. Known for their antioxidant and antimicrobial properties, these compounds are effective in food preservation and safety. Their mechanism of action includes microbial cell membrane damage, inhibition of protein synthesis, and disruption of nucleic acid metabolism.

Flavonoids disrupt microbial membranes, causing cell lysis and death. For example, flavonoids can inhibit important microbial components, such as porins in the cell membrane, which are instrumental in nutrient transport and overall microbial function (Al-Maharik et al., 2022). Nasir-Naeem et al. explained that flavonoids inhibit hydrolytic enzymes such as proteases and form complexes with proteins in the bacterial cell wall, contributing to bacterial cell death (Nasir-Naeem et al., 2021). These bonds disrupt the structural integrity of bacteria, amplifying the antimicrobial effectiveness of flavonoids through direct and indirect pathways that immobilize microbial cells.

Among plant-derived compounds, pomegranate peel extracts are rich in phenolics such as ellagitannins, which exhibit strong antioxidant properties. These compounds help delay lipid oxidation, preserve nutritional content, and extend shelf life in various food products (El-Hadary & Taha, 2020a). The antioxidant properties of flavonoids in green tea and pomegranate have been shown to neutralize free radicals, which are harmful byproducts of oxidation reactions that can accelerate spoilage (Oliveira et al., 2020). Research shows that these compounds not only inhibit the spoilage process but also contribute to the sensory quality of food, helping to maintain flavor and visual appeal during storage (El-Hadary & Taha, 2020a; Sabraoui et al., 2020). Thus, incorporating flavonoids and phenolic compounds into the food system protects against microbial contamination and chemical degradation.

In addition to their role in antimicrobial and antioxidative actions, flavonoids and phenolic compounds also exhibit anti-quorum-sensing activity, which can prevent the formation of pathogen biofilms on food surfaces. This is particularly relevant in dealing with pathogens that use biofilms to increase their resistance to antimicrobial treatments. As highlighted by Mostafa et al., phenolic compounds and flavonoids can impair quorum sensing, which is essential for biofilm development, thereby reducing the virulence of pathogenic microorganisms (Mostafa et al., 2020). In addition, flavonoids are reported to affect energy metabolism and nucleic acid synthesis in bacteria, confirming their multifaceted mechanism of action against diverse microbial threats (Al-Maharik et al., 2022; Yao et al., 2021).

Flavonoids play a role in food preservation while maintaining flavor, texture, and nutritional value through increased bioaccessibility of nutrients (Sęczyk et al., 2021). Together with phenolic compounds, flavonoids inhibit microbes and oxidation and can potentially prevent oxidative stress-related diseases and inflammation, supporting functional diets as medical prevention (Oliveira et al., 2020). Its biological potential still requires further exploration for food technology development.

Green tea (Camellia sinensis) is rich in catechins, particularly EGCG, a powerful antioxidant that prevents lipid peroxidation and extends the shelf life of fatty foods. The highest catechin content is found in green tea compared to oolong and black, depending on the processing method (Abdel-Aal et al., 2022). Besides tea, pomegranate also contains ellagic acid, a phenolic compound with high antioxidant effects that helps maintain the quality and shelf life of the fruit. Ellagic acid is also associated with protection against cancer and heart disease and has an anti-inflammatory role that supports nutritional value during storage (Tamborlin et al., 2020). Green tea extract contributes significantly to the oxidative stability of food products. Rameshrad (2017) demonstrated that incorporating 800 ppm of green tea leaf extract into sunflower oil notably delayed oxidative degradation, as indicated by reduced peroxide values and extended shelf life compared to untreated controls. Although the study did not directly quantify microbial load reduction, the potent antioxidant properties of green tea polyphenols were identified as key contributors to preservation efficacy

Cinnamon is another ingredient that is effective in enhancing food preservation. Cinnamaldehyde, the main component found in cinnamon, exhibits remarkable antibacterial and antifungal activity, thus suppressing the growth of microbes that can cause food spoilage. These antimicrobial properties of cinnamon make it an attractive natural preservative in food applications, increasing resistance to oxidation processes and extending product shelf life. The effectiveness of cinnamaldehyde has been supported by various studies, which collectively demonstrate its application in the food industry to improve product stability both microbiologically and structurally (Naik et al., 2021).

Understanding the chemical stability and interactions between these compounds in various environments is essential to maximize their effectiveness in food science. Studies have shown that environmental factors, such as pH and the presence of specific ions, can affect the stability of catechins in green tea. Maintaining an optimal pH, for example, is critical to maintaining catechin integrity during processing and storage. Similarly, the interaction of ellagic acid with other compounds present in the food matrix can affect its antioxidant ability and effectiveness. Therefore, carefully considering these factors and the strategic combination of catechins, ellagic acid, and cinnamaldehyde offers a promising approach to developing food products with enhanced oxidative stability and longer shelf life (Chen et al., 2020).

Integrating these natural compounds into the food system raises questions about their functional mechanisms. Catechins in green tea may affect fat-containing products' textural and sensory properties, crucial for consumer acceptance. Their antioxidant capabilities may act synergistically with ellagic acid and cinnamaldehyde, providing a multifaceted approach to tackling food spoilage. Research shows that extracts rich in catechins inhibit lipid oxidation and contribute to flavor retention, which is critical for maintaining food quality. Combining these natural extracts offers an innovative strategy in food preservation techniques, reminiscent of using spices and herbs to extend food life (Cabrera et al., 2021).

Evidence shows that green tea catechins, ellagic acid from pomegranate, and cinnamaldehyde have complementary antioxidant, protective, and antimicrobial roles. Their synergy supports healthier, more stable, and longer-lasting food formulations. With advances in research, these compounds have the potential to be widely adopted in modern food preservation based on scientific innovation.

1) Essential Oil

Essential oils are gaining attention as natural preservatives in the food industry due to their strong antimicrobial properties. Extracted from plants such as oregano, thyme, and rosemary, these oils contain bioactive compounds such as carvacrol, thymol, and eugenol that are effective against food pathogens such as Listeria monocytogenes, E. coli, and Salmonella spp. Their effectiveness makes them a potential alternative to synthetic preservatives (Gutiérrez-Pozo et al., 2023; György et al., 2020). Research shows some essential oils have bactericidal and bacteriostatic activity against Gram-positive and Gram-negative bacteria. György et al. (2020) reported that Salvia officinalis essential oil effectively fights Listeria monocytogenes, Staphylococcus aureus, Salmonella, and E. coli. This effectiveness comes from bioactive compounds such as carvacrol and thymol that damage microbial cell membranes, causing cell dysfunction and death (Gutiérrez-Pozo et al., 2023).

The role of essential oils as natural preservatives goes beyond their antimicrobial properties. Their application can enhance food safety, extend shelf life, and improve overall sensory quality, making them particularly attractive for application in meat products and processed foods. Essential oils' complex aromas and flavors provide additional sensory benefits and can improve consumer perception of food quality (György et al., 2020). Moreover, their natural origin aligns with the increasing consumer demand for clean-label products, driving their adoption of food preservation technologies (Ksouri et al., 2023).

Essential oils have bioactive profiles that vary between plants, affecting their effectiveness. Thymol, for example, exhibits antifungal activity that preserves food quality. Cyclodextrin encapsulation improves essential oil stability and enables controlled release while preserving quality (Serna-Escolano et al., 2020). Lemongrass oil has also been shown to have antibacterial and antioxidant activity, expanding the application of essential oils to functional foods (Hartatie et al., 2020).

Essential oils show broad potential beyond food preservation, including eugenol's antiviral properties that hold promise for health applications (Ksouri et al., 2023). Essential oils contain antimicrobial, antifungal, and antioxidant bioactive compounds, which satisfy the demand for safe and quality natural preservatives. Although they still face challenges in effectiveness, formulation, and safety, their use as alternatives to synthetic preservatives continues to grow as research progresses (Gutiérrez-Pozo et al., 2023; György et al., 2020).

Essential oils (EOs) are attracting attention as natural preservatives for meat products, mainly due to their ability to fight pathogens such as Clostridium perfringens. Their volatile compound content can damage microbial membranes and disrupt cell metabolism, which is effectively influenced by their chemical composition (Egner et al., 2023). Patchouli oil (Pogostemon cablin), in particular, shows great potential as an antimicrobial, with patchouli alcohol as the main component effective against bacteria and fungi, making it a strong candidate for improving the safety of processed meat (Duc et al., 2024).

In addition to their antimicrobial properties, essential oils are believed to affect the oxidative stability of food products. The susceptibility of meat to fat oxidation is an important concern, as oxidative spoilage can affect flavor, color, and nutritional value. Incorporating essential oils can improve the oxidative stability of foods by providing antioxidants that reduce lipid degradation (Vu et al., 2022). This dual function, providing antimicrobial and antioxidant protection, makes essential oils particularly attractive for improving the shelf life and sensory quality of meat products, especially sausages, which often contain more fat and are more susceptible to spoilage.

The prevalence of C. perfringens in meat products requires effective control measures. Studies show a high percentage of C. perfringens presence in retail meat, reaching 38% in pork and 31% in chicken (Duc et al., 2024). The pathogenic potential of this organism in food products highlights the importance of prevention strategies, including the application of essential oils. The effectiveness of essential oils in inhibiting the growth of pathogenic bacteria highlights their potential to reduce the risks associated with C. perfringens during the storage and processing of meat products.

Essential oils can synergistically increase microbial resistance, ensuring food safety over extended storage periods. In addition, developing formulations that incorporate essential oils into meat products could improve product quality, extend shelf life, and reduce reliance on synthetic preservatives. Studies focusing on optimizing the concentration and combination of essential oils may provide insights into maximizing their effectiveness in meat products (Kemala et al., 2024).

Growing consumer interest in natural preservatives provides opportunities for using essential oils in food products, which aligns with market trends favoring clean-label ingredients. Consumers increasingly favor products that avoid synthetic additives, and essential oils can be a natural alternative while providing functional benefits (Szewczuk et al., 2023). The application of essential oils addresses safety concerns related to microbial contamination and aligns with the growing demand for healthier and value-added food products.

Understanding the risk of essential oil adulteration is essential to ensure product integrity (Szewczuk et al., 2023). Misusing essential oils can impair their effectiveness and safety, decreasing antimicrobial activity. Quality control measures should be implemented to guarantee that essential oils used in food applications are genuine and unadulterated, thus maintaining their aromatic and biochemical properties. The integrity of essential oils affects their overall performance in food preservation and, therefore, requires strict standards for sourcing and application in food products.

As the demand for safer and more sustainable food preservation methods increases, essential oils become prominent candidates for inclusion in meat products. Their antimicrobial properties against pathogens such as C. perfringens and their potential to enhance oxidative stability make them invaluable for improving food safety and extending shelf life (Duc et al., 2024). Further research and development on the specific mechanisms of how essential oils act on microbial cells may further reveal their role in food preservation and strengthen their application in the food industry.

2) Natural Polysaccharides (Chitosan and Alginate)

Natural polysaccharides such as chitosan and alginate have garnered considerable attention in the food processing industry due to their multifunctional capabilities, particularly in enhancing food quality and extending shelf life. These biopolymers serve as edible coatings that help preserve moisture, freshness, and safety in perishable foods. Chitosan, derived from the deacetylation of chitin found in crustacean shells, and alginate, extracted from brown seaweed, are uniquely characterized by their film-forming properties, non-toxicity, and alignment with consumer preferences for clean-label products (Lamani & Ramaswamy, 2023; Vieira et al., 2021)

Chitosan and alginate coatings primarily form a physical barrier that significantly reduces respiration rates and minimizes moisture loss in food products. This reduction in respiration and water activity is critical, as it directly contributes to decreasing the rate of microbial spoilage—a significant concern in fresh produce and seafood. Studies have shown that chitosan-based coatings can inhibit the growth of pathogenic and spoilage microorganisms on fruits and vegetables, extending their shelf life and ensuring enhanced food safety outcomes for consumers (Vieira et al., 2021). Similarly, alginate demonstrates exceptional benefits in retaining moisture and preserving the color and overall quality of fruits, as evidenced by its application on fresh-cut pears, where it effectively prolonged shelf life by controlling fungal growth (Lamani & Ramaswamy, 2023).

The performance of these polysaccharides can often be enhanced through the formulation of composite coatings that incorporate other natural extracts, introducing synergistic effects that further boost their effectiveness. For example, research indicates that integrating ginger essential oil with alginate enhances the antimicrobial properties and overall efficacy of edible coatings applied to fresh-cut fruits, improving sensory attributes and shelf stability. Additionally, the combination of alginate with calcium salts has been successfully employed to extend the shelf life of strawberries during storage (Lamani & Ramaswamy, 2023).

Storage conditions and duration critically influence the effectiveness of these coatings in preserving food quality. Research on ready-to-eat fish products demonstrates that encapsulating seafood in composite chitosan-alginate coatings can reduce lipid oxidation during storage, thus preserving key quality attributes such as flavor, color, and nutritional composition. These coatings decelerate oxidation processes responsible for off-flavor development, which is essential for maintaining consumer acceptance of seafood products. Furthermore, malondialdehyde content has been used as an indicator to assess the extent of lipid oxidation, providing valuable insights into the shelf-life performance of polysaccharide-based coatings (Bremenkamp & Sousa-Gallagher, 2024).

In addition to their physical barrier properties, both chitosan and alginate exhibit environmental compatibility, as they are biodegradable and produced from renewable resources. This aspect aligns with the increasing global focus on sustainable food practices. Using these natural polysaccharides not only addresses food safety and quality concerns but also complements ongoing efforts to reduce the environmental impact of food packaging by offering biodegradable alternatives to synthetic materials (Vieira et al., 2021).

Consumer perception also plays an important role in accepting edible protective coatings in the food industry. Attributes such as visual appeal, texture, and flavor are critical to consumer preferences, and both chitosan and alginate coatings have generated positive responses when used appropriately. For example, the integrity of color and hardness in alginate-coated fruits has proven important in attracting consumer interest, affecting marketability and sales (Lamani & Ramaswamy, 2023). Therefore, as food scientists and researchers continue to explore the versatile applications of this natural polysaccharide, evidence supports its utility as a significant tool in modern food preservation techniques.

Chitosan, which is extracted from crustacean shells, effectively inhibits microbial growth on various food products, such as fruits and vegetables. Using a chitosan-based protective coating on strawberries demonstrated the ability of this compound to inhibit the growth of pathogens, such as E. coli and mold, thus extending the product's shelf life. Research conducted by Zárate-Moreno et al. (2023) confirmed that chitosan not only serves as a preservative but also enhances food safety without compromising the sensory quality of the product, contributing to the reduction of spoilage and maintaining fruit freshness.

3) Detoxification Mechanisms

Detoxification mechanisms of mycotoxins and heavy metals are increasingly relevant in food safety and public health discussions. Mycotoxins, secondary metabolites produced by various fungi such as Aspergillus and Penicillium species, pose a significant risk to human health through their presence in agricultural products, including grains and legumes. Lactobacillus species, especially those used in fermentation processes, have attracted attention for their potential to bind and degrade these harmful compounds, thereby mediating their toxic effects. This is supported by evidence that certain strains of Lactobacillus may have the ability to bind or degrade mycotoxins, thereby reducing their toxicity, although their effectiveness may vary depending on strain and environmental conditions (Al-Assil & Wahwah, 2024; Shanmugasundaram et al., 2023).

The mechanism that allows Lactobacillus to enhance microbial mycotoxin degradation is related to producing specific proteolytic enzymes that enhance its binding capacity through cell wall components such as peptidoglycans and polysaccharides. These biochemical interactions enable Lactobacillus to metabolize mycotoxins such as aflatoxin B1 (AFB1) and ochratoxin A (OTA) through physical adsorption and enzymatic action, significantly reducing their bioavailability in the food system (Shanmugasundaram et al., 2023).

On the other hand, heavy metals such as cadmium, lead, and mercury are another area of concern due to their widespread presence in the environment and their toxicological implications on human health. Among microorganisms that exhibit practical binding ability for heavy metals, Saccharomyces cerevisiae (baker's yeast) stands out. Research shows that this yeast utilizes unique biosorption properties facilitated mainly by functional groups in the cell wall, such as carboxyl, hydroxyl, and amino groups, which increase its binding affinity to heavy metal ions. For example, the ability of S. cerevisiae to biosorb cadmium from milk has been demonstrated, with increasing yeast concentration increasing metal uptake, indicating its application in detoxifying food products contaminated with heavy metals (Massoud et al., 2020; Parhusip et al., 2021).

The interaction between the detoxification capabilities of Lactobacillus and S. cerevisiae highlights the important role of these microorganisms in food safety, acting as natural bioremediation to reduce health risks associated with mycotoxins and heavy metals. The effectiveness of Lactobacillus in mycotoxin reduction is influenced by factors including strain specification, environmental conditions, and toxin concentration (Al-Assil & Wahwah, 2024). Studies show that variations in Lactobacillus strains can result in different levels of effectiveness in binding mycotoxins, indicating that a more granular approach is needed when considering their application in food preservation (Shanmugasundaram et al., 2023). Similarly, the effectiveness of S. cerevisiae in heavy metal biosorption suggests the need for an understanding of the biochemical dynamics at play, as some variants may show better capabilities depending on environmental exposure (Massoud et al., 2020).

Microbial detoxification offers promising ways to reduce contaminants and support broader food safety practices. Bioremediation through these microorganisms offers a cost-effective and sustainable strategy to replace or complement existing chemical detoxification methods. This drives a shift towards more environmentally friendly approaches in food production systems (Parhusip et al., 2021). Moreover, the successful application of these strategies on various food matrices confirms the need for ongoing research to optimize conditions and methodologies to maximize the detoxification potential of these beneficial microorganisms (Massoud et al., 2020; Shanmugasundaram et al., 2023).

Developing engineered strains of S. cerevisiae, such as those designed to enhance heavy metal binding through genetic modification, adds a new layer to the discourse on bioremediation. Advances such as these highlight the intersection of traditional fermentation practices with cutting-edge genetic engineering, enabling customized approaches to detoxification tailored to specific contaminants or food products. These innovations signal a future where natural and engineered microbial solutions come together to address food safety's complex challenges in a contaminated world (Parhusip et al., 2021).

Additional studies focusing on uncovering structural aspects of these microbes' interactions with toxins may identify key parameters improving detoxification efficiency. This knowledge also contributes to a better understanding of how the food matrix affects the effectiveness of this natural detoxification process (Al-Assil & Wahwah, 2024; Massoud et al., 2020).

The practical implications of these studies go beyond laboratory settings; they challenge conventional paradigms in food safety management and invite industry stakeholders to consider integrating biological strategies into their operational frameworks. Producers can explore the integration of these microorganisms during food processing to ensure safety and quality and support consumer health while meeting regulatory compliance regarding contaminants (Massoud et al., 2020). The synergy between Lactobacillus and S. cerevisiae in reducing the presence of mycotoxins and heavy metals offers a valuable opportunity to improve food safety through natural means that do not cause negative ecological impacts (Parhusip et al., 2021).

Tannins, particularly those derived from sources such as chestnut (Castanea sativa) and acacia, have attracted significant attention due to their potential to reduce heavy metal toxicity in food products. Heavy metals are widespread contaminants from agricultural practices, industrial discharges, and environmental pollution. The detrimental impact of these metals on human health emphasizes the importance of effective detoxification strategies. Tannins have been recognized for their ability to bind heavy metal ions, thereby converting them into less toxic or inert compounds that can be handled more safely by the human body (Dang et al., 2022; Sikora et al., 2022).

The mechanism by which tannins exert their protective effects involves complexation, or chelation, in which tannin molecules form stable complexes with heavy metals. This process reduces the bioavailability of these metals, facilitating their excretion from the body and preventing them from entering biological systems where they can cause oxidative stress and cellular damage (Pujiono et al., 2021). For example, tannins derived from chestnut were effective against microbial pathogens, indicating their additional role as potent antimicrobial agents (Arzola-Alavarez et al., 2020; Romani et al., 2021). Applying natural extracts such as these in food processing can provide dual benefits: improving food safety while enhancing the nutritional profile with their antioxidant properties (Arzola-Alavarez et al., 2020).

Research shows that the beneficial effects of tannins are not limited to detoxification alone; they also have antioxidant properties that can further protect cells from oxidative damage caused by heavy metals (Prodanović et al., 2023). For example, incorporating tannins into livestock diets has enhanced antioxidant defenses, which may help reduce the adverse effects of environmental toxins (Arzola-Alavarez et al., 2020). This aligns with findings showing how plant-derived tannins, such as those found in chestnut wood, have significant antioxidant activity due to their phenolic components, which can capture free radicals (Khatib et al., 2023).

Using tannins in food processing benefits human health and can resonate with the increasing consumer demand for natural and sustainable food products. The shift towards green agriculture and food safety has prompted the exploration of alternatives to synthetic preservatives. Tannins provide an attractive option, as evidenced by their ability to inhibit microbial growth while reducing heavy metal toxicity (Eckardt et al., 2025). As the industry looks to innovate with natural additives, exploring tannins derived from familiar sources such as chestnut could pave the way for developing healthier food options without compromising safety (Romani et al., 2021).

Beyond food processing, the implications of tannin utilization also extend to agricultural contexts, where they can play a role in soil health management and minimize the negative impact of heavy metals on crop production. Tannins can modulate heavy metal bioavailability by interacting with soil components, affecting plant uptake and cellular detoxification mechanisms (Dang et al., 2022). For example, research shows that plants treated with tannin-rich extracts exhibit decreased cadmium uptake, indicating the potential use of tannins in phytoremediation strategies (Pujiono et al., 2021).

Insights gained from the antioxidant capabilities of tannins reveal their postharvest applications in extending the shelf life of food products. By reducing oxidative stress, tannins can protect sensitive food components from degradation, thereby maintaining nutritional value and sensory attributes such as flavor and color (Eckardt et al., 2025). This property is increasingly beneficial in food storage and transportation, where oxidation can rapidly reduce food quality.

The multifaceted role of tannins in reducing heavy metal toxicity spans a wide range of fields, including food science, nutrition, and environmental health (Ernawati et al., 2021, 2025). Their ability to complex with metals and their inherent antioxidant properties offer multiple advantages that can improve food safety, protect public health, and align with sustainable practices (Ernawati et al., 2024; Romani et al.,

2021). As research unravels the complex interactions between tannins and heavy metals, their potential applications in various sectors are becoming increasingly clear, encouraging further innovations based on environmental sustainability and public health.

Future research needs to improve the understanding of how different sources of tannins affect the detoxification process and their effectiveness in different food matrices. Investigating the optimal conditions for tannin application, such as concentration levels, processing methods, and combination with other natural additives, will also be critical in maximizing their benefits. In addition, consumer acceptance and regulatory considerations will play an important role in determining the feasibility of tannin inclusion in mainstream food products (Arzola-Alavarez et al., 2020).

Table 1. Effectiveness of Natural Compounds in Reducing Food Contamination and Extending Shelf Life

Natural Compound	Source	Type of Inhibited Contaminant	Application Method	Effectiveness Against Contaminants	Impact on Food Quality
Chitosan	Crustacean shells	Microbial pathogens (E. coli, fungi)	Edible coating on strawberries	Inhibits microbial growth	Prolongs shelf life without compromising quality
Oregano Oil	Oregano (*Origanum vulgare*)	*Clostridium perfringens, E. coli*	Incorporated into sausages	Inhibits pathogenic growth	Enhances sensory acceptance and extends shelf life
Pomegranate Extract	Pomegranate (*Punica granatum*)	Aflatoxin (*Aspergillus flavus*)	Applied to corn	Reduces aflatoxin levels, antifungal properties	Enhances nutritional value and product quality
Flavonoids	Green tea, pomegranate	Microbial pathogens, oxidation	Added to food products	Inhibits microbial growth, and antioxidant activity	Preserves taste, color, and nutritional content

Source: Compiled and adapted from El-Hadary & Taha (2020); Abdel-Aziz et al. (Abdel-Aziz et al., 2021a) (2021); Ismail et al. (2021).(Ismail et al., 2021)

Table 1 highlights several natural compounds—chitosan, oregano oil, pomegranate extract, and flavonoids—effectively reducing food contamination and prolonging shelf life. These bioactive agents inhibit harmful contaminants such as E. coli, Clostridium perfringens, and aflatoxins while enhancing food quality by preserving flavor, color, and nutritional properties. The application methods vary, including edible coatings on fruits and direct incorporation into processed foods, offering environmentally friendly and safe alternatives for food preservation.

5. CASE STUDIES

Several case studies provide concrete evidence regarding the effectiveness of natural compounds in mitigating food contamination:

1) Chitosan Coating on Strawberries

Natural compounds, particularly chitosan, have demonstrated significant efficacy in reducing food contamination, as evidenced by studies investigating the application of chitosan coatings on fruits such as strawberries. Chitosan, a biopolymer derived from chitin, has been extensively studied for its antimicrobial properties and ability to enhance perishable goods' preservation by inhibiting pathogenic microbes, particularly Escherichia coli, on treated surfaces over time.

A study conducted by Zárate-Moreno et al. illustrates the effectiveness of chitosan coatings in preventing microbial growth on strawberries. Their research revealed that chitosan not only preserves the freshness of the fruit by inhibiting fungal development but also exhibits significant antimicrobial activity, thereby contributing to food safety by countering pathogenic organisms (Zárate-Moreno et al., 2023). These findings support the notion that chitosan is an effective barrier against contamination, extending the shelf life of strawberries without compromising their quality.

Further investigations into chitosan's properties have highlighted its ability to regulate fruit physiological processes. Ackah et al. reported that chitosan treatment enhances fruit resistance to oxidative stress by modulating reactive oxygen species (ROS) metabolism (Ackah et al., 2022). This antioxidant capability improves the overall health of the fruit, enabling it to resist microbial infections better. Thus, the dual function of chitosan as both a preservative and an antioxidant reinforces its potential as a natural strategy for mitigating foodborne pathogens.

The antimicrobial activity of chitosan has been documented in multiple studies. Varma and Vasudevan emphasized that chitosan exhibits exceptional efficacy against specific bacterial strains, including E. coli, attributing its antimicrobial effects to structural characteristics that enhance its interaction with microbial membranes (Varma & Vasudevan, 2020). Furthermore, Yan et al. demonstrated that the biocompatibility and non-toxicity of chitosan make it a promising compound for use in food preservation and medical applications, offering assurances regarding its safety for consumers (Yan et al., 2021).

In a focused study, García-García et al. examined the preservation of strawberries using chitosan coatings modified with ZnO nanostructures. Their findings indicated reduced fruit acidity levels and improved quality parameters due to decreased respiration rates and enzymatic utilization of organic acids (García-García et al., 2023). These results complement previous research, confirming that chitosan not only enhances food safety by inhibiting microbial growth but also promotes the retention of fruit quality.

2) Oregano Oil

The application of oregano oil, primarily composed of active compounds such as carvacrol and thymol, has gained recognition as an effective natural preservative in meat products, particularly sausages, for combating pathogens such as Clostridium perfringens. Studies have demonstrated that the antimicrobial properties of oregano oil are attributed mainly to its high concentrations of carvacrol and thymol, which significantly enhance its efficacy against various foodborne pathogens (Tsitsos et al., 2023). This is critical for food safety, as C. perfringens is a common meat product pathogen often linked to foodborne illnesses. Incorporating oregano oil into sausage formulations enhances microbial safety while reducing reliance on synthetic preservatives. Findings suggest that oregano oil can maintain product quality while aligning with consumer demands for natural ingredients (Alexandraki et al., 2022).

One study noted that including oregano oil in frankfurters improved sensory acceptance, indicating its dual role in preserving meat products and enhancing consumer appeal (Orozco et al., 2021). The antimicrobial efficacy of oregano oil has been supported by numerous studies, demonstrating its ability to significantly reduce undesirable microbial populations in meat products, which is crucial for extending shelf life (Alexandraki et al., 2022).

Due to its active compounds like carvacrol and thymol, Oregano oil exhibits remarkable capabilities in inhibiting foodborne pathogens, particularly C. perfringens and Escherichia coli. Tsitsos et al. (2023) confirmed that oregano oil can be a natural preservative in sausages, with proven efficacy at specific concentrations. Beyond microbial inhibition, oregano oil also enhances the sensory attributes of sausages, including flavor and aroma, which are important for consumer acceptance.

The potential synergistic effects of combining carvacrol and thymol with other natural compounds have been explored, indicating enhanced antimicrobial efficacy against foodborne pathogens. This offers a more robust strategy for food preservation (Mantzourani et al., 2025). Research consistently shows that incorporating oregano oil maintains the safety and quality of meat products and potentially enhances consumer acceptance, fulfilling both safety and marketability requirements in the processed meat sector (Alexandraki et al., 2022).

The proven effectiveness of oregano oil against pathogens such as C. perfringens and its ability to improve sensory characteristics positions it as a promising natural preservative for sausage production. Its use allows producers to meet regulatory food safety standards while addressing consumer preferences for natural products.

3) Pomegranate Extract

Pomegranate extract has demonstrated antifungal activity in reducing aflatoxin B1 contamination in maize, particularly through its phenolic constituents such as ellagic acid. Its application offers a natural approach to managing mycotoxins while improving the antioxidant profile of the product (Yu et al., 2022), underscoring the urgent need for effective control measures in agricultural practices.

Pomegranate peel extract has emerged as a promising natural solution to mitigate aflatoxin contamination in maize. In a study by Ismail et al. (2021), the application of 25% diethyl ether pomegranate peel extract reduced aflatoxin B_1 (AFB_1) levels by 78.8% after one month of storage, demonstrating its significant antifungal and detoxifying effects.

Studies have shown that pomegranate extract possesses natural antifungal properties capable of inhibiting the growth of A. flavus and reducing aflatoxin production in crops. This mechanism is linked to phytochemicals that interfere with fungal metabolic processes (Abdel-Aziz et al., 2021b). This is particularly important given that maize and soil samples with high aflatoxin levels are often found in regions with environmental conditions conducive to fungal proliferation (Jayaratne et al., 2020).

Employing pomegranate extract as a biocontrol agent provides an organic alternative to conventional mycotoxin management strategies while enhancing the safety and quality of maize-based food products. Pomegranate extract, especially from sour cherry pomace, has effectively reduced aflatoxin production by A. flavus. Its application in maize helps decrease aflatoxin contamination, which is recognized as a significant threat to human health. Abdel-Aziz et al. (2021) reported that pomegranate extract reduced aflatoxin levels and improved maize's nutritional quality by increasing its antioxidant content. Additionally, Yu et al. (2022) confirmed that bioactive compounds from pomegranate could extend maize shelf life by combating oxidation and microbial spoilage.

Additionally, its antioxidant compounds contribute to preserving the nutritional quality of maize during storage, although further studies are needed to confirm direct nutritional outcomes (Abedini et al., 2021). Using pomegranate extract aligns with current agricultural trends, emphasizing sustainable practices and safer food production strategies in response to climate change and agricultural challenges (Yu et al., 2022b).

The application of pomegranate extract in maize cultivation offers significant benefits. It functions as a natural antifungal agent to combat aflatoxin contamination while potentially enhancing the nutritional quality of the crop. This approach addresses consumer safety concerns and supports sustainable agricultural practices.

Solutions and Recommendations

1) Synergistic Formulations

The combination of various natural compounds can significantly enhance antimicrobial efficacy. Synergistic interactions among compounds such as flavonoids, essential oils, and polysaccharides can reinforce microbial inhibition mechanisms and extend the shelf life of food products. For instance, a formulation combining active components from oregano oil and flavonoids derived from green tea may offer a more effective solution for addressing microbial contamination and oxidative degradation.

2) Mild Processing Technologies

Technologies such as microencapsulation and freeze-drying have proven effective in preserving the stability of bioactive compounds. These techniques can protect natural preservatives from degradation caused by heat or moisture during food processing, thereby maintaining their quality and functional efficacy. The application of microencapsulation, particularly in processed food products, holds the potential for prolonging shelf life while enhancing the sensory quality of the final product.

3) Plant-Based Biotechnology

Developing plant varieties enriched with protective compounds offers a promising natural solution for food preservation. Innovations in biotechnology may include genetic modification or selective breeding to increase the content of bioactive compounds that inhibit pathogen growth and reduce food contamination. For example, plants with elevated levels of ellagic acid or catechins may be natural defenses against mycotoxins and other foodborne pathogens.

4) Standardization and Regulatory Frameworks

Developing safety guidelines and GRAS (Generally Recognized As Safe) standards for natural compounds is essential for their broad market acceptance. Such standards ensure that natural preservatives, while effectively reducing contamination, do not introduce new health risks. Collaborative efforts among scientists, regulatory bodies, and food industry stakeholders are vital to accelerating the adoption of natural compounds in food production systems.

5) Adaptive Sensory Formulations

Ensuring that food products retain desirable flavor and aroma profiles is critical for market acceptance of those containing natural preservatives. Formulating food products that maintain their sensory appeal while incorporating bioactive compounds can support consumer acceptance and commercial success. This approach aligns with the growing demand for healthier, natural food options without compromising taste or quality.

6) Economic Feasibility and Scalability of Natural Preservatives

Natural preservatives like pomegranate peel, green tea, and mango peel extracts show strong potential, especially when sourced from food waste, lowering costs and promoting sustainability. However, challenges remain in standardizing extraction, maintaining stability, and scaling production effectively. While lab results are promising, real-world use requires pilot studies, regulatory approval, and industry integration. Partnerships among stakeholders are key to success. Examples from low-resource settings support feasibility: a rural Indonesian bakery extended bread shelf life by three days using green tea extract, and a Tamil Nadu cooperative preserved sun-dried tomatoes for an extra week with mango peel extract using simple, low-cost methods.

Future Research Directions

1) Exploration of Novel Natural Compounds from Microbes and Deep-Sea Sources

Further investigations are required to explore the potential of novel bioactive compounds derived from underexplored microbial or marine organisms. Particular attention should be given to antimicrobial and antioxidant compounds from deep-sea microorganisms, which may present unique properties suitable for application in food preservation.

2) In-vivo studies and Clinical Trials to Assess Bioavailability and Safety

Follow-up research involving in vivo models and human clinical trials is essential to understand the bioavailability of natural compounds within the human body. These studies are also crucial for evaluating the long-term safety of such compounds, particularly when incorporated into food systems.

3) Investigation of Natural Compound Interactions with Food Matrices

Understanding how natural compounds interact with food matrices is vital to assessing their effectiveness and bioavailability within food products. These interactions may also influence food quality and safety, and thus require detailed study to optimize their application in food processing.

4) Development of Sustainable Agricultural Systems Based on Natural Compounds

Future research should also focus on developing sustainable agricultural systems incorporating natural compounds as preservatives and plant-protective agents. This approach supports environmentally friendly farming practices and produces healthier food products.

5) Evaluation of Long-Term Effects on Gut Microbiota and Consumer Health

Assessing the long-term effects of natural compound consumption on the gut microbiota and overall consumer health is imperative. This evaluation will help ensure that natural preservatives provide safety and efficacy and contribute positively to digestive health and immune function.

Limitations and Challenges in Practical Application

Despite their potential, natural preservatives face challenges in real-world use. Bioactive compounds like polyphenols are unstable when exposed to heat, light, or oxygen, reducing their effectiveness over time. Sensory issues—such as intense flavors, odors, or colors—can also affect food quality and consumer acceptance. Regulatory approval remains a barrier, as not all compounds meet GRAS standards, and global regulations vary. Overcoming these limitations requires collaboration across scientific, regulatory, and industrial sectors to ensure safety, efficacy, and commercial viability.

CONCLUSION

Applying natural compounds to mitigate food contaminants offers a transformative, effective, and sustainable approach to addressing global food safety challenges. Natural compounds—particularly flavonoids, essential oils, polysaccharides, and tannins—offer a comprehensive solution to food contamination challenges by enhancing preservation, safety, and sustainability across various food matrices.

Although challenges remain, new formulation methods and collaborations can enhance the role of natural compounds in creating safer and more sustainable food systems. With ongoing research and innovation, natural compounds hold considerable potential to replace synthetic preservatives and contribute to the evolution of a more natural and sustainable food future.

A coordinated and collaborative approach is essential to fully realize the potential of natural bioactive compounds in food preservation. Academics must continue to generate high-quality evidence and develop innovative extraction and stabilization technologies. At the same time, industry players should invest in pilot-scale applications and consumer acceptability trials. Meanwhile, policymakers play a key role in shaping regulatory frameworks that enable the safe and effective use of natural preservatives. Advancing these efforts will require cross-sector partnerships that bridge research, innovation, regulation, and market implementation. Together, these collaborations can drive the sustainable transformation of the food system toward safer, more natural preservation methods.

Furthermore, the use of plant-derived bioactive compounds for natural food preservation aligns directly with several Sustainable Development Goals (SDGs), particularly Goal 2 (Zero Hunger), Goal 3 (Good Health and Well-being), and Goal 12 (Responsible Consumption and Production). By leveraging agricultural byproducts and reducing reliance on synthetic additives, these solutions promote food security, minimize food waste, and support a more sustainable and health-oriented food system. Integrating such innovations into mainstream food processing can contribute to global efforts to achieve safer, cleaner, and more resilient food systems for future generations.

REFERENCES

Abdel-Aal, E.-S. M., Rabalski, I., Mats, L., & Rai, I. (2022). Identification and Quantification of Anthocyanin and Catechin Compounds in Purple Tea Leaves and Flakes. *Molecules (Basel, Switzerland), 27*(19), 6676. DOI: 10.3390/molecules27196676 PMID: 36235212

Abdel-Aziz, S. H., El-Esawi, M. A., Hazaa, M. M., Abdel-Aziz, H. Y., & Hassan, M. G. (2021). Antibacterial Potential of Pomegranate Peel Extracts on Escherichia coli Isolated From Benha Hospital in Egypt. *Benha Journal of Applied Sciences, 6*(3), 61–64. DOI: 10.21608/bjas.2021.169813

Abedini, M., Ghasemi-Tehrani, H., Tarrahi, M. J., & Amani, R. (2021). The effect of concentrated pomegranate juice consumption on risk factors of cardiovascular diseases in women with polycystic ovary syndrome: A randomized controlled trial. *Phytotherapy Research, 35*(1), 442–451. DOI: 10.1002/ptr.6820 PMID: 32767710

Ackah, S., Bi, Y., Xue, S., Yakubu, S., Han, Y., Zong, Y., Atuna, R. A., & Prusky, D. (2022). Postharvest chitosan treatment suppresses oxidative stress by regulating reactive oxygen species metabolism in wounded apples. *Frontiers in Plant Science, 13*, 959762. Advance online publication. DOI: 10.3389/fpls.2022.959762 PMID: 35982700

Al-Assil, B. J., & Wahwah, T. A. (2024). Control Mycotoxin Contamination in Tarasas Food Using A Lab Fermantion. *Journal of Angiotherapy, 8*(3). Advance online publication. DOI: 10.25163/angiotherapy.839544

Al-Maharik, N., Jaradat, N., Bassalat, N., Hawash, M., & Zaid, H. (2022). Isolation, Identification and Pharmacological Effects of Mandragora autumnalis Fruit Flavonoids Fraction. *Molecules (Basel, Switzerland), 27*(3), 1046. DOI: 10.3390/molecules27031046 PMID: 35164311

Alexandraki, M., Valiakos, G., Manouras, A., & Malissiova, E. (2022). Chemical, Microbiological, and Sensory Analysis Benefits Leading to Shelf-Life Extension of Minced Beef Meat Using Essential Oil of Oregano Contained in Xanthan Gum and Guar Gum Edible Coating. *Processes (Basel, Switzerland), 10*(11), 2253. DOI: 10.3390/pr10112253

Arzola-Alavarez, C., Castillo-Castillo, Y., Anderson, R., Hume, M., Ruiz-Barrera, O., Min, B., Arzola-Rubio, A., Beier, R., & Salinas-Chavira, J. (2020). Influence of Pine Bark Tannin on Bacterial Pathogens Growth and Nitrogen Compounds on Changes in Composted Poultry Litter. *Brazilian Journal of Poultry Science*, *22*(1), eRBCA-2018-0911. Advance online publication. DOI: 10.1590/1806-9061-2018-0911

Bremenkamp, I., & Sousa-Gallagher, M. J. (2024). Design and Development of an Edible Coating for a Ready-to-Eat Fish Product. *Polymers*, *16*(3), 346. DOI: 10.3390/polym16030346 PMID: 38337235

Cabrera, M., Taher, F., Llantada, A., Do, Q., Sapp, T., & Sommerhalter, M. (2021). Effect of Water Hardness on Catechin and Caffeine Content in Green Tea Infusions. *Molecules (Basel, Switzerland)*, *26*(12), 3485. DOI: 10.3390/molecules26123485 PMID: 34201178

Chen, L., Wang, W., Zhang, J., Wang, W., Ni, D., & Jiang, H. (2020). Dehydroascorbic Acid Affects the Stability of Catechins by Forming Conjunctions. *Molecules (Basel, Switzerland)*, *25*(18), 4076. DOI: 10.3390/molecules25184076 PMID: 32906587

Dang, F., Li, Y., Wang, Y., Lin, J., Du, S., & Liao, X. (2022). ZAT10 plays dual roles in cadmium uptake and detoxification in Arabidopsis. *Frontiers in Plant Science*, *13*, 994100. Advance online publication. DOI: 10.3389/fpls.2022.994100 PMID: 36110357

de Oliveira, F. L., Arruda, T. Y. P., da Silva Lima, R., Casarotti, S. N., & Morzelle, M. C. (2020). Pomegranate as a natural source of phenolic antioxidants: A review. *Journal of Food Bioactives: An Official Scientific Publication of the International Society of Nutraceuticals and Functional Foods (ISNFF)*, 10–22. DOI: 10.31665/JFB.2020.9214

Duc, H. M., Hoa, T. T. K., Ha, C. T. T., Van Hung, L., Van Thang, N., Minh Son, H., & Flory, G. A. (2024). Prevalence and Antibiotic Resistance Profile of Clostridium perfringens Isolated from Pork and Chicken Meat in Vietnam. *Pathogens (Basel, Switzerland)*, *13*(5), 400. DOI: 10.3390/pathogens13050400 PMID: 38787252

Eckardt, J., Moro, L., Colusso, E., Šket, P., Giovando, S., & Tondi, G. (2025). Comparing Hydrolysable and Condensed Tannins for Tannin Protein-Based Foams. *Polymers*, *17*(2), 153. DOI: 10.3390/polym17020153 PMID: 39861226

Egner, P., Pavlačková, J., Sedlaříková, J., Pleva, P., Mokrejš, P., & Janalíková, M. (2023). Non-Alcohol Hand Sanitiser Gels with Mandelic Acid and Essential Oils. *International Journal of Molecular Sciences*, *24*(4), 3855. DOI: 10.3390/ijms24043855 PMID: 36835267

El-Hadary, A. E., & Taha, M. (2020). Pomegranate peel methanolic-extract improves the shelf-life of edible-oils under accelerated oxidation conditions. *Food Science & Nutrition*, *8*(4), 1798–1811. DOI: 10.1002/fsn3.1391 PMID: 32328245

Enayatifard, R., Akbari, J., Babaei, A., Rostamkalaei, S. S., Hashemi, S. M., & Habibi, E. (2020, February). Anti-Microbial Potential of Nano-Emulsion form of Essential Oil Obtained from Aerial Parts of Origanum Vulgare L. as Food Additive. *Advanced Pharmaceutical Bulletin*. Advance online publication. DOI: 10.34172/apb.2021.028 PMID: 33880355

Erdem, Y. I., Kaya, S., & Gökkaya Erdem, B. (2024). Characterization and application of whey protein isolate based edible films containing clove and apricot essential oils. *Gida*, *49*(2), 299–311. DOI: 10.15237/gida.GD23137

Ernawati, K., Aslichan, A., Albar, A. S., Al Khaliel, A. T., & Al Fatich, A. S. F. (2025). Green Leadership Strategy to Encourage Pro-Environmental Behavior. In *Green Leadership Strategies for Increased Employee Engagement and Performance* (pp. 173–204). IGI Global Scientific Publishing. DOI: 10.4018/979-8-3373-2106-6.ch006

Ernawati, K., Burhan, A., Albar, A. S., Al Khaliel, A. T., & Al Fatich, A. S. F. (2024). Professional Development of Environmental Health Workers. In *Instructional Approaches for Health Professions Education* (pp. 269–294). IGI Global Scientific Publishing. DOI: 10.4018/979-8-3693-4334-0.ch010

Ernawati, K., Nadhifah, Q., Muslikha, A., Hidayat, M., Soesilo, T. E. B., Jannah, F., Widianti, D., & Yusnita, Y. (2021). Relationship of knowledge and attitude with food handling practices: A systematic review. *International Journal of Public Health Science*, *10*(2), 336. DOI: 10.11591/ijphs.v10i2.20665

Fisher, F., Africa, C., Klaasen, J., & Fisher, R. (2025). South African Medicinal Plants Traditionally Used for Wound Treatment: An Ethnobotanical Systematic Review. *Plants*, *14*(5), 818. DOI: 10.3390/plants14050818 PMID: 40094805

Gallego, M., Arnal, M., Talens, P., Toldrá, F., & Mora, L. (2020). Effect of Gelatin Coating Enriched with Antioxidant Tomato Byproducts on the Quality of Pork Meat. *Polymers*, *12*(5), 1032. DOI: 10.3390/polym12051032 PMID: 32370120

García-García, D. J., Pérez-Sánchez, G. F., Hernández-Cocoletzi, H., Sánchez-Arzubide, M. G., Luna-Guevara, M. L., Rubio-Rosas, E., Krishnamoorthy, R., & Morán-Raya, C. (2023). Chitosan Coatings Modified with Nanostructured ZnO for the Preservation of Strawberries. *Polymers*, *15*(18), 3772. DOI: 10.3390/polym15183772 PMID: 37765626

Gutiérrez-Pozo, M., Serna-Escolano, V., Giménez-Berenguer, M., Giménez, M. J., & Zapata, P. J. (2023). The Preharvest Application of Essential Oils (Carvacrol, Eugenol, and Thymol) Reduces Fungal Decay in Lemons. *Agriculture*, *13*(7), 1437. DOI: 10.3390/agriculture13071437

György, É., Laslo, É., & Csató, E. (2020). Antibacterial activity of plant extracts against listeria monocytogenes isolated from ready-to-eat salads. *Acta Universitatis Sapientiae. Alimentaria*, *13*(1), 131–143. DOI: 10.2478/ausal-2020-0008

Hartatie, E. S. (2020). Short Communication: Detection of bioactive compounds in essential oil from lemongrass cultivated in Ngantang, Malang, East Java, Indonesia. *Biodiversitas (Surakarta)*, *21*(6). Advance online publication. DOI: 10.13057/biodiv/d210659

Ibrahim, S. A., Ayivi, R. D., Zimmerman, T., Siddiqui, S. A., Altemimi, A. B., Fidan, H., Esatbeyoglu, T., & Bakhshayesh, R. V. (2021). Lactic Acid Bacteria as Antimicrobial Agents: Food Safety and Microbial Food Spoilage Prevention. *Foods*, *10*(12), 3131. DOI: 10.3390/foods10123131 PMID: 34945682

Ismail, I. A., Qari, S. H., Shawer, R., Elshaer, M. M., Dessoky, E. S., Youssef, N. H., Hamad, N. A., Abdelkhalek, A., Elsamra, I. A., & Behiry, S. I. (2021). The Application of Pomegranate, Sugar Apple, and Eggplant Peel Extracts Suppresses Aspergillus flavus Growth and Aflatoxin B1 Biosynthesis Pathway. *Horticulturae*, *7*(12), 558. DOI: 10.3390/horticulturae7120558

Jayaratne, W. M. S. C., Abeyratne, A. H. M. A. K., De Zoysa, H. K. S., Dissanayake, D. M. R. B. N., Bamunuarachchige, T. C., Waisundara, V. Y., & Chang, S. (2020). Detection and quantification of Aflatoxin B1 in corn and corn-grown soils in the district of Anuradhapura, Sri Lanka. *Heliyon*, *6*(10), e05319. DOI: 10.1016/j.heliyon.2020.e05319 PMID: 33134588

Kahrarian, Z., Mojarrab, M., & Shahbazi, Y. (2020). Investigation of Antioxidant Effects of Artemisia Sieberi Essential Oil In Iran. *Journal of Biotechnology & Bioinformatics Research*, *1–3*, 1–3. Advance online publication. DOI: 10.47363/JBBR/2020(2)117

Kemala, P., Idroes, R., Khairan, K., Ramli, M., Tallei, T. E., Helwani, Z., & Rahman, S. A. (2024). The Potent Antimicrobial Spectrum of Patchouli: Systematic Review of Its Antifungal, Antibacterial, and Antiviral Properties. *Malacca Pharmaceutics*, *2*(1), 10–17. DOI: 10.60084/mp.v2i1.156

Khatib, M., Campo, M., Bellumori, M., Cecchi, L., Vignolini, P., Innocenti, M., & Mulinacci, N. (2023). Tannins from different parts of the chestnut trunk (castanea sativa mill.): A green and effective extraction method and their profiling by high-performance liquid chromatography-diode array detector-mass spectrometry. *ACS Food Science & Technology*, *3*(11), 1903–1912. DOI: 10.1021/acsfoodscitech.3c00272

Ksouri, A., Klouz, A., Bouhaouala, B., Moussa, F., & Bezzarga, M. (2023). Docking-Based Evidence for the Potential of ImmunoDefender: A Novel Formulated Essential Oil Blend Incorporating Synergistic Antiviral Bioactive Compounds as Promising Mpro Inhibitors Against SARS-CoV-2. *Molecules*. https://doi.org/DOI: 10.20944/preprints202304.0323.v1

Lamani, N. A., & Ramaswamy, H. S. (2023). Composite Alginate–Ginger Oil Edible Coating for Fresh-Cut Pears. *Journal of Composites Science*, *7*(6), 245. DOI: 10.3390/jcs7060245

Lankanayaka, A. A. K., Matharage, D. S., Senanayake, C. M., Jayathunge, K. G. L. R., Bandara, P. C., & Manatunga, D. C. (2024). Extraction of bioactive compounds from food waste and their potential applications in the food industry: A review. *Brazilian Journal of Development*, *10*(4), e69187. DOI: 10.34117/bjdv10n4-061

Mali, R. P., Rao, P. S., & Vikhe, D. N. (2020). Wound Healing Activity of Calotropis Procera Root Bark on Diabetic Rats. *Journal of Drug Delivery and Therapeutics*, *10*(2-s), 86–89. DOI: 10.22270/jddt.v10i2-s.3955

Mantzourani, I., Daoutidou, M., & Alexopoulos, A. (2025). The Antimicrobial Effect of Thymol and Carvacrol in Combination with Organic Acids Against Foodborne Pathogens in Chicken and Beef Meat Fillets. *Microorganisms*, *13*(1), 182. DOI: 10.3390/microorganisms13010182 PMID: 39858950

Massoud, R., Sharifan, A., Khosravi-Darani, K., & Asadi, G. (2020). Cadmium Bioremoval by Saccharomyces cerevisiae in Milk. *Journal of Medical Microbiology and Infectious Diseases*, *8*(1), 29–33. DOI: 10.29252/JoMMID.8.1.29

Mostafa, I., Abbas, H. A., Ashour, M. L., Yasri, A., El-Shazly, A. M., Wink, M., & Sobeh, M. (2020). Polyphenols from Salix tetrasperma Impair Virulence and Inhibit Quorum Sensing of Pseudomonas aeruginosa. *Molecules (Basel, Switzerland)*, *25*(6), 1341. DOI: 10.3390/molecules25061341 PMID: 32187997

Naik, G., Haider, S. Z., Bhandari, U., Lohani, H., & Chauhan, N. (2021). Comparative Analysis of In vitro Antimicrobial and Antioxidant Potential of Cinnamomum tamala Extract and their Essential Oils of Two Different Chemotypes. *Agricultural Science Digest - A Research Journal*. DOI: 10.18805/ag.D-5187

Nasir-Naeem, K. O., Akande, S. A., Garba, R., & Akolade, J. O. (2021). Assessment of milling as an alternative to chemical additives in processing of jute mallow (Corchorus olitorius). *AROC in Food and Nutrition, 1*(2), 10–17. DOI: 10.53858/arocfn01021017

Ndung'u, J. K., Nguta, J. M., Mapenay, I. M., & Moriasi, G. A. (2024). A Comprehensive Review of Ethnomedicinal Uses, Phytochemistry, Pharmacology, and Toxicity of Prunus africana (Hook. F.) Kalkman from Africa. *Scientifica, 2024*, 1–17. DOI: 10.1155/2024/8862996 PMID: 38654751

Orozco, D., Alarcon-Rojo, A. D., Chavez-Mendoza, C., Luna, L., Carrillo-Lopez, L. M., & Ronquillo, O. (2021). Fat replacement by pecan nut and oregano oil and their impact on the physicochemical properties and consumer acceptability of frankfurters. *Animal Bioscience, 34*(10), 1674–1683. DOI: 10.5713/ab.20.0622 PMID: 33332948

Parhusip, A. J. N., Wijaya, T. A., & Irawati, W. (2021). *Reduction of Heavy Metals in Squid, Green Mussels, White Shrimp Using Combination of Saccharomyces cerevisiae and Candida albicans*. DOI: 10.2991/absr.k.210621.072

Prodanović, R., Nedić, S., Vujanac, I., Bojkovski, J., Nedić, S., Jovanović, L., Kirovski, D., & Borozan, S. (2023). Dietary Supplementation of Chestnut Tannins in Prepartum Dairy Cows Improves Antioxidant Defense Mechanisms Interacting with Thyroid Status. *Metabolites, 13*(3), 334. DOI: 10.3390/metabo13030334 PMID: 36984774

Pujiono, C., Taufiq Mukti, A., & Hastuti Satyantini, W. (2021). Study of Pb Heavy Metal Pollution Level on Tannin Content of Seaweed (Kappaphycus alvarezii) in Bluto and Talango Sea Waters, Sumenep, East Java. *Journal of Aquaculture and Fish Health, 10*(1), 1. DOI: 10.20473/jafh.v10i1.17088

Rameshrad, M., Razavi, B. M., & Hosseinzadeh, H. (2017). Protective effects of green tea and its main constituents against natural and chemical toxins: A comprehensive review. *Food and Chemical Toxicology, 100*, 115–137. DOI: 10.1016/j.fct.2016.11.035 PMID: 27915048

Romani, A., Simone, G., Campo, M., Moncini, L., & Bernini, R. (2021). Sweet chestnut standardized fractions from sustainable circular process and green tea extract: In vitro inhibitory activity against phytopathogenic fungi for innovative applications in green agriculture. *PLoS One, 16*(2), e0247298. DOI: 10.1371/journal.pone.0247298 PMID: 33617600

Sabraoui, T., Khider, T., Nasser, B., Eddoha, R., Moujahid, A., Benbachir, M., & Essamadi, A. (2020). Determination of Punicalagins Content, Metal Chelating, and Antioxidant Properties of Edible Pomegranate (*Punica granatum L*) Peels and Seeds Grown in Morocco. *International Journal of Food Sciences, 2020*, 1–8. DOI: 10.1155/2020/8885889 PMID: 33029486

Salih, A. M., Al-Qurainy, F., Khan, S., Nadeem, M., Tarroum, M., & Shaikhaldein, H. O. (2022). Biogenic silver nanoparticles improve bioactive compounds in medicinal plant Juniperus procera in vitro. *Frontiers in Plant Science, 13*, 962112. Advance online publication. DOI: 10.3389/fpls.2022.962112 PMID: 36226285

Sęczyk, Ł., Gawlik-Dziki, U., & Świeca, M. (2021). Influence of Phenolic-Food Matrix Interactions on In Vitro Bioaccessibility of Selected Phenolic Compounds and Nutrients Digestibility in Fortified White Bean Paste. *Antioxidants, 10*(11), 1825. DOI: 10.3390/antiox10111825 PMID: 34829697

Serna-Escolano, V., Serrano, M., Valero, D., Isabel Rodríguez-López, M., Gabaldón, J. A., Castillo, S., Valverde, J. M., Zapata, P. J., Guillén, F., & Martínez-Romero, D. (2020). Thymol Encapsulated into HP-β-Cyclodextrin as an Alternative to Synthetic Fungicides to Induce Lemon Resistance against Sour Rot Decay. *Molecules (Basel, Switzerland), 25*(18), 4348. DOI: 10.3390/molecules25184348 PMID: 32971952

Shanmugasundaram, R., Lourenco, J., Al Hakeem, W., Dycus, M. M., & Applegate, T. J. (2023). Subclinical doses of dietary fumonisins and deoxynivalenol cause cecal microbiota dysbiosis in broiler chickens challenged with Clostridium perfringens. *Frontiers in Microbiology, 14*, 1106604. Advance online publication. DOI: 10.3389/fmicb.2023.1106604 PMID: 37082176

Sikora, F., Ochmian, I., & Sobolewska, M. (2022). Quality and technological properties of flour with the addition of aesculus hippocastanum and castanea sativa. *Acta Universitatis Cibiniensis. Series E: Food Technology, 26*(1), 43–54. DOI: 10.2478/aucft-2022-0004

Szewczuk, M. A., Zych, S., Oster, N., & Karakulska, J. (2023). Activity of Patchouli and Tea Tree Essential Oils against Staphylococci Isolated from Pyoderma in Dogs and Their Synergistic Potential with Gentamicin and Enrofloxacin. *Animals (Basel), 13*(8), 1279. DOI: 10.3390/ani13081279 PMID: 37106842

Tamborlin, L., Sumere, B. R., de Souza, M. C., Pestana, N. F., Aguiar, A. C., Eberlin, M. N., Simabuco, F. M., Rostagno, M. A., & Luchessi, A. D. (2020). Characterization of pomegranate peel extracts obtained using different solvents and their effects on cell cycle and apoptosis in leukemia cells. *Food Science & Nutrition, 8*(10), 5483–5496. DOI: 10.1002/fsn3.1831 PMID: 33133551

Tsitsos, A., Economou, V., Chouliara, E., Koutouzidou, G., Arsenos, G., & Ambrosiadis, I. (2023). Effect of Chitosan and Alginate-Based Edible Membranes with Oregano Essential Oil and Olive Oil in the Microbiological, Physicochemical and Organoleptic Characteristics of Mutton. *Microorganisms, 11*(2), 507. DOI: 10.3390/microorganisms11020507 PMID: 36838470

Vaičiulienė, G., Bakutis, B., Jovaišienė, J., Falkauskas, R., Gerulis, G., & Baliukonienė, V. (2020). Origanum vulgareandthymus vulgarisextract usability to improve silage hygienic quality and reduce mycotoxin concentrations. *Journal of Microbiology and Biotechnology, 30*(8), 1149–1155. DOI: 10.4014/jmb.2003.03010 PMID: 32522971

Varma, R., & Vasudevan, S. (2020). Extraction, Characterization, and Antimicrobial Activity of Chitosan from Horse Mussel *Modiolus modiolus. ACS Omega, 5*(32), 20224–20230. DOI: 10.1021/acsomega.0c01903 PMID: 32832775

Vieira, T. M., Moldão-Martins, M., & Alves, V. D. (2021). Composite Coatings of Chitosan and Alginate Emulsions with Olive Oil to Enhance Postharvest Quality and Shelf Life of Fresh Figs (Ficus carica L. cv. 'Pingo De Mel'). *Foods, 10*(4), 718. DOI: 10.3390/foods10040718 PMID: 33805309

Vu, T. P., Gumus-Bonacina, C. E., Corradini, M. G., He, L., McClements, D. J., & Decker, E. A. (2022). Role of Solid Fat Content in Oxidative Stability of Low-Moisture Cracker Systems. *Antioxidants, 11*(11), 2139. DOI: 10.3390/antiox11112139 PMID: 36358510

Yan, D., Li, Y., Liu, Y., Li, N., Zhang, X., & Yan, C. (2021). Antimicrobial Properties of Chitosan and Chitosan Derivatives in the Treatment of Enteric Infections. *Molecules (Basel, Switzerland), 26*(23), 7136. DOI: 10.3390/molecules26237136 PMID: 34885715

Yao, L., Zhu, S., Liu, W., Manzoor, Z., Nisar, M. F., & Li, M. (2021). A comprehensive review of pharmacological and Analytical Aspects of Acacetin. *Natural Resources for Human Health, 1*(1), 8–18. DOI: 10.53365/nrfhh/141018

Yu, J., Hennessy, D. A., Tack, J., & Wu, F. (2022). Climate change will increase aflatoxin presence in US Corn. *Environmental Research Letters, 17*(5), 054017. DOI: 10.1088/1748-9326/ac6435

Zárate-Moreno, J. C., Escobar-Sierra, D. M., & Ríos-Estepa, R. (2023). Development and Evaluation of Chitosan-Based Food Coatings for Exotic Fruit Preservation. *Biotech, 12*(1), 20. DOI: 10.3390/biotech12010020 PMID: 36810447

Zolfaghari, A., Bazargani-Gilani, B., & Aghajani, N. (2023). Edible film based on corn zein containing dill extract and essential oil/β-cyclodextrin inclusion complex: Shelf life enhancement of common carp fillet. *Food Science & Nutrition, 11*(7), 4275–4288. DOI: 10.1002/fsn3.3353 PMID: 37457154

Chapter 8
Genetically-Modified Crop Risk Assessment and Biosafety: Integrative Meta Analysis and Communication Strategy

Mohammed Bashir Tijjani
https://orcid.org/0009-0009-9667-0087
University of Maiduguri, Nigeria

Adamkolo Mohammed Ibrahim
https://orcid.org/0000-0003-1662-7054
University of Maiduguri, Nigeria

Abubakar Tijjani Ibrahim
https://orcid.org/0009-0003-2410-5595
Kano State Polytechnic, Nigeria

ABSTRACT

This meta-analysis integrates findings from 52 peer-reviewed studies to assess the toxicological safety of genetically modified (GM) crops into the broader context of food-chain contamination control. By synthesising data from controlled animal feeding trials and human observational studies, the chapter demonstrates that acute toxicity remains negligible, while a minority of chronic-exposure investigations reveal subtle metabolic alterations. Employing advanced omics, particularly metabolomic and epigenetic profiling, enables detection of early biomarkers that conventional assays may miss. Framed against sustainable development goals for zero hunger, good health and well-being and responsible production, the study un-

DOI: 10.4018/979-8-3373-7052-1.ch008

derscores the necessity of robust risk-assessment frameworks. The chapter advocates a development-oriented communication strategy to translate complex toxicological evidence into actionable guidance for policymakers and practitioners. Finally, the chapter proposes a "Crop-Vigilance" system for ongoing surveillance, integrating digital tools and community engagement to mitigate potential risks. This chapter thus offers a model for uniting rigorous toxicology with food-safety innovations to safeguard public health and environmental integrity.

1. INTRODUCTION

Evaluating the safety of genetically modified (GM) crops has been pursued as a subject of considerable scientific, ethical and socio-political enquiry for several decades. Since the introduction of GM technology into agriculture, initially conceived to augment crop yield, confer resistance to pests and diseases and improve nutritional content (Amedu et al., 2025), persistent questions have arisen regarding potential long-term repercussions on human and animal health as well as on environmental systems. This paper sets out to conduct a comprehensive review of toxicological data on GM crops, employing a meta-analytical approach. The study incorporates central concepts from pharmacology and public health to critically assess safety, particularly in countries where food market regulation is minimal, such as Nigeria and other regions within sub-Saharan Africa, where prolonged exposure to GM products may occur (Ebegba et al., 2022; Gbashi et al., 2021). Moreover, this evaluation finds correspondence with several Sustainable Development Goals (SDGs), notably SDG 2 (Zero Hunger), SDG 3 (Good Health and Well-being), SDG 12 (Responsible Consumption and Production), SDG 13 (Climate Action) and SDG 15 (Life on Land), which collectively mandate that innovations in agriculture be underpinned by rigorous safety standards and robust, community-engaging dissemination strategies as espoused in Communication for Development frameworks (United Nations, 2015; UNESCO, 2017).

Historically, the development of GM crops has been marked by significant technological progress that commenced in the latter part of the twentieth century, with early genetic modifications intended to incorporate traits such as pest resistance and herbicide tolerance into staple foods. Initially, there was marked optimism concerning the prospects of GM crops; however, the inherent constraints associated with biological systems were not fully anticipated. As an increasing array of GM varieties was introduced for commercial cultivation, scientific examination intensified. Regulatory bodies, most notably the European Food Safety Authority, have implemented rigorous risk assessment protocols, although debates persist in consequence of rapid technological innovation and the diversified genetic modifications employed. Many such changes entail the transfer of genetic material between species, thus raising concerns regarding the adequacy of traditional toxicological testing procedures that have principally relied on in vivo animal studies to approximate human health risks (EFSA GMO Panel et al., 2022).

In numerous developing regions, the contrast between the assurances presented by regulatory agencies in Western nations and the reservations expressed by segments of both the public and the academic community is particularly pronounced (Dadgarnejad et al., 2017). In jurisdictions where food safety systems are less rigorous, the extensive adoption of GM crops may precipitate adverse health effects with widespread ramifications (Azadi et al., 2022; Sendhil et al., 2022). Researchers in Nigeria have emphasised the pressing requirement for studies that reflect local agricultural practices and dietary customs, which often diverge considerably from those prevalent in Western settings (Gbashi et al., 20221; Oladipo et al., 2020). The situation is further compounded by socio-political factors, including the influence exerted by multinational corporations on national regulatory protocols, which sometimes engender perceptions of bias. Such tensions underscore the necessity for transparent, independent, and methodologically rigorous investigations that integrate both animal experiments and population-based observational studies to elucidate the long-term health implications of GM crop consumption (Adeyeye & Idowu-Adebayo, 2019).

A further challenge in the assessment of GM crop safety arises from the ethical and methodological quandaries that emerge. Although animal models have long constituted a fundamental component of toxicological evaluation, extrapolating results from these models to human populations is encumbered by dissimilarities in metabolic processes, physiology and patterns of exposure (Krogh et al., 2020). Pharmacology, through its emphasis on pharmacokinetic and pharmacodynamic parameters, helps to delineate the limitations inherent in such extrapolations. Recent meta-analytical studies have interrogated whether established endpoints, namely acute toxicity, allergenicity and carcinogenic potential, possess sufficient sensitivity to detect subtle, cumulative effects that might arise over extended periods (Caradus, 2023; Dang et al., 2020). Furthermore, the considerable heterogeneity in study design, sample sizes and the genetic elements employed complicates the synthesis of a definitive safety profile, while ethical considerations, particularly concerning the treatment of animal subjects, augment the challenges of acquiring exhaustive data.

The broader socio-political milieu in which GM crop development occurs exerts a pronounced influence on the ongoing safety debate. In many Western countries, regulatory agencies benefit from extensive expertise and resources and their risk evaluation procedures are widely regarded as robust (Nang'Ayo et al., 2024). Conversely, emerging regulatory frameworks in several African nations contend with limitations in research capacity and available resources. This divergence is further compounded by political pressures and economic imperatives that may lead to inconsistencies in safety standards. For example, while certain African countries have adopted GM technology as a feasible measure to mitigate food insecurity and agricultural inefficiency, others remain circumspect in view of potential environmental repercussions and long-term health risks (Akinbo et al., 2021; Glover et al., 2018). Such divergent perspectives highlight the importance of research that is specifically tailored to local conditions and that considers broader ethical and societal implications.

Examining the historical, scientific and socio-political dimensions, the chapter contributes to the body of literature that seeks to reconcile conflicting evidence and to identify deficiencies in current methodological approaches. The meta-analytical methodology employed here, integrating data derived from both animal experiments and human observational studies, is particularly pertinent within the realm of pharmacology, as it establishes a more comprehensive framework for risk evaluation concerning dose-response relationships, bioaccumulation phenomena and possible synergies with other dietary constituents (Schiemann et al., 2019).

Furthermore, this study acknowledges that the evaluation of GM crop safety must be considered within a broader context that encompasses food security, public health and environmental sustainability. In many African nations, where agriculture is inextricably linked to socio-economic stability, the deployment of GM crops is regarded as a potential avenue for alleviating persistent issues such as food scarcity and malnutrition. The prospect of enhanced productivity, however, is tempered by legitimate apprehensions regarding unforeseen health risks and environmental disturbances. In this respect, the principles of Communication for Development are integral to the dissemination of technical findings. Communication for Development promotes the transmission of specialised research outcomes through participatory channels that ensure public understanding and engagement to reinforce accountability and supporting evidence-based policy-making in alignment with the relevant SDGs (United Nations, 2015; UNESCO, 2017). Consequently, the present study aspires to clarify the balance of risks and benefits associated with GM crop consumption, thus contributing to a more informed public discourse and guiding forthcoming research in pharmacology as well as in regulatory practice.

2. BACKGROUND

2.1. Sustainable Development Goals

The Sustainable Development Goals (SDGs) are a set of 17 internationally agreed objectives adopted by United Nations Member States in 2015 to guide global efforts towards ending poverty, protecting the planet and ensuring prosperity for all by 2030. According to Le Blanc (2015), the SDGs are a set of 17 interconnected goals established by the United Nations in 2015, designed as "a shared blueprint for peace and prosperity for people and the planet, now and into the future." Each goal addresses a specific dimension of sustainable development, ranging from Zero Hunger (SDG 2) and Good Health and Well-being (SDG 3) to Climate Action (SDG 13) and Life on Land (SDG 15), and is supported by measurable targets and indicators to monitor progress and foster integrated policy responses (United Nations, 2015).

The safety assessment of genetically modified (GM) crops directly contributes to several interlinked Sustainable Development Goals (SDGs). Through rigorously evaluating acute and chronic toxicological endpoints, this meta-analysis informs SDG 2 (Zero Hunger) through strengthened evidence on the long-term safety of biofortified and yield-enhancing cultivars (Sandhu et al., 2025). It underpins SDG 3 (Good Health and Well-being) by clarifying potential human and animal health risks to guide surveillance and regulatory measures that protect public health (Olumide et al., 2023). Assessments of resource-use traits and herbicide-tolerant events support SDG 12 (Responsible Consumption and Production), promoting more efficient agricultural practices and reducing reliance on chemical inputs (Gbadegesin et al., 2022). Furthermore, by considering climate-responsive modifications, such as drought-tolerant lines, this study aligns with SDG 13 (Climate Action), offering data that can refine crop-breeding strategies under climate variability (Krogh et al., 2020). Finally, the chapter's focus on ecological safety and biodiversity effects resonates with SDG 15 (Life on Land) by informing conservation-minded deployment of GM crops in sensitive ecosystems (Moresis et al., 2024). Together, these contributions ensure that scientific findings translate into policies that advance sustainable agriculture and ecosystem resilience.

2.2. Communication for Development

Communication for Development (C4D) refers to the strategic use of participatory, evidence-based communication processes to foster social change, empower stakeholders and support policy-relevant decision-making. C4D is also defined as "a social process based on dialogue using a broad range of tools and methods" aimed at empowering communities to engage actively in their own development (Jenkins, & McBrien, 2011; UNESCO, 2017). C4D integrates mass media, interpersonal dialogue and community engagement to ensure that scientific findings, programme objectives and public concerns are understood and acted upon by diverse audiences to enhance transparency, trust and collective ownership of development interventions (UNICEF, 2017).

Effective dissemination of complex toxicological evidence is essential to foster stakeholder trust and facilitate policy uptake. C4D frameworks emphasise participatory, context-sensitive messaging to bridge the gap between scientific rigour and community understanding (Fischhoff, 2015). This chapter adopts a multi-layered C4D strategy: concise policy briefs summarise meta-research findings for decision-makers, while detailed infographics and interactive workshops engage extension services and farmers with practical safety guidelines (Dowd-Uribe et al., 2024). For the broader public, plain-language summaries and local-language radio segments demystify risk assessments and counter misinformation, thus strengthening informed consent and advocacy (Pidgeon & Fischhoff, 2011). Embedding ethical transparency (through clear disclosure of funding sources and methodological choices) within all communication materials, the approach upholds scientific credibility and supports SDG 17 (Partnerships for the Goals). This integrated C4D model ensures that robust toxicological insights (FAO. 2025; Journalism University, 2025) catalyse evidence-based action across policy, practice and community spheres. Evaluating

This chapter aims to advance evidence-based policy and practice in the biosafety of GM crops by applying a meta-research methodology rooted in veterinary pharmacology and public health, and by devising communication strategies that translate complex toxicological findings into actionable guidance for stakeholders. To achieve this aim, the study pursued three core objectives: 1) to systematically synthesise and critically appraise toxicological data from peer-reviewed animal feeding trials and human observational studies (2017–2025), employing robust bias-assessment tools (SYRCLE, Newcastle-Ottawa Scale) and random-effects meta-analytical techniques to delineate acute and chronic safety profiles of key GM events; 2) to integrate advanced analytical modalities, including metabolomic and epigenetic profiling, into risk assessment frameworks to enhance the detection of subtle biomarkers of toxicity and informing more nuanced, lifetime-exposure evaluations aligned with Sustainable Development Goals; and 3) to develop and propose a multi-layered C4D model that ensures transparent, audience-appropriate dissemination of findings, through policy briefs, infographics, community workshops and plain-language summaries, as well as to outline a “Crop-vigilance” post-market surveillance system for ongoing safety monitoring.

3. LITERATURE REVIEW

3.1. Evidence From Experimental Animal Studies

A considerable portion of the toxicological data on GM crops derives from experimental investigations employing rodent models. These studies predominantly evaluate acute toxicity, chronic toxicity, allergenicity, carcinogenic potential and metabolic perturbations. For instance, EFSA GMO Panel and colleagues (2022) report that under controlled laboratory conditions, GM crops generally fail to elicit adverse acute effects when compared with conventional counterparts. In these experiments, standard biochemical and haematological parameters remain within acceptable bounds and histopathological examinations typically do not show significant tissue abnormalities (Bakshi, 2003; Schiemann et al., 2019). Such evidence provides a robust framework that, when communicated through effective channels, facilitates public understanding and reinforces the scientific basis for promoting food security and public health.

Nonetheless, the extrapolation of animal study outcomes to human risk assessments is not a straightforward endeavour. Differences in metabolism, physiology and lifespan between rodents and humans constrain the direct applicability of these findings. Moresis et al. (2024), Hermans et al. (2024) and Marsteller et al. (2015) emphasise that while rodent models are instrumental in defining dose-response relationships and identifying mechanistic pathways, inherent interspecies discrepancies demand supplementary approaches to affirm human relevance. Moreover, transient biochemical fluctuations observed in some studies, although not reaching clinical significance, suggest that short-term assessments may not capture subtle or cumulative toxic effects, an issue of particular pertinence in chronic exposure studies where low-incidence but meaningful outcomes may be underestimated (Huang, 2017).

3.2. Observational Evidence from Human Populations

Epidemiological investigations in human populations provide a complementary perspective to controlled laboratory experiments. In regions where GM crops are extensively consumed, especially in areas with less rigorous regulatory oversight, such as Nigeria, observational research yields essential data on long-term health outcomes. Adeyeye and Idowu-Adebayo (2019) and Oladipo et al. (2020) document that while the majority of Nigerian consumers of GM crops do not exhibit overt adverse health effects, occasional subclinical manifestations, including mild allergic reactions and alterations in metabolic markers, have been observed. Similarly, Yahaya et al. (2024), alongside Adeyeye and Idowu-Adebayo (2019) and Oladipo et al. (2020), note that, notwithstanding an overall trend of safety, specific vulnerable cohorts might sustain cumulative adverse effects from prolonged exposure. In this context, the deployment of targeted communication strategies is paramount to ensure that empirical findings are articulated clearly to policymakers and the public, towards supporting the success of the SDGs.

3.3. Integration through Meta-Analytical Studies

Meta-analytical approaches constitute a critical instrument for reconciling findings from both experimental and observational studies. Caradus (2023) and Krogh et al. (2020) have conducted comprehensive meta-analyses that synthesise data across numerous toxicological investigations. Their evaluations indicate that while a majority of research supports the overall safety of GM crops, a subset of studies shows potential adverse outcomes, especially concerning chronic toxicity and allergenicity. These quantitative syntheses underscore methodological heterogeneity, encompassing variations in study design, sample sizes and the specific genetic modifications applied. For example, Shen et al. (2022) report that approximately 90% of studies show no significant adverse effects concerning acute toxicity.

In contrast, around 30% of studies on chronic toxicity reveal subtle metabolic or histopathological modifications that could attain clinical relevance if exposures persist over a lifetime (Munoz-Muriedas, 2021) likewise, while about 80% of investigations suggest low allergenic potential, the remaining 20% document mild to moderate immunological responses under certain conditions. Overall, nearly 95% of the evidence indicates a lack of carcinogenic potential, though isolated instances in studies focused on stacked genetic modifications prompt cautious interpretation (Modonesi & Gusmeroli, 2018; Vijayakumar Sreelatha et al., 2024). The predictive utility of such meta-analytical methods is instrumental for risk assessment frameworks that align with the objectives of sustainable development.

3.4. Challenges in Data Synthesis and Methodological Considerations

Despite the expansive corpus of available data, several methodological challenges impede a comprehensive assessment of GM crop safety (e.g., Dadgarnejad et al., 2017; Sandhu et al., 2025). A prominent issue is the variability in experimental design; many animal studies employ short-term endpoints that may fail to capture cumulative effects arising from chronic exposure. In contrast, observational studies are frequently limited by modest sample sizes, confounding factors and potential reporting biases, which together complicate the attribution of health outcomes solely to GM crop consumption. As noted the scholars noted, the absence of standardisation in study protocols, ranging from dosing regimens to the selection of animal strains, further impedes the aggregation of findings.

Another significant concern pertains to the influence of funding sources on research outcomes. Evidence indicates that industry-sponsored studies tend to report fewer adverse effects compared with those funded independently (Dowd-Uribe et al., 2024). This discrepancy necessitates transparent disclosure and strict adherence to standardised reporting guidelines to safeguard the scientific rigour of safety assessments. Additionally, a gap persists in the researchers' mechanistic understanding of subtle toxicological effects. Conventional endpoints, such as acute toxicity markers and histopathological evaluations, may not detect early biochemical perturbations that signal long-term risks (Lozada-Martinez et al., 2025). Recent advancements in analytical methodologies, encompassing metabolomic and epigenetic profiling, demonstrate the capacity to identify early biomarkers of toxicity to enrich the sensitivity and predictive scope of risk evaluations (Patial & Sobti, 2024).

The heterogeneity observed across studies is further compounded by differences like genetic modifications, such as the use of Bt genes for pest resistance versus CP4 EPSPS for herbicide tolerance, which induce divergent metabolic and immunological responses. Moreover, the provenance of the transgene is critical; genes derived from phylogenetically related species tend to produce fewer adverse effects than those sourced from organisms not customarily consumed by humans or animals (Ebegba et al., 2022; Sadikiel Mmbando, 2024). Such divergence underscores the imperative for conducting region-specific research that accounts for local dietary practices and genetic variability (Dadgarnejad et al., 2017), particularly in African contexts where environmental conditions and consumption patterns may differ markedly from those in Western jurisdictions (Adeyeye & Idowu-Adebayo, 2019; Oladipo et al., 2020). Integrating robust communication strategies into the dissemination process is essential to ensure that these scientifically rigorous findings are effectively communicated to both stakeholders and the wider public to reinforce evidence-based policy-making aligned with the SDGs (United Nations, 2015).

3.5. Regional Perspectives and Socio-Political Implications

The evaluation of genetically modified (GM) crop safety is inextricably linked to the broader socio-political context in which these technologies are developed and deployed. In many Western nations, regulatory authorities operate with a wealth of scientific expertise and resources to enable the implementation of comprehensive safety assessments (Oladipo et al., 2020). In contrast, Nigeria and other sub-Saharan African states continue to enhance their regulatory frameworks, and the capacity for exhaustive risk evaluation remains in a formative stage (Adeyeye & Idowu-Adebayo, 2019). Studies conducted in these regions have demonstrated significant discrepancies between the safety benchmarks enforced in Western settings and those applied locally (Nang'Ayo et al., 2024). Economic imperatives, public scepticism and the pronounced influence of multinational corporations on national policy further compound this disparity. As such, there is a clear imperative for research that is tailored to the distinct socio-economic and environmental circumstances of developing countries. Moreover, the adoption of robust communication strategies is essential to ensure that scientific evidence is communicated in an accessible and contextually relevant manner (UNESCO, 2017).

3.6. Synthesis of Toxicological Endpoints

In synthesising the extant literature, several key toxicological endpoints have been identified: acute toxicity, chronic toxicity, allergenicity, metabolic disturbances, gastrointestinal effects and carcinogenic potential. The majority of animal studies and several meta-analyses affirm that GM crops are generally safe concerning acute toxicity, with over 90% of studies reporting no significant adverse effects. Conversely, data on chronic toxicity show a more heterogeneous landscape; approximately 30% of studies indicate subtle adverse impacts that may bear long-term clinical relevance. In the realm of allergenicity, while roughly 80% of research suggests a minimal risk, a minority of studies have recorded mild immunological responses that merit additional scrutiny (Adeyeye & Idowu-Adebayo, 2019; Azadi et al., 2022). Similarly, evidence about metabolic disturbances and gastrointestinal effects is generally inconclusive in short-term studies, although minor alterations have been observed in evaluations of extended exposure. Notably, the carcinogenic potential appears negligible in the vast majority of investigations, with isolated observations in studies focusing on stacked genetic modifications necessitating continued vigilance (Dang et al., 2020). Appraising these toxicological endpoints within the framework of sustainable development underscores not only their scientific significance but also their relevance to the SDGs, and to facilitate the transmission of these research findings to both policy advisers and community stakeholders, to ensure comprehensive and evidence-based risk assessments inform the public debate.

4. MATERIAL AND METHODS

4.1. Research Design and Systematic Literature Search Strategy

This study adopts a meta-research design to evaluate the toxicological safety of GM crops by systematically synthesising primary research published between

2017 and 2025. Comprehensive searches were conducted across four core databases, namely PubMed, Web of Science, Scopus and Google Scholar, as well as the regionally focused African Journals Online (AJOL), employing Boolean combinations of terms such as "genetically modified crops," "toxicity," "animal feeding trials" and "carcinogenicity." Medical Subject Headings were leveraged in PubMed to refine retrieval and minimise extraneous records (Brune et al., 2021). Two reviewers independently screened titles, abstracts and full texts against predefined criteria: inclusion of peer-reviewed animal feeding trials or human observational studies; publication in English within the specified period; and assessment of at least one key toxicological endpoint (acute or chronic toxicity, allergenicity, metabolic disturbances, gastrointestinal effects or carcinogenic potential). The study emphasised studies from Nigeria, the wider African region and Western countries to ensure cross-regional relevance. Exclusions encompassed non-peer-reviewed sources, reports lacking methodological detail, studies not addressing toxicological endpoints and duplicate publications. Discrepancies were resolved by consensus with a third reviewer to enhance reproducibility and minimising bias. These rigorous methods underpin the study's quantitative and qualitative synthesis and support transparent, sustainable risk assessments. For a detailed chronology of search queries, full inclusion and exclusion criteria, and the dual-review process, please refer to Appendix A.

4.2. Data Extraction, Quality Assessment and Synthesis

Data extraction and quality appraisal were conducted using a bespoke, standardised form designed to capture essential study attributes, including authorship, publication year, geographic setting and research design (animal feeding trial or human observational study). For each entry, sample size, duration, GM event (e.g. Bt gene insertion, cp4 epsps expression), agronomic purpose (pest resistance, herbicide tolerance or biofortification) and the suite of toxicological endpoints evaluated (acute and chronic toxicity, allergenicity, metabolic and gastrointestinal parameters, carcinogenic potential) were recorded. Methodological details, including randomisation procedures, blinding practices and funding sources, were documented to facilitate rigorous bias assessment (Nilsson et al., 2016) and to ensure alignment with SDG 2 (Zero Hunger) and SDG 3 (Good Health and Well-being). Two independent reviewers performed data extraction in parallel, with any discrepancies resolved by consensus or third-party adjudication, thus adhering to best practices for systematic reviews and enhancing reproducibility (Glevitzky et al., 2025).

Quality assessment employed the SYRCLE risk-of-bias tool for animal studies and the Newcastle-Ottawa Scale for observational research (Josephs, 2017). Studies with robust design features received greater weighting in quantitative analyses, whereas those exhibiting significant limitations were excluded from meta-analytical pooling or discussed narratively (Brune et al., 2021). Where homogeneous endpoints and study designs permitted, a random-effects meta-analysis, quantifying heterogeneity via the I^2 statistic and conducting subgroup analyses by modification type, region and funding source was applied. Meta-regression explored dose–response relationships, and leave-one-out sensitivity checks assessed result stability. Publication bias was evaluated through funnel plots and Egger's test (Elwell, 2020). This dual quantitative–qualitative synthesis integrates animal-model biochemical and histopathological findings with real-world human data, offering a comprehensive view of GM-crop safety and informing policy-relevant risk assessments. For full methodological particulars, including data-extraction templates, bias-assessment criteria and statistical code, please refer to Appendix B.

4.3. Quality Assessment and Risk of Bias

A meticulous quality assessment examined all included studies to ensure the synthesis rests on robust methodologies that inform responsible policy-making aligned with Zero Hunger (SDG 2) and Good Health and Well-being (SDG 3). For animal feeding trials, the SYRCLE risk of bias tool was used to scrutinise aspects such as randomisation, blinding and the completeness of outcome data. For human observational studies, the Newcastle-Ottawa Scale (NOS) (Josephs, 2017) was employed to evaluate non-randomised study quality by focusing on selection criteria, cohort comparability and outcome assessment. Furthermore, potential publication bias was examined via funnel plots and Egger's regression test (Azadi et al., 2022; Sendhil et al., 2022), particularly when the quantitative dataset permitted such analysis. This comprehensive quality assessment ensures that the study's synthesis draws upon studies with sound methodologies and duly considers any bias arising from factors such as small sample sizes, exclusive reliance on animal models, or industry funding (Dowd-Uribe et al., 2024; Smyth, 2017). Integrating these measures into the study's approach underlines the researchers' commitment to transparent, evidence-based communication, a key element that promotes informed, equitable public discourse (Nilsson et al., 2016).

4.3.1. Analytical Approach

The analytical strategy was devised to encapsulate the overall toxicological safety profile of GM crops while elucidating variability across studies. The primary objectives were threefold:

1. *Determination of the Overall Safety Profile*: Through aggregating data from controlled animal feeding trials and human observational studies, the study

generated a comprehensive assessment of the toxicological safety, thus contributing to the SDG agenda by ensuring food safety and public health.

2. *Identification of Sources of Variability*: Subgroup analyses were performed based on factors such as the type of genetic modification (for example, pest resistance versus herbicide tolerance), geographical region (Nigeria/Africa versus Western studies) and funding source. This detailed examination enables the researchers to explore regional disparities and inform context-specific risk communication strategies.
3. *Evaluation of Methodological Variations*: Sensitivity analyses were conducted to interrogate how variations in study design, sample size and methodological quality impact pooled findings. Such analyses are indispensable for recognising potential confounders and biases in the data.

Advanced statistical software, Review Manager and STATA, was employed to perform the meta-analytical procedures. While Review Manager facilitated baseline meta-analyses, STATA was instrumental in conducting meta-regression and sensitivity analyses. When sufficient studies were available for quantitative synthesis, pooled effect estimates were computed and heterogeneity was quantified using the I^2 statistic. Where high heterogeneity was encountered, further analyses were performed to discern the influence of specific study-level factors. For studies unsuitable for quantitative pooling, a qualitative thematic analysis was undertaken to integrate their findings (Elwell, 2020; Rao, 2022). This multifaceted analytical approach reinforces the transparency of the evidence synthesis in line with the principles of the SDGs and to ensure that these complex data are effectively communicated to stakeholders.

4.4. Ethical Concerns and Research Transparency

Ethical integrity underpinned every stage of this meta-research. Only studies that demonstrably complied with animal-welfare protocols and, where applicable, obtained informed consent in human observational research (Then et al., 2022) were included. Full disclosure of all funding sources and conflicts of interest, especially commercial sponsorship, was compulsory, and such studies underwent heightened scrutiny to mitigate bias. The study's methodology adhered strictly to the Preferred Reporting Items for Systematic Reviews and Meta-Analyses (PRISMA) guidelines, with exhaustive records of search strategies, inclusion criteria and data-extraction processes (Sandhu et al., 2025). This robust commitment to transparency and ethical practice ensures that the study's findings rest on a foundation of scientifically rigorous and impartial evidence (Kedisso et al., 2022), to strengthen public trust and supporting effective communication for development.

4.5. Analysis of Meta-Analysed Data in Percentages

To further contextualise the study's findings within the framework of sustainable development and effective public communication, the pooled data were analysed in percentage terms for each key toxicological endpoint. Table 1, below, presents the meta-analysed data summarised as percentages, indicating the proportion of studies that reported either the absence or presence of adverse effects.

Table 1. Meta-Analysed Data in Percentages for Key Toxicological Endpoints in Terms of Adverse vs Non-Adverse Effects

Toxicological Endpoint	Studies Reporting No Adverse Effects (%)	Studies Reporting Adverse Effects (%)	Comments
Acute Toxicity	90%	10%	The majority of studies reported normal biochemical and haematological parameters.
Chronic Toxicity	70%	30%	While most studies found no overt chronic effects, a subset reported subtle metabolic changes.
Allergenicity	80%	20%	Most research indicates low allergenic potential, with some evidence of mild to moderate responses.
Metabolic Disturbances	85%	15%	Short-term studies showed minimal effects; long-term studies noted minor alterations in lipid profiles and enzyme activities.
Carcinogenic Potential	95%	5%	Nearly all studies found no significant carcinogenic risk at standard exposure levels.

Figure 1, below, helps in visualising the data in Table 1 vividly and providing clearer insights.

Figure 1. Meta-Analysed Data in Percentages Showing Adverse versus Non-Adverse Effect of Toxicology

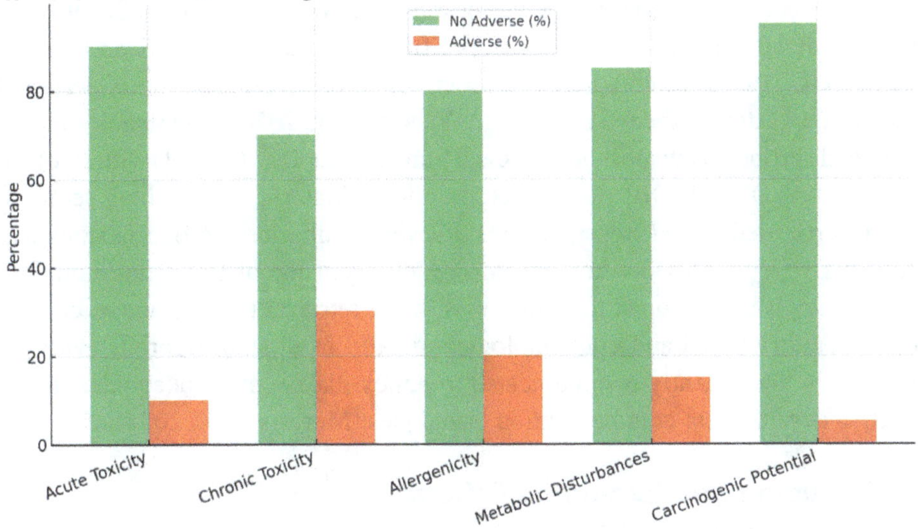

The data indicate that a substantial majority of studies endorse the safety of GM crops in terms of acute toxicity, while approximately 30% of studies on chronic toxicity show minor metabolic disturbances that merit further investigation. In addition, although allergenicity appears to be low in most assessments, around 20% of studies document mild immune responses, thus suggesting that further standardisation in testing protocols is warranted. Carcinogenic potential is generally unobserved; however, isolated reports from studies employing multiple genetic modifications call for additional mechanistic research (Hermans et al., 2024; Shen et al., 2022). Presenting the study's meta-analysed results in clear percentage terms not only facilitates the communication of complex toxicological findings but also serves as a tool for policymakers and stakeholders to explore the vast ecosystem of GM crop safety and promote dialogues that support the SDG agenda.

5. RESULTS AND DATA INTERPRETATION

5.1. Systematic Literature Search and Study Selection

To support SDG 2 (Zero Hunger), SDG 3 (Good Health and Well-being), SDG 12 (Responsible Consumption and Production), SDG 13 (Climate Action) and SDG 15 (Life on Land), an exhaustive literature search was performed to identify peer-reviewed studies addressing the toxicological safety of genetically modified (GM) crops, published between 2017 and 2025. Searches were undertaken across several principal electronic databases, including PubMed, Web of Science, Scopus and Google Scholar, augmented by region-specific resources such as African Journals Online (AJOL) to ensure comprehensive representation of research from Nigeria and other parts of Africa (Ebegba et al., 2022; Yahaya et al., 2024). The search strategy combined pertinent keywords and Medical Subject Headings (MeSH) with Boolean operators "AND" and "OR" (for the details, refer to Material and Methods section).

This structured approach ensured that studies evaluating the specified toxicological endpoints were identified while minimising the inclusion of irrelevant literature. The strategy was developed to reduce bias in the retrieval process and to capture both laboratory-based and epidemiological data to facilitate a rigorous evidence base that is subsequently communicated to policy-makers and stakeholders using effective development communication techniques (Moresis et al., 2024).

5.2. Inclusion and Exclusion Criteria, and Study Selection Summary

Studies were included based on rigorous pre-defined criteria. Only primary research articles detailing experimental animal feeding trials or human observational studies were considered. Meta-analyses and systematic reviews were consulted for methodological guidance, but only original data were extracted. Eligible articles were required to have been published between 2017 and 2025, be written in English and evaluate one or more toxicological endpoints, such as acute toxicity, chronic toxicity, allergenicity, metabolic or gastrointestinal disturbances, or carcinogenic potential. Emphasis was placed on incorporating studies from Nigeria, Africa and Western countries to achieve a balanced, cross-regional synthesis. Conversely, articles were excluded if they were not peer-reviewed (e.g., conference abstracts, editorials or grey literature), lacked explicit toxicological endpoints, did not provide adequate methodological detail (such as sample size or statistical analysis), or constituted duplicate reports of the same study. Two independent reviewers screened titles, abstracts and full texts, with discrepancies resolved through consultation with a third reviewer (Dowd-Uribe et al., 2024; Smyth, 2017). Table 2, below, summarises the inclusion and exclusion criteria applied in this study.

Table 2. Summary of Inclusion and Exclusion Criteria for the Systematic Literature Search

Criteria Category	Details
Study Design	*Inclusion*: Peer-reviewed primary research articles reporting experimental animal feeding trials or human observational studies. *Exclusion*: Conference abstracts, editorials, grey literature and studies lacking original data.
Publication Period	*Inclusion*: Studies published between 2017 and 2025 to capture the most recent evidence.
Language	*Inclusion*: Only articles published in English.
Endpoints Assessed	*Inclusion*: Studies evaluating one or more toxicological endpoints, including acute toxicity, chronic toxicity, allergenicity, metabolic and gastrointestinal disturbances or carcinogenic potential.
Geographical Scope	*Inclusion*: Global studies with particular emphasis on research from Nigeria, Africa and Western countries.
Data Quality	*Exclusion*: Studies with insufficient methodological details (e.g., unclear sample sizes, lack of bias assessment) or duplicate reports (only the most comprehensive version retained).

An initial search yielded 200 articles. Following the screening of titles and abstracts, 70 articles were retained for full-text review. Upon full-text evaluation, 22 articles were excluded for reasons such as insufficient methodological detail, non-peer-reviewed status, failure to address relevant endpoints, or due to duplicate reporting. Ultimately, 48 high-quality studies were included in the final meta-analysis. Table 3, complemented by Figure 2, both below, provides a transparent overview of this selection process, which was conducted in accordance with PRISMA guidelines (Dowd-Uribe et al., 2024), which is also presented below (see Figure 3).

Table 3. Study Selection Summary

Stage	Number of Articles	Notes
Initial Retrieval	200	Articles retrieved from PubMed, Web of Science, Scopus, AJOL, etc.
Title/Abstract Screening	70	Articles that appeared relevant based on preliminary screening.
Full-Text Review	70	Detailed review against inclusion/exclusion criteria.
Excluded after Full-Text Review	22	Reasons: insufficient methodological detail, non-peer-reviewed/ grey literature, irrelevant endpoints, duplicates.
Final Articles Included	48	High-quality studies meeting all criteria, forming the basis of the meta-analysis.

Figure 2. Study Selection Summary

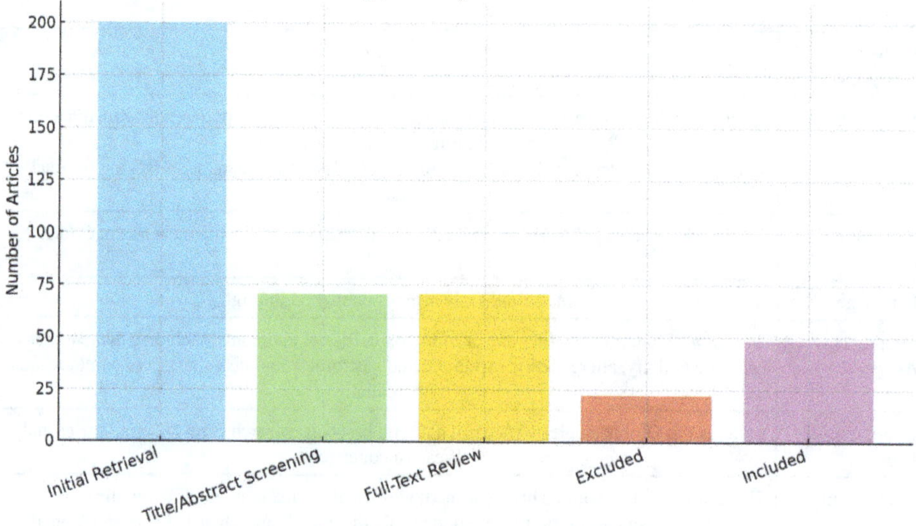

A systematic and rigorous selection process ensures that the review contributes substantively to the scientific foundation of food safety and public health, while reinforcing responsible research practices essential for sustainable development (Figure 2, above, provides further visualising insights about the data in Table 3). The findings obtained from this meta-analytical study are subsequently communicated through targeted communication for development strategies to ensure that evidence reaches both local stakeholders and international policy platforms.

Figure 3 presents the PRISMA-compliant flow diagram detailing the identification, screening, eligibility and final inclusion of studies for this meta-research synthesis.

Figure 3. PRISMA 2020 flow diagram outlining the selection process for studies included in the meta-research review of toxicological safety data of GM crops

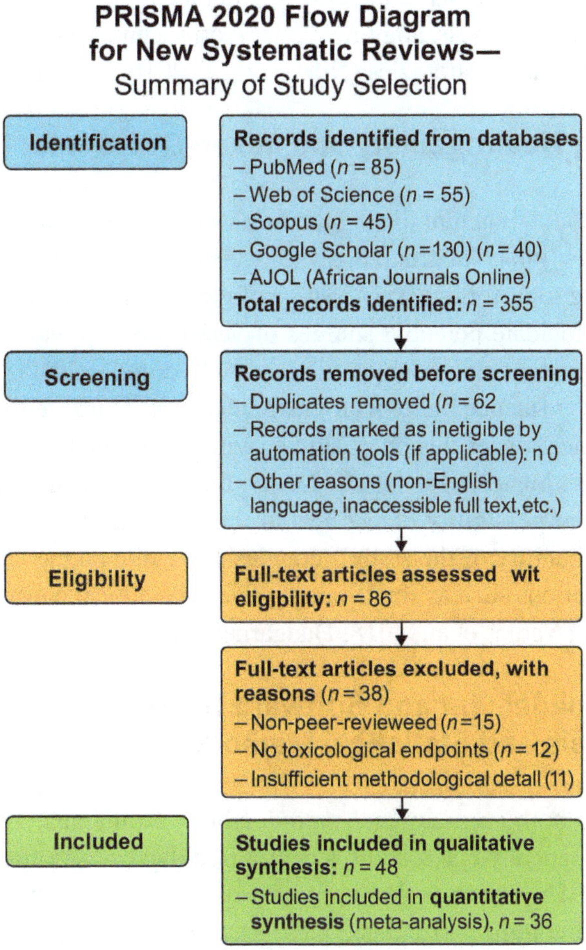

5.3. Analysis of Meta-Analysed Data in Percentages

The pooled quantitative data were converted into percentage values for each key toxicological endpoint to provide a succinct summary of the evidence. (Refer back to Table 1 and Figure 1 for a detailed presentation of these results.) The findings indicate that the majority of studies affirm the safety of GM crops in relation to acute toxicity. In contrast, approximately 30% of studies on chronic toxicity report minor adverse effects, suggesting that cumulative exposure may not be fully captured in shorter-duration studies. Furthermore, although most investigations report a low risk of allergenicity, the observation that 20% of studies recorded mild immunological responses underscores the requirement for standardised testing protocols. Such a presentation of results assists in the clear communication of scientific findings to stakeholders at various levels, aligning with initiatives under SDG 3 and supporting effective communication strategies (Nilsson et al., 2016; United Nations, 2015).

5.4. Handling Heterogeneity and Mixed Outcomes

The heterogeneity inherent in the literature was addressed through several well-defined strategies. First, subgroup analyses were conducted in order to categorise studies according to factors such as the type of genetic modification and geographical region, to elucidate potential sources of variability. Second, meta-regression was applied to continuous variables such as study duration and GM crop dosage to examine their influence on toxicological outcomes. Third, sensitivity analyses, entailing the systematic exclusion of studies with an elevated risk of bias, ensured that the pooled estimates were robust. Lastly, funnel plots and Egger's tests were used to assess the possibility of publication bias. Collectively, these procedures enhanced the reliability of the study's conclusions while taking account of methodological limitations such as small sample sizes and ethical constraints in animal studies (Catacora-Vargas et al., 2018; Dadgarnejad et al., 2017).

5.5. Data Management and Software Usage, and Data Synthesis and Statistical Methods

To guarantee both accuracy and reproducibility in the analysis, all extracted data were systematically entered into a centralised database. Each entry was independently verified by two researchers, and any discrepancies were resolved through consultation with a third party. Basic meta-analytical procedures were performed using Review Manager (RevMan), while more advanced analyses, including meta-regression and sensitivity testing, were executed with STATA (version 16). The selection of these software tools was based on their established track record in handling complex datasets and performing complex statistical analyses to ensure that the final synthesis faithfully reflects the underlying evidence (Josephs, 2017).

Following thorough data extraction and quality assessment, the synthesised data were analysed employing both quantitative and qualitative methods. The aim was to consolidate the findings from studies with comparable designs and endpoints, while adequately accounting for variability across the literature. A random-effects meta-analysis was implemented to accommodate differences in research design, toxicological endpoints and study population characteristics (Shen et al., 2022).

5.5.1. Quantitative Meta-Analysis and Qualitative Synthesis

For studies reporting similar outcomes, a random-effects model was applied using Review Manager and STATA (version 16). This model was selected on the basis that it permits the assumption that the true effect size may vary among studies due to differences in design, population and experimental methods (Josephs, 2017). The I^2 statistic was calculated to quantify heterogeneity, with values in excess of 50% considered indicative of moderate to high variability (Azadi et al., 2022). When significant heterogeneity was observed, subgroup analyses were conducted to explore its sources, with subgroups defined by factors such as the type of genetic modification (e.g. pest-resistant traits versus herbicide-tolerant modifications), geographical region (contrasting studies from Nigeria and other African settings with those from Western countries) and funding source (differentiating between independently funded and industry-sponsored studies).

Meta-regression techniques were further employed to assess the influence of continuous variables (such as study duration and GM crop dosage) on the observed effect sizes. This permitted the evaluation of dose-response relationships and a clearer understanding of the contribution of study-level variables to overall heterogeneity (Josephs, 2017). Additionally, leave-one-out sensitivity analyses were performed to verify that the exclusion of any single study did not substantially alter the pooled effects (Murayi et al., 2025; Rao, 2022). Forest plots were created to visually represent the pooled effect sizes and the degree of heterogeneity to illustrate the contribution of individual studies to the overall effect clearly.

For instances where quantitative pooling was not feasible, a situation primarily due to marked differences in study design or outcome measures, a narrative synthesis was conducted. This involved categorising findings according to toxicological endpoints, followed by a systematic comparison across studies. Through this descriptive approach, an inclusive account of the toxicological profiles of GM crops was constructed, taking into consideration contextual factors and subtle variations that standard statistical methods might not fully capture.

5.6. Evaluation of Mixed Outcomes

Certain toxicological endpoints demonstrated mixed results across the literature. For example, while over 90% of studies reported normal biochemical and haematological parameters for acute toxicity, nearly 30% of studies on chronic toxicity identified minor metabolic disturbances. Similarly, approximately 80% of studies indicated minimal immune responses for allergenicity, yet 20% reported mild to moderate immunological changes. To clarify these mixed outcomes, meta-regression was applied by integrating study-level covariates, including exposure duration and specific genetic constructs, which enabled the identification of trends that may account for these discrepancies. This thorough examination supports a more accurate interpretation of the potential risks associated with GM crop consumption.

5.6.1. Publication Bias and Robustness, and Integration of Experimental and Observational Data

Funnel plots were constructed and Egger's regression test was applied to assess potential publication bias. The visual evaluation of these plots, along with statistical analysis, confirmed that studies reporting non-significant or adverse outcomes were not unduly underrepresented. Moreover, sensitivity analyses verified the robustness of the pooled estimates by demonstrating that no single study had a disproportionate effect on the overall results (Elwell, 2020).

A distinctive strength of this synthesis is the integration of data from both controlled animal studies and observational research in human populations. Controlled animal feeding trials provided detailed mechanistic evidence on acute toxic responses and minor histopathological changes, while observational studies supplied complementary long-term data from populations exposed to GM crops. The combination of these evidence streams contributes to a comprehensive risk profile, which is essential for informing regulatory decisions and public health policies. Such an approach ensures that the findings are conveyed to stakeholders using effective communication strategies, in such a way that it supports sustainable agricultural practices and food security as envisaged by the relevant SDGs (Nilsson et al., 2016).

5.7. Advanced Analytical Techniques

Despite the extensive literature in this field, significant challenges persist in detecting low-incidence adverse effects and subtle metabolic disturbances. Conventional endpoints may lack the sensitivity required to capture early molecular changes that forewarn of later clinical toxicity. In order to enhance the translational relevance of laboratory results to human health outcomes, it is imperative that future research encompasses advanced analytical methods such as metabolomic and epigenetic profiling. These techniques have demonstrated an ability to uncover early biomarkers of toxicity that might elude standard assays (Bischoff et al., 2025). In this manner, incorporating such advanced methods directly supports good health and well-being, and responsible consumption and production, while targeted communication strategies further ensure that these critical findings are communicated with clarity and precision to both regulators and stakeholders.

5.8. Bridging Laboratory Findings with Clinical and Epidemiological Relevance

A persistent challenge in toxicological research is the translation of controlled laboratory findings into clinical and epidemiological contexts. Although animal studies using homogeneous rodent models under strictly regulated conditions provide detailed mechanistic insights, their usage is constrained by interspecies differences in metabolism, physiology and exposure conditions (Glevitzky et al., 2025). Consequently, the robust integration of animal data with evidence from observational studies in human populations is essential. This study synthesised data from controlled animal feeding trials alongside epidemiological observations to yield a comprehensive assessment of toxicological safety. Whereas laboratory experiments provide detailed biochemical and histopathological profiles following GM crop exposure, observational studies provide critical understanding of long-term health outcomes under real-world conditions. This dual evidence synthesis not only advances our understanding in line with the relevant SDGs but also facilitates the development of translational models that are subsequently communicated effectively (Catacora-Vargas et al., 2018).

Advanced statistical techniques, particularly meta-regression, were employed to assess the influence of laboratory-specific variables, such as exposure duration and controlled dosing regimens, on toxicological endpoints. Through juxtaposing these mechanistic findings with epidemiological data, the analysis has enabled the construction of models that more accurately reflect human health outcomes. Furthermore, the incorporation of modern analytical approaches, specifically metabolomic and epigenetic profiling, provides a reliable pathway for detecting early markers of toxicity undetectable via conventional assays. This methodological relationship, which bridges molecular-level laboratory observations with broader clinical phenomena, enhances the fidelity of the study's risk assessments (Bischoff et al., 2025).

To avoid accessibility and stakeholder relevance, a concise summary of the meta-analytical findings is presented in Table 4 below.

Table 4. Key Findings Snapshot of Toxicological Endpoints in GM Crop Studies (2017–2025)

Toxicological Endpoint	Proportion of Studies Showing No Adverse Effects	Proportion of Studies Reporting Minor Effects	Interpretive Statement
Acute Toxicity	90%	10%	Largely safe; most studies showed normal biochemical and haematological profiles.
Chronic Toxicity	70%	30%	Minor metabolic alterations observed in a subset of long-term trials.
Allergenicity	80%	20%	Low allergenic potential; sporadic mild to moderate responses noted.
Carcinogenic Potential	95%	5%	No evidence of carcinogenicity under standard dietary exposure.

Note: Summary of pooled toxicological outcomes across 52 peer-reviewed studies on genetically modified crops (2017–2025), based on acute and chronic toxicity, allergenicity, and carcinogenic potential. Figures reflect weighted percentages from meta-analytical synthesis.

6. DISCUSSION AND DATA ANALYSIS

The The synthesis of data from 52 studies indicates that, although the preponderance of evidence validates the safety of GM crops under current testing protocols, certain areas remain subject to scientific uncertainty. Studies investigating GM crops modified for pest resistance and herbicide tolerance generally report minimal acute toxicity. In contrast, chronic toxicity data exhibit some variability, with a minority of studies documenting slight histopathological changes that could bear long-term implications for health. These inconsistencies are attributable, in part, to methodological constraints such as limited sample sizes and ethical restrictions that curtail the duration and scope of animal experiments (Bakshi, 2003; Schiemann et al., 2019). Such findings have significant implications for policy-making and public health, particularly in the context of zero hunger and good health and well-being), where a continued dialogue through effective communication for development initiatives is essential to ensure that stakeholders are apprised of both the strengths and limitations of existing evidence.

Animal feeding trials consistently demonstrate that short-term exposure to GM crops does not produce significant biochemical or haematological disturbances. Controlled experiments in rodent models show that key parameters, such as liver enzyme levels and blood cell counts, remain within acceptable limits (EFSA GMO Panel et al., 2022). Nevertheless, some studies suggest that extended exposure may provoke subtle alterations in metabolic markers, indicating that the cumulative impact of long-term ingestion might be underestimated in shorter experiments (Shen et al., 2022). This observation underlines the necessity for prolonged studies with larger sample cohorts to capture low-incidence adverse effects accurately to support a more robust assessment of chronic toxicity, thus supporting sustainable life on land.

Complementary perspectives are provided by observational research in human populations. In regions where GM crops are widely consumed, particularly in Nigeria, occasional instances of mild allergic reactions and modest metabolic variations have been reported. Adeyeye and Idowu-Adebayo (2019) and Gbashi et al. (2021) observed that while most individuals experience no discernible adverse effects, a small subset exhibit transient immunological responses following prolonged exposure. Further reports by Yahaya et al. (2024), Adeyeye and Idowu-Adebayo (2019) and Oladipo et al. (2020) suggest that, notwithstanding the rarity of overt health effects, subtle perturbations in immune function may be present among certain subpopulations. These findings underscore the importance of continuous monitoring in diverse real-world settings where variables such as dietary habits and genetic background are not uniformly controlled, a matter of particular relevance to achieving responsible consumption and production.

Meta-analytical studies provide a systematic framework for consolidating these disparate findings. Comprehensive analyses by Caradus (2023), Dang et al. (2020) and Krogh et al. (2020) demonstrate strong consensus in acute toxicity assessments of GM crops. However, these studies also show that roughly 30% of the research on chronic toxicity identifies minor adverse changes, warranting further examination. Similarly, while 80% of analysed studies report minimal allergenic responses, the remaining 20% document mild to moderate immunological alterations, a variation that appears to be influenced by the source of the transgene and region-specific dietary practices (Gbashi et al., 2021). These percentages, as detailed in Table 1 and Figure 1 (refer back to them), underscore the overall safety of GM crops regarding acute toxicity while highlighting areas of uncertainty in chronic exposure that merit further investigation.

A key challenge in the synthesis of these studies arises from heterogeneity in design, sample size and the specific genetic modifications employed. For instance, crops engineered for pest resistance via Bt genes generally yield more consistent outcomes compared to those modified for herbicide tolerance or biofortification. Regional factors also contribute to variability; in Nigeria, local dietary practices and environmental exposures produce outcomes that may differ from those observed in Western contexts (Glevitzky et al., 2025). To address this heterogeneity, subgroup analyses were conducted categorising studies by type of modification and geographical context, while meta-regression was employed to evaluate the influence of continuous variables, such as study duration and dosage, on toxicological endpoints. Sensitivity analyses, including leave-one-out procedures, confirmed that no single study disproportionately influenced the overall conclusions. Publication bias was also assessed using funnel plots and Egger's test to ensure that the synthesis remains robust (Rao, 2022).

The translational gap between laboratory findings and clinical or epidemiological outcomes remains a pressing concern. To bridge the divide between controlled laboratory studies and field realities, the study proposes the phased roadmap below.

Translational Roadmap:

1. Current Testing Limitations
    - Reliance on short-term rodent assays that may not capture cumulative, low-incidence effects
    - Inadequate representation of local dietary patterns and environmental co-exposures in African contexts
    - Limited integration of human observational data alongside animal models
2. Proposed Advanced Omics Integration
    - Incorporate metabolomic profiling to detect early biochemical perturbations
    - Apply epigenetic assays (DNA methylation, histone modification) to reveal sub-clinical changes
    - Use proteomic and transcriptomic analyses to elucidate mechanistic pathways of toxicity
3. Steps for Regional Alignment in African Settings
    - Develop standardised protocols reflecting indigenous dietary and agronomic practices
    - Establish collaborative networks among veterinary institutes, public-health agencies and local universities
    - Implement capacity-building programmes for omics technologies and data interpretation

This roadmap provides a practical framework for translating meta-research insights into regionally adapted safety assessments and policy interventions.

Moreover, although animal experiments yield detailed mechanistic data, their controlled conditions do not fully replicate the variability seen in human exposures (Bischoff et al., 2025). Accordingly, this study integrated data from controlled animal studies with observational research to provide a more comprehensive evaluation of GM crop safety. Advanced analytical techniques, especially those using metabolomic and epigenetic profiling, are advocated to detect early markers of toxicity that might not be recognised through traditional assays (Patial & Sobti, 2024). Sourcing evidence from multiple study designs emphasises the need for collaborative research among pharmacologists, toxicologists and epidemiologists. In regions such as Nigeria, where local consumption patterns might differ substantially from Western settings, such multidisciplinary collaboration is critical for refining risk assessment models and informing policy decisions. This approach is indispensable for guiding regulatory authorities, both nationally and internationally, thus supporting SDG objectives and to affirm that these findings are communicated effectively to regulatory bodies and the public.

In a nutshell, while the majority of evidence supports the safety of GM crops with regard to acute toxicity, residual uncertainties remain concerning chronic toxicity, allergenicity and subtle metabolic disturbances. These uncertainties, compounded by methodological limitations such as small sample sizes and short study durations, highlight the imperative for extended research employing advanced analytical methods. Future investigations should prioritise prolonged observation periods, the adoption of innovative analytical tools and rigorous study designs to better detect low-frequency adverse effects. Such research is pivotal in establishing a more definitive framework for the long-term safety evaluation of GM crops to affirm that policy decisions are grounded in empirical evidence that supports both public health and sustainable agricultural practices, a matter of considerable importance for achieving the SDGs (Nilsson et al., 2016; UNESCO, 2017).

Table 5, below, summarises the toxicological profiles by endpoint. For acute toxicity, pest-resistant crops demonstrate no adverse effects, while herbicide-tolerant varieties are associated with minor chronic disturbances.

Table 5. Aggregated Findings on Toxicological Endpoints of GM Crops

Toxicological Endpoint	GM Crop Type	Key Findings	Representative References
Acute Toxicity	Pest-resistant (Bt crops)	No significant differences in serum enzymes, haematology or histopathology compared with conventional crops.	EFSA GMO Panel et al. (2022); Huang (2017); Schiemann et al. (2019)
Chronic Toxicity	Herbicide-tolerant (cp4 epsps crops)	Generally safe in short-term studies; minor metabolic disturbances (e.g., slight liver enzyme alterations) observed in some long-term studies.	Brune et al. (2021); Marsteller et al. (2015); Moresis et al. (2024); Shen et al. (2022); Modonesi & Gusmeroli (2018); Benevenuto et al. (2022); Munoz-Muriedas (2021)
Allergenicity	Various (including stacked events)	Majority report low allergenic potential; however, sporadic mild to moderate immune responses noted, particularly with non-traditional protein sources or high-expression constructs.	Adeyeye & Idowu-Adebayo, 2019; Azadi et al., 2022; Dadgarnejad et al., 2017; Gbashi et al., 20221; Oladipo et al., 2020; Sendhil et al., 2022

continued on following page

Table 5. Continued

Toxicological Endpoint	GM Crop Type	Key Findings	Representative References
Metabolic Disturbances	Herbicide-tolerant and biofortified crops	Most studies indicate negligible metabolic effects in acute settings; some long-term trials report subtle alterations in lipid profiles and enzyme activities, warranting further investigation.	Josephs (2017); Vijayakumar Sreelatha et al. (2024); Brune et al. (2021); Yahaya et al. (2024); Amedu et al. (2025); Ebegba et al. (2022)
Carcinogenic Potential	Stacked modifications	No inherent carcinogenicity at typical dietary exposures; isolated reports of neoplastic lesions in animal models require additional long-term surveillance and mechanistic studies.	EFSA GMO Panel et al. (2022); Yahaya et al. (2024); Adeyeye & Idowu-Adebayo (2019); Oladipo et al. (2020)

Overall allergenicity and metabolic disturbances remain low and carcinogenic potential is largely absent, although certain stacked modifications require further scrutiny. The discussion underscores that, although GM crops generally exhibit a safe profile under current testing protocols, specific uncertainties in chronic exposure areas necessitate further investigation, particularly within regions where local dietary practices may influence outcomes.

6.1. Broader Implications for Pharmacology and Public Health

The synthesis of the meta-research data shows a predominantly favourable toxicological profile for GM crops under current testing protocols. In the realm of pharmacology, especially veterinary pharmacology to affirm that livestock feed derived from GM crops is safe is paramount for animal welfare and productivity. For example, although most studies demonstrate normal acute toxicity markers in such feeds, the detection of subtle metabolic disturbances in a fraction of studies suggests that prolonged exposure might eventually impact livestock health. This potential for low-incidence adverse effects is particularly significant in regions such as Nigeria, where variations in local dietary practices and environmental exposures may further influence health outcomes (Gbashi et al., 2021). Such findings underscore the need for a continual, evidence-based monitoring system akin to pharmacovigilance, an independent 'Crop-Vigilance' mechanism designed to collect, analyse and rapidly communicate safety data. This approach directly aligns with a number of SDGs, especially good health and well-being, life on land, and responsible consumption and production, while robust food-security-related communication strategies are employed to ensure that these findings are disseminated effectively to regulatory bodies and the public (Amedu et al., 2025; Yahaya et al., 2024).

For human health, while the acute safety of GM crops is well demonstrated, uncertainties persist regarding the long-term effects of chronic exposure. Observational research from high-consumption regions, particularly in Nigeria, indicates that even minor subclinical effects might accumulate over time, potentially leading to significant health risks (Oladipo et al., 2020). Advanced omics techniques, such as metabolomic and epigenetic profiling, have proven their capacity to detect early molecular alterations that precede overt clinical toxicity (Lozada-Martinez et al., 2025). Through incorporating these advanced analytical methodologies, future research can provide a more detailed and predictive risk assessment that not only informs public health policy but also underpins effective communication strategies. This translational approach contributes to the broader goals on zero hunger and good health and well-being to ensure that continuously emerging scientific findings are conveyed to both stakeholders and communities through targeted development communication initiatives (Nilsson et al., 2016).

The implications of these findings extend into the policy arena, where independent, stringent regulatory oversight is essential. This meta-analysis review advocates for standardised testing protocols that integrate conventional toxicological endpoints with cutting-edge analytical techniques and mandates that safety assessments be subject to independent audits and transparent reporting. Interdisciplinary collaborations among pharmacologists, toxicologists and epidemiologists are crucial for developing robust risk assessment models that accurately capture the continuously changing relationship between experimental and observational data. Moreover, effective dissemination of these findings through well-structured communication for development strategies acts as a tool for informed decision-making to facilitate a resilient framework for food system integrity and public health. This approach not only addresses the existing scientific uncertainties but also promotes a transparent, evidence-based dialogue with policymakers and the public, an imperative for realising the targets outlined under the climate action and life on land goals (Elwell, 2020).

Hence, while the aggregated evidence robustly supports the acute safety of GM crops, persistent uncertainties in chronic toxicity, allergenicity and subtle metabolic disturbances necessitate further investigation. Extended-duration studies employing advanced analytical techniques are essential to refine our understanding of long-term safety. Incorporating data from controlled animal experiments and real-world observational studies provides a comprehensive risk profile that is indispensable for guiding regulatory practices and public health policies. Through embedding these scientific findings within a framework of proactive communication strategies and aligning them with the relevant SDGs, the chapter can promote informed decision-making and sustainable agricultural practices across diverse regions, particularly in emerging markets such as Nigeria.

6.1.1. Significance of Findings

The outcomes of this systematic review and meta-analysis hold considerable significance for contemporary pharmacology, public health policy and the regulation of agri-food biotechnology. The comprehensive synthesis of data drawn from both controlled animal experiments and observational studies yields a robust evidence base that clarifies the toxicological profile of GM crops, with particular relevance for global as well as regional (notably African and Nigerian) agricultural consumption.

The consistent observation that over 80% of studies report normal acute toxicity parameters is reassuring, especially given the increasing incorporation of GM crops in animal feed. For pharmacology practitioners and animal nutritionists, this finding implies that short-term exposure to widely deployed GM crops does not compromise systemic physiological function, a conclusion that buttresses current dietary guidelines incorporating GM-derived ingredients.

However, the detection of minor metabolic and immunological alterations in a subset of studies focused on chronic exposure invites further pharmacodynamic scrutiny. Although these modifications do not typically reach clinical thresholds, their recurring presence suggests that subtle physiological perturbations may accumulate with prolonged exposure, particularly within species characterised by extended life cycles. This issue bears significant implications for food-producing animals, as long-term health directly affects meat and dairy quality, reproductive performance and economic viability. Moreover, the study's focus on African data, especially from Nigeria, reveals a critical knowledge gap regarding indigenous long-term studies in tropical pharmacological environments. Such findings underscore the necessity for local regulatory authorities, including bodies like the National Food, Drug, Administration and Control Agency (NAFDAC) and the National Biosafety Management Agency, to cultivate locally contextualised evidence that supports or recalibrates current biosafety frameworks.

The deployment of robust statistical techniques, such as meta-regression and subgroup analysis, to elucidate heterogeneity and identify bias further enhances the methodological rigour of the study. This rigorous approach advocates for harmonisation of experimental protocols and stricter adherence to OECD toxicity testing guidelines in future research.

Ultimately, this systematic review contributes significantly to pharmacological discourse on GM crop safety by bridging the gap between laboratory toxicology and field-based risk assessment. The findings are expected to guide pharmacologists, development communication specialists, regulatory bodies and policymakers in making more empirically substantiated decisions regarding GM crop usage to reinforce food system integrity in line with the SDGs. The effective communication of these findings through relevant channels will further empower stakeholders to ensure that policy and practice are underpinned by robust, science-based evidence.

Table 5 (refer back to it) encapsulates the toxicological profiles of GM crops, delineated by endpoint. For acute toxicity, pest-resistant varieties exhibit virtually no adverse effects, whereas herbicide-tolerant modifications are associated with minor chronic disturbances. Overall, metrics related to allergenicity and metabolic changes remain low, and carcinogenic potential is negligible, with occasional signals in studies on stacked modifications warranting further investigation. These results collectively indicate that GM crops generally exhibit a safe profile under controlled experimental conditions. Nonetheless, the observed discrepancies in chronic toxicity and allergenicity underscore the need for further targeted and methodologically robust research to fully elucidate long-term safety implications.

This advanced synthesis emphasises the relationship between cutting-edge analytical techniques and translational research models, while also reinforcing the critical role of effective Communication for Development strategies in disseminating research findings. Such an approach not only supports progressive public health and agricultural policies but also contributes to the global agenda outlined by the relevant SDGs (United Nations, 2015).

7. FUTURE RESEARCH DIRECTIONS

Despite its value, this meta-analysis is limited by considerable heterogeneity in study designs, small sample sizes and reliance on short-term animal trials. Interspecies differences further restrict the direct translation of laboratory findings to human populations, while limited longitudinal data impede robust conclusions regarding cumulative toxic effects. Regional disparities, particularly between Western and Nigerian contexts, underscore the need for more locally focused studies. Future research must prioritise extended-duration investigations and incorporate advanced omics technologies to capture subtle molecular changes. Strengthening interdisciplinary collaboration will be vital in developing better risk assessment models that underpin sustainable public health and agricultural policies globally.

Additionally, to advance GM-crop safety science, future work must embrace truly interdisciplinary frameworks. The chapter recommends integrating high-throughput omics technologies, like metabolomics, proteomics and epigenomics, with pharmacovigilance-style surveillance to detect emergent adverse signals. Coupling these analytical modalities with C4D strategies will ensure that complex data are distilled into actionable messages for diverse stakeholders. Collaborative networks between veterinary pharmacologists, toxicologists, epidemiologists and social scientists will be essential to co-design locally relevant studies, strengthen laboratory-to-field translation and ultimately inform evidence-based policy interventions.

8. CONCLUSION

This meta-research study unequivocally demonstrates that, under existing testing protocols, GM crops exhibit a robust safety profile with most studies to ensure normal acute toxicity markers. Nonetheless, a continuously shifting interaction among evidence shows that chronic exposure may induce minor histopathological and metabolic alterations in a proportion of cases. Highlighting of the phased Translational Roadmap (refer back to section 6.0) has substantially bridged the divide between controlled laboratory assessments and real-world outcomes, especially in regions such as Nigeria. Through integrating both experimental and epidemiological evidence, the study provides a robust understanding of the nuanced risk profile of GM crops, reinforcing the necessity for ongoing surveillance. This synthesis, underpinned by advanced analytical methods, serves as an instrument for future risk assessments that must incorporate emerging omics technologies to enhance predictive capability. Furthermore, the alignment with multiple Sustainable Development Goals underscores the broader societal relevance and public health implications of the study's findings. The study aims to promote dialogue among regulatory authorities, pharmacologists and community stakeholders through robust and transparent dissemination. In essence, while the overall safety of GM crops is well supported, the observed challenges warrant a serious commitment to extended and interdisciplinary research efforts that continuously refine risk assessments and inform sustainable regulatory practices.

8.1. Limitations

Despite rigorous methodology, several constraints temper the study's conclusions. First, significant heterogeneity across studies, e.g., in design, endpoints and statistical power, limits the precision of pooled estimates and underscores the need for standardised testing protocols. Second, long-term toxicological data from many African regions remain scarce, impeding region-specific risk assessments and underlining the necessity of locally driven observational studies. Third, potential biases arising from industry-funded research may skew outcomes towards favourable safety profiles; although stringent disclosure requirements and sensitivity analyses were implemented, residual commercial influence cannot be entirely excluded. Finally, the detection of low-incidence adverse effects is inherently challenging within the time frames and sample sizes of existing animal trials. These constraints may obscure subtle or cumulative health impacts, particularly under real-world exposure scenarios. Transparently acknowledging these limitations reinforces the study's commitment to scientific integrity and highlight priorities for future research endeavours.

8.2. Recommendations

Based on the study's findings, the chapter recommends establishing an independent 'Crop-Vigilance' system to continuously monitor GM crop safety in both animal feed and human food. Future studies should integrate advanced analytical methods, such as metabolomics and epigenetics, to detect early biomarkers and refine risk evaluations. Regulatory agencies must enforce strict, unbiased oversight while stimulating interdisciplinary collaboration among pharmacologists, toxicologists and epidemiologists. In parallel, tailored communication strategies should be implemented to ensure that complex scientific findings are accessible to all stakeholders. Such proactive oversight and transparent communication are essential for realising global public health and food security objectives.

Moreover, the following are some policy engagement action recommendations for regulators:

- Standardise toxicological testing protocols for GM crops across regulatory jurisdictions.
- Mandate full disclosure of funding sources and potential conflicts of interest.
- Support region-specific long-term observational studies in African and low-regulation contexts.
- Integrate advanced omics-based screening methods (e.g., metabolomics and epigenetics) in risk assessments.
- Establish a transparent, publicly accessible GM crop database with post-market surveillance reports.
- Facilitate multi-sectoral "Crop-Vigilance" platforms for ongoing food safety monitoring and reporting.
- Promote evidence-informed public communication strategies tailored to local contexts.

REFERENCES

Adeyeye, S. A. O., & Idowu-Adebayo, F. (2019). Genetically modified and biofortified crops and food security in developing countries: A review. *Nutrition & Food Science*, *49*(5), 978–986. DOI: 10.1108/NFS-12-2018-0335

Akinbo, O., Obukosia, S., Ouedraogo, J., Sinebo, W., Savadogo, M., Timpo, S., Mbabazi, R., Maredia, K., Makinde, D., & Ambali, A. (2021). Commercial release of genetically modified crops in Africa: Interface between biosafety regulatory systems and varietal release systems. *Frontiers in Plant Science*, *12*, 605937. DOI: 10.3389/fpls.2021.605937

Amedu, J., Adediji, A., Miracle, N., Anthony, A., Adeyemi, P., Ahmed, R., Atsumbe, S., Costly, M., Majekodunmi, A., Balogun, O., Akinpelu, O., Borgbara, K., Olufowobi, O., Tahir, H. J., Aroworamimo, L., & Asagbra, A. (2025). What's in (y) our food?– Occurrence of GM-containing foods on the Nigerian market and compliance with national regulations. *Journal of Genetic Engineering and Biotechnology*, *23*(2), 100481. DOI: 10.1016/j.jgeb.2025.100481

Azadi, H., Taheri, F., Ghazali, S., Moghaddam, S. M., Siamian, N., Goli, I., & Witlox, F. (2022). Genetically modified crops in developing countries: Savior or traitor? *Journal of Cleaner Production*, *371*, 133296. DOI: 10.1016/j.jclepro.2022.133296

Bakshi, A. (2003). Potential adverse health effects of genetically modified crops. *Journal of Toxicology and Environmental Health. Part B, Critical Reviews*, *6*(3), 211–225. DOI: 10.1080/10937400306469

Benevenuto, R. F., Venter, H. J., Zanatta, C. B., Nodari, R. O., & Agapito-Tenfen, S. Z. (2022). Alterations in genetically modified crops assessed by omics studies: Systematic review and meta-analysis. *Trends in Food Science & Technology*, *120*, 325–337. DOI: 10.1016/j.tifs.2022.01.002

Bischoff, N. S., Bussi, M. R., Van Breda, S. G., Jolani, S., Sijm, D. T., de Kok, T. M., & Briedé, J. J. (2025). Food-grade titanium dioxide exposure between age groups and in global regions: A systematic review and meta-analysis. *Critical Reviews in Food Science and Nutrition*, 1–11. DOI: 10.1080/10408398.2025.2467823

Brune, P., Chakravarthy, S., Graser, G., Mathesius, C. A., McClain, S., Petrick, J. S., & Goodwin, L. (2021). Core and supplementary studies to assess the safety of genetically modified (GM) plants used for food and feed. *Journal of Regional Science*, *9*(1), 45–60. https://orcid.org/0000-0001-8245-998X

Caradus, J. R. (2023). Intended and unintended consequences of genetically modified crops-myth, fact and/or manageable outcomes? *New Zealand Journal of Agricultural Research*, *66*(6), 519–619. DOI: 10.1080/00288233.2022.2141273

Catacora-Vargas, G., Binimelis, R., Myhr, A. I., & Wynne, B. (2018). Socio-economic research on genetically modified crops: A study of the literature. *Agriculture and Human Values*, *35*(2), 489–513. DOI: 10.1007/s10460-017-9842-4

Dadgarnejad, M., Kouser, S., & Moslemi, M. (2017). Genetically modified foods: Promises, challenges and safety assessments. *Applied Food Biotechnology*, *4*(4), 193–202. DOI: 10.22037/afb.v4i4.17244

Dang, C., & Wang, F., LU, Z., & YE, G. (2020). Application of Meta-analysis in the Safety Assessments of Transgenic Crops. *Zhongguo Shengwu Fangzhi Xuebao*, *36*(1), 17–23. DOI: 10.16409/j.cnki.2095-039x.2019.05.012

Dowd-Uribe, B., Rock, J. S., Spreadbury, T., Chiril, P., & Uminsky, D. (2024). Bridging the gap? Public-private partnerships and genetically modified crop development for smallholder farmers in Africa. *Plants, People, Planet*, *6*(2), 437–451. DOI: 10.1002/ppp3.10453

Ebegba, R., Bello, S., & Onwude, J. (2022). Status of Nigeria's biosafety regulation. In *Biosafety and bioethics in biotechnology* (pp. 1-18). CRC Press. DOI: 10.1201/9781003179177-1

Elwell, D. A. (2020). Industry-influenced evidence: Bias, conflict, and manipulation in scientific evidence. *BCL Rev.*, *61*, 2155.

FAO. (2025). *Communication for Development*. Food and Agricultural Organization. Retrieved from https://www.fao.org/communication-for-development/en/

Gbadegesin, L. A., Ayeni, E. A., Tettey, C. K., Uyanga, V. A., Aluko, O. O., Ahiakpa, J. K., Okoye, C. O., Mbadianya, J. I., Adekoya, M. A., Aminu, R. O., Oyawole, F. P., & Odufuwa, P. (2022). GMOs in Africa: Status, adoption and public acceptance. *Food Control*, *141*, 109193. DOI: 10.1016/j.foodcont.2022.109193

Gbashi, S., Adebo, O., Adebiyi, J. A., Targuma, S., Tebele, S., Areo, O. M., Olopade, B., Odukoya, J. O., & Njobeh, P. (2021). Food safety, food security and genetically modified organisms in Africa: A current perspective. *Biotechnology & Genetic Engineering Reviews*, *37*(1), 30–63. DOI: 10.1080/02648725.2021.1940735

Glevitzky, M., Glevitzky, I., Mucea- tef, P., Popa, M., Dumitrel, G. A., & Vică, M. L. (2025). Integrated risk framework (IRF)—Interconnection of the Ishikawa Diagram with the enhanced HACCP system in risk assessment for the sustainable food industry. *Sustainability (Basel)*, *17*(2), 536. DOI: 10.3390/su17020536

Glover, B., Akinbo, O., Savadogo, M., Timpo, S., Lemgo, G., Sinebo, W., Akile, S., Obukosia, S., Ouedraogo, J., Ndomondo-Sigonda, M., Koch, M., Makinde, D., & Ambali, A. (2018). Strengthening regulatory capacity for gene drives in Africa: Leveraging NEPAD's experience in establishing regulatory systems for medicines and GM crops in Africa. *BMC Proceedings, 12*(S8, Suppl 8), 11. DOI: 10.1186/s12919-018-0108-y

Hermans, E., Toelen, J., Ventrella, D., De Schaepdrijver, L., Turner, K., Croubels, S., & Devreese, M. (2024). Application of preclinical juvenile animal models. In Gasthuys, E., Allegaert, K., Dossche, L., & Turner, M. (Eds.), *Essentials of translational pediatric drug development: From past needs to future opportunities* (pp. 189–212). Academic Press. DOI: 10.1016/B978-0-323-88459-4.00009-2

Huang, K. (2017). *Safety assessment of genetically modified foods.* Springer Singapore. DOI: 10.1007/978-981-10-3488-6

Jenkins, H., & McBrien, J. (2011). Communicating for change: Development Communication and social transformation. *Communication, Culture & Critique*, *4*(3), 299–316. DOI: 10.1111/j.1468-2885.2010.01275.x

Josephs, J. J. (2017). *Perceptions of validity: How knowledge is created, transformed and used in bio-agricultural technology safety testing for the development of government policies and regulations* (doctoral dissertation, College of Arts, Humanities, and Social Sciences), Nova Southeastern University.

Journalism University. (2025). "Understanding development communication: Concepts and goals". Available at: https://journalism.university/development-journalism-for-social-change/development-communication-concepts-goals/

Kedisso, E. G., Maredia, K., Guenthner, J., & Koch, M. (2022). Commercialization of genetically modified crops in Africa: Opportunities and challenges. *African Journal of Biotechnology*, *21*(5), 188–197. DOI: 10.5897/AJB2021.17434

Krogh, P. H., Kostov, K., & Damgaard, C. F. (2020). The effect of *Bt* crops on soil invertebrates: A systematic review and quantitative meta-analysis. *Transgenic Research*, *29*(5-6), 487–498. DOI: 10.1007/s11248-020-00213-y

Le Blanc, D. (2015). Towards integration at last? The Sustainable Development Goals as a network of targets. *Sustainable Development (Bradford)*, *23*(3), 176–187. DOI: 10.1002/sd.1582

Lozada-Martinez, I. D., Hernandez-Paz, D. A., Fiorillo-Moreno, O., Picón-Jaimes, Y. A., & Bermúdez, V. (2025). Meta-research in biomedical investigation: Gaps and opportunities based on meta-research publications and global indicators in health, science, and human development. *Publications / MDPI, 13*(1), 7. DOI: 10.3390/publications13010007

Marsteller, N., Bøgh, K. L., Goodman, R. E., & Epstein, M. M. (2015). A review of animal models used to evaluate potential allergenicity of genetically modified organisms (GMOs). *Drug Discovery Today. Disease Models, 17*, 81–88. DOI: 10.1016/j.ddmod.2016.11.001

Mmbando, G. S. (2023). The legal aspect of the current use of genetically modified organisms in Kenya, Tanzania, and Uganda. *GM Crops and Food: Biotechnology in Agriculture and the Food Chain, 14*(1), 1–12. DOI: 10.1080/21645698.2023.2208999

Modonesi, C., & Gusmeroli, F. (2018). Evidence-based science or science-based evidence? The GM crops between false myths and ecological systems. *Organisms. The Journal of Biological Sciences, 2*(1), 63–72. DOI: 10.13133/2532-5876_3.13

Moresis, A., Restivo, L., Bromilow, S., Flik, G., Rosati, G., Scorrano, F., Tsoory, M., O'Connor, E. C., Gaburro, S., & Bannach-Brown, A. (2024). A minimal metadata set (MNMS) to repurpose nonclinical in vivo data for biomedical research. *Lab Animal, 53*(3), 67–79. DOI: 10.1038/s41684-024-01335-0

Munoz-Muriedas, J. (2021). Large scale meta-analysis of preclinical toxicity data for target characterisation and hypotheses generation. *PLoS One, 16*(6), e0252533. DOI: 10.1371/journal.pone.0252533

Murayi, J. A., Evenson, E., Britton, C., Gehred, A., & Goday, P. S. (2025). Clinical effects of pediatric commercial food-based formulas: A systematic review. *Journal of Pediatric Gastroenterology and Nutrition, 80*(3), 501–509. DOI: 10.1002/jpn3.12450

Nang'Ayo, F., Simiyu-Wafukho, S., & Oikeh, S. O. (2014). Regulatory challenges for GM crops in developing economies: The African experience. *Transgenic Research, 23*(6), 1049–1055. DOI: 10.1007/s11248-014-9805-0

Nilsson, M., Griggs, D., & Visbeck, M. (2016). Policy: Map the interactions between Sustainable Development Goals. *Nature, 534*(7607), 320–322. DOI: 10.1038/534320a

Ogwu, M. C., Thompson, O. P., Kosoe, E. A., Mumuni, E., Akolgo-Azupogo, H., Obahiagbon, E. G., Izah, S. C., & Imarhiagbe, O. (2024). Factors driving the acceptance of genetically modified food crops in Ghana. *Food Safety and Health, 2*(1), 158–168. DOI: 10.1002/fsh3.12031

Oladipo, O. H., Ibrahim, R. R., Adeboye, S. E., & Kuiper, H. (2020). Readiness of the Nigerian public for the introduction of genetically modified crops into the food market. *African Journal of Biotechnology*, *19*(7), 426–438. DOI: 10.5897/AJB2020.17136

Ongu, I., Olayide, P., Alexandersson, E., Mugwanya Zawedde, B., & Eriksson, D. (2023). Biosafety regulatory frameworks in Kenya, Nigeria, Uganda and Sweden and their potential impact on international R&D collaborations. *GM Crops and Food: Biotechnology in Agriculture and the Food Chain*, *14*(1), 1–17. DOI: 10.1080/21645698.2023.2194221

Patial, R., & Sobti, R. C. (2024). Exploring the impact of meta-analysis in scientific research: A review. *Medinformatics*, *00*(00), 1–11. DOI: 10.47852/bonviewMEDIN42022447

Rao, A. (2022). Industry-funded research and bias in food science. *Quantitative Marketing and Economics*, *20*(1), 39–67. DOI: 10.1007/s11129-021-09244-z

Sadikiel Mmbando, G. (2024). The adoption of genetically modified crops in Africa: The public's current perception, the regulatory obstacles, and ethical challenges. *GM Crops and Food: Biotechnology in Agriculture and the Food Chain*, *15*(1), 185–199. DOI: 10.1080/21645698.2024.2345401

Sandhu, R., Chaudhary, N., Shams, R., & Dash, K. K. (2025). Genetically modified crops and sustainable development: Navigating challenges and opportunities. *Food Science and Biotechnology*, *34*(2), 307–323. DOI: 10.1007/s10068-024-01669-y

Schiemann, J., Dietz-Pfeilstetter, A., Hartung, F., Kohl, C., Romeis, J., & Sprink, T. (2019). Risk assessment and regulation of plants modified by modern biotechniques: Current status and future challenges. *Annual Review of Plant Biology*, *70*(1), 699–726. DOI: 10.1146/annurev-arplant-050718-100025

Schrenk, D., Bignami, M., Bodin, L., Chipman, J. K., del Mazo, J., Grasl-Kraupp, B., Hoogenboom, L. R., Leblanc, J.-C., Nebbia, C. S., Nielsen, E., Ntzani, E., Petersen, A., Sand, S., Schwerdtle, T., Vleminckx, C., Wallace, H., Rose, M., Cottrill, B., Lundebye, A. K., & Hogstrand, C.EFSA GMO Panel on Contaminants in the Food Chain (CONTAM). (2022). Decontamination process for dioxins and dioxin-like PCBs from fish oil and vegetable oils and fats by a physical process with activated carbon. *EFSA Journal*, *20*(9), e07524. DOI: 10.2903/j.efsa.2022.7524

Sendhil, R., Nyika, J., Yadav, S., Mackolil, J., Workie, E., Ragupathy, R., & Ramasundaram, P. (2022). Genetically modified foods: Bibliometric analysis on consumer perception and preference. *GM Crops and Food: Biotechnology in Agriculture and the Food Chain*, *13*(1), 65–85. DOI: 10.1080/21645698.2022.2038525

Shen, C., Yin, X. C., Jiao, B. Y., Li, J., Jia, P., Zhang, X. W., Cheng, X.-H., Ren, J.-X., Lan, H.-D., Hou, W.-B., Fang, M., Li, X., Fei, Y.-T., Robinson, N., & Liu, J. P. (2022). Evaluation of adverse effects/events of genetically modified food consumption: A systematic review of animal and human studies. *Environmental Sciences Europe*, *34*(1), 1–33. DOI: 10.1186/s12302-021-00578-9

Smyth, S. J. (2017). Genetically modified crops, regulatory delays, and international trade. *Food and Energy Security*, *6*(2), 78–86. DOI: 10.1002/fes3.100

Then, C., Miyazaki, J., & Bauer-Panskus, A. (2022). Deficiencies in the risk assessment of genetically engineered Bt cowpea approved for cultivation in Nigeria: A critical review. *Plants*, *11*(3), 380. DOI: 10.3390/plants11030380

UNESCO. (2017). Communication for Development: Harnessing information for sustainable development. Retrieved from https://en.unesco.org/themes/communication

United Nations. (2015). The 17 Goals - Sustainable Development. Available at: https://sdgs.un.org/goals

Vijayakumar Sreelatha, H., Patel, S., & Nagarajan, P. (2024). The Key Concepts of Animal Models. In V. Sreelatha, H., Patel, S. & Nagarajan, P. (Eds.), *Animal models in research: Principles and practice* (pp. 3–16). Singapore: Springer Nature Singapore. DOI: 10.1007/978-981-97-0048-6_1

Yahaya, U., Suleiman, R. A., Hussaini, Y., Hamidu, S. T., Odey, B. O., & Adaaja, B. O. (2024). Environmental risk and biosafety of genetically modified plants. *Nigerian Journal of Biotechnology*, *41*(2), 104–108. DOI: 10.4314/njb.v41i2.9

ADDITIONAL READING

Adegbaju, M. S., Ajose, T., Adegbaju, I. E., Omosebi, T., Ajenifujah-Solebo, S. O., Falana, O. Y., Shittu, O. B., Adetunji, C. O., & Akinbo, O. (2024). Genetic engineering and genome editing technologies as catalyst for Africa's food security: The case of plant biotechnology in Nigeria. *Frontiers in Genome Editing*, *6*, 1398813. DOI: 10.3389/fgeed.2024.1398813

Chowdhury, A., Kabir, K. H., Asafo-Agyei, E. K., & Abdulai, A. R. (2024). Participatory and community-based approach in combating agri-food misinformation: A scoping review. *Advancements in Agricultural Development*, *5*(2), 81–104. DOI: 10.37433/aad.v5i2.349

Chukwugozie, D. C., Victoria, A. F., Onawo, U. G., Kawino, D. A., Mazi, I. M., & Onyeaka, H. (2025). Assessing the impact of new agricultural technologies on food safety in Sub-Saharan Africa. *Modern Agriculture*, *3*(1), e70011. DOI: 10.1002/moda.70011

Eneh, C. A., & Sowunmi, K. O. (2024). Human nutritive health, food security and animal production: Issues and concerns for genetically engineered organisms and technological innovations. *TechnoScience Review*, *15*(1), 30–43.

Hu, Y., & Chen, Y. (2025). Genetic modification technology and food security: Opportunities, challenges and response strategies. *Global Academic Frontiers*, *3*(1), 11–27. DOI: 10.5281/zenodo.15074505

Kabir, K. H., Chowdhury, A., & Edet, U. (2024). Data-driven digital participation in agri-food context: Why should C4D and CfSC scholars and practitioners pay attention to information disorder? In Chowdhury, A., & Gow, G. A. (Eds.), *Digital communication for agricultural and rural development* (pp. 51–70). Routledge. DOI: 10.4324/9781003282075-6

Meyer, H. (2011). Systemic risks of genetically modified crops: The need for new approaches to risk assessment. *Environmental Sciences Europe*, *23*(1), 1–11. DOI: 10.1186/2190-4715-23-7

Muzhinji, N., & Ntuli, V. (2021). Genetically modified organisms and food security in Southern Africa: Conundrum and discourse. *GM Crops and Food: Biotechnology in Agriculture and the Food Chain*, *12*(1), 25–35. DOI: 10.1080/21645698.2020.1794489

Nezhmetdinova, F. T., Guryleva, M. E., & Kardanova, L. D. (2025). Risks and safety standards for the use of genetically modified organisms in the context of food security. In *International Scientific and Practical Conference "Agriculture and Food Security: Technology, Innovation, Markets, Human Resources*" (FIES 2024) (Vol. 161, p. 00030). EDP Sciences. DOI: 10.1051/bioconf/202516100030

Smyth, S. J., Phillips, P. W., & Kerr, W. A. (2015). Food security and the evaluation of risk. *Global Food Security*, *4*, 16–23. DOI: 10.1016/j.gfs.2014.08.001

Warda, M., Tekin, S., Khafaga, N., Sengul, E., Çelebi, F., & Laçin, B. B. (2025). Comprehensive insights into genetically modified foods: Technological advances, nutritional benefits, therapeutic applications, and health implications. *Food Safety and Health*, *00*(3), 1–22. DOI: 10.1002/fsh3.70011

KEY TERMS AND DEFINITIONS

Acute Toxicity: The immediate harmful effects that can occur from a single or short-term exposure to a substance.

Allergenicity: The capacity of a substance to trigger allergic reactions in people, potentially causing symptoms like rashes, sneezing or breathing difficulties.

Chronic Toxicity: The adverse effects that may develop gradually after repeated or long-term exposure to a substance.

Communication for Development (C4D): This is the process of transforming detailed scientific findings into clear, accessible messages that can be easily understood by everyone, from community members and farmers to policy-makers. It ensures that important research outcomes and technical data are communicated in everyday language, enabling informed decision-making and practical actions to improve public health and sustainability.

Crop-Vigilance: A monitoring system designed to continuously oversee the safety of GM crops in the food supply to ensure that any unexpected effects are promptly detected.

Epigenetic Profiling: The investigation of modifications in gene activity that do not involve changes to the underlying DNA sequence but can influence how cells behave.

Genetically Modified (GM) Crops: Crops whose genetic material has been deliberately altered through scientific techniques to enhance traits such as pest resistance, yield or nutritional value.

Metabolomic Profiling: A modern analytical technique used to measure and evaluate the small molecules within cells to detect changes that might indicate early signs of harm.

Risk Assessment: The systematic process of determining the likelihood and severity of potential harm linked to a substance or technology.

Sustainable Development Goals (SDGs): A collection of global targets set by the United Nations aimed at improving aspects of societal, environmental and economic well-being for all countries.

Toxicological Safety: The assurance that a substance or product, like GM crops, does not cause harm or pose health risks when consumed or used.

APPENDIX A

Material and Methods

Research Design and Systematic Literature Search Strategy

This study employs a meta-research approach to meticulously synthesise and critically evaluate the toxicological safety of genetically modified (GM) crops.

A comprehensive literature search was systematically performed to identify peer-reviewed studies published between 2017 and 2025 that examine the toxicological endpoints associated with GM crops. Searches were executed across multiple electronic databases, including PubMed, Web of Science, Scopus and Google Scholar, as well as region-specific platforms such as African Journals Online (AJOL). The choice of these databases was dictated by their extensive coverage of biomedical, agricultural and toxicological research to ensure a holistic perspective on GM crop safety from both international (e.g., Sendhil et al., 2022) and African (e.g., Yahaya et al., 2024) vantage points.

Search queries were carefully formulated using Boolean operators to combine key terms and relevant subject headings. One primary query used in PubMed was:

```
("genetically modified crops" OR "GM crops") AND ("toxicolo-
gy" OR "toxicity" OR "safety") AND ("animal feeding trials" OR
"in vivo" OR "observational studies") AND ("acute toxicity" OR
"chronic toxicity" OR "allergenicity" OR "metabolic disturbanc-
es" OR "carcinogenicity").
```

Usage of Medical Subject Headings (MeSH) further refined the search to limit the retrieval of extraneous publications. This systematic approach minimised selection bias and optimised the capture of pertinent literature (Brune et al., 2021).

Inclusion and Exclusion Criteria

Stringent inclusion and exclusion criteria were applied to ensure that only high-quality studies were incorporated into the analysis. Table A1, below, summarises these criteria.

Inclusion Criteria:

i. *Study Design*: Only primary research articles reporting experimental animal feeding trials or human observational studies were considered. While meta-analyses and systematic reviews were appraised for background and methodological understanding, only original data from primary studies were synthesised.

ii. *Publication Period*: Articles published between 2017 and 2025 were selected to reflect current research and contemporary methodological rigour in GM crop safety evaluation.
iii. *Language*: Studies published in English were incorporated to ensure the necessary language proficiency for detailed critical analysis.
iv. *Endpoints Assessed*: Studies had to evaluate at least one key toxicological endpoint: acute toxicity, chronic toxicity, allergenicity, metabolic disturbances, gastrointestinal effects, or carcinogenic potential.
v. *Geographical Scope*: Although the search was global, emphasis was placed on studies from Nigeria, the broader African region and Western countries to achieve a balanced, cross-regional synthesis.

Exclusion Criteria:

i. *Non-Peer-Reviewed Sources*: Publications that had not undergone rigorous peer review (e.g., conference abstracts, grey literature, editorials) were excluded.
ii. *Irrelevant Outcomes*: Studies focusing solely on agronomic performance or those not directly assessing toxicological endpoints were omitted.
iii. *Insufficient Data*: Articles lacking sufficient methodological details, such as sample size or statistical analysis, were not considered.
iv. *Duplicate Publications*: In instances where multiple outputs emerged from the same study cohort, the most comprehensive and current version was retained.

Table A1. Summary of Inclusion and Exclusion Criteria for the Systematic Literature Search

Criteria Category	**Details**
Study Design	Inclusion: Peer-reviewed primary research articles reporting experimental animal feeding trials or human observational studies. Exclusion: Conference abstracts, editorials, grey literature and studies lacking original data.
Publication Period	Inclusion: Studies published between 2017 and 2025 to capture the most recent evidence.
Language	Inclusion: Only articles published in English.
Endpoints Assessed	Inclusion: Studies evaluating one or more toxicological endpoints, including acute toxicity, chronic toxicity, allergenicity, metabolic and gastrointestinal disturbances, or carcinogenic potential.

continued on following page

Table A1. Continued

Criteria Category	Details
Geographical Scope	Inclusion: Global studies with particular emphasis on research from Nigeria, Africa and Western countries.
Data Quality	Exclusion: Studies with insufficient methodological details (e.g., unclear sample sizes, lack of bias assessment) or duplicate reports (only the most comprehensive version retained).

These criteria were applied independently by two reviewers, with discrepancies resolved through consultation with a third reviewer. This methodological rigour minimised subjective bias and ensured the integrity and replicability of the data synthesis. The procedures adopted not only reinforce scientific transparency but also align with sustainable development and Communication for Development principles to facilitate the translation of findings into effective policy and public health strategies (Dowd-Uribe et al., 2024).

APPENDIX B

Data Extraction, Quality Assessment and Synthesis

A comprehensive, standardised data extraction form was developed to systematically record the salient characteristics of each study identified in the literature search. This form was designed to capture crucial information, including author details, year of publication, geographical origin, study design, sample size, study duration, the specific GM crop investigated, details of genetic modifications (for example, Bt gene insertion or cp4 epsps expression), and the toxicological endpoints assessed. Further, methodological elements, namely randomisation procedures, blinding methods and funding sources, were documented to enable a meticulous appraisal of potential bias (Nilsson et al., 2016). The addition of these details not only anchors the research within the framework of responsible evidence synthesis but also aligns with the imperatives of zero hunger and good health and well-being goals by supporting robust, policy-relevant risk assessments. Moreover, employing effective communication techniques ensures that findings are communicated clearly to stakeholders and decision-makers to cultivate public understanding and transparency (United Nations, 2015; UNESCO, 2017).

The extraction form facilitated the categorisation of each study under several distinct domains:

1. *Study Characteristics*: This domain encompasses authorship, publication year and geographical context, thus providing a temporal and spatial overview of the research environment. Details regarding the study design, whether an experimental animal feeding trial or a human observational study, alongside sample size and duration, were recorded to capture variables influencing statistical reliability.
2. *Intervention Details*: For every study, the specific nature of the GM crop and the genetic modifications implemented (such as the insertion of Bt genes for insect resistance or the use of CP4 EPSPS for herbicide tolerance) were noted. The intended agronomic objective, be it enhanced pest control, improved nutritional content, or herbicide tolerance, was documented to contextualise the resultant toxicological outcomes.
3. *Toxicological Endpoints*: Data extraction focused on a range of endpoints. For example, acute toxicity was evaluated via biochemical markers, including serum enzyme levels and haematological parameters, whereas chronic toxicity was assessed based on histopathological findings and metabolic alterations. Moreover,

immunological assays for allergenicity, evaluations of gastrointestinal function and evidence suggestive of carcinogenic potential were carefully recorded.

4. *Methodological Quality*: Information on the adequacy of randomisation, blinding practices and sample size was extracted, together with details of funding sources. This approach emphasised the identification of any potential commercial influence that might skew study outcomes and was essential to the transparent dissemination of research findings in line with the principles of responsible consumption and production.

Data extraction was independently carried out by two reviewers. Their parallel approach ensured accuracy and consistency in data collation, with any discrepancies resolved through consensus or consultation with a third reviewer. This dual-review process is emblematic of best practice in systematic reviews and supports the overall integrity of the synthesis (Glevitzky et al., 2025).

For the quality appraisal, the study employed established assessment tools. The SYRCLE risk of bias tool was used for animal feeding trials, enabling a rigorous evaluation of random assignment, blinding and the completeness of outcome reporting. In parallel, human observational studies were appraised using the Newcastle-Ottawa Scale (NOS) (Josephs, 2017), with particular attention to the selection of study groups, comparability across cohorts and the rigour of outcome assessment. Studies demonstrating superior methodological design received greater weighting in the synthesis, whereas those manifesting significant limitations were either excluded from quantitative analyses or discussed separately in the narrative review (Brune et al., 2021).

Synthesised data were subsequently managed using a combination of quantitative and qualitative techniques. In instances where studies shared similar design parameters and reported analogous endpoints, a random-effects meta-analysis was performed. This approach was chosen to account for the inherent variability among studies in terms of design, population characteristics and measured endpoints (Benevenuto et al., 2022). Statistical heterogeneity was quantified using the I^2 statistic, with values exceeding 50% interpreted as indicative of moderate to high heterogeneity. Where substantial heterogeneity was detected, subgroup analyses were conducted based on variables such as the genetic modification type (e.g., pest-resistance versus herbicide tolerance), geographical region (African versus Western studies) and funding source (independent versus industry-funded studies).

In addition, meta-regression analyses were performed to evaluate the impact of continuous variables such as study duration and dosage of GM crop consumption on the observed toxicological outcomes. These analyses allowed the researchers to explore dose-response relationships and delineate sources of variability. Sensitivity analyses, including leave-one-out procedures, were undertaken to test the robustness of the findings and publication bias was examined through funnel plot construction and Egger's test to identify any suppression of null or adverse findings (Elwell, 2020).

The fusion of data from controlled animal studies with observational research enabled the construction of a complex overview of the toxicological profile of GM crops. While animal studies elucidated biochemical and histopathological changes following GM crop ingestion, human observational studies contributed significantly to long-term health outcomes under real-world conditions. This synthesised analysis is pivotal for developing a balanced perspective regarding potential health risks associated with prolonged GM crop consumption, which is essential for guiding SDG-related policy and practice (Nilsson et al., 2016).

Furthermore, recognising the limitations inherent in current methodologies, including small sample sizes and ethical constraints that restrict the duration of animal studies, the study's analysis acknowledges reduced statistical power that may hinder the detection of low-incidence adverse effects (Josephs, 2017; Moresis et al., 2024). Additionally, the extrapolation from rodent models to human populations remains challenging owing to intrinsic interspecies differences in metabolism and immune response. To address these challenges, the study advocates for the integration of advanced analytical techniques, such as metabolomic and epigenetic profiling, which may reveal early biomarkers of toxicity currently obscured by conventional methods (Lozada-Martinez et al., 2025).

Throughout the process, ethical rigour was upheld. Full disclosure of funding sources and any potential conflicts of interest was mandated for study inclusion to ensure that the synthesis remained free from undue commercial influence. The entire process, from data extraction to statistical analysis, was executed in strict accordance with PRISMA guidelines to enhance methodological transparency and laying a robust foundation for future research.

Finally, data management was consolidated using a centralised database. The subsequent analyses were conducted using Review Manager (RevMan) for the meta-analytical component and STATA (version 16) for advanced meta-regression and sensitivity analyses. Detailed documentation of all analytical decisions was maintained to provide a clear audit trail to ensure that the study's findings are robust, replicable and delivered in a manner that empowers stakeholders, supporting the communication of scientific evidence essential for achieving the SDGs (Jenkins, & McBrien, 2011).

Chapter 9
Indigenous and Traditional Knowledge Systems in Food Safety and Contaminant Management

Matthew Chidozie Ogwu
https://orcid.org/0000-0001-6054-1667
Appalachian State University, USA

Aliu Olugbemiga Ojo
https://orcid.org/0000-0003-4942-7454
Auchi Polytechnic, Nigeria

ABSTRACT

This chapter discusses the indispensable role of Indigenous and Traditional Knowledge (ITK) systems in ensuring food safety and managing contaminants within diverse socio-ecological landscapes. Through an interdisciplinary analysis of examples from Africa, Asia, the Americas, and the Pacific, this study highlights a range of time-honored practices, including fermentation, herbal preservation, natural pest management, biocultural indicators, and traditional storage techniques, that Indigenous communities have developed to prevent contamination and maintain food quality. The chapter highlights how these practices are rooted in holistic worldviews that link ecological health with cultural, ethical, and spiritual dimensions of food. It further investigates how ITK systems support community-based environmental monitoring, toxin risk assessment, and adaptive responses to environmental change. Through interdisciplinary analysis, the chapter assesses the complementarities and

DOI: 10.4018/979-8-3373-7052-1.ch009

frictions between ITK and modern scientific paradigms, calling for co-produced solutions that respect Indigenous epistemologies. Critical legal and ethical issues such as food sovereignty, intellectual property rights, and the protection of traditional knowledge are addressed as essential for equitable governance. The chapter argues for the revitalization, documentation, and meaningful integration of ITK into contemporary food safety frameworks to advance environmental justice, cultural resilience, and sustainable food systems.

1. INTRODUCTION

Indigenous and Traditional Knowledge (ITK) systems refer to a cumulative body of knowledge, practices, and beliefs that are developed over time through cultural transmission within various Indigenous communities (Lam et al., 2020). These systems embody a profound understanding of local ecosystems, food sources, and sustainable practices developed through generations of experience and adaptation to specific environmental conditions (Swiderska et al., 2022). The significance of ITK in the context of sustainable food systems cannot be overstated. Historically, Indigenous populations have cultivated biodiversity through meticulous management practices that ensure food security while fostering ecological balance (Mercer et al., 2009).

The realm of food safety and contaminants integrates both traditional and modern perspectives, where ITK can play a crucial role in enhancing food quality and safety standards. Traditional food practices often incorporate natural food preservation techniques and knowledge about local species that modern food safety measures may overlook (Ikhajiagbe et al., 2021; Pilnik et al., 2023). Moreover, understanding how contaminants affect local diets through traditional dietary connections can improve overall health outcomes and even prevent foodborne illnesses (Robin and Hart, 2025). One of the compelling rationales for integrating ITK with scientific approaches is the potential for a more comprehensive and nuanced understanding of food safety and ecological health. As outlined by various researchers, bridging the gap between Indigenous knowledge and contemporary scientific methods can lead to more comprehensive and culturally relevant food safety protocols (Bohensky and Maru, 2011). For example, Indigenous methodologies often employ holistic approaches that consider the interconnectedness of species and ecosystems, thereby enhancing modern scientific frameworks that focus on singular aspects of health (Berkes & Berkes, 2009).

In contemporary food systems, a prevalent challenge exists in standardizing practices, leading to the marginalization of ITK. Standardized food safety protocols may overlook or undervalue the intricate knowledge possessed by Indigenous communities, potentially leading to the loss of valuable biodiversity and sustainable practices (Weiler et al., 2014). What becomes evident is a stark contrast in epistemologies; Western scientific paradigms often compartmentalize knowledge, while Indigenous approaches frequently embrace contextual and holistic understandings of ecosystems (Spencer et al., 2021). Such a disconnect is usually exacerbated by systemic factors, including colonialism and environmental racism, which create barriers to the integration of ITK in policymaking and food systems management (Robin et al., 2020). Food sovereignty movements have countered these challenges by advocating for the inclusion of Indigenous perspectives in shaping food systems (Jernigan et al., 2021). Scholars argue that addressing long-term systemic problems related to food insecurity requires a reconceptualization of governance structures that include ITK as an essential component (Robin, 2019). Collaborative governance that respects Indigenous rights and cultural values can facilitate a more sustainable food system, potentially leading to improved health outcomes and increased food access for Indigenous peoples (Jernigan et al., 2021).

The integration of ITK with scientific knowledge presents unique opportunities for managing ecological systems, particularly in the face of climate change and food insecurity issues. As noted in various studies, effective collaboration emerges when both systems of knowledge operate on a platform of mutual respect, with methodologies designed to address shared concerns, such as health and sustainability (Scholes, 2019). This partnership acknowledges the wisdom embedded in ITK regarding sustainable food production practices, which can be leveraged in conjunction with scientific advancements (Kom et al., 2024). For example, incorporating Indigenous knowledge regarding crop rotation and pest management can significantly enhance food production resilience against climate fluctuations (Kom et al., 2024).

Additionally, research has shown that bridging ITK with contemporary science through interdisciplinary approaches leads to innovative solutions for food security (Ullah, 2024). For instance, employing Indigenous knowledge of rainfall prediction can play a significant role in agricultural planning, helping farmers time their planting optimally based on environmental indicators rather than relying solely on traditional meteorological data. This research emphasizes the necessity of honoring the contextual, local knowledge of Indigenous people, who have historically adapted to environmental changes over centuries (Osawaru et al., 2014; Kom et al., 2024). Preservation of biodiversity is not merely an ecological imperative; it is also central to many Indigenous knowledge systems that intrinsically recognize the value of varied species for food security and health. The utilization of Indigenous agricultural practices, such as mixed cropping and the cultivation of traditional food plants, supports genetic diversity and promotes sustainable agrarian methods (Nnamani et al., 2017). As highlighted in the literature, ITK systems often contain critical information on local species that modern agriculture may overlook due to its focus on productivity over biodiversity (Pilnik et al., 2023; Ogwu, 2023).

Furthermore, Indigenous practices embody agroecological principles that align with modern sustainability goals. For instance, practices such as shifting cultivation and traditional pest management strategies can be harmonized with contemporary ecological knowledge to provide pathways for sustainable agricultural practices that respect both cultural heritage and environmental health (Swiderska et al., 2022). This fusion of knowledge supports a more resilient agrarian framework that can withstand the challenges posed by climate change and global food demands.

This chapter aims to explore the critical role that Indigenous and traditional knowledge (ITK) systems play in ensuring food safety and managing contaminants within diverse cultural and ecological contexts. Drawing on interdisciplinary perspectives and grounded case studies from around the world, including Africa, the Americas, Asia, and the Pacific, this work highlights the sophisticated practices employed by Indigenous communities to monitor, prevent, and respond to foodborne hazards and environmental contaminants. The chapter aims to illuminate how local ecological knowledge, customary norms, and traditional technologies have historically safeguarded food systems through culturally embedded strategies, including fermentation, herbal preservation, pest management, and environmental sensing. Furthermore, it investigates the synergies and tensions between ITK and modern scientific approaches, advocating for integrative frameworks that respect Indigenous sovereignty and knowledge co-production. This chapter contributes to broader discussions on food sovereignty, environmental justice, and the decolonization of food safety governance by highlighting ethical concerns, policy implications, and community-driven innovations. It provides a starting point for reconsidering how conventional ecological knowledge can inform more resilient, inclusive, and sustainable approaches to contaminant management in the face of global challenges.

2. CULTURAL AND ETHICAL FOUNDATIONS OF FOOD SAFETY

The intersection of indigenous and traditional knowledge systems with food safety is a significant field of inquiry, particularly when considering cultural cosmologies and ethical foundations that vary widely across different societies (Osawaru & Ogwu 2020). Indigenous knowledge systems are rooted in the cultural and spiritual practices that shape understanding of food ethics, contamination, and overall health. These systems blend practical knowledge about food safety with broader cosmological values that connect the land, community, and sustenance derived from nature. Critical to this discussion are the cultural interpretations of purity, contamination, and health as they relate to food systems across various indigenous groups, including those in Africa, the Andes, Asia, and Indigenous North America.

In many indigenous cultures, cosmological perspectives inform the ethical foundations of food safety. For instance, the Inca in the Andes emphasized a harmonious relationship with nature, which is reflected in agricultural practices that prioritize the purity of food sources and consider the spiritual dimensions of food production and consumption (Malli et al., 2023). Similar views can be found in many African cultures where ancestral connections play a vital role in food safety practices, seeing food not merely as sustenance but as a vessel carrying spiritual significance. Here, purity and contamination are often interpreted through a lens of cultural respect and reverence (Kachapulula et al., 2017). Practices that honor the land and ecosystem are integral to maintaining food safety, as they are believed to protect both consumers and the community's heritage (Sherriff et al., 2022).

Definitions of what constitutes contamination and purity can vary widely among cultures, often leading to conflicting understandings of food safety. For example, while certain Western scientific frameworks emphasize the absence of pathogens or chemicals as indicators of safe food, many indigenous cultures consider the use of specific plants or traditional preservation methods, which may not align with modern safety protocols, as essential. This divergence underscores the complex interplay between conventional food safety practices and traditional knowledge (Ogwu et al., 2016; Gbashi et al., 2019). In Ayurveda, a traditional system of medicine from India, notions of food purity extend beyond biological safety to encompass the energetic quality of food and its effects on health, integrating physical and spiritual dimensions into food safety considerations (Adekoya et al., 2017; Oladeji et al., 2024; Ogwu & Ogunsola 2024).

African food systems face significant challenges related to contamination, particularly with mycotoxins such as aflatoxins that substantially impact agricultural yields and nutritional security. With maize as a key staple, the contamination of food crops poses not just health risks but also economic challenges, connecting issues of contamination to broader socio-economic implications (Nji et al., 2022). Nonetheless, indigenous practices, such as the timing of planting and harvesting, often align with lunar cycles or other natural indicators, offering promising methods to mitigate such risks while respecting cultural traditions (Nji et al., 2022; Osei-Kwarteng et al., 2024). These approaches underscore an ethical stance on food safety embedded in local ecosystems and knowledge (Wu and Güçlü, 2012).

Furthermore, ethical considerations surrounding food safety in indigenous contexts are often linked to broader societal issues, such as the legacies of colonialism that have disrupted traditional food systems and imposed external frameworks of understanding (Malli et al., 2023). Such colonial effects can lead to disconnection from traditional food sources and may introduce contaminants associated with industrialized food systems. The reliance on processed foods, frequently contaminated, highlights the tension between modern convenience and traditional wisdom (Roche & Tidou, 2008). Thus, Indigenous communities advocate for a return to traditional systems that value the interconnectedness of food, health, and cultural identity, representing an act of resistance against ongoing colonial influences (Abass et al., 2017). In Indigenous North America, many tribes continue to engage in traditional hunting and gathering, reflecting a cosmology that regards land and its animals as relatives deserving of respect. Their ethical stance aims to ensure that food sources remain clean and abundant, honoring practices that embody gratitude toward nature (Sherriff et al., 2022). This illustrates a complex interplay between food safety, cultural identity, and environmental stewardship. Evidence suggests that minimizing contamination is closely aligned with cultural rituals recognizing the sanctity of food as a life-giving force (Nji & Mwanza, 2024). Analyzing diverse indigenous perspectives reveals that food safety transcends mere practical concern, embodying a multifaceted concept grounded in cultural values, community integrity, and a reciprocal relationship with the environment. Practices and beliefs regarding food safety across Africa, the Andes, Asia, and Indigenous North America challenge conventional paradigms of food safety, advocating for an integrative approach that respects the complexities of cultural identity and environmental interdependence (Abass et al., 2017; Omotayo et al., 2019).

3. TRADITIONAL KNOWLEDGE AND PRACTICES THAT SAFEGUARD FOOD

Indigenous communities worldwide possess rich traditional knowledge and practices that significantly contribute to food preparation, preservation, and safety (Obahiagbon & Ogwu 2024). This body of knowledge has been passed down through generations. It encompasses various techniques, such as food fermentation, the use of natural antimicrobial agents, seasonal harvesting, and traditional methods of pest and disease control (Izah et al., 2024). Each of these techniques encapsulates the cultural heritage of these communities and plays a crucial role in ensuring food security and safety. Table 1 presents selected ITK practices from diverse regions that contribute to the protection of food systems against biological, chemical, and environmental contaminants. These practices reflect centuries of ecological understanding, cultural innovation, and adaptive strategies that ensure food safety, quality, and resilience. They illustrate how local communities integrate environmental cues, spiritual values, and empirical knowledge to prevent spoilage, manage pests, store food safely, and detect contamination, and highlight the ongoing relevance of ITK in contemporary efforts toward sustainable and just food systems.

Fermentation is a prime example of traditional food preservation practiced by numerous indigenous groups. The process involves the conversion of carbohydrates to alcohol or organic acids through the action of microorganisms under anaerobic conditions (Msumuzzaman et al., 2021; Evivie et al., 2020; 2022). For instance, the production of 'picungan bandeng', a fermented fish dish in Banten, Indonesia, showcases a blend of traditional knowledge and local resources that have been sustained through socio-cultural practices in food preparation (Khastini et al., 2023). This method not only extends the shelf life of food but also enhances the nutritional value and flavor profile of the resulting products (Abajue & Ogwu 2024). Various indigenous communities, such as those in northeast India, utilize local animal and plant resources to create a wide range of fermented foods, which play a pivotal role in their diets and food security (Kadirvel et al., 2018; Ovuru et al., 2023). Moreover, the fermentation techniques employed by these communities reflect their deep understanding of microbial processes and the interplay between ingredients. In addition to fermentation, the utilization of natural antimicrobial agents such as specific herbs and spices remains a foundational aspect of traditional food preparation (Omorotionmwan et al., 2019). For example, spices like turmeric and garlic are widely recognized for their antimicrobial properties and are commonly used in various cuisines worldwide to inhibit the growth of pathogens (Wana & Kudama, 2022). Indigenous cultures inherently understand the importance of these natural preservatives, utilizing them for flavor and food safety. The practice of smoking meats and fish, prevalent among various tribes, further illustrates this knowledge. The smoke acts as a preservative, imparting flavor while reducing moisture content and creating a barrier to spoilage. This method, along with the cultural significance of the smoked products, exemplifies the multifaceted role of food preservation within indigenous diets.

The importance of seasonal harvesting and storage knowledge is critical. Indigenous communities have developed strategies to harvest food sustainably and store it effectively (Ogwu et al., 2018; 2023). Food is often harvested during the peak season to ensure optimal taste and nutritional content, and then preserved through methods such as drying or curing (Abdul-Mutalib et al., 2012). The knowledge of seasonal cycles is closely linked with traditional agricultural practices, where indigenous farmers employ methods that avoid overexploitation of resources. This aspect of traditional knowledge is crucial for climate resilience, as the ability to adapt to changing environmental conditions can significantly impact food security (Ogwu et al., 2024).

Traditional pest and disease control methods represent another significant area where indigenous knowledge excels. These practices often rely on natural remedies derived from locally available plants, rather than synthetic pesticides, which can be harmful to the environment and human health. Ethnobotanical studies reveal the diverse range of plant species utilized by indigenous communities for pest management, highlighting their in-depth understanding of local ecosystems (Suwardi et al., 2020). Some groups may employ specific plants that repel pests or attract beneficial insects, thereby reducing the need for chemical inputs. Such practices also preserve biodiversity and promote ecological balance, making them increasingly relevant in discussions surrounding sustainable agriculture. The interplay between these practices forms a holistic system that is the backbone of indigenous food cultures. The combination of fermentation techniques, the use of antimicrobial agents, seasonal knowledge, and natural pest control exemplifies a comprehensive approach to food preparation that maximizes safety, flavor, and nutritional value while maintaining a minimal ecological footprint.

However, recent evidence suggests that as globalization intensifies, the transmission of this vital knowledge is declining, posing threats to food diversity and cultural identity (Ahmad & Pieroni, 2016). Reviving and continuing these practices is essential not only for safeguarding traditional knowledge systems but also for enhancing food security in modern contexts. There is a pressing need for research that integrates traditional knowledge systems with contemporary practices to foster sustainable food systems. Indigenous methods enhance the resilience of local food systems by promoting biodiversity and sustainable agricultural practices. These practices must be valued and integrated into contemporary food systems to ensure the preservation of cultural heritage while promoting sustainable health benefits for future generations.

Table 1. Some Traditional Knowledge and Practices That Safeguard Food Safety Across Indigenous Communities

Practice Type	Region/Community	Function	Food Safety Outcome
Fermentation of Foods	Northeast India, Indonesia (Banten), South Africa	Natural preservation through microbial activity	Inhibits spoilage, extends shelf life, and reduces pathogens
Herbal Preservation	Ethiopia, West Africa, South Asia	Use of spices (turmeric, garlic, neem) with antimicrobial properties	Prevents microbial contamination, inhibits fungal growth
Natural Pest Management	Vhembe District (South Africa), West Africa	Use of botanical repellents and intercropping	Reduces chemical pesticide use, lowers aflatoxin risks
Traditional Storage Techniques	Sahelian regions, Ghana, Indigenous Americas	Use of mud silos, gourds, and elevated platforms	Minimizes moisture, controls pests, and prevents mold

continued on following page

Table 1. Continued

Practice Type	Region/Community	Function	Food Safety Outcome
Biocultural Indicators	Amazon, Ghana	Visual signs (e.g., leaf discoloration, soil texture) to detect contamination	Early warning for spoilage or toxic exposure
Ethnopedology	Latin America	Indigenous classification of soils	Helps identify heavy metal risks and nutrient deficiencies
Cultural Burning (Fire Management)	Australia (Aboriginal communities)	Land management to enhance fertility and reduce contaminants	Prevents accumulation of toxic biomass, supports soil health
Seasonal Harvesting Knowledge	Ghana, Zambia	Timing aligned with climate and moisture cycles	Reduces the risk of mycotoxin formation in crops
Cosmological Food Governance	Inca (Andes), Cree Nation (Canada)	Ethical harvesting based on lunar cycles and rituals	Maintains ecological balance, ensures "clean" and sacred food
Participatory Monitoring	Greenland, Canada	Community-based surveillance of environmental change	Detects contaminant trends in water, soil, and wildlife
Women's Knowledge Transmission	Northern Ghana, Pacific Islands	Intergenerational transfer of food safety, water handling, and hygiene	Preserves community food health norms, empowers food sovereignty

4. MONITORING AND MANAGING CONTAMINANTS USING INDIGENOUS AND TRADITIONAL KNOWLEDGE SYSTEMS

The utilization of Indigenous and traditional knowledge systems in monitoring and managing food spoilage and environmental contamination offers a rich tapestry of insights that integrates local cultural practices with contemporary understanding of ecological risks. Indigenous communities have long demonstrated their ability to identify local indicators of food spoilage and contamination, utilizing a range of traditional practices that reflect their deep connection to the land and natural resources (Izah et al., 2022; Richard et al., 2022). The value of such knowledge is increasingly recognized in scientific research, where it serves as a bridge to sustainable management practices and holistic health approaches.

Local indicators of food spoilage often include sensory perceptions, such as changes in color, texture, and smell of food products, which are rooted in generations of observational learning within these communities (Table 2). For instance, certain leaves and natural substances are commonly employed as indicators of spoilage or toxins in foods, leveraging their biochemical properties as alerts for potential risks (Williams et al., 2023). These sensory and ecological cues based on sight, smell, taste, texture, and environmental interactions serve as culturally embedded, experience-based tools for assessing food quality and detecting potential contamination. They reflect a deep understanding of local ecosystems and are critical for maintaining food safety in the absence of modern diagnostic technologies. Elders in Indigenous societies often serve as repositories of essential knowledge, guiding younger generations on the appropriate practices and signs to observe when determining food safety (Dawoe et al., 2012). The practical application of these indicators can be bolstered by modern science, which seeks to validate and quantify these observations through empirical measures (Wang et al., 2021).

Table 2. Local Indicators of Food Spoilage in Indigenous and Traditional Knowledge Systems

Indicator Type	Observation	Community/Region	Spoilage Cue
Visual	Leaf color changes upon food contact	Amazon Basin	Detection of chemical spoilage or toxicity
Color	Discoloration of grains, meats, or fermented items	West Africa, Southeast Asia	Fungal growth, oxidation, and microbial activity
Texture	Slimy or unusually soft consistency	Global (multiple communities)	Bacterial spoilage or enzymatic degradation
Smell	Sour, rancid, or foul odor	Ghana, South Africa, and the Andean regions	Fermentation gone wrong or protein degradation
Taste (cautiously used)	Bitter, overly sour, metallic taste	Select African and Andean communities	Early detection of mycotoxins or chemical leaching
Behavioral (animal/insect)	Absence of ants or flies around food	Indigenous Australia, Ghana	Suggests chemical contamination
Sound	Absence of popping sound in dry grains when bitten	Sahelian and South Asian communities	Indicates moisture and fungal growth

Moreover, community-based surveillance methods provide a framework through which Indigenous populations can actively engage in monitoring their local environments for contaminants. This often involves participatory approaches that harness the insights of local farmers and foragers who have extensive experiential knowledge about soil, water, and crop health (Omari et al., 2018; du Bray et al., 2023). For instance, community-led initiatives to map soil quality indicators have shown that local knowledge can correlate closely with scientific measures, promoting mutual understanding and respect between traditional wisdom and modern agricultural science (Barrios & Trejo, 2003). Such collaborative efforts reinforce the significance of ethnopedology the study of how cultures understand soil, which plays a crucial role in the discourse of sustainable agriculture and environmental stewardship (Barrios et al., 2006).

Biocultural indicators serve as a critical intersection between environmental health and cultural practices. The health of soil and water is often directly perceived through traditional cultivation practices that emphasize biodiversity and ecological balance, reflecting Indigenous philosophies (Country et al., 2024). Indicators such as the presence of specific plant species or the behavioral patterns of local wildlife are emblematic of ecosystem health and can provide valuable warnings about potential contamination (Smout et al., 2017; Ogwu et al., 2024). The integration of these traditional ecological indicators into a modern framework facilitates a more comprehensive understanding of environmental health, enabling communities to preemptively address contamination issues before they escalate (Sultan et al., 2019).

Case studies illustrating the practical application of traditional knowledge in food safety are abundant across various cultures. For example, Indigenous communities in South America utilize specific local species of leaves that change color when in contact with spoiled fruits or other perishable items, providing an immediate visual cue for food safety (Williams et al., 2023). Such empirical techniques exemplify the adaptive capacity of Indigenous practices in responding to the contemporary challenges of food safety and spoilage in a rapidly changing environmental landscape. Furthermore, the ongoing discourse surrounding decolonization of food regulatory frameworks underscores the necessity of recognizing and integrating traditional knowledge in assessing dietary safety, thus reinforcing the cultural value of traditional food systems within modern regulatory contexts (Williams et al., 2023).

The development of participatory techniques in soil management inspired by Indigenous practices marks a significant innovation in agricultural practices, enhancing sustainability and resilience among farmers. Collaborative research between scientific and Indigenous communities has underscored the importance of integrating local insights with technical knowledge to achieve better soil health outcomes (Omari et al., 2018). For example, narratives surrounding the seasonal use of fire in traditional land management not only serve to understand ecological relationships but also to develop modern fire management protocols that honor Indigenous traditional practices while addressing contemporary societal needs (Country et al., 2024).

Furthermore, modern analytics tools enhance the collaboration between community-based practices and scientific validation. The employment of machine learning algorithms for analyzing traditional practices highlights the potential of combining local wisdom with high technology, offering data-driven methods for evaluating the safety and quality of food products (Wang et al., 2021). Such approaches may streamline food safety assessments while appreciating the context-specific practices rooted in Indigenous culture. The longevity of traditional food safety practices, such as fermented and preserved products, also bears witness to the health benefits of Indigenous dietary regimes in enhancing food safety (Filip & Fink, 2017). Traditional techniques tend to anticipate potential spoilage and contamination, thereby ensuring community resilience in the face of ecological changes. As documented, the methods of food preservation not only safeguard nutritional value but also enhance the overall safety of food products while maintaining cultural heritage (Filip & Fink, 2017). The blending of Indigenous and traditional knowledge systems with modern scientific practices holds promise for enhancing community resilience against food spoilage and environmental contamination. By valuing local insights, we can foster a more comprehensive understanding of soil, water, and crop health that incorporates traditional ecological knowledge (Ikhajiagbe et al, 2021). The journey toward recognizing and integrating these vital perspectives is essential for building sustainable systems of food safety and environmental management.

5. FOOD CONTAMINANT SOURCES AND RISK PERCEPTION IN INDIGENOUS COMMUNITIES

Various factors, including sources of food contaminants, community perceptions of food safety, and the effects of industrial activities on traditional food systems, significantly shape the management and consumption of traditional foods among Indigenous communities. Figure 1 show common sources of food contamination encountered in local and traditional food systems. Contaminants may originate from animals, water, raw ingredients, utensils, packaging materials, and environmental particles such as dust and soil. Recognizing these factors and sources is essential for identifying risks and implementing community-based practices that safeguard food safety and public health. These factors intertwine in complex ways, affecting food sovereignty and security, which are foundational to the health and cultural identity of these populations.

Figure 1. Diverse local sources of food contamination

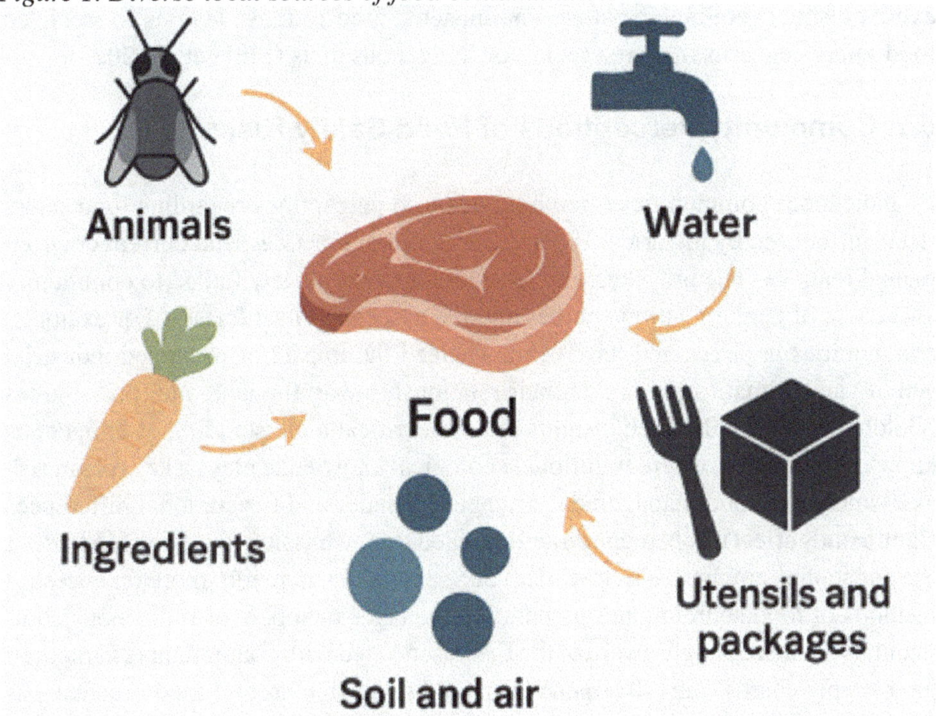

5.1. Natural vs. Anthropogenic Contaminant Sources

Contaminants in the food supply can originate from both natural and anthropogenic sources. Heavy metals, for instance, typically enter the food chain through agricultural practices and industrial activities, which may leach contaminants into the soil and water systems (Ikhajiagbe & Ogwu 2020). A review detailing the transfer of heavy metals through contaminated soil reveals that anthropogenic activities, such as mining and waste disposal, significantly contribute to heavy metal accumulation in agricultural settings (Abdullahi et al., 2023). Likewise, contamination of food products can also stem from environmental pollutants, such as pesticides and fertilizers, which are used extensively in industrial agriculture (Thompson & Darwish, 2019). The impact of these contaminants is compounded when traditional food systems are stressed by industrialization, resulting in food insecurity and health risks associated with consuming contaminated food (Goldhar et al., 2010). Moreover, specific anthropogenic contaminants, such as perchlorate, are released into the environment through agricultural runoff and can accumulate in food products, particularly in fruits and vegetables. Contaminants also differ in their mobility and bioavailability, depending on their source. Pollutants from industrial discharge often exhibit distinct geographic patterns in impacted food systems, leading to localized food safety concerns that are crucial for Indigenous diets (Milia et al., 2023).

5.2. Community Perceptions of Food Safety Risks

Indigenous communities often hold nuanced perceptions regarding food safety risks, influenced by historical experiences, cultural practices, and current environmental realities. The prevalence of food insecurity is closely linked to community awareness of contamination sources and the associated health risks. For example, communities in Greenland have acknowledged the impact of increased industrial activity and climate stressors on their relationship with the land and food sources (Goldhar et al., 2010). Such insights inform a critical understanding of Indigenous knowledge systems, where traditional ecological knowledge plays a key role in risk assessment and food management strategies. Gender and generational differences significantly affect risk perception and knowledge transmission regarding food safety. Recent studies emphasize that while older generations may rely more on ancestral methods of food sourcing and preparation, younger members of Indigenous communities are increasingly aware of the risks associated with contaminants in modern food supply chains. This divergence can lead to tension around food practices, as younger community members may advocate for food safety measures and adopt new strategies to mitigate risk (Sampson et al., 2021).

5.3. Effects of Mining and Agriculture on Traditional Food Systems

Mining and agricultural practices have profound implications for traditional food systems and the livelihoods of communities. Empirical evidence from various regions, including India and Ghana, demonstrates how mining activities reduce agricultural productivity and contribute to local food insecurity (Yadav et al., 2019; Ofosu et al., 2020; Schueler et al., 2011). The disruption of land use by mining diminishes arable land and leads to significant cultural losses as traditional practices are abandoned in favor of economic survival strategies influenced by mining revenues. These changes often displace local farmers and threaten food sovereignty, thereby eroding cultural identity tied to food practices (Ofosu et al., 2020; Schueler et al., 2011). Moreover, the interrelation between mining and agricultural health is often exacerbated by environmental degradation. For instance, the release of heavy metals and toxic effluents into agricultural fields can severely compromise the quality and safety of food produced in these areas, directly impacting Indigenous diets that rely on local foods (Asante et al., 2023; Gao and Chen, 2023). When traditional food sources are contaminated, communities face stark choices between consuming potentially unsafe local foods and purchasing expensive alternatives, further entrenching food insecurity (Yadav et al., 2019; Yiridomoh, 2021).

5.4. Gendered and Generational Knowledge Transmission

The transmission of knowledge about food safety and traditional practices is often gendered and influenced by generational divides within Indigenous communities. Women, who historically have played crucial roles in food gathering, preparation, and education, are central to maintaining traditional food practices and improving family food security (Sampson et al., 2021; Williams et al., 2020). Their insights into sustainable practices, particularly in the context of shifting environmental conditions resulting from industrial pollution and climate change, are invaluable. However, as younger generations migrate towards urban areas or adopt more Western lifestyles, there is a risk that vital traditional knowledge may be lost (Sampson et al., 2021). Furthermore, community efforts to revitalize traditional knowledge may differ based on gender, with women often spearheading initiatives aimed at managing traditional food systems and advocating for improved food policies (Sampson et al., 2021; Erinle et al., 2021). The dynamics of food practice changes can thus reflect broader social changes, with women often serving as custodians of cultural heritage and advocates for food sovereignty.

6. BRIDGING KNOWLEDGE SYSTEMS: SCIENTIFIC VALIDATION AND INTEGRATIVE APPROACHES

The intersection of ITK and scientific validation provides a foundation for enhancing contemporary food safety practices. To effectively bridge these two realms, it is essential to explore how traditional food safety methods, particularly those identified through the antimicrobial properties of native plants, can be integrated into modern protocols. Research has demonstrated that training food handlers to enhance their understanding of hygiene practices can significantly influence food safety outcomes, thereby supporting the potential for co-production of knowledge between traditional and scientific approaches (Mahunu et al., 2024). For example, Jores et al. (201) found that although there was not a direct statistical correlation between knowledge and food safety practices among street food vendors, knowledge acquired through training is crucial for improving food safety and hygiene practices, highlighting the value of continuous training.

Moreover, Malavi et al. (2021), established that training food handlers on food safety principles led to significant behavior changes, indicating that education strengthens the relationship between knowledge and practice in food safety contexts, particularly in food production facilities in Kenya. This aligns with Adesokan et al. (2015), who provided evidence that food safety training correlates with improved knowledge and behaviors among food service workers, highlighting the necessity of ongoing professional development in food safety. Conversely, Aquino et al. (2021) showed that food safety training does not always enhance attitudes toward food safety practices, suggesting a need for re-evaluating training methodologies to ensure their effectiveness. Integrating traditional knowledge of food safety into modern training programs could enhance educational efforts. Traditional methods often involve the use of local plants noted for their antimicrobial properties, as highlighted by Chen et al. (2018), who have emphasized the potential benefits of biopesticides derived from native plants in enhancing food safety. The historical knowledge embedded in these traditional practices can offer insights into natural methods for preventing foodborne illnesses and improving food preservation through culturally informed approaches.

In addressing the challenges and opportunities of co-producing knowledge, exploring how indigenous approaches can be validated scientifically holds promise. A study by Ferk et al. (2015) indicated that educational levels correlate with food safety attitudes and behaviors, suggesting that integrating ITK with formal education could enhance collective knowledge about food safety. It is essential to consider diverse educational and cultural backgrounds when designing food safety programs, as seen in studies assessing university students' food safety knowledge and practices (Akabanda et al., 2017). Such considerations can help facilitate effective comparative analyses and adaptation of educational resources, addressing both technical and culturally relevant needs. Challenges also arise when attempting to synthesize traditional knowledge and modern science, particularly regarding the acceptance and adaptation of these approaches within broader food safety protocols. Numerous studies highlight the gap between knowledge and practice, showing that even with significant training, many food handlers fail to properly implement food safety protocols (Tegegne & Phyo, 2017; Al-Slaihat et al., 2024). This gap necessitates a tailored approach to training programs that accommodates traditional practices while incorporating robust scientific validation to enhance comprehension and execution among food handlers.

As organizations and governments enhance food safety frameworks in response to evolving challenges, such as foodborne illnesses, it is crucial to include interdisciplinary strategies that leverage insights from both traditional and contemporary methods. For instance, practical improvements in food safety knowledge, attitudes, and practices can be achieved by combining culturally relevant education with formal scientific knowledge (Brown et al., 2014). This fosters a comprehensive understanding of food safety that resonates with diverse populations and their respective practices. Partnerships between traditional knowledge holders and scientists can facilitate knowledge exchange while strengthening validated protocols. Engaging local communities in the co-creation of knowledge regarding food safety can lead to more effective interventions and improved practices. Collaborative models have been proposed to bridge the divide between ITK and scientific approaches, emphasizing the need for mutual respect, acknowledgment of diverse knowledge creation processes, and participatory frameworks that prioritize local contexts. The opportunities for integrating traditional food safety methods with modern protocols are substantial, and addressing these collaboratively is essential for achieving better food safety outcomes globally. As research continues to illuminate the benefits of such integrative approaches, skepticism towards traditional practices in scientific circles may begin to decline. Understanding that food safety is not merely a technical issue but also a cultural one is crucial in ensuring that food safety education and training effectively address the needs of diverse populations while promoting safe food practices.

7. GOVERNANCE, RIGHTS, AND ETHICS IN INDIGENOUS FOOD SAFETY SYSTEMS

The intricate web of governance, rights, and ethics surrounding Indigenous food safety systems encompasses various elements, notably Indigenous rights, food sovereignty, the protection of traditional knowledge, and customary law. Indigenous rights and food sovereignty are foundational elements of self-governance, emphasizing the inherent right of Indigenous peoples to control their land, resources, and food systems. This notion of food sovereignty extends beyond mere access to food; it reflects a broader calling for Indigenous communities to reclaim and affirm their rights to make decisions about their cultural and ecological heritage (Loring & Gerlach, 2015; Whyte, 2018). The implementation of the United Nations Declaration on the Rights of Indigenous Peoples serves as a crucial framework to uphold these rights, recognizing the necessity for Indigenous peoples to play a primary role in matters affecting their food systems (Loring & Gerlach, 2015). Legal frameworks can empower Indigenous communities to restore autonomy over food production, distribution, and consumption practices, thus reshaping their relationships with local ecosystems (Young et al., 2024).

In this context, traditional ecological knowledge plays an integral role in the governance of food systems among Indigenous communities. This knowledge, often passed down through generations, provides critical insights into sustainable practices that promote food security (Phillipps et al., 2022). For Indigenous peoples, the protection of such knowledge is vital as it embodies their cultural identity and sustains their unique relationship with the environment (Lovo et al., 2021). Acknowledging and integrating Indigenous knowledge into food safety governance frameworks can enhance the effectiveness of food systems by infusing them with culturally relevant practices that have been honed over millennia (Kuhnlein, 2014). Furthermore, the enforcement of customary law within these frameworks not only respects Indigenous knowledge but also legitimizes traditional practices that may conflict with Western legal systems (Tuominen et al., 2023). The recognition and protection of traditional knowledge are closely tied to discussions of intellectual property and benefit-sharing mechanisms. The potential for biocolonialism necessitates the establishment of fair frameworks that ensure Indigenous communities benefit from their genetic resources and traditional knowledge (Soares et al., 2023; Ogwu et al., 2024). Emerging frameworks, such as the CARE Principles for Indigenous Data Governance, emphasize the importance of establishing collective benefit agreements that respect the rights of Indigenous peoples while promoting ethical engagement in research involving their cultural heritage (Carroll et al., 2022). These principles reflect a growing recognition of the need for Indigenous authority to control access to their knowledge and resources, which is crucial for promoting equitable partnerships in research and agricultural practices (Carroll et al., 2022; Onoabhagbe et al., 2024).

The role of customary law in regulating food safety is particularly salient in contexts where Indigenous peoples face external pressures from state and federal regulations that may not align with their traditional practices (Popken et al., 2023). Customary laws, grounded in culturally specific traditions and practices, protect the food systems upon which many Indigenous communities rely, reinforcing a collective responsibility to sustain these systems for future generations (Ban et al., 2019). Exploring the intersection of customary law with national and international regulations can illuminate pathways for inclusivity in food governance systems that reflect the values and needs of Indigenous communities (Onomu, 2023). The interplay between these various systems invites critical questions about ethical engagement in research and policymaking related to Indigenous food systems. Ethical guidelines must be implemented that honor the rights of Indigenous peoples and ensure their meaningful participation in decision-making processes affecting their food sovereignty (Burchill et al., 2022). Engaging in dialogue, building trust, and respecting the unique cultural contexts of Indigenous communities are integral aspects of this engagement (Bond et al., 2016). Ethical research that incorporates Indigenous perspectives not only contributes to more robust knowledge production but also fosters an environment in which Indigenous voices are prioritized and respected (Burchill et al., 2022).

To advance Indigenous food sovereignty further, it is essential to understand the contemporary and historical contexts that shape food systems today. Colonial histories have often undermined Indigenous food practices, leading to a dependence on industrial food sources that do not align with the cultural dietary needs of these communities (Robin, 2019). Addressing these inequities requires a commitment to revitalizing traditional foods and practices in a way that acknowledges their significance to health and well-being (Farfán et al., 2021; Izah et al., 2023). Moreover, promoting self-governance in food systems provides an avenue for healing and recovery, allowing Indigenous communities to draw upon their cultural heritage as a source of resilience (Popken et al., 2023). The movement toward food sovereignty is not merely an Indigenous issue; it intersects with broader societal actions aimed at dismantling oppressive structures such as colonialism and capitalism that impact marginalized communities worldwide (Tuominen et al., 2023). This intersectionality highlights the importance of solidarity between Indigenous movements and other social justice struggles, advocating for systemic changes that acknowledge the rights and needs of diverse groups (Phillipps et al., 2022). Progress in governance, rights, and ethics surrounding Indigenous food safety systems demands continual examination of new and existing policy frameworks. Advocates must remain vigilant in monitoring how policies are crafted and implemented to ensure they do not inadvertently perpetuate colonial legacies or hinder Indigenous self-determination (Weiler et al., 2014). By reinforcing Indigenous governance structures, decision-making processes can be better aligned with the cultural values and traditional knowledge of Indigenous communities, leading to more effective and ethical food safety systems (Robin & Hart, 2025).

8. INDIGENOUS RESPONSES TO FOOD CONTAMINATION AND SAFETY CHALLENGES

Indigenous communities around the globe face numerous challenges related to food contamination and safety, particularly when impacted by industrial activities such as mining and oil extraction. Such events can severely undermine traditional food systems, causing detrimental effects on health, culture, and the environment. For example, significant contamination from mining has been reported in regions such as the Amazon and the Canadian boreal forests, where Indigenous peoples have had to navigate toxic pollutants that compromise not only their food sources but also their way of life. Communities engage in various responses, including legal actions, grassroots movements, and the revival of traditional ecological practices, to mitigate the effects of contamination and reclaim their rights to safe food environments (Phillipps et al., 2022; Levkoe et al., 2021). Among the Indigenous knowledge systems (IKS) addressing food safety, traditional grain storage practices in sub-Saharan Africa offer valuable insights into aflatoxin prevention, which is crucial for food safety in these regions. Knowledge regarding traditional methods of grain storage, such as the use of nature-based techniques and community validation systems, can significantly reduce the risk of mycotoxin contamination. Such practices are being rejuvenated through intergenerational knowledge transfer and community-led initiatives that affirm cultural identity while addressing contemporary food safety issues (Neufeld et al., 2017). These Indigenous methods not only preserve nutritional integrity but also enhance resilience against the changing dynamics of global food systems.

The role of customary governance in ensuring food safety cannot be understated, especially in the Pacific Islands, where traditional structures are crucial for managing local resources. Customary institutions have been integral in enforcing local rules regarding food production and safety, promoting sustainable practices that safeguard both land and marine environments. Communities have revitalized long-standing customs, blending them with modern approaches to establish policy frameworks that prioritize traditional knowledge in food governance. This blending portrays a holistic approach that respects cultural practices while enhancing food sovereignty (Thow et al., 2011; Sievert et al., 2019).

In Canada, Indigenous-led movements are gaining momentum in addressing food contamination issues, particularly those related to industrial pollution. By employing a rights-based approach, these communities work to reclaim access to traditional lands and resources. They actively participate in negotiating land-use agreements and assessing environmental impacts associated with energy development projects. These efforts underscore the importance of Indigenous leadership in environmental stewardship and highlight the necessity of integrating Indigenous traditional knowledge into national policy frameworks (Levkoe et al., 2021; Bunkley et al., 2023). This approach addresses the systemic inequities faced by Indigenous communities in both the enjoyment of their rights and the protection of food sources from contamination. Similarly, in the Amazon basin, Indigenous responses to food safety challenges posed by environmental degradation resulting from oil spill incidents demonstrate their resilience and proactive engagement. Indigenous groups utilize both legal frameworks and community activism to safeguard their environment. They advocate for the enforcement of international treaties and national regulations that promise environmental justice and food safety while leveraging their traditional ecological knowledge to adapt and mitigate the impacts of contamination on their food systems (Rhodes et al., 2020). The use of participatory governance has proven effective in fostering collaboration among Indigenous communities, environmental organizations, and governmental agencies, thereby leading to comprehensive strategies to address these pressing food safety challenges.

Policy reforms are essential for integrating Indigenous traditional knowledge (ITK) into national food safety frameworks. Countries are beginning to recognize the value of ITK in supporting food security and safety initiatives. For instance, moving from marginalization of Indigenous practices to their inclusion in mainstream agricultural policies can yield significant benefits. Such reforms advocate for equity and sustainability by aligning modern food safety regulations with traditional practices that have proven effective over time. Evidence from various studies indicates a marked improvement in food safety and community health outcomes when inclusive policies are adopted (Foale et al., 2010; Thow et al., 2011). Such integration is necessary to steer food policies toward a sustainable future that prioritizes the health of communities and ecosystems, reiterating the importance of a value-driven approach in addressing food safety (Thow et al., 2011). This work underscores the adaptive capacity of Indigenous communities to protect and transfer traditional knowledge about food safety across generations. Such continuity is vital for ensuring that future generations maintain an understanding of sustainable practices and their applications in mitigating the effects of contamination on food resources (Neufeld et al., 2017).

9. SUSTAINING INDIGENOUS KNOWLEDGE SYSTEMS FOR DETECTING CONTAMINANTS IN FOODS: CHALLENGES AND TRANSFORMATIVE PATHWAYS

The task of sustaining ITK systems for detecting contaminants in food encompasses numerous complex challenges and transformative pathways that are critical for the future of food sovereignty and sustainability for Indigenous Peoples. These challenges are exacerbated by the marginalization and continuous erosion of ITK, necessitating robust documentation, revitalization, and intergenerational transmission of this knowledge. The marginalization of Indigenous Knowledge Systems is a historical and ongoing issue that impedes their effective utilization within contemporary governance and scientific frameworks. Several studies have documented the realities faced by Indigenous communities regarding their traditional food practices and how colonial narratives have often overshadowed their knowledge systems in policymaking processes (Vijayan et al., 2022; Swiderska et al., 2022; Malli et al., 2023). This institutional neglect has led to a significant gap in the integration of ITK into modern food safety regulations, thereby reinforcing systemic inequalities (Heaney et al., 2024; Domingo et al., 2023). The consequences are especially pronounced concerning food insecurity, as Indigenous populations often find themselves at the intersection of inadequate support from governmental policies that fail to acknowledge their unique needs and contributions to food systems (Sherriff et al., 2022; Domingo et al., 2023; Kom et al., 2024; Ajani et al., 2013).

There is a pressing need for proper documentation and revitalization of IK. Documentation serves not merely as a record of practices but as a living repository of knowledge that is essential for intergenerational transmission. The wisdom encapsulated in Indigenous practices offers ecological insights crucial for modern agricultural practices, such as resilience to climate change and managing pest infestations (Kom et al., 2024; Ajani et al., 2013; Coté, 2016). Programs that engage elders and community members in workshops and other participative formats have been shown to be effective in passing on food knowledge, enhancing the cultural identity of younger generations while simultaneously addressing food sovereignty issues (Phillipps et al., 2022; Neufeld et al., 2017; Levkoe et al., 2021). The challenges of utilizing ITK in contemporary food systems must also be viewed in light of socio-economic factors that directly impact the application of these knowledge systems. Indigenous practices related to crop rotation, rainfall prediction, and pest management are often not supported by formal education systems, leading to the underutilization of valuable ecological knowledge (Ajani et al., 2013; Kom et al., 2024). To remedy this, capacity-building initiatives are imperative. They should foster partnerships between Indigenous and non-Indigenous scientists, promoting participatory research methodologies that recognize and validate ITK, thus enhancing the efficacy and reach of food safety frameworks (Michnik et al., 2021; Malcom, 2024; Persaud et al., 2025).

Capacity building within Indigenous communities is essential. This encompasses developing skills not only in the documentation and application of ITK but also in advocacy and policy reform. The complexities of integrating Indigenous perspectives into national food safety frameworks are compounded by historical traumas and current socio-political dynamics that frequently marginalize Indigenous voices in policymaking (Swiderska et al., 2022; Malli et al., 2023; Domingo et al., 2023). It is therefore vital that policy frameworks include provisions for the participatory engagement of Indigenous peoples in all stages of decision-making processes related to food systems (McCarter et al., 2014; Johnson-Jennings et al., 2020; Malcom, 2024). Doing so aligns with a decolonizing approach that emphasizes the importance of Indigenous agency and governance in food sovereignty initiatives.

Furthermore, policy reforms must include mechanisms that specifically support and reinforce the benefits inherent in ITK. A key recommendation emerging from the literature is to align national food safety policies with international frameworks, such as the United Nations Declaration on the Rights of Indigenous Peoples (UNDRIP), which advocates for the recognition of Indigenous rights and self-determination (Malcom, 2024; Swiderska et al., 2022; Coté, 2016). Such alignment not only offers a recognizable pathway for Indigenous communities to assert control over their food systems but also emphasizes the cultural responsibilities that accompany these rights (Coté, 2016; Levkoe et al., 2021; Domingo et al., 2023). As Indigenous food systems cornerstones recognize climate resilience and sustainability, they are fundamentally tied to the health and wellness of Indigenous communities (Swiderska et al., 2022; Malli et al., 2023). Addressing gaps in food security requires considering the intersectionality of health, economic, and agricultural policies that directly impact Indigenous peoples' access to traditional and contemporary food systems (Sherriff et al., 2022; Deen et al., 2024; Kom et al., 2024). An integrative approach that combines ITK with contemporary science offers the potential for innovative solutions to food safety challenges, particularly for detecting and mitigating food contaminants.

Incorporating ITK in food safety challenges also means extending beyond mere contamination detection. This includes promoting traditional food preservation methods, which are vital for maintaining the nutritional integrity of food and are often overlooked in modern culinary practices (Malli et al., 2023; Masekoameng & Molotja, 2023; Heaney et al., 2024). Educational initiatives that impart these food preservation techniques can empower community members while ensuring that cultural practices are honored and passed down (Johnson-Jennings et al., 2020; Michnik et al., 2021). As we explore the transformative pathways for sustaining Indigenous Knowledge Systems in food safety, it becomes clear that enhancing cooperation among various stakeholders' Indigenous communities, scientists, policymakers, and educators will be instrumental in creating inclusive frameworks that celebrate and integrate diverse knowledge systems (Vijayan et al., 2022; Muller, 2018; Domingo et al., 2023). The essence of these collaborative efforts will not only advocate for the recognition and revitalization of ITK, but also promote its continued development. Still, it will also work towards dismantling the foundations of systemic inequities that have long disenfranchised Indigenous Peoples. These multifaceted challenges necessitate a concerted, respectful, and persistent approach to reestablishing and sustaining the integrity and efficacy of Indigenous food systems, particularly in light of contemporary food safety concerns.

10. CONCLUSION

Indigenous and traditional knowledge (ITK) systems provide deeply rooted, place-based frameworks for ensuring food safety and managing contaminants that have evolved over generations through observation, adaptation, and cultural continuity. As this chapter has demonstrated, Indigenous communities have developed sophisticated methods for food preservation, environmental monitoring, and risk assessment that reflect a holistic understanding of ecosystem health, spiritual beliefs, and communal well-being. These practices ranging from fermentation and the use of natural preservatives to biocultural indicators, are not only scientifically valid but also often more sustainable, context-appropriate, and resilient than many industrial approaches. Yet, despite their demonstrated value, ITK systems continue to be marginalized in formal food safety governance and research agendas. This exclusion reflects more profound legacies of colonialism, structural inequality, and epistemic injustice that undermine Indigenous agency and erode traditional food systems. As such, the path forward must prioritize the respectful integration of Indigenous knowledge into broader contaminant management strategies rather than assimilation. This includes legal recognition, ethical engagement, and the co-production of knowledge that honors Indigenous rights, sovereignty, and intellectual traditions. Empowering Indigenous communities to lead in food safety initiatives, supporting intergenerational knowledge transmission, and reforming policy frameworks to include customary institutions are essential steps toward more inclusive and sustainable food governance. In an era marked by intensifying environmental crises and growing concerns over food system vulnerabilities, drawing on the strengths of ITK offers not only practical solutions but also a transformative shift toward justice, resilience, and ecological balance.

REFERENCES

Abajue, M. C., & Ogwu, M. C. (2024). Nutritional Status of Indigenous and Traditional Edible Insects: Challenges and Limitations. In Ogwu, M. C., Izah, S. C., & Ntuli, N. R. (Eds.), *Food Safety and Quality in the Global South* (pp. 711–729). Springer. DOI: 10.1007/978-981-97-2428-4_23

Abass, A., Awoyale, W., Sulyok, M., & Alamu, E. (2017). Occurrence of regulated mycotoxins and other microbial metabolites in dried cassava products from Nigeria. *Toxins*, *9*(7), 207. DOI: 10.3390/toxins9070207

Abdul-Mutalib, N., Abdul-Rashid, M., Mustafa, S., Nordin, S., Hamat, R., & Osman, M. (2012). Knowledge, Attitudes, and Practices Regarding Food Hygiene and Sanitation of Food Handlers in Kuala Pilah, Malaysia. *Food Control*, *27*(2), 289–293. DOI: 10.1016/j.foodcont.2012.04.001

Abdullahi, N., Igwe, E. C., Dandago, M. A., Saní, A., & Umar, N. B. (2023). A review of heavy metals accumulation and control in active agricultural soil. *Turkish Journal of Food and Agriculture Sciences*, *5*(1), 7–18. DOI: 10.53663/turjfas.1247846

Adekoya, I., Njobeh, P., Obadina, A., Chilaka, C., Okoth, S., Boevre, M., & Saeger, S. (2017). Awareness and prevalence of mycotoxin contamination in selected Nigerian fermented foods. *Toxins*, *9*(11), 363. DOI: 10.3390/toxins9110363

Adesokan, H., Akinseye, V., & Adesokan, G. (2015). Food safety training is associated with improved knowledge and behaviours among foodservice establishments' workers. *International Journal of Food Sciences*, *2015*, 1–8. DOI: 10.1155/2015/328761

Ahmad, K., & Pieroni, A. (2016). Folk knowledge of wild food plants among the tribal communities of Thakht-e-Sulaiman Hills, north-west Pakistan. *Journal of Ethnobiology and Ethnomedicine*, *12*(1), 17. Advance online publication. DOI: 10.1186/s13002-016-0090-2

Ajani, E., Mgbenka, R., & Okeke, M. (2013). Utilization of Indigenous Knowledge as a Strategy for Climate Change Adaptation among Farmers in Sub-Saharan Africa: Implications for Policy. *Asian Journal of Agricultural Extension Economics and Sociology*, *2*(1), 23–40. DOI: 10.9734/AJAEES/2013/1856

Akabanda, F., Hlortsi, E., & Owusu-Kwarteng, J. (2017). Food Safety Knowledge, Attitudes, and Practices of Institutional Food Handlers in Ghana. *BMC Public Health*, *17*(1), 40. Advance online publication. DOI: 10.1186/s12889-016-3986-9

Al-Slaihat, A., Saleh, M., Al-Awadi, A., Almomani, H., Alsakran, S., Alsheikh, D., & Lee, Y. (2024). Food safety knowledge, attitudes, and practices (KAP) among dairy retail vendors in AmmaJordandan... https://doi.org/DOI: 10.20944/preprints202405.1041.v1

Aquino, H., Yap, T., Lacap, J., Tuazon, G., & Flores, M. (2021). Food Safety Knowledge, Attitudes, Practices, and Training of Fast-Food Restaurant Food Handlers: A Moderation Analysis. *British Food Journal*, *123*(12), 3824–3840. DOI: 10.1108/BFJ-01-2021-0026

Asante, K., Abugre, S., Akoto, D., & Derkyi, N. (2023). Cassava (Manihot esculenta) yield, nutrition, and heavy metal bioaccumulation responses to circular economy-based innovations in a mining-degraded landscape. *International Journal of Plant and Soil Science*, *35*(22), 840–850. DOI: 10.9734/ijpss/2023/v35i224195

Ban, N., Wilson, E., & Neasloss, D. (2019). Historical and Contemporary Indigenous Marine Conservation Strategies in the North Pacific. *Conservation Biology*, *34*(1), 5–14. DOI: 10.1111/cobi.13432

Barrios, E., Delve, R., Bekunda, M., Mowo, J., Agunda, J., Ramisch, J., Trejo, M. T., & Thomas, R. (2006). Indicators of soil quality: A south–south development of a methodological guide for linking local and technical knowledge. *Geoderma*, *135*, 248–259. DOI: 10.1016/j.geoderma.2005.12.007

Barrios, E., & Trejo, M. (2003). Implications of Local Soil Knowledge for Integrated Soil Management in Latin America. *Geoderma*, *111*(3-4), 217–231. DOI: 10.1016/S0016-7061(02)00265-3

Berkes, F., & Berkes, M. (2009). Ecological complexity, fuzzy logic, and holism in indigenous knowledge. *Futures*, *41*(1), 6–12. DOI: 10.1016/j.futures.2008.07.003

Bohensky, E., & Maru, Y. (2011). Indigenous knowledge, science, and resilience: What have we learned from a decade of international literature on "integration"? *Ecology and Society*, *16*(4), art6. Advance online publication. DOI: 10.5751/ES-04342-160406

Bond, C., Foley, W., & Askew, D. (2016). "it puts a human face on the researched" – a qualitative evaluation of an indigenous health research governance model. *Australian and New Zealand Journal of Public Health*, *40*, S89–S95. DOI: 10.1111/1753-6405.12422

Brown, L., Le, B., Wong, M., Reimann, D., Nicholas, D., Faw, B., Davis, E., & Selman, C. (2014). Restaurant manager and worker food safety certification and knowledge. *Foodborne Pathogens and Disease*, *11*(11), 835–843. DOI: 10.1089/fpd.2014.1787

Bunkley, N., McCool, J., & Garton, K. (2023). A review of international trade and investment agreements and nutrition policy space in the Pacific. *Frontiers in Nutrition*, *10*, 1208542. Advance online publication. DOI: 10.3389/fnut.2023.1208542

Burchill, L., Kotevski, A., Duke, D., Ward, J., Prictor, M., Lamb, K., & Kennedy, M. (2022). Ethics guidelines use and indigenous governance and participation in aboriginal and torres strait islander health research: A national survey. *The Medical Journal of Australia*, *218*(2), 89–93. DOI: 10.5694/mja2.51757

Carroll, S., Plevel, R., Jennings, L., Garba, I., Sterling, R., Cordova-Marks, F., Hiratsuka, V., Hudson, M., & Garrison, N. (2022). Extending the care principles from tribal research policies to benefit sharing in genomic research. *Frontiers in Genetics*, *13*, 1052620. Advance online publication. DOI: 10.3389/fgene.2022.1052620

Chen, Y., Ji, H., Chen, L., Jiang, R., & Wu, Y. (2018). Food Safety Knowledge, Attitudes, and Behavior among Dairy Plant Workers in Beijing, Northern China. *International Journal of Environmental Research and Public Health*, *15*(1), 63. DOI: 10.3390/ijerph15010063

Coté, C. (2016). "Indigenizing" food sovereignty. Revitalizing Indigenous Food Practices and Ecological Knowledges in Canada and the United States. *Humanities (Washington)*, *5*(3), 57. DOI: 10.3390/h5030057

Country, M., Davis, J., Simmons, J., Snelson, S., Channell, V., Haynes, K., & Dosseto, A. (2024). Quantitative assessment of the effect of agency-led prescribed burns and cultural burns on soil properties in southeastern Australia. *Fire (Basel, Switzerland)*, *7*(3), 75. DOI: 10.3390/fire7030075

Dawoe, E., Quashie-Sam, J., Isaac, M., & Oppong, S. (2012). Exploring farmers' local knowledge and perceptions of soil fertility and management in the Ashanti region of Ghana. *Geoderma*, *179-180*, 96–103. DOI: 10.1016/j.geoderma.2012.02.015

Deen, C., Sherriff, S., Shelling, M., Gall, A., Cubillo, B., Morenga, L., & Matthews, V. (2024). Measuring indigenous food security—A case for indigenous designed tools. *Health Promotion Journal of Australia*, *36*(1), e945. Advance online publication. DOI: 10.1002/hpja.945

Domingo, A., Yessis, J., Charles, K., Skinner, K., & Hanning, R. (2023). Integrating knowledge and action: learnings from an implementation program for food security and food sovereignty with First Nations communities in Canada. DOI: 10.21203/rs.3.rs-2801301/v1

du Bray, M., Sanchez, A., & Ogwu, M. (2023). Time Immemorial: Water Governance Approaches and Indigenous Rights. River Field Studies Network, QUBES Educational Resources. DOI: 10.25334/2HPW-H547

Erinle, K. O., Ogwu, M. C., Evivie, S. E., Zaheer, M. S., Ogunyemi, S. O., & Adeniran, S. O. (2021). Impacts of COVID-19 on agriculture and food security in developing countries: Potential mitigation strategies. *Perspectives in Agriculture, Veterinary Science, Nutrition and Natural Resources*, *16*(16), 1–16. DOI: 10.1079/PAVSNNR202116016

Evivie, S. E., Ogwu, M. C., Abdelazez, A., Bian, X., Liu, F., Li, B., & Huo, G. (2020). Suppressive effects of *Streptococcus thermophilus* KLDS 3.1003 on some foodborne pathogens revealed through *in vitro, in vivo* and genomic insights. *Food & Function*, *11*(7), 6573–6587. DOI: 10.1039/D0FO01218A

Evivie, S. E., Ogwu, M. C., Abdelazez, A., Bian, X., Liu, F., Li, B., & Huo, G. (2021). Suppressive effects of *streptococcus thermophilus* KLDS 3.1003 on some foodborne pathogens revealed through *in vitro, in vivo* and genomic insights. *Food & Function*, *12*(7), 3280. DOI: 10.1039/D1FO90021H

Farfán, J., Chaux, J., & Torres, D. (2021). Food autonomy: Decolonial perspectives for indigenous health and buen vivir. *Global Health Promotion*, *28*(3), 50–58. DOI: 10.1177/1757975920984206

Ferk, C., Calder, B., & Camire, M. (2015). Assessing the food safety knowledge of University of mainMainedents. *Journal of Food Science Education*, *15*(1), 14–22. DOI: 10.1111/1541-4329.12076

Filip, S., & Fink, R. (2017). Hazard analysis and critical control points of traditional vs industrial cremeschnitte production. *Acta Alimentaria*, *46*(1), 9–16. DOI: 10.1556/066.2017.46.1.2

Foale, S., Cohen, P., Januchowski-Hartley, S., Wenger, A., & Macintyre, M. (2010). Tenure and taboos: Origins and implications for fisheries in the Pacific. *Fish and Fisheries*, *12*(4), 357–369. DOI: 10.1111/j.1467-2979.2010.00395.x

Gao, J., & Chen, Z. (2023). Application of association rule mining algorithm based on 5 G technology in information management system. *Scalable Computing Practice and Experience*, *24*(2), 139–149. DOI: 10.12694/scpe.v24i2.2150

Gbashi, S., Madala, N., Saeger, S., Boevre, M., Adekoya, I., Adebo, O., … Njobeh, P. (2019). The Socioeconomic Impact of Mycotoxin Contamination in Africa. DOI: 10.5772/intechopen.79328

Goldhar, C., Ford, J., & Berrang-Ford, L. (2010). Prevalence of Food Insecurity in a Greenlandic Community and the Importance of Social, Economic, and Environmental Stressors. *International Journal of Circumpolar Health, 69*(3), 285–303. DOI: 10.3402/ijch.v69i3.17616

Heaney, D., Padilla-Zakour, O., & Chen, C. (2024). Processing and preservation technologies to enhance indigenous food sovereignty, nutrition security, and health equity in North America. *Frontiers in Nutrition, 11*, 1395962. Advance online publication. DOI: 10.3389/fnut.2024.1395962

Ikhajiagbe, B., Atoe, R., Ogwu, M. C., & Loveniers, P.-J. (2021). Changes in *Telfaria occidentalis* leaf morphology, quality and phytochemical composition under different local preservation regimes. *Vegetos, 31*(1), 29–36. DOI: 10.1007/s42535-021-00188-z

Ikhajiagbe, B., & Ogwu, M. C. (2020). Hazard quotient, microbial diversity and plant composition of spent crude oil polluted-soil. *Beni-Suef University Journal of Basic and Applied Sciences, 9*(26), 26. Advance online publication. DOI: 10.1186/s43088-020-00052-0

Ikhajiagbe, B., Ogwu, M. C., Ogochukwu, O. F., Odozi, E. B., Adekunle, I. J., & Omage, Z. E. (2021). The place of neglected and underutilized legumes in human nutrition and protein security. *Critical Reviews in Food Science and Nutrition*. Advance online publication. DOI: 10.1080/10408398.2020.1871319

Izah, S. C., Aigberua, A. O., & Ogwu, M. C. (2022). Trace element composition of *Gallus gallus domesticus* eggs and health risks associated with their consumption in Port Harcourt, Nigeria. *Journal of Food Safety and Hygiene, 8*(3), 202–222.

Izah, S. C., Odubo, T. C., Ogwu, M. C., & Hait, M. (2024). Aspects of Microorganisms in the Food Industry. In Ogwu, M. C., Izah, S. C., & Ntuli, N. R. (Eds.), *Food Safety and Quality in the Global South* (pp. 399–425). Springer. DOI: 10.1007/978-981-97-2428-4_13

Izah, S. C., Richard, G., Stanley, H. O., Sawyer, W. E., Ogwu, M. C., & Uwaeme, O. R. (2023). Integrating the One Health Approach and Statistical Analysis for Sustainable Aquatic Ecosystem Management and Trace Contamination Mitigation. *ES Food and Agroforestry, 14*, 1012. DOI: 10.30919/esfaf1012

Johnson-Jennings, M., Jennings, D., Paul, K., & Little, M. (2020). Identifying needs and uses of digital indigenous food knowledge and practices for an indigenous food wisdom repository. *Alternative*, *16*(4), 290–299. DOI: 10.1177/1177180120954446

Jores, D., Arif, M., & Rahman, M. (2018). Factors Associated with Food Hygiene Practices among Street Food Vendors in Padawan, Sarawak. *Borneo Journal of Resource Science and Technology*, *8*(1), 56–65. DOI: 10.33736/bjrst.824.2018

Kachapulula, P., Akello, J., Bandyopadhyay, R., & Cotty, P. (2017). Aflatoxin contamination of groundnut and maize in Zambia: Observed and potential concentrations. *Journal of Applied Microbiology*, *122*(6), 1471–1482. DOI: 10.1111/jam.13448

Kadirvel, G., Banerjee, B., Meitei, S., Doley, S., Sen, A., & Muthukumar, M. (2018). Market Potential and Opportunities for Commercialization of Traditional Meat Products in the Northeast Hill Region of India. *Veterinary World*, *11*(2), 118–124. DOI: 10.14202/vetworld.2018.118-124

Khastini, R., Maryani, N., Haryati, S., Rahmah, R., Sa'ban, A., Aisyah, H., Fadillah, T. N., & Nadia, S. (2023). Understanding the Traditional Knowledge of Picungan Bandeng, a Fermented Fish Food, and Its Sustainability in Banten Province. *IOP Conference Series. Earth and Environmental Science*, *1241*(1), 012076. DOI: 10.1088/1755-1315/1241/1/012076

Kom, Z., Nicolau, M., & Nenwiini, S. (2024). The use of indigenous knowledge systems and practices to enhance food security in Vhembe District, South Africa. *Agricultural Research*, *13*(3), 599–612. DOI: 10.1007/s40003-024-00716-8

Kuhnlein, H. (2014). Food system sustainability for health and well-being of indigenous peoples. *Public Health Nutrition*, *18*(13), 2415–2424. DOI: 10.1017/S1368980014002961

Lam, D., Hinz, E., Lang, D., Tengö, M., Wehrden, H., & Martín-López, B. (2020). Indigenous and local knowledge in sustainability transformations research: A literature review. *Ecology and Society*, *25*(1), art3. Advance online publication. DOI: 10.5751/ES-11305-250103

Levkoe, C., McLaughlin, J., & Strutt, C. (2021). Mobilizing Networks and Relationships through Indigenous Food Sovereignty: The Indigenous Food Circle's Response to the COVID-19 Pandemic in Northwestern Ontario. *Frontiers in Communication*, *6*, 672458. Advance online publication. DOI: 10.3389/fcomm.2021.672458

Loring, P., & Gerlach, S. (2015). Searching for progress on food security in the North American North: A research synthesis and meta-analysis of the peer-reviewed literature + supplementary appendix (see article tools). *Arctic*, *68*(3), 380. DOI: 10.14430/arctic4509

Lovo, E., Woodward, L., Larkins, S., Preston, R., & Baba, U. (2021). Indigenous knowledge around the ethics of human research from the Oceania region: A scoping literature review. *Philosophy, Ethics, and Humanities in Medicine; PEHM*, *16*(1), 12. Advance online publication. DOI: 10.1186/s13010-021-00108-8

Mahunu, G. K., Osei-Kwarteng, M., Ogwu, M. C., & Afoakwah, N. A. (2024). Safe Food Handling Techniques to Prevent Microbial Contamination. In Ogwu, M. C., Izah, S. C., & Ntuli, N. R. (Eds.), *Food Safety and Quality in the Global South* (pp. 427–461). Springer. DOI: 10.1007/978-981-97-2428-4_14

Malavi, D., Abong, G., & Muzhingi, T. (2021). Effect of food safety training on behavior change of food handlers: A case of orange-fleshed sweetpotato purée processing in Kenya. *Food Control*, *119*, 107500. DOI: 10.1016/j.foodcont.2020.107500

Malcom, A. (2024). Integrating indigenous knowledge into livestock policy formulation. *International Journal of Livestock Policy*, *3*(1), 30–42. DOI: 10.47941/ijlp.1962

Malli, A., Monteith, H., Hiscock, C., Smith, E., Fairman, K., Galloway, T., & Mashford-Pringle, A. (2023). Impacts of Colonization on Indigenous Food Systems in Canada and the United States: A Scoping Review. *BMC Public Health*, *23*(1), 2105. Advance online publication. DOI: 10.1186/s12889-023-16997-7

Masekoameng, M., & Molotja, M. (2023). Contribution of indigenous food preservation and processing practices to food security of rural households in Sekhukhune District of Limpopo Province, South Africa. *African Journal of Food, Agriculture, Nutrition and Development*, *23*(122), 24115–24135. DOI: 10.18697/ajfand.122.21705

Masumuzzaman, M., Evivie, S. E., Ogwu, M. C., Bailiang, L., Jin-Cheng, D., Wan, L., Guicheng, H., Fei, L., & Song, W. (2021). Genomic and *in vitro* properties of the dairy *Streptococcus thermophilus* SMQ-301 strain against selected pathogens. *Food & Function*, *12*(8), 7017–7028. DOI: 10.1039/D0FO02951C

McCarter, J., Gavin, M., Baereleo, S., & Love, M. (2014). The Challenges of Maintaining Indigenous Ecological Knowledge. *Ecology and Society*, *19*(3), art39. Advance online publication. DOI: 10.5751/ES-06741-190339

Mercer, J., Kelman, I., Taranis, L., & Suchet-Pearson, S. (2009). Framework for integrating indigenous and scientific knowledge for disaster risk reduction. *Disasters*, *34*(1), 214–239. DOI: 10.1111/j.1467-7717.2009.01126.x

Michnik, K., Thompson, S., and Beardy, B. (2021). Moving your body, soul, and heart to share and harvest food. Canadian Food Studies / La Revue Canadienne Des Études Sur L Alimentation, 8(2). DOI: 10.15353/cfs-rcea.v8i2.446

Milia, A., Buonocunto, F., Mali, M., Giandomenico, S., Leo, A., Spada, L., Ferraro, L., & Giordano, L. (2023). Natural vs. Anthropogenic Metals for Reconstructing the Source-to-Sink Pathway (Naples Bay, Eastern Tyrrhenian Sea). *Frontiers in Earth Science (Lausanne)*, *11*, 1225658. Advance online publication. DOI: 10.3389/feart.2023.1225658

Muller, M. (2018). Promoting or protecting traditional knowledges? tensions in the resurgence of indigenous food practices on vancouver island. *International Indigenous Policy Journal*, *9*(4). Advance online publication. DOI: 10.18584/iipj.2018.9.4.4

Neufeld, H., Richmond, C., & Centre, S. (2017). Impacts of place and social spaces on traditional food systems in southwestern ontario. *International Journal of Indigenous Health*, *12*(1), 93–115. DOI: 10.18357/ijih112201716903

Nji, Q., Babalola, O., Ekwomadu, T., Nleya, N., & Mwanza, M. (2022). Six main contributing factors to high levels of mycotoxin contamination in african foods. *Toxins*, *14*(5), 318. DOI: 10.3390/toxins14050318

Nji, Q., Babalola, O., & Mwanza, M. (2022). Aflatoxins in maize: Can their occurrence be effectively managed in africa in the face of climate change and food insecurity? *Toxins*, *14*(8), 574. DOI: 10.3390/toxins14080574

Nji, Q., & Mwanza, M. (2024). Three-year multi-mycotoxin analysis of south african commercial maize from three provinces. *Frontiers in Fungal Biology*, *5*, 1426782. Advance online publication. DOI: 10.3389/ffunb.2024.1426782

Nnamani, C., Ajayi, S., Oselebe, H., Atkinson, C., Igboabuchi, A., & Ezigbo, E. (2017). Sphenostylis stenocarpa (ex. A. Rich.) harms., a fading genetic resource in a changing climate: Prerequisite for conservation and sustainability. *Plants*, *6*(3), 30. DOI: 10.3390/plants6030030

Obahiagbon, E. G., & Ogwu, M. C. (2024). Organic Food Preservatives: The Shift Towards Natural Alternatives and Sustainability in the Global South's Markets. In Ogwu, M. C., Izah, S. C., & Ntuli, N. R. (Eds.), *Food Safety and Quality in the Global South* (pp. 299–329). Springer. DOI: 10.1007/978-981-97-2428-4_10

Ofosu, G., Dittmann, A., Sarpong, D., & Botchie, D. (2020). Socio-economic and environmental implications of artisanal and small-scale mining (ASM) on agriculture and livelihoods. *Environmental Science & Policy*, *106*, 210–220. DOI: 10.1016/j.envsci.2020.02.005

Ogwu, M. C. (2023). Local Food Crops in Africa: Sustainable Utilization, Threats, and Traditional Storage Strategies. In *Sustainable Utilization and Conservation of Africa's Biological Resources and Environment.* Springer. DOI: 10.1007/978-981-19-6974-4_13

Ogwu, M. C., Ahana, C. M., & Osawaru, M. E. (2018). Sustainable food production in Nigeria: A case study for Bambara groundnut (*Vigna subterranean* (L.) Verdc. Fabaceae). *Journal of Energy and Natural Resources Management, 1*, 68–77.

Ogwu, M. C., Izah, S. C., & Ntuli, N. R. (2024). *Food Safety and Quality in the Global South.* Springer Nature. DOI: 10.1007/978-981-97-2428-4

Ogwu, M. C., Izah, S. C., Ntuli, N. R., & Odubo, T. C. (2024). Food Security Complexities in the Global South. In Ogwu, M. C., Izah, S. C., & Ntuli, N. R. (Eds.), *Food Safety and Quality in the Global South* (pp. 3–33). Springer. DOI: 10.1007/978-981-97-2428-4_1

Ogwu, M. C., & Ogunsola, O. A. (2024). Physicochemical Methods of Food Preservation to Ensure Food Safety and Quality. In Ogwu, M. C., Izah, S. C., & Ntuli, N. R. (Eds.), *Food Safety and Quality in the Global South* (pp. 263–298). Springer. DOI: 10.1007/978-981-97-2428-4_9

Ogwu, M. C., Osawaru, M. E., Obadiaru, O., & Itsepopo, I. C. (2016). Nutritional composition of some uncommon indigenous spices and vegetables. *International Conference on Natural Resource Development and Utilization (4th Edition): Positioning small and medium enterprises (SMEs) for optimal utilization of available resources and economic growth.* Raw Materials Research and Development Council (Federal Ministry of Science and Technology), Maitama, Abuja Nigeria.

Ogwu, M. C., Osawaru, M. E., & Owie, M. O. (2023). Effects of Storage at Room Temperature on the Food Components of Three Cocoyam Species (*Colocasia esculenta, Xanthosoma atrovirens, and X. sagittifolium*). *Food Studies, 13*(2), 59–83. DOI: 10.18848/2160-1933/CGP/v13i02/59-83

Oladeji, O. A., Karigidi, K. O., & Ogwu, M. C. (2024). Indices for Monitoring and Measuring the Physicochemical Properties of Safe and Quality Food. In Ogwu, M. C., Izah, S. C., & Ntuli, N. R. (Eds.), *Food Safety and Quality in the Global South* (pp. 123–150). Springer. DOI: 10.1007/978-981-97-2428-4_5

Omari, R., Bellingrath-Kimura, S., Addo, E., Oikawa, Y., & Fujii, Y. (2018). Exploring farmers' indigenous knowledge of soil quality and fertility management practices in selected farming communities of the guinea savannah agro-ecological zone of ghana. *Sustainability (Basel), 10*(4), 1034. DOI: 10.3390/su10041034

Omorotionmwan, F. O.-O., Ogwu, H. I., & Ogwu, M. C. (2019). Antibacterial characterization and bacteria composition of pineapple (*Ananas comosus* [Linn.] Merr.) peel and pulp. *Food and Health*, *5*(1), 1–11. DOI: 10.3153/FH19001

Omotayo, O., Omotayo, A., Mwanza, M., & Babalola, O. (2019). Prevalence of mycotoxins and their consequences on human health. *Toxicological Research*, *35*(1), 1–7. DOI: 10.5487/TR.2019.35.1.001

Onoabhagbe, O., Ogwu, M. C., & Ikhajiagbe, B. (2024). Germination characteristics of *Sorghum bicolor* (L.) Moench. under different pH regimes after chemo-priming. *Vegetos*, *37*(5), 1876–1886. Advance online publication. DOI: 10.1007/s42535-024-00909-0

Onomu, A. (2023). Pitfalls and potential pathways to commercialization of indigenous food crops, fruits, and vegetables in Africa. *Asian Journal of Agriculture and Rural Development*, *13*(1), 25–38. DOI: 10.55493/5005.v13i1.4716

Osawaru, M. E., & Ogwu, M. C. (2020). Survey of plant and plant products in local markets within Benin City and environs. In Filho, L. W., Ogugu, N., Ayal, D., Adelake, L., & da Silva, I. (Eds.), *African Handbook of Climate Change Adaptation* (pp. 1–24). DOI: 10.1007/978-3-030-42091-8_159-1

Osawaru, M. E., Ogwu, M. C., & Aigbefue, D. (2014). Survey of ornamental gardens in five local government areas of Southern Edo State Nigeria. *The Bioscientist*, *2*(1), 87–102.

Osei-Kwarteng, M., Ogwu, M. C., Mahunu, G. K., & Afoakwah, N. A. (2024). Post-harvest Food Quality and Safety in the Global South: Sustainable Management Perspectives. In Ogwu, M. C., Izah, S. C., & Ntuli, N. R. (Eds.), *Food Safety and Quality in the Global South* (pp. 151–195). Springer. DOI: 10.1007/978-981-97-2428-4_6

Ovuru, K. F., Izah, S. C., Ogidi, O. I., Imarhiagbe, I., & Ogwu, M. C. (2023). Slaughterhouse Facilities in Developing Nations: Sanitation and Hygiene Practices, Microbial Contamination and Sustainable Management System. *Food Science and Biotechnology*. Advance online publication. DOI: 10.1007/s10068-023-01406-x

Persaud, J., Wannamaker, K., Stark, K., Lambert, C., Harrison, C., & Keller, N. (2025). Decolonizing education: Advancing indigenous student success through culturally responsive practices in Ontario. *Alternative*, *21*(2), 264–273. DOI: 10.1177/11771801251340657

Phillipps, B., Skinner, K., Parker, B., & Neufeld, H. (2022). An intersectionality-based policy analysis examining the complexities of access to wild game and fish for urban indigenous women in northwestern Ontario. *Frontiers in Communication*, *6*, 762083. Advance online publication. DOI: 10.3389/fcomm.2021.762083

Pilnik, M., Argentim, T., Kinupp, V., Haverroth, M., & Ming, L. (2023). Traditional botanical knowledge: Food plants from the Huni Kuĩ indigenous people, Acre, Western Brazilian Amazon. Rodriguésia, 74. https://doi.org/DOI: 10.1590/2175-7860202374016

Popken, L., Griffin, P., Coté, C., & Angel, É. (2023). Indigenous Food Sovereignty through Resurgent Self-Governance: Centering Nuuchah-nulth Principles in Sea Otter Management in Canada. *Ecology and Society*, *28*(2), art12. Advance online publication. DOI: 10.5751/ES-13702-280212

Rhodes, L., Smith, K., Murray, J., Nishimura, T., & Finch, S. (2020). Ciguatera fish poisoning: The risk from an Aotearoa/New Zealand perspective. *Toxins*, *12*(1), 50. DOI: 10.3390/toxins12010050

Richard, G., Izah, S. C., & Ogwu, M. C. (2022). Implications of Artisanal Crude Oil Refining on Sustainable Food Production in the Niger Delta Region of Nigeria. *Journal of Environmental Bioremediation and Toxicology*, *5*(2), 69–77. DOI: 10.54987/jebat.v5i2.775

Robin, T. (2019). Our Hands at Work: Indigenous Food Sovereignty in Western Canada. *Journal of Agriculture, Food Systems, and Community Development*, 1–15. DOI: 10.5304/jafscd.2019.09B.007

Robin, T., Dennis, M., & Hart, M. (2020). Feeding indigenous people in Canada. *International Social Work*, *65*(4), 652–662. DOI: 10.1177/0020872820916218

Robin, T., & Hart, M. (2025). Cree food knowledge and being well. *International Journal of Environmental Research and Public Health*, *22*(2), 181. DOI: 10.3390/ijerph22020181

Roche, H., & Tidou, A. (2008). First ecotoxicological assessment assay in a hydroelectric reservoir: The Lake Taabo (côte d'Ivoire). *Bulletin of Environmental Contamination and Toxicology*, *82*(3), 322–326. DOI: 10.1007/s00128-008-9572-9

Sampson, D., Cely-Santos, M., Gemmill-Herren, B., Babin, N., Bernhart, A., Kerr, R., & Wittman, H. (2021). Food sovereignty and rights-based approaches strengthen food security and nutrition across the globe: A systematic review. *Frontiers in Sustainable Food Systems*, *5*, 686492. Advance online publication. DOI: 10.3389/fsufs.2021.686492

Scholes, C. (2019). Educational modules for increasing indigenous australian students' involvement in chemistry. *Journal of Chemical Education*, *96*(9), 1914–1921. DOI: 10.1021/acs.jchemed.9b00207

Schueler, V., Kuemmerle, T., & Schröder, H. (2011). Impacts of Surface Gold Mining on Land Use Systems in Western Ghana. *Ambio*, *40*(5), 528–539. DOI: 10.1007/s13280-011-0141-9

Sherriff, S., Kalucy, D., Tong, A., Naqvi, N., Nixon, J., Eades, S., Ingram, T., Slater, K., Dickson, M., Lee, A., & Muthayya, S. (2022). Murradambirra dhangaang (make food secure): Aboriginal community and stakeholder perspectives on food insecurity in urban and regional Australia. *BMC Public Health*, *22*(1), 1066. Advance online publication. DOI: 10.1186/s12889-022-13202-z

Sievert, K., Lawrence, M., Naika, A., & Baker, P. (2019). Processed Foods and Nutrition Transition in the Pacific: Regional Trends, Patterns, and Food System Drivers. *Nutrients*, *11*(6), 1328. DOI: 10.3390/nu11061328

Smout, F., Skerratt, L., Butler, J., Johnson, C., Congdon, B., & Thompson, R. (2017). The hookworm Ancylostoma ceylanicum: An emerging public health risk in Australian tropical rainforests and Indigenous communities. *One Health*, *3*, 66–69. DOI: 10.1016/j.onehlt.2017.04.002

Soares, G., Hedges, J., Sethi, S., Poirier, B., & Jamieson, L. (2023). From biocolonialism to emancipation: Considerations on ethical and culturally respectful omics research with indigenous Australians. *Medicine, Health Care, and Philosophy*, *26*(3), 487–496. DOI: 10.1007/s11019-023-10151-1

Spencer, J., Maxwell, D., Erickson, K., Wall, D., Nicholas-Figueroa, L., Pratt, K., & Shultz, G. (2021). Cultural relevance in chemistry education: Snow chemistry and the iñupiaq community. *Journal of Chemical Education*, *99*(1), 363–372. DOI: 10.1021/acs.jchemed.1c00480

Sultan, K., Ali, T., Fahmy, N., & El-Shibiny, A. (2019). Using millimeter-waves for rapid detection of pathogenic bacteria in food based on bacteriophage. *Engineering Reports*, *1*(1), e12026. Advance online publication. DOI: 10.1002/eng2.12026

Suwardi, A., Navia, Z., Harmawan, T., Syamsuardi, S., & Mukhtar, E. (2020). Ethnobotany and Conservation of Indigenous Edible Fruit Plants in South Aceh, Indonesia. *Biodiversitas (Surakarta)*, *21*(5). Advance online publication. DOI: 10.13057/biodiv/d210511

Swiderska, K., Argumedo, A., Wekesa, C., Ndalilo, L., Song, Y., Rastogi, A., & Ryan, P. (2022). Indigenous Peoples' Food Systems and Biocultural Heritage: Addressing Indigenous Priorities through Decolonial and Interdisciplinary Research Approaches. *Sustainability (Basel)*, *14*(18), 11311. DOI: 10.3390/su141811311

Tegegne, H., & Phyo, H. (2017). Food safety knowledge, attitude, and practices of meat handler in abattoir and retail meat shops of Jigjiga town, Ethiopia. DOI: 10.15167/2421-4248/jpmh2017.58.4.737

Thompson, L., & Darwish, W. (2019). Environmental chemical contaminants in food: Review of a global problem. *Journal of Toxicology*, *2019*, 1–14. DOI: 10.1155/2019/2345283

Thow, A., Heywood, P., Schultz, J., Quested, C., Jan, S., & Colagiuri, S. (2011). Trade and the nutrition transition: Strengthening policy for health in the Pacific. *Ecology of Food and Nutrition*, *50*(1), 18–42. DOI: 10.1080/03670244.2010.524104

Tuominen, I., Ballardini, R., Mähönen, J., & Pihlajarinne, T. (2023). Protecting and accessing indigenous peoples' digital cultural heritage through sustainable governance and IPR structures - the case of Sámi culture. *Arctic Review on Law and Politics*, *14*. Advance online publication. DOI: 10.23865/arctic.v14.5809

Ullah, M. (2024). NGOs' role in sustaining indigenous knowledge in rural Bangladesh: Agriculture, healthcare, and disaster management. *South Asian Journal of Social Sciences and Humanities*, *5*(1), 79–102. DOI: 10.48165/sajssh.2024.5106

Vijayan, D., Ludwig, D., Rybak, C., Kaechele, H., Hoffmann, H., Schönfeldt, H., Mbwana, H. A., Rivero, C. V., & Löhr, K. (2022). Indigenous knowledge in food system transformations. *Communications Earth & Environment*, *3*(1), 213. Advance online publication. DOI: 10.1038/s43247-022-00543-1

Wana, H., & Kudama, G. (2022). Do traditional food commodities transform food security? evidence from indigenous knowledge in West part of Ethiopia. DOI: 10.21203/rs.3.rs-1687653/v1

Wang, X., Bouzembrak, Y., Lansink, A., & Fels-Klerx, H. (2021). Application of machine learning to the monitoring and prediction of food safety: A review. *Comprehensive Reviews in Food Science and Food Safety*, *21*(1), 416–434. DOI: 10.1111/1541-4337.12868

Weiler, A., Hergesheimer, C., Brisbois, B., Wittman, H., Yassi, A., & Spiegel, J. (2014). Food sovereignty, food security, and health equity: A meta-narrative mapping exercise. *Health Policy and Planning*, *30*(8), 1078–1092. DOI: 10.1093/heapol/czu109

Whyte, K. (2018). Food sovereignty, justice, and indigenous peoples. DOI: 10.1093/oxfordhb/9780199372263.013.34

Williams, L., Jones, M., & Wright, P. (2023). Decolonising food regulatory frameworks: Importance of recognising traditional culture when assessing dietary safety of traditional foods. *The Proceedings of the Nutrition Society*, *82*(4), 454–467. DOI: 10.1017/S0029665123003026

Williams, P., Sikutshwa, L., & Shackleton, S. (2020). Acknowledging indigenous and local knowledge to facilitate collaboration in landscape approaches—Lessons from a systematic review. *Land (Basel)*, *9*(9), 331. DOI: 10.3390/land9090331

Wu, F., & Güçlü, H. (2012). Aflatoxin regulations in a network of global maize trade. *PLoS One*, *7*(9), e45151. DOI: 10.1371/journal.pone.0045151

Yadav, A., Bhagat, R., & Yadav, V. (2019). Does coal mining affect household food security? An Empirical Study of Dhanbad, India. *Agriculture & Food Security*, *8*(1), 7. Advance online publication. DOI: 10.1186/s40066-018-0244-7

Yiridomoh, G. (2021). "Illegal" gold mining operations in Ghana: Implication for climate-smart agriculture in northwestern Ghana. *Frontiers in Sustainable Food Systems*, *5*, 745317. Advance online publication. DOI: 10.3389/fsufs.2021.745317

Young, L., Shukla, S., & Wilson, T. (2024). Indigenous values and perspectives for strengthening food security and sovereignty: Learning from a community-based case study of misko-ziibiing (bloodvein river First Nation), Manitoba, Canada. *Frontiers in Sustainable Food Systems*, *8*, 1321231. Advance online publication. DOI: 10.3389/fsufs.2024.1321231

Compilation of References

Abass, A., Awoyale, W., Sulyok, M., & Alamu, E. (2017). Occurrence of regulated mycotoxins and other microbial metabolites in dried cassava products from Nigeria. *Toxins*, *9*(7), 207. DOI: 10.3390/toxins9070207

Abbasi, A., Hashemi, M., Kafil, H. S., Astamal, M. A., Lahouty, M., Tajani, A. G., Hosseini, H., & Nasirifar, S. Z. (2024). a critical review on the bioavailability promotion of the food bioactive compounds: Nano lipid carriers perspective. *Ulum-i Daruyi*. Advance online publication. DOI: 10.34172/PS.2024.11

Abd Rahim, M. H., Hazrin-Chong, N. H., Harith, H. H., Wan, W. A., & Sukor, R. (2023). Roles of fermented plant-, dairy-and meat-based foods in the modulation of allergic responses. *Food Science and Human Wellness*, *12*(3), 691–701. DOI: 10.1016/j.fshw.2022.09.002

Abdel-Aal, E.-S. M., Rabalski, I., Mats, L., & Rai, I. (2022). Identification and Quantification of Anthocyanin and Catechin Compounds in Purple Tea Leaves and Flakes. *Molecules (Basel, Switzerland)*, *27*(19), 6676. DOI: 10.3390/molecules27196676 PMID: 36235212

Abdel-Aziz, S. H., El-Esawi, M. A., Hazaa, M. M., Abdel-Aziz, H. Y., & Hassan, M. G. (2021). Antibacterial Potential of Pomegranate Peel Extracts on Escherichia coli Isolated From Benha Hospital in Egypt. *Benha Journal of Applied Sciences*, *6*(3), 61–64. DOI: 10.21608/bjas.2021.169813

Abdul Khalil, Tye, Leh, Saurabh, Ariffin, Mohammad Fizree, & Suriani. (2018). Cellulose reinforced biodegradable polymer composite film for packaging applications. *Bionanocomposites for packaging applications.* 49-69. .DOI: 10.1007/978-3-319-67319-6_3

Abdullahi, N., Igwe, E. C., Dandago, M. A., Sani, A., & Umar, N. B. (2023). A review of heavy metals accumulation and control in active agricultural soil. *Turkish Journal of Food and Agriculture Sciences*, *5*(1), 7–18. DOI: 10.53663/turjfas.1247846

Abdul-Mutalib, N., Abdul-Rashid, M., Mustafa, S., Nordin, S., Hamat, R., & Osman, M. (2012). Knowledge, Attitudes, and Practices Regarding Food Hygiene and Sanitation of Food Handlers in Kuala Pilah, Malaysia. *Food Control, 27*(2), 289–293. DOI: 10.1016/j.foodcont.2012.04.001

Abedini, M., Ghasemi-Tehrani, H., Tarrahi, M. J., & Amani, R. (2021). The effect of concentrated pomegranate juice consumption on risk factors of cardiovascular diseases in women with polycystic ovary syndrome: A randomized controlled trial. *Phytotherapy Research, 35*(1), 442–451. DOI: 10.1002/ptr.6820 PMID: 32767710

Abthiyar, L. (2023). *Effect of UV-C treatment on sanitised and unsanitised ready-to-eat leafy green vegetables, produced in New Zealand: a thesis presented in partial fulfilment of the requirements for the degree of Master of Food Technology, Massey University, Albany Campus, New Zealand* (Doctoral dissertation, Massey University).

Acevedo-Fani, A., & Singh, H. (2021). Biopolymer interactions during gastric digestion: Implications for nutrient delivery. *Food Hydrocolloids, 116*, 106644. DOI: 10.1016/j.foodhyd.2021.106644

Acevedo-Fani, A., Soliva-Fortuny, R., & Martín-Belloso, O. (2017). Nanoemulsions as edible coatings. *Current Opinion in Food Science, 15*, 43–49. DOI: 10.1016/j.cofs.2017.06.002

Ackah, S., Bi, Y., Xue, S., Yakubu, S., Han, Y., Zong, Y., Atuna, R. A., & Prusky, D. (2022). Postharvest chitosan treatment suppresses oxidative stress by regulating reactive oxygen species metabolism in wounded apples. *Frontiers in Plant Science, 13*, 959762. Advance online publication. DOI: 10.3389/fpls.2022.959762 PMID: 35982700

Addisu, S., & Wajera, A. (2024). Climate change and variability impacts on food security of smallholder farmers in Assosa district, Ethiopia. DOI: 10.21203/rs.3.rs-4187640/v1

Adeagbo, O., Ojo, T., & Adetoro, A. (2021). Understanding the determinants of climate change adaptation strategies among smallholder maize farmers in southwest Nigeria. *Heliyon, 7*(2), e06231. DOI: 10.1016/j.heliyon.2021.e06231

Adeghizadeh-Yazdi, Habibi, Kamali, & Banaei. (2019). Application of edible anSd biodegradable starch-based films in food packaging: A systematic review and meta-analysis. *Current research in nutrition and food science journal, 7*(3), 624-637.

Adekoya, I., Njobeh, P., Obadina, A., Chilaka, C., Okoth, S., Boevre, M., & Saeger, S. (2017). Awareness and prevalence of mycotoxin contamination in selected Nigerian fermented foods. *Toxins, 9*(11), 363. DOI: 10.3390/toxins9110363

Adenubi, A. O., Oduroye, A. P., Mosud, O., & Akanni, A. (2024). Cultivating the future: Nanotechnology's green revolution in agriculture. *International Journal of Research in Education Humanities and Commerce, 05*(02), 16–28. DOI: 10.37602/IJREHC.2024.5202

Adesokan, H., Akinseye, V., & Adesokan, G. (2015). Food safety training is associated with improved knowledge and behaviours among foodservice establishments' workers. *International Journal of Food Sciences, 2015*, 1–8. DOI: 10.1155/2015/328761

Adeyeye, S. A. O. (2019). Food packaging and nanotechnology: Safeguarding consumer health and safety. *Nutrition & Food Science, 49*(6), 1164–1179. DOI: 10.1108/NFS-01-2019-0020

Adeyeye, S. A. O., & Idowu-Adebayo, F. (2019). Genetically modified and biofortified crops and food security in developing countries: A review. *Nutrition & Food Science, 49*(5), 978–986. DOI: 10.1108/NFS-12-2018-0335

Adhikari, S., Kelkar, V., Kumar, R., & Halden, R. U. (2022). Methods and challenges in the detection of microplastics and nanoplastics: A mini-review. *Polymer International, 71*(5), 543–551. DOI: 10.1002/pi.6348

Agoh, E. C., Ogwu, M. C., Chukwuemeka, O. S., & Ekeledo, P. I. (2024). Environmental and human health effects of cassava processing and processing waste. In *Plant Biology, Sustainability and Climate change, Sustainable Cassava.* Academic Press. DOI: 10.1016/B978-0-443-21747-0.00001-1

Ahangarkolaee, S. S., & Gorton, M. (2020). The effects of perceived regulatory efficacy, ethnocentrism and food safety concern on the demand for organic food. *International Journal of Consumer Studies, 45*(2), 273–286. DOI: 10.1111/ijcs.12619

Ahmad, K., & Pieroni, A. (2016). Folk knowledge of wild food plants among the tribal communities of Thakht-e-Sulaiman Hills, north-west Pakistan. *Journal of Ethnobiology and Ethnomedicine, 12*(1), 17. Advance online publication. DOI: 10.1186/s13002-016-0090-2

Ahmed, S., Akther, S., & Hossain, M. S. (2024). Electrolyzed Water-Based Technology for Fresh Fruits and Vegetables. In *Sustainable Postharvest Technologies for Fruits and Vegetables* (pp. 293–303). CRC Press. DOI: 10.1201/9781003370376-25

Aigberua, A. O., Izah, S. C., & Aigberua, A. A. (2023). Occurrence, source delineation, and health hazard of polycyclic aromatic hydrocarbons in tissues of *Sarotherodon melanotheron* and *Chrysichthys nigrodigitatus* from Okulu River, Nigeria. *Environmental Monitoring and Assessment, 195*(3), 364. DOI: 10.1007/s10661-023-10970-y

Aigberua, A. O., Izah, S. C., & Isaac, I. U. (2018). Level and health risk assessment of heavy metals in selected seasonings and culinary condiments used in Nigeria. *Biological Evidence*, *8*(2), 6–15. DOI: 10.5376/be.2018.08.0002

Aigberua, A. O., Izah, S. C., & Richard, G. (2021). Hazard analysis of trace metals in muscle of *Sarotherodon melanotheron* and *Chrysichthys nigrodigitatus* from Okulu River, Rivers State, Nigeria. *Journal of Environmental Health and Sustainable Development*, *6*(3), 1340–1356.

Aigberua, A. O., Izah, S. C., & Richard, G. (2021). Hazard Analysis of Trace Metals in Muscle of Sarotherodon melanotheron and Chrysichthys nigrodigitatus from Okulu River, Rivers State, Nigeria. *Journal of Environmental Health and Sustainable Development.*, *6*(3), 1340–1356.

Ajani, E., Mgbenka, R., & Okeke, M. (2013). Utilization of Indigenous Knowledge as a Strategy for Climate Change Adaptation among Farmers in Sub-Saharan Africa: Implications for Policy. *Asian Journal of Agricultural Extension Economics and Sociology*, *2*(1), 23–40. DOI: 10.9734/AJAEES/2013/1856

Ajayeoba, T. A., & Ijabadeniyi, O. A. (2025). Transforming Food for the Future: Precision Fermentation as a Key to Sustainability, Nutrition, and Health. *Nutrition and Health (Berkhamsted, Hertfordshire)*.

Akabanda, F., Hlortsi, E., & Owusu-Kwarteng, J. (2017). Food Safety Knowledge, Attitudes, and Practices of Institutional Food Handlers in Ghana. *BMC Public Health*, *17*(1), 40. Advance online publication. DOI: 10.1186/s12889-016-3986-9

Akil, L. (2021). Trends of foodborne diseases in Mississippi: Association with racial and economic disparities. *Diseases (Basel, Switzerland)*, *9*(4), 83. DOI: 10.3390/diseases9040083

Akil, L., & Ahmad, H. (2019). Quantitative risk assessment model of human salmonellosis resulting from consumption of broiler chicken. *Diseases (Basel, Switzerland)*, *7*(1), 19. DOI: 10.3390/diseases7010019

Akil, L., & Ahmad, H. (2023). Socioeconomic Impacts of the COVID-19 Pandemic on Foodborne Illnesses in the United States. *European Journal of Environment and Public Health*, *7*(2), em0128. Advance online publication. DOI: 10.29333/ejeph/12585

Akil, L., Ahmad, H., & Reddy, R. (2014). Effects of Climate Change on Salmonella Infections. *Foodborne Pathogens and Disease*, *11*(12), 974–980. DOI: 10.1089/fpd.2014.1802

Akinbo, O., Obukosia, S., Ouedraogo, J., Sinebo, W., Savadogo, M., Timpo, S., Mbabazi, R., Maredia, K., Makinde, D., & Ambali, A. (2021). Commercial release of genetically modified crops in Africa: Interface between biosafety regulatory systems and varietal release systems. *Frontiers in Plant Science*, *12*, 605937. DOI: 10.3389/fpls.2021.605937

Al Amin, M., Sobhani, Z., Liu, Y., Dharmaraja, R., Chadalavada, S., Naidu, R., Chalker, J. M., & Fang, C. (2020). Recent advances in the analysis of per- and poly-fluoroalkyl substances (PFAS)—A review. *Environmental Technology & Innovation*, *19*, 100879. DOI: 10.1016/j.eti.2020.100879

Alamri, M. S., Qasem, A. A., Mohamed, A. A., Hussain, S., Ibraheem, M. A., Shamlan, G., & Qasha, A. S. (2021). Food packaging's materials: A food safety perspective. *Saudi Journal of Biological Sciences*, *28*(8), 4490–4499. DOI: 10.1016/j.sjbs.2021.04.047 PMID: 34354435

Al-Assil, B. J., & Wahwah, T. A. (2024). Control Mycotoxin Contamination in Tarasas Food Using A Lab Fermantion. *Journal of Angiotherapy*, *8*(3). Advance online publication. DOI: 10.25163/angiotherapy.839544

Alexandraki, M., Valiakos, G., Manouras, A., & Malissiova, E. (2022). Chemical, Microbiological, and Sensory Analysis Benefits Leading to Shelf-Life Extension of Minced Beef Meat Using Essential Oil of Oregano Contained in Xanthan Gum and Guar Gum Edible Coating. *Processes (Basel, Switzerland)*, *10*(11), 2253. DOI: 10.3390/pr10112253

Alexandre, E. M., Pinto, C. A., Moreira, S. A., Pintado, M., & Saraiva, J. A. (2019). Nonthermal food processing/preservation technologies. In *Saving food* (pp. 141–169). Academic Press. DOI: 10.1016/B978-0-12-815357-4.00005-5

Alexandri, M., Kachrimanidou, V., Papapostolou, H., Papadaki, A., & Kopsahelis, N. (2022). Sustainable food systems: The case of functional compounds towards the development of clean label food products. *Foods*, *11*(18), 2796. DOI: 10.3390/foods11182796 PMID: 36140924

Ali, A. A., Altemimi, A. B., Alhelfi, N., & Ibrahim, S. A. (2020). Application of biosensors for detection of pathogenic food bacteria: A review. *Biosensors (Basel)*, *10*(6), 58. DOI: 10.3390/bios10060058 PMID: 32486225

Allende, A., McEvoy, J. L., Luo, Y., Artes, F., & Wang, C. Y. (2006). Effectiveness of two-sided UV-C treatments in inhibiting natural microflora and extending the shelf-life of minimally processed 'Red Oak Leaf' lettuce. *Food Microbiology*, *23*(3), 241–249. DOI: 10.1016/j.fm.2005.04.009 PMID: 16943010

Al-Maharik, N., Jaradat, N., Bassalat, N., Hawash, M., & Zaid, H. (2022). Isolation, Identification and Pharmacological Effects of Mandragora autumnalis Fruit Flavonoids Fraction. *Molecules (Basel, Switzerland)*, *27*(3), 1046. DOI: 10.3390/molecules27031046 PMID: 35164311

Al-Moghazy, M., & El-Sayed, H. S. (2023). Antimicrobial-prebiotic: Novel dual approach of pomegranate peel extract in vitro and in food system. *Biocatalysis and Agricultural Biotechnology*, *49*, 102664. DOI: 10.1016/j.bcab.2023.102664

Alqarni, A. M. (2024). Analytical Methods for the Determination of Pharmaceuticals and Personal Care Products in Solid and Liquid Environmental Matrices: A Review. *Molecules (Basel, Switzerland)*, *29*(16), 3900. DOI: 10.3390/molecules29163900 PMID: 39202981

Alrobaish, W. S., Vlerick, P., Luning, P. A., & Jacxsens, L. (2020). Food safety governance in Saudi Arabia: Challenges in control of imported food. *Journal of Food Science*, *86*(1), 16–30. DOI: 10.1111/1750-3841.15552

Al-Slaihat, A., Saleh, M., Al-Awadi, A., Almomani, H., Alsakran, S., Alsheikh, D., & Lee, Y. (2024). Food safety knowledge, attitudes, and practices (KAP) among dairy retail vendors in AmmaJordandan... https://doi.org/DOI: 10.20944/preprints202405.1041.v1

Altemimi, A., Aziz, S. N., Al-Hilphy, A. R., Lakhssassi, N., Watson, D. G., & Ibrahim, S. A. (2019). Critical review of radio-frequency (RF) heating applications in food processing. *Food Quality and Safety*, *3*(2), 81–91. DOI: 10.1093/fqsafe/fyz002

Amedu, J., Adediji, A., Miracle, N., Anthony, A., Adeyemi, P., Ahmed, R., Atsumbe, S., Costly, M., Majekodunmi, A., Balogun, O., Akinpelu, O., Borgbara, K., Olufowobi, O., Tahir, H. J., Aroworamimo, L., & Asagbra, A. (2025). What's in (y) our food?– Occurrence of GM-containing foods on the Nigerian market and compliance with national regulations. *Journal of Genetic Engineering and Biotechnology*, *23*(2), 100481. DOI: 10.1016/j.jgeb.2025.100481

Ameta, S. K., Rai, A. K., Hiran, D., Ameta, R., & Ameta, S. C. (2020). Use of nanomaterials in food science. In *Biogenic nano-particles and their use in agro-ecosystems* (pp. 457–488). Springer Singapore. DOI: 10.1007/978-981-15-2985-6_24

Amor-Gutiérrez, O., Selvolini, G., Fernández-Abedul, M. T., de la Escosura-Muñiz, A., & Marrazza, G. (2020). Folding-based electrochemical aptasensor for the determination of β-lactoglobulin on poly-L-lysine modified graphite electrodes. *Sensors (Basel)*, *20*(8), 2349. DOI: 10.3390/s20082349 PMID: 32326088

Andrady, A. L. (2011). Microplastics in the marine environment. *Marine Pollution Bulletin, 62*(8), 1596–1605. DOI: 10.1016/j.marpolbul.2011.05.030 PMID: 21742351

Ansari, M. A. (2023). Nanotechnology in food and plant science: Challenges and future prospects. *Plants, 12*(13), 2565. DOI: 10.3390/plants12132565 PMID: 37447126

Anwar, R. S., Ahmed, R. R., Streimikiene, D., Streimikis, J., & Zamek, D. (2024). Securing food futures: The interplay of safety governance, hygiene, supplier beliefs and consumer engagement. *British Food Journal, 127*(2), 758–778. DOI: 10.1108/BFJ-09-2024-0882

Aquino, H., Yap, T., Lacap, J., Tuazon, G., & Flores, M. (2021). Food Safety Knowledge, Attitudes, Practices, and Training of Fast-Food Restaurant Food Handlers: A Moderation Analysis. *British Food Journal, 123*(12), 3824–3840. DOI: 10.1108/BFJ-01-2021-0026

Arficho, T. S., & Kebede, A. H. (2019). Microbiological quality and safety of ready-to-eat foods from restaurants and food establishments in Yirgalem town, Southern Ethiopia. *Research Square*. DOI: 10.21203/rs.2.13735/v1

Arienzo, A., Coff, C., & Barling, D. (2008). The European Union and the Regulation of Food Traceability: From Risk Management to Informed Choice? In Coff, C., Barling, D., Korthals, M., & Nielsen, T. (Eds.), *Ethical Traceability and Communicating Food* (Vol. 15, pp. 23–42). Springer Netherlands. DOI: 10.1007/978-1-4020-8524-6_2

Arsène, M. M. J., Davares, A. K. L., Viktorovna, P. I., Andreevna, S. L., Sarra, S., Khelifi, I., & Sergueïevna, D. M. (2022). The public health issue of antibiotic residues in food and feed: Causes, consequences, and potential solutions. *Veterinary World*, 662–671. DOI: 10.14202/vetworld.2022.662-671 PMID: 35497952

Arzola-Alavarez, C., Castillo-Castillo, Y., Anderson, R., Hume, M., Ruiz-Barrera, O., Min, B., Arzola-Rubio, A., Beier, R., & Salinas-Chavira, J. (2020). Influence of Pine Bark Tannin on Bacterial Pathogens Growth and Nitrogen Compounds on Changes in Composted Poultry Litter. *Brazilian Journal of Poultry Science, 22*(1), eRBCA-2018-0911. Advance online publication. DOI: 10.1590/1806-9061-2018-0911

Asante, K., Abugre, S., Akoto, D., & Derkyi, N. (2023). Cassava (Manihot esculenta) yield, nutrition, and heavy metal bioaccumulation responses to circular economy-based innovations in a mining-degraded landscape. *International Journal of Plant and Soil Science, 35*(22), 840–850. DOI: 10.9734/ijpss/2023/v35i224195

Assogba, M. F., Afé, O. H. I., Ahouansou, R. H., Anihouvi, D. G. H., Kpoclou, Y. E., Djago, D., Douny, C., Igout, A., Mahillon, J., Hounhouigan, D. J., Scippo, M.-L., & Anihouvi, V. B. (2021). Performances of the barrel kiln used in cottage industry for fish processing and effects on physicochemical characteristics and safety of smoked fish products. *Journal of the Science of Food and Agriculture*, *102*(2), 851–861. DOI: 10.1002/jsfa.11421

Assunção, R., Martins, C., Viegas, S., Viegas, C., Jakobsen, L., Pires, S., & Alvito, P. (2018). Climate change and the health impact of aflatoxins exposure in Portugal – an overview. *Food Additives & Contaminants. Part A, Chemistry, Analysis, Control, Exposure & Risk Assessment*, *35*(8), 1610–1621. DOI: 10.1080/19440049.2018.1447691

Atalay, Y., Gebeyehu, N., & Gelaw, K. (2025). Food hygiene practice and associated factors among food handlers working in food establishments in sub-Saharan Africa: A systematic review and meta-analysis. *Primary Health Care Research and Development*, *26*, e13. Advance online publication. DOI: 10.1017/S146342362500009X

Awlqadr, F. H., Altemimi, A. B., Omar, A. M. A., Saeed, M. N., Qadir, S. A., Faraj, A. M., Al-Manhel, A. J. A., Salih, T. H., Hesarinejad, M. A., & Vieira, I. R. S. (2025, June). Advancing Sustainability in Fruit and Vegetable Packaging: The Role of Nanotechnology in Food Preservation. *eFood*, *6*(3), e70060. Advance online publication. DOI: 10.1002/efd2.70060

Awuchi, C. G., Ondari, E. N., Ogbonna, C. U., Upadhyay, A. K., Baran, K., Okpala, C. O. R., Korzeniowska, M., & Guiné, R. P. (2021). Mycotoxins affecting animals, foods, humans, and plants: Types, occurrence, toxicities, action mechanisms, prevention, and detoxification strategies—A revisit. *Foods*, *10*(6), 1279. DOI: 10.3390/foods10061279 PMID: 34205122

Azadi, H., Taheri, F., Ghazali, S., Moghaddam, S. M., Siamian, N., Goli, I., & Witlox, F. (2022). Genetically modified crops in developing countries: Savior or traitor? *Journal of Cleaner Production*, *371*, 133296. DOI: 10.1016/j.jclepro.2022.133296

Azangoo, M., Taherkordi, A., Blech, J. O., & Vyatkin, V. (2021). Digital twin-assisted controlling of AGVs in flexible manufacturing environments. In *2021 IEEE 30th International Symposium on Industrial Electronics (ISIE)* (pp. 1–7). IEEE. DOI: 10.1109/ISIE45552.2021.9576361

Bai, X., Ga, L., & Ai, J. (2023). A fluorescent biosensor based on carbon quantum dots and single-stranded dna for the detection of escherichia coli. *Analyst*, *148*(16), 3892–3898. DOI: 10.1039/D3AN01024D PMID: 37462388

Bakshi, A. (2003). Potential adverse health effects of genetically modified crops. *Journal of Toxicology and Environmental Health. Part B, Critical Reviews*, *6*(3), 211–225. DOI: 10.1080/10937400306469

Baldos, U., & Hertel, T. (2014). Global food security in 2050: The role of agricultural productivity and climate change. *The Australian Journal of Agricultural and Resource Economics*, *58*(4), 554–570. DOI: 10.1111/1467-8489.12048

Banerjee, R., Verma, A. K., Naveena, B. M., & Kulkarni, V. V. (2018). Processing, Storage, and Transportation of Meat and Meat Products. In *Animal Sourced Foods for Developing Economies* (pp. 77–117). CRC Press. DOI: 10.1201/9780429398575-5

Ban, N., Wilson, E., & Neasloss, D. (2019). Historical and Contemporary Indigenous Marine Conservation Strategies in the North Pacific. *Conservation Biology*, *34*(1), 5–14. DOI: 10.1111/cobi.13432

Baranyi, J., & da Silva, N. B. (2017). The use of predictive models to optimize risk of decisions. *International Journal of Food Microbiology*, *240*, 19–23. DOI: 10.1016/j.ijfoodmicro.2016.10.016

Barboza, L. G. A., Dick Vethaak, A., Lavorante, B. R. B. O., Lundebye, A.-K., & Guilhermino, L. (2018). Marine microplastic debris: An emerging issue for food security, food safety and human health. *Marine Pollution Bulletin*, *133*, 336–348. DOI: 10.1016/j.marpolbul.2018.05.047 PMID: 30041323

Barrios, E., Delve, R., Bekunda, M., Mowo, J., Agunda, J., Ramisch, J., Trejo, M. T., & Thomas, R. (2006). Indicators of soil quality: A south–south development of a methodological guide for linking local and technical knowledge. *Geoderma*, *135*, 248–259. DOI: 10.1016/j.geoderma.2005.12.007

Barrios, E., & Trejo, M. (2003). Implications of Local Soil Knowledge for Integrated Soil Management in Latin America. *Geoderma*, *111*(3-4), 217–231. DOI: 10.1016/S0016-7061(02)00265-3

Barry, V., Winquist, A., & Steenland, K. (2013). Perfluorooctanoic Acid (PFOA) Exposures and Incident Cancers among Adults Living Near a Chemical Plant. *Environmental Health Perspectives*, *121*(11–12), 1313–1318. DOI: 10.1289/ehp.1306615 PMID: 24007715

Basak, G., Sharma, B., Parul, S., Jain, U., Mishra, R. P., & Srivastava, M. K. (2021). Strategies for food safety: A contemporary approach. *Journal of Entomology and Zoology Studies*, *9*(1), 117–122.

Battilani, P., Toscano, P., Fels-Klerx, H., Moretti, A., Leggieri, M., Brera, C., & Robinson, T. (2016). Aflatoxin B1 contamination in maize in Europe increases due to climate change. *Scientific Reports, 6*(1), 24328. Advance online publication. DOI: 10.1038/srep24328

Beard, C., Eisen, R., Barker, C., Garofalo, J., Hahn, M., Hayden, M., … Schramm, P. (2016). Ch. 5: Vectorborne Diseases. The Impacts of Climate Change on Human Health in the United States: A Scientific Assessment. DOI: 10.7930/J0765C7V

Béné, C., Fanzo, J., Prager, S. D., Achicanoy, H. A., Mapes, B. R., Alvarez Toro, P., & Bonilla Cedrez, C. (2020). Global drivers of food system (un)sustainability: A multi-country correlation analysis. *PLoS One, 15*(4), e0231071. DOI: 10.1371/journal.pone.0231071

Benevenuto, R. F., Venter, H. J., Zanatta, C. B., Nodari, R. O., & Agapito-Tenfen, S. Z. (2022). Alterations in genetically modified crops assessed by omics studies: Systematic review and meta-analysis. *Trends in Food Science & Technology, 120*, 325–337. DOI: 10.1016/j.tifs.2022.01.002

Ben, Y., Fu, C., Hu, M., Liu, L., Wong, M. H., & Zheng, C. (2019). Human health risk assessment of antibiotic resistance associated with antibiotic residues in the environment: A review. *Environmental Research, 169*, 483–493. DOI: 10.1016/j.envres.2018.11.040 PMID: 30530088

Berkes, F., & Berkes, M. (2009). Ecological complexity, fuzzy logic, and holism in indigenous knowledge. *Futures, 41*(1), 6–12. DOI: 10.1016/j.futures.2008.07.003

Bharathiraja, S., Suriya, J., Krishnan, M., Manivasagan, P., & Kim, S. K. (2017). Production of enzymes from agricultural wastes and their potential industrial applications. In *Advances in food and nutrition research* (Vol. 80, pp. 125–148). Academic Press. DOI: 10.1016/bs.afnr.2016.11.003

Białk-Bielińska, A., Kumirska, J., Borecka, M., Caban, M., Paszkiewicz, M., Pazdro, K., & Stepnowski, P. (2016). Selected analytical challenges in the determination of pharmaceuticals in drinking/marine waters and soil/sediment samples. *Journal of Pharmaceutical and Biomedical Analysis, 121*, 271–296. DOI: 10.1016/j.jpba.2016.01.016 PMID: 26818066

Bintsis, T. (2017). Foodborne pathogens. *AIMS Microbiology, 3*(3), 529–563. DOI: 10.3934/microbiol.2017.3.529

Birania, S., Kumar, S., Kumar, N., Attkan, A. K., Panghal, A., Rohilla, P., & Kumar, R. (2022). Advances in development of biodegradable food packaging material from agricultural and agro-industry waste. *Journal of Food Process Engineering*, *45*(1), e13930. DOI: 10.1111/jfpe.13930

Bischoff, N. S., Bussi, M. R., Van Breda, S. G., Jolani, S., Sijm, D. T., de Kok, T. M., & Briedé, J. J. (2025). Food-grade titanium dioxide exposure between age groups and in global regions: A systematic review and meta-analysis. *Critical Reviews in Food Science and Nutrition*, 1–11. DOI: 10.1080/10408398.2025.2467823

Biswas, S., Gogoi, P., Wilkinson, M., Deka, P., Das, V., Goswami, K., & Sandilya, S. P. (2025). Evaluating Regulatory Approaches to Emerging Pollutants. In Dey, S., & Bhattacharya, S. (Eds.), *Biotechnological Interventions in the Removal of Emerging Pollutants* (pp. 19–36). Springer Nature Singapore. DOI: 10.1007/978-981-97-9922-0_2

Bodenhamer, W. T., Jackowski, G., & Davies, E. (2004). Inventors; Toxin Alert Inc, assignee. Surface binding of an immunoglobulin to a flexible polymer using a water soluble varnish matrix. United States patent US 6, 692, 973. 02.17.

Bohensky, E., Butler, J., & Mitchell, D. (2011). Scenarios for knowledge integration: Exploring ecotourism futures in Milne Bay, Papua New Guinea. *Journal of Marine Biology*, *2011*, 1–11. DOI: 10.1155/2011/504651

Bohensky, E., & Maru, Y. (2011). Indigenous knowledge, science, and resilience: What have we learned from a decade of international literature on "integration"? *Ecology and Society*, *16*(4), art6. Advance online publication. DOI: 10.5751/ES-04342-160406

Bolouri, P., Salami, R., Kouhi, S., Kordi, M., Asgari Lajayer, B., Hadian, J., & Astatkie, T. (2022). Applications of essential oils and plant extracts in different industries. *Molecules (Basel, Switzerland)*, *27*(24), 8999. DOI: 10.3390/molecules27248999 PMID: 36558132

Bolumar, T., Orlien, V., Sikes, A., Aganovic, K., Bak, K. H., Guyon, C., Stübler, A. S., de Lamballerie, M., Hertel, C., & Brüggemann, D. A. (2021). High-pressure processing of meat: Molecular impacts and industrial applications. *Comprehensive Reviews in Food Science and Food Safety*, *20*(1), 332–368. DOI: 10.1111/1541-4337.12670 PMID: 33443800

Bond, C., Foley, W., & Askew, D. (2016). "it puts a human face on the researched" – a qualitative evaluation of an indigenous health research governance model. *Australian and New Zealand Journal of Public Health*, *40*, S89–S95. DOI: 10.1111/1753-6405.12422

Booth, J. (1982). Antibiotic residues in milk. *In Practice*, *4*(4), 100–109. DOI: 10.1136/inpract.4.4.100 PMID: 7186890

Borgå, K., Saloranta, T., & Ruus, A. (2010). Simulating climate change-induced alterations in bioaccumulation of organic contaminants in an Arctic marine food web. *Environmental Toxicology and Chemistry*, *29*(6), 1349–1357. DOI: 10.1002/etc.159

Borriello, L., Scivicco, M., Cacciola, N. A., Esposito, F., Severino, L., & Cirillo, T. (2023). Microplastics, a Global Issue: Human Exposure through Environmental and Dietary Sources. *Foods*, *12*(18), 3396. DOI: 10.3390/foods12183396 PMID: 37761106

Bouwmeester, H., Hollman, P. C. H., & Peters, R. J. B. (2015). Potential Health Impact of Environmentally Released Micro- and Nanoplastics in the Human Food Production Chain: Experiences from Nanotoxicology. *Environmental Science & Technology*, *49*(15), 8932–8947. DOI: 10.1021/acs.est.5b01090 PMID: 26130306

Brambilla, G., & Testa, C. (2014). Food safety/food security aspects related to the environmental release of pharmaceuticals. *Chemosphere*, *115*, 81–87. DOI: 10.1016/j.chemosphere.2014.01.024 PMID: 24602346

Bremenkamp, I., & Sousa-Gallagher, M. J. (2024). Design and Development of an Edible Coating for a Ready-to-Eat Fish Product. *Polymers*, *16*(3), 346. DOI: 10.3390/polym16030346 PMID: 38337235

Briassoulis, D. (2004). An overview on the mechanical behaviour of biodegradable agricultural films. *Journal of Polymers and the Environment*, *12*(2), 65–81. DOI: 10.1023/B:JOOE.0000010052.86786.ef

Briassoulis, H. (2004). The institutional complexity of environmental policy and planning problems: The example of Mediterranean desertification. *Journal of Environmental Planning and Management*, *47*(1), 115–135. DOI: 10.1080/0964056042000189835

Brito, S. D. C., Bresolin, J. D., Sivieri, K., & Ferreira, M. D. (2019). Low-density polyethylene films incorporated with silver nanoparticles to promote antimicrobial efficiency in food packaging. *Food Science & Technology International*, *26*(4), 353–366. DOI: 10.1177/1082013219894202 PMID: 31870192

Brody, Bugusu, Han, Sand, & McHugh. (2008). Scientific status summary: innovative food packaging solutions. .DOI: 10.1111/j.1750-3841.2008.00933.x

Brown, L., Le, B., Wong, M., Reimann, D., Nicholas, D., Faw, B., Davis, E., & Selman, C. (2014). Restaurant manager and worker food safety certification and knowledge. *Foodborne Pathogens and Disease*, *11*(11), 835–843. DOI: 10.1089/fpd.2014.1787

Brune, P., Chakravarthy, S., Graser, G., Mathesius, C. A., McClain, S., Petrick, J. S., & Goodwin, L. (2021). Core and supplementary studies to assess the safety of genetically modified (GM) plants used for food and feed. *Journal of Regional Science*, *9*(1), 45–60. https://orcid.org/0000-0001-8245-998X

Brunn, H., Arnold, G., Körner, W., Rippen, G., Steinhäuser, K. G., & Valentin, I. (2023). PFAS: Forever chemicals—persistent, bioaccumulative and mobile. Reviewing the status and the need for their phase out and remediation of contaminated sites. *Environmental Sciences Europe*, *35*(1), 20. DOI: 10.1186/s12302-023-00721-8

Buck, R. C., Franklin, J., Berger, U., Conder, J. M., Cousins, I. T., De Voogt, P., Jensen, A. A., Kannan, K., Mabury, S. A., & Van Leeuwen, S. P. (2011). Perfluoroalkyl and polyfluoroalkyl substances in the environment: Terminology, classification, and origins. *Integrated Environmental Assessment and Management*, *7*(4), 513–541. DOI: 10.1002/ieam.258 PMID: 21793199

Bunkley, N., McCool, J., & Garton, K. (2023). A review of international trade and investment agreements and nutrition policy space in the Pacific. *Frontiers in Nutrition*, *10*, 1208542. Advance online publication. DOI: 10.3389/fnut.2023.1208542

Burchill, L., Kotevski, A., Duke, D., Ward, J., Prictor, M., Lamb, K., & Kennedy, M. (2022). Ethics guidelines use and indigenous governance and participation in aboriginal and torres strait islander health research: A national survey. *The Medical Journal of Australia*, *218*(2), 89–93. DOI: 10.5694/mja2.51757

Burt, S. (2004). Essential oils: Their antibacterial properties and potential applications in foods—a review. *International Journal of Food Microbiology*, *94*(3), 223–253. DOI: 10.1016/j.ijfoodmicro.2004.03.022 PMID: 15246235

Cabello, F. C., Godfrey, H. P., Buschmann, A. H., & Dölz, H. J. (2016). Aquaculture as yet another environmental gateway to the development and globalisation of antimicrobial resistance. *The Lancet. Infectious Diseases*, *16*(7), e127–e133. DOI: 10.1016/S1473-3099(16)00100-6 PMID: 27083976

Cabrera, M., Taher, F., Llantada, A., Do, Q., Sapp, T., & Sommerhalter, M. (2021). Effect of Water Hardness on Catechin and Caffeine Content in Green Tea Infusions. *Molecules (Basel, Switzerland)*, *26*(12), 3485. DOI: 10.3390/molecules26123485 PMID: 34201178

Cachon, R., Girardon, P., & Voilley, A. (Eds.). (2019). *Gases in agro-food processes*. Academic Press.

Caradus, J. R. (2023). Intended and unintended consequences of genetically modified crops-myth, fact and/or manageable outcomes? *New Zealand Journal of Agricultural Research*, *66*(6), 519–619. DOI: 10.1080/00288233.2022.2141273

Carroll, S., Plevel, R., Jennings, L., Garba, I., Sterling, R., Cordova-Marks, F., Hiratsuka, V., Hudson, M., & Garrison, N. (2022). Extending the care principles from tribal research policies to benefit sharing in genomic research. *Frontiers in Genetics*, *13*, 1052620. Advance online publication. DOI: 10.3389/fgene.2022.1052620

Cassini, A., Hathaway, S., Havelaar, A., Koopmans, M., Koutsoumanis, K., Messens, W., Müller-Seitz, G., Nørrung, B., Rizzi, V., & Scheutz, F. (2016). Microbiological risk assessment. *EFSA Journal*, *14*(S1). Advance online publication. DOI: 10.2903/j.efsa.2016.s0507

Castro-Mayorga, J. L., Fabra, M. J., Cabedo, L., & Lagaron, J. M. (2016). On the use of the electrospinning coating technique to produce antimicrobial polyhydroxyalkanoate materials containing in situ-stabilized silver nanoparticles. *Nanomaterials (Basel, Switzerland)*, *7*(1), 4. DOI: 10.3390/nano7010004 PMID: 28336838

Catacora-Vargas, G., Binimelis, R., Myhr, A. I., & Wynne, B. (2018). Socio-economic research on genetically modified crops: A study of the literature. *Agriculture and Human Values*, *35*(2), 489–513. DOI: 10.1007/s10460-017-9842-4

Chammem, N., Issaoui, M., De Almeida, A. I. D., & Delgado, A. M. (2018). Food crises and food safety incidents in European Union, United States, and Maghreb area: Current risk communication strategies and new approaches. *Journal of AOAC International*, *101*(4), 923–938. DOI: 10.5740/jaoacint.17-0446

Chandorkar, S. (2023). Climate change and food safety. DOI: 10.57772/intechopen.112575

Chang, E. T., Adami, H.-O., Boffetta, P., Cole, P., Starr, T. B., & Mandel, J. S. (2014). A critical review of perfluorooctanoate and perfluorooctanesulfonate exposure and cancer risk in humans. Critical Reviews in Toxicology, 44(sup1), 1–81. DOI: 10.3109/10408444.2014.905767

Cheng, M., McCarl, B., & Fei, C. (2022). Climate change and livestock production: A literature review. *Atmosphere (Basel)*, *13*(1), 140. DOI: 10.3390/atmos13010140

Chen, J., Ying, G.-G., & Deng, W.-J. (2019). Antibiotic Residues in Food: Extraction, Analysis, and Human Health Concerns. *Journal of Agricultural and Food Chemistry*, *67*(27), 7569–7586. DOI: 10.1021/acs.jafc.9b01334 PMID: 31198037

Chen, L., Wang, W., Zhang, J., Wang, W., Ni, D., & Jiang, H. (2020). Dehydroascorbic Acid Affects the Stability of Catechins by Forming Conjunctions. *Molecules (Basel, Switzerland), 25*(18), 4076. DOI: 10.3390/molecules25184076 PMID: 32906587

Chen, M., Haq, S., Ahmed, K., Hussain, A., & Ahmed, M. (2021). The Link Between Climate Change, Food Security, and Fertility: The Case of Bangladesh. *PLoS One, 16*(10), e0258196. DOI: 10.1371/journal.pone.0258196

Chen, T., Wang, L., & Wang, J. (2017). Transparent assessment of the supervision information in China's food safety: A fuzzy-ANP comprehensive evaluation method. *Journal of Food Quality, 2017*, 1–14. DOI: 10.1155/2017/4878926

Chen, Y., Ji, H., Chen, L., Jiang, R., & Wu, Y. (2018). Food Safety Knowledge, Attitudes, and Behavior among Dairy Plant Workers in Beijing, Northern China. *International Journal of Environmental Research and Public Health, 15*(1), 63. DOI: 10.3390/ijerph15010063

Chen, Y., Li, H., Dou, H., Wen, H., & Dong, Y. (2023). Prediction and visual analysis of food safety risk based on TabNet-GRA. *Foods, 12*(16), 3113. DOI: 10.3390/foods12163113

Chen, Z., Li, H., Jia, W., Liu, X., Li, Z., Wen, F., Zheng, N., Jiang, J., & Xu, D. (2017). Bivalent aptasensor based on silver-enhanced fluorescence polarization for rapid detection of lactoferrin in milk. *Analytical Chemistry, 89*(11), 5900–5908. DOI: 10.1021/acs.analchem.7b00261 PMID: 28467701

Chen, Z., Pan, M., Cheng, C., Luo, J., & Deng, X. (2024). Water disinfection: Advances in photocatalysis and piezo/triboelectric catalysis with progressively enhanced energy utilization. *SusMat, 4*(5), e232. DOI: 10.1002/sus2.232

Chiellini, E., Barghini, A., Cinelli, P., & Ilieva, V. I. (2008). Overview of environmentally compatible polymeric materials for food packaging. In *Environmentally compatible food packaging* (pp. 371–395). Woodhead Publishing. DOI: 10.1533/9781845694784.3.371

Choudhury, O., Fairoza, N., Sylla, I., & Das, A. (2019). A blockchain framework for managing and monitoring data in multi-site clinical trials. *arXiv preprint arXiv:1902.03975.*

Chugh, G., Siddique, K. H., & Solaiman, Z. M. (2021). Nanobiotechnology for agriculture: Smart technology for combating nutrient deficiencies with nanotoxicity challenges. *Sustainability (Basel), 13*(4), 1781. DOI: 10.3390/su13041781

Chu, Z., Wang, H., & Dong, B. (2024). Research on Food Preservation Based on Antibacterial Technology: Progress and Future Prospects. *Molecules (Basel, Switzerland), 29*(14), 3318. DOI: 10.3390/molecules29143318 PMID: 39064897

Ciecierska, M., Dasiewicz, K., & Wołosiak, R. (2023). Methods of minimizing polycyclic aromatic hydrocarbon content in homogenized smoked meat sausages using different casings and variants of meat-fat raw material. *Foods, 12*(22), 4120. DOI: 10.3390/foods12224120

Ciecierska, M., Komorowska, U., Bryła, M., & Roszko, M. (2025). Formation of polycyclic aromatic hydrocarbons on grilled pork neck loins as affected by different marinades and grill types. *Foods, 14*(10), 1673. DOI: 10.3390/foods14101673

Cissé, G. (2019). Food-borne and water-borne diseases under climate change in low- and middle-income countries: Further efforts are needed to reduce environmental health exposure risks. *Acta Tropica, 194*, 181–188. DOI: 10.1016/j.actatropica.2019.03.012

Clayton, S., Devine-Wright, P., Swim, J., Bonnes, M., Steg, L., Whitmarsh, L., & Carrico, A. (2016). Expanding the role for psychology in addressing environmental challenges. *The American Psychologist, 71*(3), 199–215. DOI: 10.1037/a0039482

Coban, H. B. (2020). Organic acids as antimicrobial food agents: Applications and microbial productions. *Bioprocess and Biosystems Engineering, 43*(4), 569–591. DOI: 10.1007/s00449-019-02256-w PMID: 31758240

Codex Alimentarius Commission. (2020). Codex Pesticides Residues in Food Online Database. Food and Agriculture Organization of the United Nations. Retrieved from https://www.fao.org/fao-who-codexalimentarius/codex-texts/dbs/pestres/en/

Commeau, N., Cornu, M., Albert, I., Denis, J. B., & Parent, E. (2011). Hierarchical Bayesian models to assess between- and within-batch variability of pathogen contamination in food. *Risk Analysis, 32*(3), 395–415. DOI: 10.1111/j.1539-6924.2011.01699.x

Connolly-Boutin, L., & Smit, B. (2015). Climate Change, Food Security, and Livelihoods in Sub-Saharan Africa. *Regional Environmental Change, 16*(2), 385–399. DOI: 10.1007/s10113-015-0761-x

Connors, J., Safayet, M., Rosenheim, N., & Watson, M. (2022). Assessing changes in food pantry access after extreme events. *Agriculture and Human Values, 40*(2), 619–634. DOI: 10.1007/s10460-022-10373-8

Cortés-Sánchez, A. D. J., Diaz-Ramírez, M., Torres-Ochoa, E., Espinosa-Chaurand, L. D., Rayas-Amor, A. A., Cruz-Monterrosa, R. G., Aguilar-Toalá, J. E., & Salgado-Cruz, M. D. L. P. (2024). Processing, quality and elemental safety of fish. *Applied Sciences (Basel, Switzerland), 14*(7), 2903. DOI: 10.3390/app14072903

Coté, C. (2016). "Indigenizing" food sovereignty. Revitalizing Indigenous Food Practices and Ecological Knowledges in Canada and the United States. *Humanities (Washington), 5*(3), 57. DOI: 10.3390/h5030057

Country, M., Davis, J., Simmons, J., Snelson, S., Channell, V., Haynes, K., & Dosseto, A. (2024). Quantitative assessment of the effect of agency-led prescribed burns and cultural burns on soil properties in southeastern Australia. *Fire (Basel, Switzerland), 7*(3), 75. DOI: 10.3390/fire7030075

Cox, K. D., Covernton, G. A., Davies, H. L., Dower, J. F., Juanes, F., & Dudas, S. E. (2019). Human Consumption of Microplastics. *Environmental Science & Technology, 53*(12), 7068–7074. DOI: 10.1021/acs.est.9b01517 PMID: 31184127

Currie, A., Farber, J. M., Nadon, C., Sharma, D., Whitfield, Y., Gaulin, C., Galanis, E., Bekal, S., Flint, J., Tschetter, L., Pagotto, F., Lee, B., Jamieson, F., Badiani, T., MacDonald, D., Ellis, A., May-Hadford, J., McCormick, R., Savelli, C., & Sierpinska, U. (2015). Multi-province listeriosis outbreak linked to contaminated deli meat consumed primarily in institutional settings, Canada, 2008. *Foodborne Pathogens and Disease, 12*(8), 645–652. DOI: 10.1089/fpd.2015.1939

Currie, S. D., Wang, J.-S., & Tang, L. (2024). Impacts of PFAS Exposure on Neurodevelopment: A Comprehensive Literature Review. *Environments (Basel, Switzerland), 11*(9), 188. DOI: 10.3390/environments11090188

Curulli, A. (2020). Nanomaterials in electrochemical sensing area: Applications and challenges in food analysis. *Molecules (Basel, Switzerland), 25*(23), 5759. DOI: 10.3390/molecules25235759 PMID: 33297366

Cverenkárová, K., Valachovičová, M., Mackuľak, T., Žemlička, L., & Bírošová, L. (2021). Microplastics in the Food Chain. *Life (Chicago, Ill.), 11*(12), 1349. DOI: 10.3390/life11121349 PMID: 34947879

D'Amato, G., & Akdiş, C. (2020). Global warming, climate change, air pollution, and allergies. *Allergy, 75*(9), 2158–2160. DOI: 10.1111/all.14527

D'Amore, T., Smaoui, S., & Varzakas, T. (2025). Chemical food safety in Europe under the spotlight: Principles, regulatory framework and roadmap for future directions. *Foods, 14*(9), 1628. DOI: 10.3390/foods14091628

Dadgarnejad, M., Kouser, S., & Moslemi, M. (2017). Genetically modified foods: Promises, challenges and safety assessments. *Applied Food Biotechnology*, *4*(4), 193–202. DOI: 10.22037/afb.v4i4.17244

Dahiya, D., Manuel, J. V., & Nigam, P. S. (2021). An overview of bioprocesses employing specifically selected microbial catalysts for γ-aminobutyric acid production. *Microorganisms*, *9*(12), 2457. DOI: 10.3390/microorganisms9122457 PMID: 34946060

Dai, L., Cheng, T., Duan, C., Zhao, W., Zhang, W., Zou, X., & Ni, Y. (2019). 3D printing using plant-derived cellulose and its derivatives: A review. *Carbohydrate Polymers*, *203*, 71–86. DOI: 10.1016/j.carbpol.2018.09.027 PMID: 30318237

Dang, C., & Wang, F., LU, Z., & YE, G. (2020). Application of Meta-analysis in the Safety Assessments of Transgenic Crops. *Zhongguo Shengwu Fangzhi Xuebao*, *36*(1), 17–23. DOI: 10.16409/j.cnki.2095-039x.2019.05.012

Dang, F., Li, Y., Wang, Y., Lin, J., Du, S., & Liao, X. (2022). ZAT10 plays dual roles in cadmium uptake and detoxification in Arabidopsis. *Frontiers in Plant Science*, *13*, 994100. Advance online publication. DOI: 10.3389/fpls.2022.994100 PMID: 36110357

Darwish, W. S., Eldaly, E. A., El-Abbasy, M. T., Ikenaka, Y., Nakayama, S., & Ishizuka, M. (2013). Antibiotic residues in food: The African scenario (No. Supplement). Graduate School of Veterinary Medicine, Hokkaido University. DOI: 10.14943/jjvr.61.suppl.s13

Das, A. K. (2023). *Microbial determination methods in pharmaceutical and food quality control* (Doctoral dissertation, Brac University).

Dasgupta, S., & Robinson, E. (2022). Attributing changes in food insecurity to a changing climate. *Scientific Reports*, *12*(1), 4709. Advance online publication. DOI: 10.1038/s41598-022-08696-x

David, G., Gontard, N., & Angellier-Coussy, H. (2019). Mitigating the impact of cellulose particles on the performance of biopolyester-based composites by gas-phase esterification. *Polymers*, *11*(2), 200. DOI: 10.3390/polym11020200 PMID: 30960185

Dawoe, E., Quashie-Sam, J., Isaac, M., & Oppong, S. (2012). Exploring farmers' local knowledge and perceptions of soil fertility and management in the Ashanti region of Ghana. *Geoderma*, *179-180*, 96–103. DOI: 10.1016/j.geoderma.2012.02.015

Dawson, A. L., Kawaguchi, S., King, C. K., Townsend, K. A., King, R., Huston, W. M., & Bengtson Nash, S. M. (2018). Turning microplastics into nanoplastics through digestive fragmentation by Antarctic krill. *Nature Communications*, *9*(1), 1001. DOI: 10.1038/s41467-018-03465-9 PMID: 29520086

de Mesquita Souza Saraiva, M., Lim, K., do Monte, D. F. M., Givisiez, P. E. N., Alves, L. B. R., de Freitas Neto, O. C., Kariuki, S., Júnior, A. B., de Oliveira, C. J. B., & Gebreyes, W. A. (2022). Antimicrobial resistance in the globalized food chain: A One Health perspective applied to the poultry industry. *Brazilian Journal of Microbiology : Publication of the Brazilian Society for Microbiology*, *53*(1), 465–486. DOI: 10.1007/s42770-021-00635-8 PMID: 34775576

de Oliveira, F. L., Arruda, T. Y. P., da Silva Lima, R., Casarotti, S. N., & Morzelle, M. C. (2020). Pomegranate as a natural source of phenolic antioxidants: A review. *Journal of Food Bioactives: An Official Scientific Publication of the International Society of Nutraceuticals and Functional Foods (ISNFF)*, 10–22. DOI: 10.31665/JFB.2020.9214

Death, C., Bell, C., Champness, D., Milne, C., Reichman, S., & Hagen, T. (2021). Per- and polyfluoroalkyl substances (PFAS) in livestock and game species: A review. *The Science of the Total Environment*, *774*, 144795. DOI: 10.1016/j.scitotenv.2020.144795 PMID: 33609849

Deen, C., Sherriff, S., Shelling, M., Gall, A., Cubillo, B., Morenga, L., & Matthews, V. (2024). Measuring indigenous food security—A case for indigenous designed tools. *Health Promotion Journal of Australia*, *36*(1), e945. Advance online publication. DOI: 10.1002/hpja.945

De-la-Torre, G. E. (2020). Microplastics: An emerging threat to food security and human health. *Journal of Food Science and Technology*, *57*(5), 1601–1608. DOI: 10.1007/s13197-019-04138-1 PMID: 32327770

Deshmukh, A. R., Aloui, H., Khomlaem, C., Negi, A., Yun, J. H., Kim, H. S., & Kim, B. S. (2021). Biodegradable films based on chitosan and defatted Chlorella biomass: Functional and physical characterization. *Food Chemistry*, *337*, 127777. DOI: 10.1016/j.foodchem.2020.127777 PMID: 32799163

Devlieghere, F., Vermeiren, L., & Debevere, J. (2004). New preservation technologies: Possibilities and limitations. *International Dairy Journal*, *14*(4), 273–285. DOI: 10.1016/j.idairyj.2003.07.002

Devra, V. (2022). Applications of metal nanoparticles in the agri-food sector. Egyptian Journal of Agricultural Research. DOI: 10.21608/ejar.2022.102565.1164

DeWitt, J. C., Blossom, S. J., & Schaider, L. A. (2019). Exposure to perfluoroalkyl and polyfluoroalkyl substances leads to immunotoxicity: Epidemiological and toxicological evidence. *Journal of Exposure Science & Environmental Epidemiology, 29*(2), 148–156. DOI: 10.1038/s41370-018-0097-y PMID: 30482935

Diamanti-Kandarakis, E., Bourguignon, J.-P., Giudice, L. C., Hauser, R., Prins, G. S., Soto, A. M., Zoeller, R. T., & Gore, A. C. (2009). Endocrine-Disrupting Chemicals: An Endocrine Society Scientific Statement. *Endocrine Reviews, 30*(4), 293–342. DOI: 10.1210/er.2009-0002 PMID: 19502515

Díaz-Montes, E. (2022). Polysaccharide-based biodegradable films: An alternative in food packaging. *Polysaccharide., 3*(4), 761–775. DOI: 10.3390/polysaccharides3040044

Díaz-Montes, E., & Castro-Muñoz, R. (2021). Trends in chitosan as a primary biopolymer for functional films and coatings manufacture for food and natural products. *Polymers, 13*(5), 767. DOI: 10.3390/polym13050767 PMID: 33804445

Ding, S., Hu, H., Yue, X., Feng, K., Gao, X., Dong, Q., Yang, M., Tamer, U., Huang, G., & Zhang, J. (2022). A fluorescent biosensor based on quantum dot–labeled streptavidin and poly-l-lysine for the rapid detection of Salmonella in milk. *Journal of Dairy Science, 105*(4), 2895–2907. DOI: 10.3168/jds.2021-21229 PMID: 35181133

Dobrucka, R., & Ankiel, M. (2018). Possible applications of metal nanoparticles in antimicrobial food packaging. *Journal of Food Safety, 39*(2), e12617. Advance online publication. DOI: 10.1111/jfs.12617

Domingo, A., Yessis, J., Charles, K., Skinner, K., & Hanning, R. (2023). Integrating knowledge and action: learnings from an implementation program for food security and food sovereignty with First Nations communities in Canada. DOI: 10.21203/rs.3.rs-2801301/v1

Domingo, J. L., & Nadal, M. (2017). Per- and Polyfluoroalkyl Substances (PFASs) in Food and Human Dietary Intake: A Review of the Recent Scientific Literature. *Journal of Agricultural and Food Chemistry, 65*(3), 533–543. DOI: 10.1021/acs.jafc.6b04683 PMID: 28052194

Domínguez-Manzano, J., Olmo-Ruiz, C., Bautista-Gallego, J., Arroyo-López, F. N., Garrido-Fernández, A., & Jiménez-Díaz, R. (2012). Biofilm formation on abiotic and biotic surfaces during Spanish style green table olive fermentation. *International Journal of Food Microbiology, 157*(2), 230–238. DOI: 10.1016/j.ijfoodmicro.2012.05.011 PMID: 22656327

Dong, H., Wang, X., Niu, X., Zeng, J., Zhou, Y., Suona, Z., Yuan, Y., & Chen, X. (2023). Overview of analytical methods for the determination of microplastics: Current status and trends. *Trends in Analytical Chemistry*, *167*, 117261. DOI: 10.1016/j.trac.2023.117261

Dove, M., Balasubramanian, A., & Narayanan, B. (2020). Transparency as a way of attaining quality, safety and optimal food purchases. *SocioEconomic Challenges*, *4*(4), 48–62. DOI: 10.21272/sec.4(4).48-62.2020

Dowd-Uribe, B., Rock, J. S., Spreadbury, T., Chiril, P., & Uminsky, D. (2024). Bridging the gap? Public-private partnerships and genetically modified crop development for smallholder farmers in Africa. *Plants, People, Planet*, *6*(2), 437–451. DOI: 10.1002/ppp3.10453

Drago, E., Campardelli, R., Pettinato, M., & Perego, P. (2020). Innovations in smart packaging concepts for food: An extensive review. *Foods*, *9*(11), 1628. DOI: 10.3390/foods9111628

du Bray, M., Sanchez, A., & Ogwu, M. (2023). Time Immemorial: Water Governance Approaches and Indigenous Rights. River Field Studies Network, QUBES Educational Resources. DOI: 10.25334/2HPW-H547

Duc, H. M., Hoa, T. T. K., Ha, C. T. T., Van Hung, L., Van Thang, N., Minh Son, H., & Flory, G. A. (2024). Prevalence and Antibiotic Resistance Profile of Clostridium perfringens Isolated from Pork and Chicken Meat in Vietnam. *Pathogens (Basel, Switzerland)*, *13*(5), 400. DOI: 10.3390/pathogens13050400 PMID: 38787252

Duchenne-Moutien, R., & Neetoo, H. (2021). Climate change and emerging food safety issues: A review. *Journal of Food Protection*, *84*(11), 1884–1897. DOI: 10.4315/JFP-21-141

Duncan, T. V., Khan, S. A., Patri, A. K., & Wiggins, S. (2024). Regulatory Science Perspective on the Analysis of Microplastics and Nanoplastics in Human Food. *Analytical Chemistry*, *96*(11), 4343–4358. DOI: 10.1021/acs.analchem.3c05408 PMID: 38452774

Duque-Acevedo, M., Belmonte-Ureña, L. J., Cortés-García, F. J., & Camacho-Ferre, F. (2020). Agricultural waste: Review of the evolution, approaches and perspectives on alternative uses. *Global Ecology and Conservation*, *22*, e00902. DOI: 10.1016/j.gecco.2020.e00902

Durán, N., & Marcato, P. D. (2012). Nanobiotechnology perspectives. role of nanotechnology in the food industry: A review. *International Journal of Food Science & Technology*, *48*(6), 1127–1134. DOI: 10.1111/ijfs.12027

Ebegba, R., Bello, S., & Onwude, J. (2022). Status of Nigeria's biosafety regulation. In *Biosafety and bioethics in biotechnology* (pp. 1-18). CRC Press. DOI: 10.1201/9781003179177-1

Ebi, K. (2011). Climate change and health risks: Assessing and responding to them through 'adaptive management.'. *Health Affairs (Project Hope), 30*(5), 924–930. DOI: 10.1377/hlthaff.2011.0071

Eckardt, J., Moro, L., Colusso, E., Šket, P., Giovando, S., & Tondi, G. (2025). Comparing Hydrolysable and Condensed Tannins for Tannin Protein-Based Foams. *Polymers, 17*(2), 153. DOI: 10.3390/polym17020153 PMID: 39861226

Eckelman, M., & Sherman, J. (2018). Estimated global disease burden from U.S. health care sector greenhouse gas emissions. *American Journal of Public Health, 108*(S2), S120–S122. DOI: 10.2105/AJPH.2017.303846

EFSA Panel on Contaminants in the Food Chain (CONTAM). (2014). Scientific Opinion on Chloramphenicol in food and feed. *EFSA Journal, 12*(11). Advance online publication. DOI: 10.2903/j.efsa.2014.3907

Egboduku, W. O., Ogwu, M. C., Egboduku, T., Enujeke, E. C., Edokpaiawe, S., & Golohor, O. M. (2024). Sustainable cassava processing techniques to eliminate cyanogenic glycosides. In *Plant Biology, Sustainability and Climate Change, Sustainable Cassava.* Academic Press. DOI: 10.1016/B978-0-443-21747-0.00007-2

Egner, P., Pavlačková, J., Sedlaříková, J., Pleva, P., Mokrejš, P., & Janalíková, M. (2023). Non-Alcohol Hand Sanitiser Gels with Mandelic Acid and Essential Oils. *International Journal of Molecular Sciences, 24*(4), 3855. DOI: 10.3390/ijms24043855 PMID: 36835267

Ehrlich, H., Shaala, L. A., Youssef, D. T., Żółtowska-Aksamitowska, S., Tsurkan, M., Galli, R., & Jesionowski, T. (2018). Discovery of chitin in skeletons of non-verongiid Red Sea demosponges. *PLoS One, 13*(5), e0195803. DOI: 10.1371/journal.pone.0195803 PMID: 29763421

Eissa, S., Tlili, C., Mounir, B. A., & Kanoun, O. (2022). Editorial: Nanomaterials-based electrochemical biosensors. *Frontiers in Bioengineering and Biotechnology, 10*, 1091592. Advance online publication. DOI: 10.3389/fbioe.2022.1091592 PMID: 36588931

El-Hadary, A. E., & Taha, M. (2020). Pomegranate peel methanolic-extract improves the shelf-life of edible-oils under accelerated oxidation conditions. *Food Science & Nutrition, 8*(4), 1798–1811. DOI: 10.1002/fsn3.1391 PMID: 32328245

Eliasson, A. C. (2004). Starch-lipid interactions and their relevance in food products AC Eliasson and M. Wahlgren, Lund University, Sweden. *Starch in food: Structure, function and applications.* 441.

El-Khishin, M., Gooneratne, R., and Hussain, M. (2017). Microbial Safety of Foods in the Supply Chain and Food Security Advanced in Food Technology and Nutritional Sciences - Open Journal, 3(1), 22-32. DOI: 10.17140/AFTNSOJ-3-141

Elwell, D. A. (2020). Industry-influenced evidence: Bias, conflict, and manipulation in scientific evidence. *BCL Rev., 61*, 2155.

Enayatifard, R., Akbari, J., Babaei, A., Rostamkalaei, S. S., Hashemi, S. M., & Habibi, E. (2020, February). Anti-Microbial Potential of Nano-Emulsion form of Essential Oil Obtained from Aerial Parts of Origanum Vulgare L. as Food Additive. *Advanced Pharmaceutical Bulletin.* Advance online publication. DOI: 10.34172/apb.2021.028 PMID: 33880355

Endersen, L., O'Mahony, J., Hill, C., Ross, R. P., McAuliffe, O., & Coffey, A. (2014). Phage therapy in the food industry. *Annual Review of Food Science and Technology, 5*(1), 327–349. DOI: 10.1146/annurev-food-030713-092415 PMID: 24422588

Enova, A., Singh, R., Aga, D., Mariano, S., Villanoy, C., & Jaraula, C. (2024). Modeling pharmaceutical contaminant transport in an urban coastal estuarine system: The case of Manila Bay, Philippines. *ACS ES&T Water, 5*(1), 156–167. DOI: 10.1021/acsestwater.4c00736

Erdem, Y. I., Kaya, S., & Gökkaya Erdem, B. (2024). Characterization and application of whey protein isolate based edible films containing clove and apricot essential oils. *Gida, 49*(2), 299–311. DOI: 10.15237/gida.GD23137

Erdoğan, N. Ö., Uslu, B., & Aydoğdu Tığ, G. (2023). Development of an electrochemical biosensor utilizing a combined aptamer and MIP strategy for the detection of the food allergen lysozyme. *Microchimica Acta, 190*(12), 471. DOI: 10.1007/s00604-023-06054-w PMID: 37975892

Erdogan, S. (2024). On the impact of natural resources on environmental sustainability in African countries: A comparative approach based on the EKC and LCC hypotheses. *Resources Policy, 88*, 104492. DOI: 10.1016/j.resourpol.2023.104492

Erhunmwunse, N. O., Pajiah, T. J., & Ogwu, M. C. (2024). Microplastics as Water Pollutants and Sustainable Management Strategies. In Izah, S. C., Ogwu, M. C., Loukas, A., & Hamidifar, H. (Eds.), *Water Crises and Sustainable Management in the Global South*. Springer., DOI: 10.1007/978-981-97-4966-9_8

Erinle, K. O., Ogwu, M. C., Evivie, S. E., Zaheer, M. S., Ogunyemi, S. O., & Adeniran, S. O. (2021). Impacts of COVID-19 on agriculture and food security in developing countries: Potential mitigation strategies. *Perspectives in Agriculture, Veterinary Science, Nutrition and Natural Resources*, *16*(16), 1–16. DOI: 10.1079/PAVSNNR202116016

Ernawati, K., Aslichan, A., Albar, A. S., Al Khaliel, A. T., & Al Fatich, A. S. F. (2025). Green Leadership Strategy to Encourage Pro-Environmental Behavior. In *Green Leadership Strategies for Increased Employee Engagement and Performance* (pp. 173–204). IGI Global Scientific Publishing. DOI: 10.4018/979-8-3373-2106-6.ch006

Ernawati, K., Burhan, A., Albar, A. S., Al Khaliel, A. T., & Al Fatich, A. S. F. (2024). Professional Development of Environmental Health Workers. In *Instructional Approaches for Health Professions Education* (pp. 269–294). IGI Global Scientific Publishing. DOI: 10.4018/979-8-3693-4334-0.ch010

Ernawati, K., Nadhifah, Q., Muslikha, A., Hidayat, M., Soesilo, T. E. B., Jannah, F., Widianti, D., & Yusnita, Y. (2021). Relationship of knowledge and attitude with food handling practices: A systematic review. *International Journal of Public Health Science*, *10*(2), 336. DOI: 10.11591/ijphs.v10i2.20665

Eslami, A., Gholami, Z., Nargesi, S., Rostami, B., & Avazpour, M. (2017). Evaluation of microbial contamination of ready-to-eat foods (pizza, frankfurters, sausages) in the city of Ilam. *Environmental Health Engineering and Management*, *4*(2), 117–122. DOI: 10.15171/EHEM.2017.16

Esmaeili, Y., Paidari, S., Baghbaderani, S. A., Nateghi, L., Al-Hassan, A. A., & Ariffin, F. (2022). Essential oils as natural antimicrobial agents in postharvest treatments of fruits and vegetables: A review. *Journal of Food Measurement and Characterization*, *16*(1), 507–522. DOI: 10.1007/s11694-021-01178-0

Europe, P. (2015). An analysis of European plastics production, demand and waste data. *Plastics–the facts*, 147.

Evers, E. G., Berk, P. A., Horneman, M. L., van Leusden, F. M., & de Jonge, R. (2014). A quantitative microbiological risk assessment for *Campylobacter* in petting zoos. *Risk Analysis*, *34*(9), 1618–1638. DOI: 10.1111/risa.12197

Evivie, S. E., Ogwu, M. C., Abdelazez, A., Bian, X., Liu, F., Li, B., & Huo, G. (2020). Suppressive effects of *Streptococcus thermophilus* KLDS 3.1003 on some foodborne pathogens revealed through *in vitro, in vivo* and genomic insights. *Food & Function*, *11*(7), 6573–6587. DOI: 10.1039/D0FO01218A

Evivie, S. E., Ogwu, M. C., Abdelazez, A., Bian, X., Liu, F., Li, B., & Huo, G. (2021). Suppressive effects of *streptococcus thermophilus* KLDS 3.1003 on some foodborne pathogens revealed through *in vitro, in vivo* and genomic insights. *Food & Function, 12*(7), 3280. DOI: 10.1039/D1FO90021H

Evivie, S. E., Ogwu, M. C., Ebabhamiegbebho, P. A., Abel, E. S., Imaren, J. O., & Igene, J. O. (2020). Packaging and the Nigerian food industry: challenges and opportunities. In Ogunlade, C. A., Adeleke, K. M., & Oladejo, M. T. (Eds.), *Food Technology and Culture in Africa* (pp. 28–99). Reamsworth Publishing.

Falloon, P., Bebber, D., Dalin, C., Ingram, J., Mitchell, D., Hartley, T., Johnes, P. J., Newbold, T., Challinor, A. J., Finch, J., Galdos, M. V., Petty, C., Cornforth, R., Bhunnoo, R., Pope, E., Enow, A., Borrion, A., Waterson, A., MacNeill, K., & Houldcroft, A. (2022). What do changing weather and climate shocks and stresses mean for the UK food system? *Environmental Research Letters, 17*(5), 051001. DOI: 10.1088/1748-9326/ac68f9

Fang, C., Luo, Y., & Naidu, R. (2023). Microplastics and nanoplastics analysis: Options, imaging, advancements and challenges. *Trends in Analytical Chemistry, 166*, 117158. DOI: 10.1016/j.trac.2023.117158

FAO. (2025). *Communication for Development.* Food and Agricultural Organization. Retrieved from https://www.fao.org/communication-for-development/en/

Farfán, J., Chaux, J., & Torres, D. (2021). Food autonomy: Decolonial perspectives for indigenous health and buen vivir. *Global Health Promotion, 28*(3), 50–58. DOI: 10.1177/1757975920984206

Farré, M. L., Pérez, S., Kantiani, L., & Barceló, D. (2008). Fate and toxicity of emerging pollutants, their metabolites and transformation products in the aquatic environment. *Trends in Analytical Chemistry, 27*(11), 991–1007. DOI: 10.1016/j.trac.2008.09.010

Fatema, K. N., Biswas, M. R., Bang, S. H., Cho, K. Y., & Oh, W. C. (2020). Electro analytical characteristic of a novel biosensor designed with graphene–polymer-based quaternary and mesoporous nanomaterials. *Bulletin of Materials Science, 12*(43), 1–3.

Fatta-Kassinos, D., Vasquez, M. I., & Kümmerer, K. (2011). Transformation products of pharmaceuticals in surface waters and wastewater formed during photolysis and advanced oxidation processes – Degradation, elucidation of byproducts and assessment of their biological potency. *Chemosphere, 85*(5), 693–709. DOI: 10.1016/j.chemosphere.2011.06.082 PMID: 21835425

Favas, C., Cresta, C., Whelan, E., Smith, K., Manger, M., Chandrasenage, D., Singhkumarwong, A., Kawasaki, J., Moreno, S., & Goudet, S. (2024). Exploring food system resilience to the global polycrisis in six asian countries. *Frontiers in Nutrition, 11*, 1347186. Advance online publication. DOI: 10.3389/fnut.2024.1347186

FDA. (2023). Food Ingredients & Packaging. Available online: https://www.fda.gov/food/food-ingredients-packaging

Fels-Klerx, H., Liu, C., & Battilani, P. (2016). Modeling climate change impacts on mycotoxin contamination. *World Mycotoxin Journal, 9*(5), 717–726. DOI: 10.3920/WMJ2016.2066

Feng, Z., Xu, S., Zhang, S., & Sun, M. (2017, May). From the holistic governance perspective to study multidimensional analysis of China's food safety supervision system. In *2017 International Conference on Education, Economics and Management Research (ICEEMR 2017)* (pp. 56–62). Atlantis Press. DOI: 10.2991/iceemr-17.2017.14

Fenta, L., & Kebede, A. (2019). *Effect of climate change on food- and waterborne disease outbreaks: a mini review*. FSQM. DOI: 10.7176/FSQM/88-02

Ferk, C., Calder, B., & Camire, M. (2015). Assessing the food safety knowledge of University of mainMainedents. *Journal of Food Science Education, 15*(1), 14–22. DOI: 10.1111/1541-4329.12076

Ferrone, E., Araneo, R., Notargiacomo, A., Pea, M., & Rinaldi, A. (2019). Zno nanostructures and electrospun zno–polymeric hybrid nanomaterials in biomedical, health, and sustainability applications. *Nanomaterials (Basel, Switzerland), 9*(10), 1449. DOI: 10.3390/nano9101449 PMID: 31614707

Filip, S., & Fink, R. (2017). Hazard analysis and critical control points of traditional vs industrial cremeschnitte production. *Acta Alimentaria, 46*(1), 9–16. DOI: 10.1556/066.2017.46.1.2

Fisher, F., Africa, C., Klaasen, J., & Fisher, R. (2025). South African Medicinal Plants Traditionally Used for Wound Treatment: An Ethnobotanical Systematic Review. *Plants, 14*(5), 818. DOI: 10.3390/plants14050818 PMID: 40094805

Fisher, N., & Chen, C. (2011). Interdisciplinary approaches for addressing marine contamination issues. *Environmental Conservation, 38*(2), 187–198. DOI: 10.1017/S0376892911000269

Foale, S., Cohen, P., Januchowski-Hartley, S., Wenger, A., & Macintyre, M. (2010). Tenure and taboos: Origins and implications for fisheries in the Pacific. *Fish and Fisheries, 12*(4), 357–369. DOI: 10.1111/j.1467-2979.2010.00395.x

Fraceto, L. F., Grillo, R., de Medeiros, G. A., Scognamiglio, V., Rea, G., & Bartolucci, C. (2016). Nanotechnology in agriculture: Which innovation potential does it have? *Frontiers in Environmental Science*, *4*. Advance online publication. DOI: 10.3389/fenvs.2016.00020

Frank-Kamenetskii, D. (2012). *Plasma: the fourth state of matter*. Springer Science & Business Media.

Freitas, C. M. P., Coimbra, J. S. R., Souza, V. G. L., & Sousa, R. C. S. (2021). Structure and applications of pectin in food, biomedical, and pharmaceutical industry: A review. *Coatings*, *11*(8), 922. DOI: 10.3390/coatings11080922

Fuertes, G., Soto, I., Carrasco, R., Vargas, M., Sabattin, J., & Lagos, C. (2016). Intelligent packaging systems: Sensors and nanosensors to monitor food quality and safety. *Journal of Sensors*, *2016*, 1–8. DOI: 10.1155/2016/4046061

Fu, H., Bai, Z., Li, P., Feng, X., Hu, X., Song. And, X., & Chen, L. (2023). Molecular imprinted electrochemical sensor for ovalbumin detection based on boronate affinity and signal amplification approach. *Food Chemistry*, *409*, 135292. DOI: 10.1016/j.foodchem.2022.135292 PMID: 36584533

Fu, K., Zhang, H., Guo, Y., Li, J., Nie, H., Song, X., Xu, K., Wang, J., & Zhao, C. (2020). Rapid and selective recognition of Vibrio parahaemolyticus assisted by perfluorinated alkoxysilane modified molecularly imprinted polymer film. *RSC Advances*, *10*(24), 14305–14312. DOI: 10.1039/D0RA00306A PMID: 35498485

Fussell, R. J., Garcia Lopez, M., Mortimer, D. N., Wright, S., Sehnalova, M., Sinclair, C. J., Fernandes, A., & Sharman, M. (2014). Investigation into the Occurrence in Food of Veterinary Medicines, Pharmaceuticals, and Chemicals Used in Personal Care Products. *Journal of Agricultural and Food Chemistry*, *62*(17), 3651–3659. DOI: 10.1021/jf4052418 PMID: 24495205

Fu, Y., & Dudley, E. G. (2021). Antimicrobial-coated films as food packaging: A review. *Comprehensive Reviews in Food Science and Food Safety*, *20*(4), 3404–3437. DOI: 10.1111/1541-4337.12769 PMID: 34056844

Gaggero, E., López-Muñoz, M. J., Paganini, M. C., Arencibia, A., Bertinetti, S., Fernández De Paz, N., & Calza, P. (2023). Mercury and Organic Pollutants Removal from Aqueous Solutions by Heterogeneous Photocatalysis with ZnO-Based Materials. *Molecules (Basel, Switzerland)*, *28*(6), 2650. DOI: 10.3390/molecules28062650 PMID: 36985622

Gagnon, P., Sheedy, C., Rousseau, A., Bourgeois, G., & Chouinard, G. (2015). Integrated assessment of climate change impact on surface runoff contamination by pesticides. *Integrated Environmental Assessment and Management, 12*(3), 559–571. DOI: 10.1002/ieam.1706

Gaines, L. G. T. (2023). Historical and current usage of per- and polyfluoroalkyl substances (PFAS): A literature review. *American Journal of Industrial Medicine, 66*(5), 353–378. DOI: 10.1002/ajim.23362 PMID: 35614869

Galanakis, C. M. (2024). The future of food. *Foods, 13*(4), 506. DOI: 10.3390/foods13040506 PMID: 38397483

Gallego, M., Arnal, M., Talens, P., Toldrá, F., & Mora, L. (2020). Effect of Gelatin Coating Enriched with Antioxidant Tomato Byproducts on the Quality of Pork Meat. *Polymers, 12*(5), 1032. DOI: 10.3390/polym12051032 PMID: 32370120

Gallego-Ramírez, C., Chica, E., & Rubio-Clemente, A. (2024). Combination of Biochar and Advanced Oxidation Processes for the Sustainable Elimination of Pharmaceuticals in Water. *Sustainability (Basel), 16*(23), 10761. DOI: 10.3390/su162310761

Galus, S., Arik Kibar, E. A., Gniewosz, M., & Kraśniewska, K. (2020). Novel materials in the preparation of edible films and coatings—A review. *Coatings, 10*(7), 674. DOI: 10.3390/coatings10070674

Gao, H., Yan, C., Wu, W., & Li, J. (2020). Application of microfluidic chip technology in food safety sensing. *Sensors (Basel), 20*(6), 1792. DOI: 10.3390/s20061792 PMID: 32213909

Gao, J., & Chen, Z. (2023). Application of association rule mining algorithm based on 5 G technology in information management system. *Scalable Computing Practice and Experience, 24*(2), 139–149. DOI: 10.12694/scpe.v24i2.2150

Gao, J., Sun, Y., Lu, Y., & Li, L. (2014). Impact of ambient humidity on child health: A systematic review. *PLoS One, 9*(12), e112508. DOI: 10.1371/journal.pone.0112508

Gao, X., Zuo, Q., Fu, X., Song, L., Cen, M., & Wu, J. (2023). Association between prenatal exposure to per- and polyfluoroalkyl substances and neurodevelopment in children: Evidence based on birth cohort. *Environmental Research, 236*, 116812. DOI: 10.1016/j.envres.2023.116812 PMID: 37536558

García, G., Carlin, M., & Cano, R. D. J. (2025). Holobiome harmony: Linking environmental sustainability, agriculture, and human health for a thriving planet and one health. *Microorganisms, 13*(3), 514. DOI: 10.3390/microorganisms13030514 PMID: 40142407

García-García, D. J., Pérez-Sánchez, G. F., Hernández-Cocoletzi, H., Sánchez-Arzubide, M. G., Luna-Guevara, M. L., Rubio-Rosas, E., Krishnamoorthy, R., & Morán-Raya, C. (2023). Chitosan Coatings Modified with Nanostructured ZnO for the Preservation of Strawberries. *Polymers*, *15*(18), 3772. DOI: 10.3390/polym15183772 PMID: 37765626

Garcia-Gutierrez, E., Mayer, M. J., Cotter, P. D., & Narbad, A. (2019). Gut microbiota as a source of novel antimicrobials. *Gut Microbes*, *10*(1), 1–21. DOI: 10.1080/19490976.2018.1455790 PMID: 29584555

Gavai, A., Bouzembrak, Y., Mu, W., Martin, F., Kaliyaperumal, R., van Soest, J., Choudhury, A., Heringa, J., Dekker, A., & Marvin, H. (2022). A federated learning approach to data sharing in a food supply chain. *Research Square*. DOI: 10.21203/rs.3.rs-2350301/v1

Gayán, E., Serrano, M. J., Pagán, R., Álvarez, I., & Condón, S. (2015). Environmental and biological factors influencing the UV-C resistance of Listeria monocytogenes. *Food Microbiology*, *46*, 246–253. DOI: 10.1016/j.fm.2014.08.011 PMID: 25475293

Gayathri, D., Soundarya, R., & Prashantkumar, C. (2024). Various facets of nanotechnology in food processing [review]. *International Journal of Functional Nutrition*, *5*(1), 4. Advance online publication. DOI: 10.3892/ijfn.2024.38

Gbadegesin, L. A., Ayeni, E. A., Tettey, C. K., Uyanga, V. A., Aluko, O. O., Ahiakpa, J. K., Okoye, C. O., Mbadianya, J. I., Adekoya, M. A., Aminu, R. O., Oyawole, F. P., & Odufuwa, P. (2022). GMOs in Africa: Status, adoption and public acceptance. *Food Control*, *141*, 109193. DOI: 10.1016/j.foodcont.2022.109193

Gbashi, S., Madala, N., Saeger, S., Boevre, M., Adekoya, I., Adebo, O., … Njobeh, P. (2019). The Socioeconomic Impact of Mycotoxin Contamination in Africa. DOI: 10.5772/intechopen.79328

Gbashi, S., Adebo, O., Adebiyi, J. A., Targuma, S., Tebele, S., Areo, O. M., Olopade, B., Odukoya, J. O., & Njobeh, P. (2021). Food safety, food security and genetically modified organisms in Africa: A current perspective. *Biotechnology & Genetic Engineering Reviews*, *37*(1), 30–63. DOI: 10.1080/02648725.2021.1940735

Gbashi, S., Adelusi, O., & Njobeh, P. (2024). Insights from modeling sixteen years of climatic and fumonisin patterns in maize in South Africa. *Scientific Reports*, *14*(1), 11643. Advance online publication. DOI: 10.1038/s41598-024-60904-y

Geissdoerfer, M., Savaget, P., Bocken, N. M. P., & Hultink, E. J. (2017). The Circular Economy – A new sustainability paradigm? *Journal of Cleaner Production*, *143*, 757–768. DOI: 10.1016/j.jclepro.2016.12.048

Ghisi, R., Vamerali, T., & Manzetti, S. (2019). Accumulation of perfluorinated alkyl substances (PFAS) in agricultural plants: A review. *Environmental Research, 169*, 326–341. DOI: 10.1016/j.envres.2018.10.023 PMID: 30502744

Ghosh, A. K., Ullah, A. S., & Kubo, A. (2019). Hidden Markov model-based digital twin construction for futuristic manufacturing systems. *Artificial Intelligence for Engineering Design, Analysis and Manufacturing, 33*(3), 317–331. DOI: 10.1017/S089006041900012X

Giesy, J. P., & Kannan, K. (2001). Global distribution of perfluorooctane sulfonate in wildlife. *Environmental Science & Technology, 35*(7), 1339–1342. DOI: 10.1021/es001834k PMID: 11348064

Giesy, J. P., & Kannan, K. (2002). Peer Reviewed: Perfluorochemical Surfactants in the Environment. *Environmental Science & Technology, 36*(7), 146A–152A. DOI: 10.1021/es022253t PMID: 11999053

Giri, S., Lamichhane, G., Khadka, D., & Devkota, H. P. (2024). Microplastics contamination in food products: Occurrence, analytical techniques and potential impacts on human health. *Current Research in Biotechnology, 7*, 100190. DOI: 10.1016/j.crbiot.2024.100190

Givanoudi, S., Cornelis, P., Rasschaert, G., Wackers, G., Iken, H., Rolka, D., Yongabi, D., Robbens, J., Schöning, M. J., Heyndrickx, M., & Wagner, P. (2021). Selective Campylobacter detection and quantification in poultry: A sensor tool for detecting the cause of a common zoonosis at its source. *Sensors and Actuators. B, Chemical, 332*, 129484. DOI: 10.1016/j.snb.2021.129484

Glevitzky, M., Glevitzky, I., Mucea- tef, P., Popa, M., Dumitrel, G. A., & Vică, M. L. (2025). Integrated risk framework (IRF)—Interconnection of the Ishikawa Diagram with the enhanced HACCP system in risk assessment for the sustainable food industry. *Sustainability (Basel), 17*(2), 536. DOI: 10.3390/su17020536

Glover, B., Akinbo, O., Savadogo, M., Timpo, S., Lemgo, G., Sinebo, W., Akile, S., Obukosia, S., Ouedraogo, J., Ndomondo-Sigonda, M., Koch, M., Makinde, D., & Ambali, A. (2018). Strengthening regulatory capacity for gene drives in Africa: Leveraging NEPAD's experience in establishing regulatory systems for medicines and GM crops in Africa. *BMC Proceedings, 12*(S8, Suppl 8), 11. DOI: 10.1186/s12919-018-0108-y

Godde, C., Mason-D'Croz, D., Mayberry, D., Thornton, P., & Herrero, M. (2021). Impacts of climate change on the livestock food supply chain: A review of the evidence. *Global Food Security, 28*, 100488. DOI: 10.1016/j.gfs.2020.100488

Gohil, K., Chandarana, K. A., Chandwani, S., Jawahar, G., & Amaresan, N. (2022). Production, cost analysis, and marketing of acetic acid (Vinegar). In *Industrial microbiology based entrepreneurship: Making money from microbes* (pp. 93–115). Springer Nature Singapore. DOI: 10.1007/978-981-19-6664-4_8

Goldhar, C., Ford, J., & Berrang-Ford, L. (2010). Prevalence of Food Insecurity in a Greenlandic Community and the Importance of Social, Economic, and Environmental Stressors. *International Journal of Circumpolar Health, 69*(3), 285–303. DOI: 10.3402/ijch.v69i3.17616

Gómez-Arribas, L. N., Benito-Peña, E., Hurtado-Sanchez, M. D., & Moreno-Bondi, M. C. (2018). Biosensing based on nanoparticles for food allergens detection. *Sensors (Basel), 18*(4), 1087. DOI: 10.3390/s18041087 PMID: 29617319

Gonçalves, S. F., Lee, S. H. I., Khaneghah, A. M., & Oliveira, C. A. F. (2020). Enzyme-based approaches to control microbial biofilms in dairy processing environments: A review. *Quality Assurance and Safety of Crops & Foods, 12*(SP1), 50–58. DOI: 10.15586/qas.v12SP1.828

González-González, R. B., Sharma, P., Singh, S. P., Américo-Pinheiro, J. H. P., Parra-Saldívar, R., Bilal, M., & Iqbal, H. M. N. (2022). Persistence, environmental hazards, and mitigation of pharmaceutically active residual contaminants from water matrices. *The Science of the Total Environment, 821*, 153329. DOI: 10.1016/j.scitotenv.2022.153329 PMID: 35093347

Grace, D. (2023). Burden of foodborne disease in low-income and middle-income countries and opportunities for scaling food safety interventions. *Food Security, 15*(6), 1475–1488. DOI: 10.1007/s12571-023-01391-3

Graham, F., Paradis, L., Bégin, P., Paradis, J., Babin, Y., & Des Roches, A. (2014). Risk of allergic reaction and sensitization to antibiotics in foods. *Annals of Allergy, Asthma & Immunology, 113*(3), 329–330. DOI: 10.1016/j.anai.2014.06.029 PMID: 25168224

Griffith, C., Livesey, K., & Clayton, D. (2010). The assessment of food safety culture. *British Food Journal, 112*(4), 439–456. DOI: 10.1108/00070701011034448

Gumienna, M., & Górna, B. (2021). Antimicrobial food packaging with biodegradable polymers and bacteriocins. *Molecules (Basel, Switzerland), 26*(12), 3735. DOI: 10.3390/molecules26123735 PMID: 34207426

Gummadi, S. N. (2024). Advances in the applications of Bacteriophages and phage products against food-contaminating bacteria. *Critical Reviews in Microbiology, 50*(5). PMID: 37861086

Guo, X. (2024). Research progress on the detection of foodborne pathogens based on aptamer recognition. *Microchimica Acta, 191*(6), 318. DOI: 10.1007/s00604-024-06375-4 PMID: 38727855

Guo, X., Broeze, J., Groot, J., Axmann, H., & Vollebregt, M. (2020). A worldwide hotspot analysis on food loss and waste, associated greenhouse gas emissions, and protein losses. *Sustainability (Basel), 12*(18), 7488. DOI: 10.3390/su12187488

Guo, X., Lin, H., Xu, S., & He, L. (2022). Recent Advances in Spectroscopic Techniques for the Analysis of Microplastics in Food. *Journal of Agricultural and Food Chemistry, 70*(5), 1410–1422. DOI: 10.1021/acs.jafc.1c06085 PMID: 35099960

Guo, Z. (2025). Research progress of nanomaterials as food packaging materials. *Applied and Computational Engineering, 137*(1), 18–25. DOI: 10.54254/2755-2721/2025.21001

Gutiérrez-Pozo, M., Serna-Escolano, V., Giménez-Berenguer, M., Giménez, M. J., & Zapata, P. J. (2023). The Preharvest Application of Essential Oils (Carvacrol, Eugenol, and Thymol) Reduces Fungal Decay in Lemons. *Agriculture, 13*(7), 1437. DOI: 10.3390/agriculture13071437

György, É., Laslo, É., & Csató, E. (2020). Antibacterial activity of plant extracts against listeria monocytogenes isolated from ready-to-eat salads. *Acta Universitatis Sapientiae. Alimentaria, 13*(1), 131–143. DOI: 10.2478/ausal-2020-0008

Hadjigeorgiou, A., Soteriades, E. S., Philalithis, A., Psaroulaki, A., Tselentis, Y., & Gikas, A. (2012). National food safety systems in the European Union: A comparative survey. *International Journal of Food Studies, 2*(1). Advance online publication. DOI: 10.7455/ijfs/2.1.2013.a8

Halden, R. U. (2010). Plastics and health risks. *Annual Review of Public Health, 31*(1), 179–194. DOI: 10.1146/annurev.publhealth.012809.103714 PMID: 20070188

Hamaideh, S., Olaimat, A., Al-Holy, M., Ababneh, A., Shahbaz, H., Abughoush, M., Al-Nabulsi, A., Osaili, T., Ayyash, M., & Holley, R. (2024). The influence of technological shifts in the food chain on the emergence of foodborne pathogens: An overview. *Applied Microbiology (Basel, Switzerland), 4*(2), 594–606. DOI: 10.3390/applmicrobiol4020041

Hardy, A., Benford, D., Halldorsson, T., Jeger, M. J., Knutsen, H. K., More, S., Naegeli, H., Noteborn, H., Ockleford, C., Ricci, A., Rychen, G., Schlatter, J. R., Silano, V., Solecki, R., Turck, D., Younes, M., Chaudhry, Q., Cubadda, F., Gott, D., & Mortensen, A.EFSA Scientific Committee. (2018). Guidance on risk assessment of the application of nanoscience and nanotechnologies in the food and feed chain: Part 1, human and animal health. *EFSA Journal*, *16*(7). Advance online publication. DOI: 10.2903/j.efsa.2018.5327 PMID: 32625968

Harris-Glenville, F., & Cloos, P. (2024). "I think they should give primary health care a little more priority". The primary health care in Caribbean SIDS: What can be said about adaptation to the changing climate? The case of Dominica—a qualitative study. *BMC Primary Care*, *25*(1), 65. Advance online publication. DOI: 10.1186/s12875-024-02311-w

Harris, S. J., Cormican, M., & Cummins, E. (2012). Antimicrobial Residues and Antimicrobial-Resistant Bacteria: Impact on the Microbial Environment and Risk to Human Health—A Review. *Human and Ecological Risk Assessment*, *18*(4), 767–809. DOI: 10.1080/10807039.2012.688702

Hartatie, E. S. (2020). Short Communication: Detection of bioactive compounds in essential oil from lemongrass cultivated in Ngantang, Malang, East Java, Indonesia. *Biodiversitas (Surakarta)*, *21*(6). Advance online publication. DOI: 10.13057/biodiv/d210659

Hasegawa, T., Fujimori, S., Havlík, P., Valin, H., Bodirsky, B., Doelman, J., Fellmann, T., Kyle, P., Koopman, J. F. L., Lotze-Campen, H., Mason-D'Croz, D., Ochi, Y., Pérez Domínguez, I., Stehfest, E., Sulser, T. B., Tabeau, A., Takahashi, K., Takakura, J., van Meijl, H., & Witzke, P. (2018). Risk of increased food insecurity under a stringent global climate change mitigation policy. *Nature Climate Change*, *8*(8), 699–703. DOI: 10.1038/s41558-018-0230-x

Heaney, D., Padilla-Zakour, O., & Chen, C. (2024). Processing and preservation technologies to enhance indigenous food sovereignty, nutrition security, and health equity in North America. *Frontiers in Nutrition*, *11*, 1395962. Advance online publication. DOI: 10.3389/fnut.2024.1395962

Heidari, T., Sadighara, P., Oskoei, V., & Zeinali, T. (2023). Significance of climate change in food safety and hygiene. *Jundishapur J Health Sci*, *15*(4). Advance online publication. DOI: 10.5812/jjhs-141415

Heinze, El Seoud, Koschella, Heinze, El Seoud, & Koschella. (2018). Structure and properties of cellulose and its derivatives. *Cellulose derivatives: synthesis, structure, and properties*, 39-172. DOI: 10.1007/978-3-319-73168-1_2

Hermans, E., Toelen, J., Ventrella, D., De Schaepdrijver, L., Turner, K., Croubels, S., & Devreese, M. (2024). Application of preclinical juvenile animal models. In Gasthuys, E., Allegaert, K., Dossche, L., & Turner, M. (Eds.), *Essentials of translational pediatric drug development: From past needs to future opportunities* (pp. 189–212). Academic Press. DOI: 10.1016/B978-0-323-88459-4.00009-2

Hernando, M., Mezcua, M., Fernandezalba, A., & Barcelo, D. (2006). Environmental risk assessment of pharmaceutical residues in wastewater effluents, surface waters and sediments. *Talanta*, *69*(2), 334–342. DOI: 10.1016/j.talanta.2005.09.037 PMID: 18970571

He, X., Deng, H., & Hwang, H. M. (2019). The current application of nanotechnology in food and agriculture. *Yao Wu Shi Pin Fen Xi*, *27*(1), 1–21. DOI: 10.1016/j.jfda.2018.12.002 PMID: 30648562

He, X., & Hwang, H. M. (2016). Nanotechnology in food science: Functionality, applicability, and safety assessment. *Yao Wu Shi Pin Fen Xi*, *24*(4), 671–681. DOI: 10.1016/j.jfda.2016.06.001 PMID: 28911604

Heyder, M., Hollmann-Hespos, T., & Theuvsen, L. (2010). Agribusiness Firm Reactions to Regulations: The Case of Investments in Traceability Systems. *International Journal on Food System Dynamics*, *1*(2), 133–142. DOI: 10.18461/ijfsd.v1i2.125

Hong, S. P., Mohd-Naim, N. F., Keasberry, N. A., & Ahmed, M. U. (2022). Electrochemical Detection of β-Lactoglobulin Allergen Using Titanium Dioxide/Carbon Nanochips/Gold Nanocomposite-based Biosensor. *Electroanalysis*, *34*(4), 684–691. DOI: 10.1002/elan.202100207

Hoover, E., Renauld, M., Edelstein, M., & Brown, P. (2015). Social science collaboration with environmental health. *Environmental Health Perspectives*, *123*(11), 1100–1106. DOI: 10.1289/ehp.1409283

Hopewell, J., Dvorak, R., & Kosior, E. (2009). Plastics recycling: Challenges and opportunities. *Philosophical Transactions of the Royal Society of London. Series B, Biological Sciences*, *364*(1526), 2115–2126. DOI: 10.1098/rstb.2008.0311 PMID: 19528059

Huang, K. (2017). *Safety assessment of genetically modified foods*. Springer Singapore. DOI: 10.1007/978-981-10-3488-6

Huang, S., Wang, G., Lei, D., & Yan, Y. (2022). Toward digital validation for rapid product development based on digital twin: A framework. *International Journal of Advanced Manufacturing Technology*, *119*(3–4), 2509–2523. DOI: 10.1007/s00170-021-08475-4

Huang, X., Swain, D., & Hall, A. (2020). Future precipitation increase from very high-resolution ensemble downscaling of extreme atmospheric river storms in California. *Science Advances*, *6*(29), eaba1323. Advance online publication. DOI: 10.1126/sciadv.aba1323

Huang, Y., Zhang, W., Ruan, G., Li, X., Cong, Y., Du, F., & Li, J. (2018). Reduced graphene oxide-hybridized polymeric high-internal phase emulsions for highly efficient removal of polycyclic aromatic hydrocarbons from water matrix. *Langmuir*, *34*(12), 3661–3668. DOI: 10.1021/acs.langmuir.8b00005

Hu, J., Tang, F., Jiang, Y. Z., & Liu, C. (2020). Rapid screening and quantitative detection of Salmonella using a quantum dot nanobead-based biosensor. *Analyst*, *145*(6), 2184–2190. DOI: 10.1039/D0AN00035C PMID: 32101227

Hume, M. E. (2011). Historic perspective: Prebiotics, probiotics, and other alternatives to antibiotics. *Poultry Science*, *90*(11), 2663–2669. DOI: 10.3382/ps.2010-01030 PMID: 22010256

Hyde, R., Hoflund, A. B., & Pautz, M. C. (2016). One HACCP, two approaches: Experiences with and perceptions of the Hazard Analysis and Critical Control Points (HACCP) food safety management systems in the US and the EU. *Administration & Society*, *48*(8), 962–987. DOI: 10.1177/0095399714548266

Ibrahim, S. A., Ayivi, R. D., Zimmerman, T., Siddiqui, S. A., Altemimi, A. B., Fidan, H., Esatbeyoglu, T., & Bakhshayesh, R. V. (2021). Lactic Acid Bacteria as Antimicrobial Agents: Food Safety and Microbial Food Spoilage Prevention. *Foods*, *10*(12), 3131. DOI: 10.3390/foods10123131 PMID: 34945682

Ikhajiagbe, B., Atoe, R., Ogwu, M. C., & Loveniers, P.-J. (2021). Changes in *Telfaria occidentalis* leaf morphology, quality and phytochemical composition under different local preservation regimes. *Vegetos*, *31*(1), 29–36. DOI: 10.1007/s42535-021-00188-z

Ikhajiagbe, B., & Ogwu, M. C. (2020). Hazard quotient, microbial diversity and plant composition of spent crude oil polluted-soil. *Beni-Suef University Journal of Basic and Applied Sciences*, *9*(26), 26. Advance online publication. DOI: 10.1186/s43088-020-00052-0

Ikhajiagbe, B., Ogwu, M. C., Ogochukwu, O. F., Odozi, E. B., Adekunle, I. J., & Omage, Z. E. (2021). The place of neglected and underutilized legumes in human nutrition and protein security. *Critical Reviews in Food Science and Nutrition*. Advance online publication. DOI: 10.1080/10408398.2020.1871319

Ilias, S., Zakir, M. A. W. M., & Abd Almisreb, A. (2021). Review of the Effectiveness of UV-C for Disinfection of High Touch Objects. *Southeast Europe Journal of Soft Computing*, *10*(1), 30–37.

Imarhiagbe, O., & Ogwu, M. C. (2022). Sacred Groves in the Global South: A Panacea for Sustainable Biodiversity Conservation. In *Biodiversity in Africa: Potentials, Threats and Conservation.* Springer. DOI: 10.1007/978-981-19-3326-4_20

Imarhiagbe, O., Ogwu, M. C., Ikponmwosa, B. O., Mukah, F. E., Akemu, S. E., & Ohiaba, E. E. (2024). Threats to cassava cultivation, production, and processing: Global status and sustainable management strategies. In *Plant Biology, Sustainability and Climate Change, Sustainable Cassava.* Academic Press. DOI: 10.1016/B978-0-443-21747-0.00016-3

Imarhiagbe, O., Onyeukwu, I. I., Egboduku, W., Mukah, F. E., & Ogwu, M. C. (2022). Forest Conservation Strategies in Africa: Historical Perspectives, Status and Sustainable Avenues for Progress. In *Biodiversity in Africa: Potentials, Threats and Conservation.* Springer. DOI: 10.1007/978-981-19-3326-4_21

Inbaraj, B. S., & Chen, B. H. (2016). Nanomaterial-based sensors for detection of foodborne bacterial pathogens and toxins as well as pork adulteration in meat products. *Yao Wu Shi Pin Fen Xi*, *24*(1), 15–28. DOI: 10.1016/j.jfda.2015.05.001 PMID: 28911398

Ineyougha, E. R., Orutugu, L. A., & Izah, S. C. (2015). Assessment of Microbial Quality of Smoked *Trachurus trachurus* sold in some Markets of Three South-South States of Nigeria. *International Journal of Food Research*, *2*, 16–23.

Ingegno, M., Zianni, R., Della Rovere, I., Chiappinelli, A., Nardelli, V., Casamassima, F., Calitri, A., Quinto, M., Nardiello, D., & Iammarino, M. (2024). Development of a highly sensitive method based on QuEChERS and GC–MS/MS for the determination of polycyclic aromatic hydrocarbons in infant foods. *Frontiers in Nutrition*, *11*, 1403541. Advance online publication. DOI: 10.3389/fnut.2024.1403541

Islam, F., Imran, A., Nosheen, F., Fatima, M., Arshad, M. U., Afzaal, M., Ijaz, N., Noreen, R., Mehta, S., Biswas, S., Rasool, I. F. U., Aslam, M. A., Usman, I., Zahra, S. M., Segueni, N., & Amer Ali, Y. (2023). Functional roles and novel tools for improving-oxidative stability of polyunsaturated fatty acids: A comprehensive review. *Food Science & Nutrition*, *11*(6), 2471–2482. DOI: 10.1002/fsn3.3272 PMID: 37324849

Ismail, I. A., Qari, S. H., Shawer, R., Elshaer, M. M., Dessoky, E. S., Youssef, N. H., Hamad, N. A., Abdelkhalek, A., Elsamra, I. A., & Behiry, S. I. (2021). The Application of Pomegranate, Sugar Apple, and Eggplant Peel Extracts Suppresses Aspergillus flavus Growth and Aflatoxin B1 Biosynthesis Pathway. *Horticulturae*, *7*(12), 558. DOI: 10.3390/horticulturae7120558

Issa-Zacharia, A. (2024). Application of slightly acidic electrolyzed water as a potential sanitizer in the food industry. *Journal of Food Quality*, *2024*(1), 5559753. DOI: 10.1155/2024/5559753

Iwegbue, C. M., Agadaga, H., Bassey, F. I., Overah, L. C., Tesi, G. O., & Nwajei, G. E. (2015). Concentrations and profiles of polycyclic aromatic hydrocarbons in some commercial brands of tea-, coffee-, and cocoa-based food drinks in Nigeria. *International Journal of Food Properties*, *18*(10), 2124–2133. DOI: 10.1080/10942912.2014.908906

Iyiola, A. O., Izah, S. C., Morya, S., Akinsorotan, A., & Ogwu, M. C. (2023). Owalla Reservoir in South-western Nigeria: Assessment of Fish Distribution, Biological Diversity, and Water Quality Index. *Indonesian Journal of Agricultural Research*, *6*(02), 106–125. DOI: 10.32734/injar.v6i2.12207

Iyiola, A. O., Kolawole, A. S., Setufe, S. B., Bilikoni, J., Ofori, E., & Ogwu, M. C. (2024). Fish as a Sustainable Biomonitoring Tool in Aquatic Environments. In Izah, S. C., Ogwu, M. C., & Hamidifar, H. (Eds.), *Biomonitoring of Pollutants in the Global South*. Springer., DOI: 10.1007/978-981-97-1658-6_12

Izah, S.C., Kigigha, L.T., & Anene, E.K. (2016b). Bacteriological Quality Assessment of *Malus domestica* Borkh and *Cucumis sativus* L. in Yenagoa Metropolis, Bayelsa state, Nigeria. British Journal of Applied Research, 1(2), 5-7.

Izah, S. C., & Aigberua, A. O. (2020). Microbial and heavy metal hazard analysis of edible tomatoes (*Lycopersicon esculentum*) in Port Harcourt, Nigeria. *Toxicology and Environmental Health Sciences*, *12*(4), 371–380. DOI: 10.1007/s13530-020-00060-8

Izah, S. C., Aigberua, A. O., & Ogwu, M. C. (2022). Trace element composition of *Gallus gallus domesticus* eggs and health risks associated with their consumption in Port Harcourt, Nigeria. *Journal of Food Safety and Hygiene*, *8*(3), 202–222.

Izah, S. C., Aigberua, A. O., & Richard, G. (2022). Concentration, source, and health risk of trace metals in some liquid herbal medicine sold in Nigeria. *Biological Trace Element Research*, *200*(6), 3009–3022. DOI: 10.1007/s12011-021-02879-9

Izah, S. C., Aseiba, E. R., & Orutugu, L. A. (2015b). Microbial quality of polythene packaged sliced fruits sold in major markets of Yenagoa Metropolis, Nigeria. *Point Journal of Botany and Microbiology Research, 1*(3), 30–36.

Izah, S. C., Etebu, E. N., Aigberua, A. O., Odubo, T. C., & Iniamagha, I. (2022). A meta-analysis of microbial contaminants in selected ready-to-eat foods in Bayelsa State, Nigeria: Public health implications and risk-reduction strategies. *Hygiene and Environmental Health Advances, 4*, 100017. DOI: 10.1016/j.heha.2022.100017

Izah, S. C., Kigigha, L. T., Aseibai, E. R., Okowa, I. P., & Orutugu, L. A. (2016a). Advances in preservatives and condiments used in zobo (a food-drink) production. *Biotechnological Research, 2*(3), 104–119.

Izah, S. C., Nurmakhanova, A., Ogwu, M. C., Toktarbay, Z., Umirbaeva, Z., Ussen, K., Koibasova, L., Nazarbekova, S., Tynybekov, B., & Guo, Z. (2025). Public health risks associated with antibiotic residues in poultry food products. *Journal of Agriculture and Food Research, 21*, 101815. DOI: 10.1016/j.jafr.2025.101815

Izah, S. C., Orutugu, L. A., & Kigigha, L. T. (2015a). A review of the quality assessment of zobo drink consumed in Nigeria. ASIO Journal of Microbiology. *Food Science and Biotechnology Innovations, 1*(1), 34–44.

Izah, S. C., Richard, G., Stanley, H. O., Sawyer, W. E., Ogwu, M. C., & Uwaeme, O. R. (2023). Integrating the One Health Approach and Statistical Analysis for Sustainable Aquatic Ecosystem Management and Trace Contamination Mitigation. *ES Food and Agroforestry, 14*, 1012. DOI: 10.30919/esfaf1012

Izah, S. C., Stanley, H. O., Richard, G., Sawyer, W. E., Uwaeme, O. R., & Sylva, L. (2024). Source and health risks of trace metals in *Clarias batrachus* and *Chrysichthys nigrodigitatus* from surface waters in Bayelsa State, Nigeria: A probabilistic model. *Frontiers in Sustainable Food Systems, 8*, 1419143. DOI: 10.3389/fsufs.2024.1419143

Izah, S. C., Uzoekwe, S. A., & Aigberua, A. O. (2021). Source, geochemical spreading and risks of trace metals in particulate matter 2.5 within a gas flaring area in Bayelsa State, Nigeria. *Advances in Environmental Technology, 7*(2), 101–118.

Jagtap, S., Trollman, H., Trollman, F., Garcia-Garcia, G., Parra-López, C., Duong, L., Martindale, W., Munekata, P. E. S., Lorenzo, J. M., Hdaifeh, A., Hassoun, A., Salonitis, K., & Afy-Shararah, M. (2022). The Russia-Ukraine Conflict: Its Implications for the Global Food Supply Chains. *Foods, 11*(14), 2098. DOI: 10.3390/foods11142098

Jahan, N., & Rahman, M. M. (2021). Photonic crystal fiber based biosensor for pseudomonas bacteria detection: A simulation study. *IEEE Access,* 9, 42206-15. DOI: 10.1109/ACCESS.2021.3063691

Jahnke, A., & Berger, U. (2009). Trace analysis of per- and polyfluorinated alkyl substances in various matrices—How do current methods perform? *Journal of Chromatography. A, 1216*(3), 410–421. DOI: 10.1016/j.chroma.2008.08.098 PMID: 18817914

Jambeck, J. R., Geyer, R., Wilcox, C., Siegler, T. R., Perryman, M., Andrady, A., Narayan, R., & Law, K. L. (2015). Plastic waste inputs from land into the ocean. *Science, 347*(6223), 768–771. DOI: 10.1126/science.1260352 PMID: 25678662

Jamieson, M., Trowbridge, A., Raffa, K., & Lindroth, R. (2012). Consequences of climate warming and altered precipitation patterns for plant-insect and multitrophic interactions. *Plant Physiology, 160*(4), 1719–1727. DOI: 10.1104/pp.112.206524

Jan, B., Shams, R., Rizvi, Q. E. H., & Manzoor, A. (2021). Ohmic heating technology for food processing: A review of recent developments. *Journal of Postharvest Technology, 9*(1), 20–34.

Janevska, D., Gospavić, R., Pacholewicz, E., & Popov, V. (2010). Application of an HACCP–QMRA approach for managing the impact of climate change on food quality and safety. *Food Research International, 43*(7), 1915–1924. DOI: 10.1016/j.foodres.2010.01.025

Jayaratne, W. M. S. C., Abeyratne, A. H. M. A. K., De Zoysa, H. K. S., Dissanayake, D. M. R. B. N., Bamunuarachchige, T. C., Waisundara, V. Y., & Chang, S. (2020). Detection and quantification of Aflatoxin B1 in corn and corn-grown soils in the district of Anuradhapura, Sri Lanka. *Heliyon, 6*(10), e05319. DOI: 10.1016/j.heliyon.2020.e05319 PMID: 33134588

Jenkins, H., & McBrien, J. (2011). Communicating for change: Development Communication and social transformation. *Communication, Culture & Critique, 4*(3), 299–316. DOI: 10.1111/j.1468-2885.2010.01275.x

Jiang, C., Shaw, K., Upperman, C., Blythe, D., Mitchell, C., Murtugudde, R., Sapkota, A. R., & Sapkota, A. (2015). Climate change, extreme events, and the increased risk of salmonellosis in Maryland, USA: Evidence for coastal vulnerability. *Environment International, 83*, 58–62. DOI: 10.1016/j.envint.2015.06.006

Jiang, D., Feng, Z., Jiang, H., Cao, H., Xiang, X., & Wang, L. (2024). 3D bio-printing-based vascular-microtissue electrochemical biosensor for fish parvalbumin detection. *Food Chemistry*, *18*, 138799. DOI: 10.1016/j.foodchem.2024.138799 PMID: 38401313

Jiang, D., Ge, P., Wang, L., Jiang, H., Yang, M., Yuan, L., Ge, Q., Fang, W., & Ju, X. (2019). A novel electrochemical mast cell-based paper biosensor for the rapid detection of milk allergen casein. *Biosensors & Bioelectronics*, *130*, 299–306. DOI: 10.1016/j.bios.2019.01.050 PMID: 30776617

Jiang, D., Jiang, H., & Wang, L. (2020). A novel paper-based capacitance mast cell sensor for evaluating peanut allergen protein Arah2. *Food Analytical Methods*, *13*(10), 1993–2001. DOI: 10.1007/s12161-020-01769-5

Jimenez, A., Fabra, M. J., Talens, P., & Chiralt, A. (2012). Edible and biodegradable starch films: A review. *Food and Bioprocess Technology*, *5*(6), 2058–2076. DOI: 10.1007/s11947-012-0835-4

Jin, C., Levi, R., Liang, Q., Renegar, N., Springs, S., Zhou, J., & Zhou, W. (2021). Testing at the source: Analytics-enabled risk-based sampling of food supply chains in China. *Management Science*, *67*(5), 2985–2996. DOI: 10.1287/mnsc.2020.3839

Jin, M., Wang, X., Ren, T., Wang, J., & Shan, J. (2021). Microplastics contamination in food and beverages: Direct exposure to humans. *Journal of Food Science*, *86*(7), 2816–2837. DOI: 10.1111/1750-3841.15802 PMID: 34146409

Jin, N., Song, Y., Ma, R., Li, J., Li, G., & Zhang, D. (2022). Characterization and identification of microplastics using Raman spectroscopy coupled with multivariate analysis. *Analytica Chimica Acta*, *1197*, 339519. DOI: 10.1016/j.aca.2022.339519 PMID: 35168726

Jin, Y., & Adhikari, A. (2025). Emerging and Innovative Technologies for the Sanitization of Fresh Produce: Advances, Mechanisms, and Applications for Enhancing Food Safety and Quality. *Foods*, *14*(11), 1924. DOI: 10.3390/foods14111924 PMID: 40509452

Jo, Robby, Kim, Lee, Lee, & Park. (2021). Reusable biosensor-based polymer dot-coated electrode surface for wireless detection of bacterial contamination. *Sensors and Actuators B: Chemical.* 1, 346:130503. DOI: 10.1016/j.snb.2021.130503

Johnson-Jennings, M., Jennings, D., Paul, K., & Little, M. (2020). Identifying needs and uses of digital indigenous food knowledge and practices for an indigenous food wisdom repository. *Alternative*, *16*(4), 290–299. DOI: 10.1177/1177180120954446

Jores, D., Arif, M., & Rahman, M. (2018). Factors Associated with Food Hygiene Practices among Street Food Vendors in Padawan, Sarawak. *Borneo Journal of Resource Science and Technology*, *8*(1), 56–65. DOI: 10.33736/bjrst.824.2018

Josephs, J. J. (2017). *Perceptions of validity: How knowledge is created, transformed and used in bio-agricultural technology safety testing for the development of government policies and regulations* (doctoral dissertation, College of Arts, Humanities, and Social Sciences), Nova Southeastern University.

Journalism University. (2025). "Understanding development communication: Concepts and goals". Available at: https://journalism.university/development-journalism-for-social-change/development-communication-concepts-goals/

Kachapulula, P., Akello, J., Bandyopadhyay, R., & Cotty, P. (2017). Aflatoxin contamination of groundnut and maize in Zambia: Observed and potential concentrations. *Journal of Applied Microbiology*, *122*(6), 1471–1482. DOI: 10.1111/jam.13448

Kadirvel, G., Banerjee, B., Meitei, S., Doley, S., Sen, A., & Muthukumar, M. (2018). Market Potential and Opportunities for Commercialization of Traditional Meat Products in the Northeast Hill Region of India. *Veterinary World*, *11*(2), 118–124. DOI: 10.14202/vetworld.2018.118-124

Kadirvel, V., Palanisamy, Y., & Ganesan, N. D. (2024). Active packaging system—An overview of recent advances for enhanced food quality and safety. *Packaging Technology & Science*, *38*(2), 145–162. DOI: 10.1002/pts.2863

Kaewprachu, P., Osako, K., & Rawdkuen, S. (2018). Effects of plasticizers on the properties of fish myofibrillar protein film. *Journal of Food Science and Technology*, *55*(8), 3046–3055. DOI: 10.1007/s13197-018-3226-7 PMID: 30065414

Kahrarian, Z., Mojarrab, M., & Shahbazi, Y. (2020). Investigation of Antioxidant Effects of Artemisia Sieberi Essential Oil In Iran. *Journal of Biotechnology & Bioinformatics Research*, *1–3*, 1–3. Advance online publication. DOI: 10.47363/JBBR/2020(2)117

Kalita, J. J., Sharma, P., & Bora, U. (2023). Recent developments in application of nucleic acid aptamer in food safety. *Food Control*, *145*, 109406. DOI: 10.1016/j.foodcont.2022.109406

Kampers, F. W. H. (2011). Opportunities for nanotechnologies in food production. *Acta Horticulturae*, (916), 113–117. DOI: 10.17660/ActaHortic.2011.916.11

Kang, Shi, Sun, Dan, Liang, Zhang, Su, Wang, & Zhang. (2022). Magnetic Nano separation technology for efficient control of Microorganisms and Toxins in Foods: A Review. *Journal of Agricultural and Food Chemistry*, 70(51), 16050-68. .DOI: 10.1021/acs.jafc.2c07132

Kannan, N. (2023). An analysis of the climate change effects on pesticide vapor drift from ground-based pesticide applications to cotton. *Scientific Reports*, *13*(1), 9740. Advance online publication. DOI: 10.1038/s41598-023-36941-4

Kanu, W., & Onyekwere, I. (2024). Perceptions of climate change-related disasters and impact on household food security in rural farm households in Imo State, Nigeria. *Journal of Agriculture and Food Sciences*, *21*(2), 121–134. DOI: 10.4314/jafs.v21i2.9

Kappenstein, O., Vieth, B., Luch, A., & Pfaff, K. (2012). Toxicologically relevant phthalates in food. *Molecular, Clinical and Environmental Toxicology: Volume 3. Environmental Toxicology*, *101*, 87–106. DOI: 10.1007/978-3-7643-8340-4_4 PMID: 22945567

Katsigiannis, A. S., Bayliss, D. L., & Walsh, J. L. (2022). Cold plasma for the disinfection of industrial food-contact surfaces: An overview of current status and opportunities. *Comprehensive Reviews in Food Science and Food Safety*, *21*(2), 1086–1124. DOI: 10.1111/1541-4337.12885 PMID: 35075786

Kattwinkel, M., Kühne, J., Foit, K., & Liess, M. (2011). Climate change, agricultural insecticide exposure, and risk for freshwater communities. *Ecological Applications*, *21*(6), 2068–2081. DOI: 10.1890/10-1993.1

Kaur, P., Devgan, K., Kumar, N., Kaur, A., Kumar, M., & Sandhu, K. (2021). Quality retention and shelf-life prolongation of cucumbers (Cucumis sativus L.) under different cool storage systems with passive modified atmosphere bulk packaging. *Packaging Technology & Science*, *34*(9), 567–578. DOI: 10.1002/pts.2595

Kedisso, E. G., Maredia, K., Guenthner, J., & Koch, M. (2022). Commercialization of genetically modified crops in Africa: Opportunities and challenges. *African Journal of Biotechnology*, *21*(5), 188–197. DOI: 10.5897/AJB2021.17434

Kemala, P., Idroes, R., Khairan, K., Ramli, M., Tallei, T. E., Helwani, Z., & Rahman, S. A. (2024). The Potent Antimicrobial Spectrum of Patchouli: Systematic Review of Its Antifungal, Antibacterial, and Antiviral Properties. *Malacca Pharmaceutics*, *2*(1), 10–17. DOI: 10.60084/mp.v2i1.156

Kennedy, G. L., & Symons, J. M. (2015). Carcinogenicity of Perfluoroalkyl Compounds. In DeWitt, J. C. (Ed.), *Toxicological Effects of Perfluoroalkyl and Polyfluoroalkyl Substances* (pp. 265–304). Springer International Publishing. DOI: 10.1007/978-3-319-15518-0_12

Kerry, O'Grady, & Hogan. (2006). Past, current and potential utilisation of active and intelligent packaging systems for meat and muscle-based products: A review. *Meat science*, 74(1), 113-30. DOI: 10.1016/j.meatsci.2006.04.024

Khachatryan, G., & Khachatryan, K. (2019). Starch based nanocomposites as sensors for heavy metals-detection of Cu2+ and Pb2+ ions. *International Agrophysics*, *33*(1), 121–126. DOI: 10.31545/intagr/104414

Khastini, R., Maryani, N., Haryati, S., Rahmah, R., Sa'ban, A., Aisyah, H., Fadillah, T. N., & Nadia, S. (2023). Understanding the Traditional Knowledge of Picungan Bandeng, a Fermented Fish Food, and Its Sustainability in Banten Province. *IOP Conference Series. Earth and Environmental Science*, *1241*(1), 012076. DOI: 10.1088/1755-1315/1241/1/012076

Khatib, M., Campo, M., Bellumori, M., Cecchi, L., Vignolini, P., Innocenti, M., & Mulinacci, N. (2023). Tannins from different parts of the chestnut trunk (castanea sativa mill.): A green and effective extraction method and their profiling by high-performance liquid chromatography-diode array detector-mass spectrometry. *ACS Food Science & Technology*, *3*(11), 1903–1912. DOI: 10.1021/acsfoodscitech.3c00272

Khiari, R., Zemni, H., Maury, C., & Mihoubi, D. (2020). Modeling desorption isotherms and thermodynamic properties of Italia grapes. *Journal of Food Processing and Preservation*, *44*(10), e14731. DOI: 10.1111/jfpp.14731

Khodabandeh Lu, N., Alamdari, H., Rashedi, H., Yazdian, F., Pourmadadi, M., Rahdar, A., & Romanholo Ferreira, L. F. (2025). Photocatalytic Bioreactors in Water Treatment: Mechanisms, Challenges, and Innovations in Microbial Inactivation. *Water, Air, and Soil Pollution*, *236*(5), 1–28. DOI: 10.1007/s11270-025-07919-x

Kigigha, L. T., Berezi, J., & Izah, S. C. (2017). Bacteriological quality assessment of meat pie sold in Yenagoa metropolis, Nigeria. *EC Nutrition*, *6*(6), 189–195.

Kigigha, L. T., Berezi, J., & Izah, S. C. (2017a). Bacteriological Quality Assessment of Meat Pie Sold in Yenagoa Metropolis, Nigeria. *EC Nutrition*, *6*(6), 189–195.

Kigigha, L. T., Ebieto, L. O., & Izah, S. C. (2017). Health risk assessment of heavy metal in smoked *Trachurus trachurus* sold in Yenagoa, Bayelsa State, Nigeria. *International Journal of Healthcare and Medical Sciences*, *3*(9), 62–69.

Kigigha, L. T., Igoya, U. O. S., & Izah, S. C. (2016). Microbiological quality assessment of unpeeled groundnut sold in Yenagoa metropolis, Nigeria. *International Journal of Innovative Biochemistry & Microbiology Research*, *4*(4), 11–22.

Kigigha, L. T., Igoya, U. O. S., & Izah, S. C. (2016). Microbiological Quality Assessment Of Unpeeled Groundnut Sold in Yenagoa Metropolis, Nigeria. *International Journal of Innovative Biochemistry & Microbiology Research*, *4*(4), 11–22.

Kigigha, L. T., Izah, S. C., & Kpea, T. B. (2015b). Microbiological quality of fermented Cassava Flakes (Gari) sold in Yenagoa, Metropolis, Nigeria. *Bulletin of Advanced Scientific Research*, *01*(07), 157–160.

Kigigha, L. T., Nyenke, P., & Izah, S. C. (2018). Health risk assessment of selected heavy metals in gari (cassava flake) sold in some major markets in Yenagoa metropolis, Nigeria. *MOJ Toxicology*, *4*(2), 47–52.

Kigigha, L. T., Opusunju, I. C., & Izah, S. C. (2017b). Assessment of Bacteriological Quality of Puff-Puff Sold in Amassoma, Bayelsa State, Nigeria. *Bulletin of Trends in Biological Sciences*, *1*(1), 18–22.

Kigigha, L. T., Ovunda, H. O., & Izah, S. C. (2015a). Microbiological quality assessment of suya sold in Yenagoa Metropolis, Nigeria. *Journal of Advances in Biological and Basic Research*, *01*(05), 106–109.

Kigigha, L. T., Samson, G. A., Izah, S. C., & Aseibai, E. R. (2018). Microbial Assessment of Zobo Drink Sold in Some Locations in Yenagoa Metropolis, Nigeria. *EC Nutrition*, *13*(7), 470–476.

Kim, G., Cha, D., Park, C., Lee, G., Jin, C., Lee, D., Suh, M.-S., Ahn, J.-B., Min, S.-K., Hong, S.-Y., & Kang, H. (2018). Future Changes in Extreme Precipitation Indices over Korea. *International Journal of Climatology*, *38*(S1). Advance online publication. DOI: 10.1002/joc.5414

Kim, N. N., Kim, W. J., & Kang, S. S. (2019a). Anti-biofilm effect of crude bacteriocin derived from Lactobacillus brevis DF01 on Escherichia coli and Salmonella Typhimurium. *Food Control*, *98*, 274–280. DOI: 10.1016/j.foodcont.2018.11.004

Kim, S. S., Park, S. H., Kim, S. H., & Kang, D. H. (2019b). Synergistic effect of ohmic heating and UV-C irradiation for inactivation of Escherichia coli O157: H7, Salmonella Typhimurium and Listeria monocytogenes in buffered peptone water and tomato juice. *Food Control*, *102*, 69–75. DOI: 10.1016/j.foodcont.2019.03.011

Klinger, C., Żółtowska-Aksamitowska, S., Wysokowski, M., Tsurkan, M. V., Galli, R., Petrenko, I., & Ehrlich, H. (2019). Express method for isolation of ready-to-use 3D chitin scaffolds from Aplysina archeri (Aplysineidae: Verongiida) Demosponge. *Marine Drugs, 17*(2), 131. DOI: 10.3390/md17020131 PMID: 30813373

Knipe, H., Temperton, B., Lange, A., Bass, D., & Tyler, C. R. (2021). Probiotics and competitive exclusion of pathogens in shrimp aquaculture. *Reviews in Aquaculture, 13*(1), 324–352. DOI: 10.1111/raq.12477

Knorr, D., Froehling, A., Jaeger, H., Reineke, K., Schlueter, O., & Schoessler, K. (2011). Emerging technologies in food processing. *Annual Review of Food Science and Technology, 2*(1), 203–235. DOI: 10.1146/annurev.food.102308.124129 PMID: 22129381

Kogovšek, U., Kovač, B., & Babnik, K. (2017). Perceptions of nanotechnology utilization in food: A pilot study. *Journal of Engineering & Processing Management, 9*(1), 10–14. DOI: 10.7251/JEPM1709010K

Kokkinos, C., Angelopoulou, M., Economou, A., Prodromidis, M., Florou, A., Haasnoot, W., Petrou, P., & Kakabakos, S. (2016). Lab-on-a-membrane foldable devices for duplex drop-volume electrochemical biosensing using quantum dot tags. *Analytical Chemistry, 88*(13), 6897–6904. DOI: 10.1021/acs.analchem.6b01625 PMID: 27257985

Kom, Z., Nicolau, M., & Nenwiini, S. (2024). The use of indigenous knowledge systems and practices to enhance food security in Vhembe District, South Africa. *Agricultural Research, 13*(3), 599–612. DOI: 10.1007/s40003-024-00716-8

Korkala, E., Hugg, T., & Jaakkola, J. (2014). Awareness of climate change and the dietary choices of young adults in Finland: A population-based cross-sectional study. *PLoS One, 9*(5), e97480. DOI: 10.1371/journal.pone.0097480

Koseki, S. (2016). Predictive modeling for estimation of bacterial behavior from farm to table. *Food Safety (Tokyo, Japan), 4*(2), 33–44. DOI: 10.14252/foodsafetyfscj.2016006

Kosoe, E. A., & Ogwu, M. C. (2024). Sustainable cassava processing: processes, techniques, wastes, and waste streams. In *Plant Biology, Sustainability and Climate Change, Sustainable Cassava.* Academic Press. DOI: 10.1016/B978-0-443-21747-0.00011-4

Kosoe, E. A., Achana, G. T. W., & Ogwu, M. C. (2023). Meta-Evaluation of the One Health Implication on Food Systems of Agrochemical Use. In *One Health Implications of Agrochemicals and their Sustainable Alternatives*. Springer. DOI: 10.1007/978-981-99-3439-3_14

Koutchma, T. (2014). *Food Plant Safety: UV Applications for Food and Non-Food Surfaces*. Elsevier.

Koutchma, T. (2019). *Ultraviolet light in food technology: principles and applications*. CRC press. DOI: 10.1201/9780429244414

Krauss, W. (2015). Anthropology in the Anthropocene: Sustainable Development, Climate Change and Interdisciplinary Research. In Greschke, H., & Tischler, J. (Eds.), *Grounding Global Climate Change* (pp. 59–76). Springer Netherlands. DOI: 10.1007/978-94-017-9322-3_4

Krishnamurthy, K., Tewari, J. C., Irudayaraj, J., & Demirci, A. (2010). Microscopic and spectroscopic evaluation of inactivation of Staphylococcus aureus by pulsed UV light and infrared heating. *Food and Bioprocess Technology*, *3*(1), 93–104. DOI: 10.1007/s11947-008-0084-8

Kroeker, K., Micheli, F., Gambi, M., & Martz, T. (2011). Divergent ecosystem responses within a benthic marine community to ocean acidification. *Proceedings of the National Academy of Sciences of the United States of America*, *108*(35), 14515–14520. DOI: 10.1073/pnas.1107789108

Krogh, P. H., Kostov, K., & Damgaard, C. F. (2020). The effect of *Bt* crops on soil invertebrates: A systematic review and quantitative meta-analysis. *Transgenic Research*, *29*(5-6), 487–498. DOI: 10.1007/s11248-020-00213-y

Ksouri, A., Klouz, A., Bouhaouala, B., Moussa, F., & Bezzarga, M. (2023). Docking-Based Evidence for the Potential of ImmunoDefender: A Novel Formulated Essential Oil Blend Incorporating Synergistic Antiviral Bioactive Compounds as Promising Mpro Inhibitors Against SARS-CoV-2. *Molecules*. https://doi.org/DOI: 10.20944/preprints202304.0323.v1

Kuhnlein, H. (2014). Food system sustainability for health and well-being of indigenous peoples. *Public Health Nutrition*, *18*(13), 2415–2424. DOI: 10.1017/S1368980014002961

Kumar, A., Kumar, V., Arsenov, D., Thakur, M., Kumar, A., Khokhar, A., Seth, C. S., & Kumar, R. (2024). The science of food safety and their health impacts. *Journal of Geochemical Exploration*, *267*, 107596. DOI: 10.1016/j.gexplo.2024.107596

Kumar, A., Singh, G. A., Gourkhede, P. H., Goutam, P. K., Laxman, T., Pandey, S. K., & Singh, P. (2023). Revolutionizing agriculture: A comprehensive review of nanotechnology applications. *International Journal of Environment and Climate Change, 13*(11), 3586–3603. DOI: 10.9734/ijecc/2023/v13i113534

Kumar, M. N. R. (2000). A review of chitin and chitosan applications. *Reactive & Functional Polymers, 46*(1), 1–27. DOI: 10.1016/S1381-5148(00)00038-9

Kumar, N., Kaur, P., & Bhatia, S. (2017). Advances in bio-nanocomposite materials for food packaging: A review. *Nutrition & Food Science, 47*(4), 591–606. DOI: 10.1108/NFS-11-2016-0176

Kumar, R., Manna, C., Padha, S., Verma, A., Sharma, P., Dhar, A., Ghosh, A., & Bhattacharya, P. (2022). Micro(nano)plastics pollution and human health: How plastics can induce carcinogenesis to humans? *Chemosphere, 298*, 134267. DOI: 10.1016/j.chemosphere.2022.134267 PMID: 35301996

Küníli, İ. E., Çolakoğlu, S., & Çolakoğlu, F. (2020). Levels of PAHs, PCBs, and toxic metals in *Ruditapes philippinarum* and *Donax trunculus* in Marmara Sea, Turkey. *Journal of the Science of Food and Agriculture, 101*(3), 1167–1173. DOI: 10.1002/jsfa.10728

Kuzukiran, O., Simsek, I., Filazi, A., & Yurdakok-Dikmen, B. (2022). Perfluorooctane sulfonate and perfluorooctanoic acid. In *Reproductive and Developmental Toxicology* (pp. 815–831). Elsevier. DOI: 10.1016/B978-0-323-89773-0.00041-2

Kwiatkowski, C. F., Andrews, D. Q., Birnbaum, L. S., Bruton, T. A., DeWitt, J. C., Knappe, D. R. U., Maffini, M. V., Miller, M. F., Pelch, K. E., Reade, A., Soehl, A., Trier, X., Venier, M., Wagner, C. C., Wang, Z., & Blum, A. (2020). Scientific Basis for Managing PFAS as a Chemical Class. *Environmental Science & Technology Letters, 7*(8), 532–543. DOI: 10.1021/acs.estlett.0c00255 PMID: 34307722

Kwon, J.-H., Kim, J.-W., Pham, T. D., Tarafdar, A., Hong, S., Chun, S.-H., Lee, S.-H., Kang, D.-Y., Kim, J.-Y., Kim, S.-B., & Jung, J. (2020). Microplastics in Food: A Review on Analytical Methods and Challenges. *International Journal of Environmental Research and Public Health, 17*(18), 6710. DOI: 10.3390/ijerph17186710 PMID: 32942613

Lake, I. (2017). Food-borne disease and climate change in the United Kingdom. *Environmental Health : A Global Access Science Source, 16*(S1), 117. Advance online publication. DOI: 10.1186/s12940-017-0327-0

Lake, I., Hooper, L., Abdelhamid, A., Bentham, G., Boxall, A., Draper, A., Fairweather-Tait, S., Hulme, M., Hunter, P. R., Nichols, G., & Waldron, K. (2012). Climate Change and Food Security: Health Impacts in Developed Countries. *Environmental Health Perspectives, 120*(11), 1520–1526. DOI: 10.1289/ehp.1104424

Lamani, N. A., & Ramaswamy, H. S. (2023). Composite Alginate–Ginger Oil Edible Coating for Fresh-Cut Pears. *Journal of Composites Science, 7*(6), 245. DOI: 10.3390/jcs7060245

Lambert, S., Scherer, C., & Wagner, M. (2017). Ecotoxicity testing of microplastics: Considering the heterogeneity of physicochemical properties. *Integrated Environmental Assessment and Management, 13*(3), 470–475. DOI: 10.1002/ieam.1901 PMID: 28440923

Lambert, S., & Wagner, M. (2018). Microplastics Are Contaminants of Emerging Concern in Freshwater Environments: An Overview. In Wagner, M., & Lambert, S. (Eds.), *Freshwater Microplastics* (Vol. 58, pp. 1–23). Springer International Publishing. DOI: 10.1007/978-3-319-61615-5_1

Lam, D., Hinz, E., Lang, D., Tengö, M., Wehrden, H., & Martín-López, B. (2020). Indigenous and local knowledge in sustainability transformations research: A literature review. *Ecology and Society, 25*(1), art3. Advance online publication. DOI: 10.5751/ES-11305-250103

Lam, W. R., & Liu, Y. (2020). Tackling Small and Medium-Sized Enterprises (SMEs) Financing in China. *Annals of Economics and Finance, 21*(1).

Lankanayaka, A. A. K., Matharage, D. S., Senanayake, C. M., Jayathunge, K. G. L. R., Bandara, P. C., & Manatunga, D. C. (2024). Extraction of bioactive compounds from food waste and their potential applications in the food industry: A review. *Brazilian Journal of Development, 10*(4), e69187. DOI: 10.34117/bjdv10n4-061

Lapworth, D. J., Baran, N., Stuart, M. E., & Ward, R. S. (2012). Emerging organic contaminants in groundwater: A review of sources, fate, and occurrence. *Environmental Pollution, 163*, 287–303. DOI: 10.1016/j.envpol.2011.12.034 PMID: 22306910

Lassa, J., Teng, P., Caballero-Anthony, M., & Shrestha, M. (2018). Revisiting emergency food reserve policy and practice under disaster and extreme climate events. *International Journal of Disaster Risk Science, 10*(1), 1–13. DOI: 10.1007/s13753-018-0200-y

Lau, T. C., Chan, M. W., Tan, H. P., & Kwe, C. L. (2013). Functional food: A growing trend among the health conscious. *Asian Social Science, 9*(1), 198. DOI: 10.5539/ass.v9n1p198

Le Blanc, D. (2015). Towards integration at last? The Sustainable Development Goals as a network of targets. *Sustainable Development (Bradford), 23*(3), 176–187. DOI: 10.1002/sd.1582

Lee, L., Tung, S., Gan, W., & Kaur, S. (2022). Knowledge, attitude, and perception of climate change and dietary choices in a predominantly Chinese university student population in Klang Valley. *Malaysian Applied Biology, 51*(3), 37–45. DOI: 10.55230/mabjournal.v51i3.2275

Leggieri, M., Giorni, P., Pietri, A., & Battilani, P. (2019). Aspergillus flavus and Fusarium verticillioides interaction: Modeling the impact on mycotoxin production. *Frontiers in Microbiology, 10*, 2653. Advance online publication. DOI: 10.3389/fmicb.2019.02653

Leite, Y., Duque, T., Santos, J., & Santos, E. (2025). Potential residual pesticide consumption: A stratified analysis of Brazilian families. *Journal of Xenobiotics, 15*(2), 37. DOI: 10.3390/jox15020037

Levkoe, C., McLaughlin, J., & Strutt, C. (2021). Mobilizing Networks and Relationships through Indigenous Food Sovereignty: The Indigenous Food Circle's Response to the COVID-19 Pandemic in Northwestern Ontario. *Frontiers in Communication, 6*, 672458. Advance online publication. DOI: 10.3389/fcomm.2021.672458

Liao, W., Badri, W., Dumas, E., Ghnimi, S., Elaïssari, A., Saurel, R., & Gharsallaoui, A. (2021). Nanoencapsulation of essential oils as natural food antimicrobial agents: An overview. *Applied Sciences (Basel, Switzerland), 11*(13), 5778. DOI: 10.3390/app11135778

Li, C., Zwiers, F., Zhang, X., Chen, G., Lu, J., Li, G., Norris, J., Tan, Y., Sun, Y., & Liu, M. (2019). Larger increases in more extreme local precipitation events as climate warms. *Geophysical Research Letters, 46*(12), 6885–6891. DOI: 10.1029/2019GL082908

Li, F., Liu, J., Ren, J., Cao, X., & Zhu, Y. (2016). Numerical investigation of airborne contaminant transport under different vortex structures in the aircraft cabin. *International Journal of Heat and Mass Transfer, 96*, 287–295. DOI: 10.1016/j.ijheatmasstransfer.2016.01.004

Li, J. H., Yu, W. J., Lai, Y. H., & Ko, Y. C. (2012). Major food safety episodes in Taiwan: Implications for the necessity of international collaboration on safety assessment and management. *The Kaohsiung Journal of Medical Sciences, 28*(7S). Advance online publication. DOI: 10.1016/j.kjms.2012.05.004

Li, L., Shen, Y., Yuan, Z., & Li, J. (2018). Analysis and research of food safety risk assessment based on data mining. In *Proceedings of the 2018 International Conference on Big Data Engineering and Technology* (pp. 68–72). DOI: 10.1145/3297730.3297748

Lindner, L. F. (2008). Regulating food safety: The power of alignment and drive towards convergence. *Innovation (Abingdon)*, *21*(2), 133–143. DOI: 10.1080/13511610802214487

Lin, W., & Liang, J. (2024). Regulatory decentralization and food safety: Evidence from China. *American Journal of Agricultural Economics*, *107*(1), 54–80. DOI: 10.1111/ajae.12484

Li, R., Yue, J., Zhu, F., Zhou, J., & Liu, X. (2025). Application progress of carbon quantum dot composites in fluorescent detection of food safety. *Journal of Food Science*, *90*(5), e70299. Advance online publication. DOI: 10.1111/1750-3841.70299 PMID: 40433944

Li, T., Luo, Y., & Zhao, T. (2024). Research on enhancing the efficiency of food safety sampling inspections in China based on Pareto's law. *Journal of the Science of Food and Agriculture*, *105*(1), 235–243. DOI: 10.1002/jsfa.13822

Liu, C., Hofstra, N., & Franz, E. (2016). Impacts of climate and management variables on the contamination of preharvest leafy greens with Escherichia coli. *Journal of Food Protection*, *79*(1), 17–29. DOI: 10.4315/0362-028X.JFP-15-255

Liu, Q., Chen, Z., Chen, Y., Yang, F., Yao, W., & Xie, Y. (2021). Microplastics and Nanoplastics: Emerging Contaminants in Food. *Journal of Agricultural and Food Chemistry*, *69*(36), 10450–10468. DOI: 10.1021/acs.jafc.1c04199 PMID: 34473500

Liu, X., Bu, Y., Li, M., & Li, J. (2021). Is interdisciplinary collaboration research more disruptive than monodisciplinary research? *Proceedings of the Association for Information Science and Technology*, *58*(1), 264–272. DOI: 10.1002/pra2.454

Liu, Y., Ahmed, S., Sameen, D. E., Lu, Y., Wang, R., Dai, J., & Qin, W. (2021). A review of cellulose and its derivatives in biopolymer-based for food packaging application. *Trends in Food Science & Technology*, *112*, 532–546. DOI: 10.1016/j.tifs.2021.04.016

Li, Y., Tao, L., Wang, Q., Wang, F., Li, G., & Song, M. (2023). Potential Health Impact of Microplastics: A Review of Environmental Distribution, Human Exposure, and Toxic Effects. *Environment & Health*, *1*(4), 249–257. DOI: 10.1021/envhealth.3c00052 PMID: 39474495

Loecke, T., Burgin, A., Riveros-Iregui, D., Ward, A., Thomas, S., Davis, C., & Clair, M. (2017). Weather whiplash in agricultural regions drives deterioration of water quality. *Biogeochemistry, 133*(1), 7–15. DOI: 10.1007/s10533-017-0315-z

Lohita, B., & Srijaya, M. (2024). Novel technologies for shelf-life extension of food products as a competitive advantage: A review. *Food production, diversity, and safety under climate change*, 285-306.

Lohmann, R. (2017). Microplastics are not important for the cycling and bioaccumulation of organic pollutants in the oceans—But should microplastics be considered POPs themselves? *Integrated Environmental Assessment and Management, 13*(3), 460–465. DOI: 10.1002/ieam.1914 PMID: 28440937

Lorenzen, M., Weiser, A., Pieper, R., Lahrssen-Wiederholt, M., & Numata, J. (2021). Introducing the Rapid Alert Supply Network Extractor (RASNEX) tool to mine supply chain information from food and feed contamination notifications in Europe. *PLoS One, 16*(7), e0254301. DOI: 10.1371/journal.pone.0254301

Loretz, M., Stephan, R., & Zweifel, C. (2011). Antibacterial activity of decontamination treatments for cattle hides and beef carcasses. *Food Control, 22*(3-4), 347–359. DOI: 10.1016/j.foodcont.2010.09.004

Loring, P., & Gerlach, S. (2015). Searching for progress on food security in the North American North: A research synthesis and meta-analysis of the peer-reviewed literature + supplementary appendix (see article tools). *Arctic, 68*(3), 380. DOI: 10.14430/arctic4509

Lovo, E., Woodward, L., Larkins, S., Preston, R., & Baba, U. (2021). Indigenous knowledge around the ethics of human research from the Oceania region: A scoping literature review. *Philosophy, Ethics, and Humanities in Medicine; PEHM, 16*(1), 12. Advance online publication. DOI: 10.1186/s13010-021-00108-8

Lozada-Martinez, I. D., Hernandez-Paz, D. A., Fiorillo-Moreno, O., Picón-Jaimes, Y. A., & Bermúdez, V. (2025). Meta-research in biomedical investigation: Gaps and opportunities based on meta-research publications and global indicators in health, science, and human development. *Publications / MDPI, 13*(1), 7. DOI: 10.3390/publications13010007

Loza, K., Sengstock, C., Chernousova, S., Köller, M., & Epple, M. (2014). The predominant species of ionic silver in biological media is colloidally dispersed nanoparticulate silver chloride. *RSC Advances, 4*(67), 35290. DOI: 10.1039/C4RA04764H

Lozano, A. M. C., & Arias, M. D. C. (2009). Residuos de fármacos en alimentos de origen animal: Panorama actual en Colombia. *Revista Colombiana de Ciencias Pecuarias*, *21*(1), 15. DOI: 10.17533/udea.rccp.324262

Macdiarmid, J., & Whybrow, S. (2019). Nutrition from a climate change perspective. *The Proceedings of the Nutrition Society*, *78*(3), 380–387. DOI: 10.1017/S0029665118002896

Madanayake, N. H., Hossain, A., & Adassooriya, N. M. (2021). Nanobiotechnology for agricultural sustainability, and food and environmental safety. *Quality Assurance and Safety of Crops & Foods*, *13*(1), 20–36. DOI: 10.15586/qas.v13i1.838

Mafe, A. N., & Büsselberg, D. (2024). Mycotoxins in Food: Cancer Risks and Strategies for Control. *Foods*, *13*(21), 3502. DOI: 10.3390/foods13213502

Mahmudatusa'adah, A., & Sudewi, S. (2017). Hazard analysis critical control point (HACCP) in sweet potato's liquid sugar. *Innovation of Vocational Technology Education*, *12*(2). Advance online publication. DOI: 10.17509/invotec.v12i2.6203

Mahunu, G. K., Osei-Kwarteng, M., Ogwu, M. C., & Afoakwah, N. A. (2024). Safe food handling techniques to prevent microbial contamination. In *Food safety and quality in the global south* (pp. 427–461). Springer Nature Singapore. DOI: 10.1007/978-981-97-2428-4_14

Mainero Rocca, L., Gentili, A., Pérez-Fernández, V., & Tomai, P. (2017). Veterinary drugs residues: A review of the latest analytical research on sample preparation and LC-MS based methods. *Food Additives & Contaminants. Part A, Chemistry, Analysis, Control, Exposure & Risk Assessment*, 1–19. DOI: 10.1080/19440049.2017.1298846 PMID: 28278127

Malavi, D., Abong, G., & Muzhingi, T. (2021). Effect of food safety training on behavior change of food handlers: A case of orange-fleshed sweetpotato purée processing in Kenya. *Food Control*, *119*, 107500. DOI: 10.1016/j.foodcont.2020.107500

Malchi, T., Maor, Y., Tadmor, G., Shenker, M., & Chefetz, B. (2014). Irrigation of Root Vegetables with Treated Wastewater: Evaluating Uptake of Pharmaceuticals and the Associated Human Health Risks. *Environmental Science & Technology*, *48*(16), 9325–9333. DOI: 10.1021/es5017894 PMID: 25026038

Malcom, A. (2024). Integrating indigenous knowledge into livestock policy formulation. *International Journal of Livestock Policy*, *3*(1), 30–42. DOI: 10.47941/ijlp.1962

Malhi, G., Kaur, M., & Kaushik, P. (2021). Impact of climate change on agriculture and its mitigation strategies: A review. *Sustainability (Basel)*, *13*(3), 1318. DOI: 10.3390/su13031318

Malhotra, B. D., Srivastava, S., & Augustine, S. (2015). Biosensors for food toxin detection: Carbon nanotubes and graphene. *Proceedings of the Materials Research Society, 1725*, mrsf14–1725. DOI: 10.1557/opl.2015.165

Mali, R. P., Rao, P. S., & Vikhe, D. N. (2020). Wound Healing Activity of Calotropis Procera Root Bark on Diabetic Rats. *Journal of Drug Delivery and Therapeutics, 10*(2-s), 86–89. DOI: 10.22270/jddt.v10i2-s.3955

Malli, A., Monteith, H., Hiscock, C., Smith, E., Fairman, K., Galloway, T., & Mashford-Pringle, A. (2023). Impacts of Colonization on Indigenous Food Systems in Canada and the United States: A Scoping Review. *BMC Public Health, 23*(1), 2105. Advance online publication. DOI: 10.1186/s12889-023-16997-7

Mamun, A. A., Prasetya, T. A. E., Dewi, I. R., & Ahmad, M. (2023). Microplastics in human food chains: Food becoming a threat to health safety. *The Science of the Total Environment, 858*, 159834. DOI: 10.1016/j.scitotenv.2022.159834 PMID: 36461575

Manatunga, D. C., Dassanayake, R. S., & Liyanage, R. N. (2022). Novel acumens into biodegradation: impact of nanomaterials and their contribution. DOI: 10.5772/intechopen.98771

Mandal, R., Mohammadi, X., Wiktor, A., Singh, A., & Pratap Singh, A. (2020). Applications of pulsed light decontamination technology in food processing: An overview. *Applied Sciences (Basel, Switzerland), 10*(10), 3606. DOI: 10.3390/app10103606

Manikandan, V. S., Adhikari, B., & Chen, A. (2018). Nanomaterial based electrochemical sensors for the safety and quality control of food and beverages. *Analyst, 143*(19), 4537–4554. DOI: 10.1039/C8AN00497H PMID: 30113611

Mani-López, E., García, H. S., & López-Malo, A. (2012). Organic acids as antimicrobials to control Salmonella in meat and poultry products. *Food Research International, 45*(2), 713–721. DOI: 10.1016/j.foodres.2011.04.043

Mantzourani, I., Daoutidou, M., & Alexopoulos, A. (2025). The Antimicrobial Effect of Thymol and Carvacrol in Combination with Organic Acids Against Foodborne Pathogens in Chicken and Beef Meat Fillets. *Microorganisms, 13*(1), 182. DOI: 10.3390/microorganisms13010182 PMID: 39858950

Marsili, D., Pasetto, R., Iavarone, I., Fazzo, L., Zona, A., & Comba, P. (2021). Fostering environmental health literacy in contaminated sites: National and local experience in Italy from a public health and equity perspective. *Frontiers in Communication, 6*, 697547. Advance online publication. DOI: 10.3389/fcomm.2021.697547

Marsteller, N., Bøgh, K. L., Goodman, R. E., & Epstein, M. M. (2015). A review of animal models used to evaluate potential allergenicity of genetically modified organisms (GMOs). *Drug Discovery Today. Disease Models, 17*, 81–88. DOI: 10.1016/j.ddmod.2016.11.001

Martínez-Megías, C., Mentzel, S., Fuentes-Edfuf, Y., Moe, S., & Rico, A. (2022). A Bayesian network approach to assess the influence of climate change and pesticide use practices on the ecological risks of pesticides in a protected Mediterranean wetland. https://doi.org/DOI: 10.1101/2022.12.15.520567

Martin, P., Wen, Y., Woods, A., Fayida, T., & Hobbs, S. (2024). Exploring pesticide transport, groundwater, and environmental justice in a changing climate: A community-engaged research approach. *Environmental Research Letters, 19*(9), 094009. DOI: 10.1088/1748-9326/ad67f3

Martirosyan, A., & Schneider, Y. J. (2014). Engineered nanomaterials in food: Implications for food safety and consumer health. *International Journal of Environmental Research and Public Health, 11*(6), 5720–5750. DOI: 10.3390/ijerph110605720 PMID: 24879486

Martischang, R., Godycki-Ćwirko, M., Kowalczyk, A., Kosiek, K., Turjeman, A., Babich, T., Shiber, S., Leibovici, L., Von Dach, E., Harbarth, S., & Huttner, A. (2021). Risk factors for treatment failure in women with uncomplicated lower urinary tract infection. *PLoS One, 16*(8), e0256464. DOI: 10.1371/journal.pone.0256464

Marvin, H. J., Janssen, E. M., Bouzembrak, Y., Hendriksen, P. J., & Staats, M. (2016). Big data in food safety: An overview. *Critical Reviews in Food Science and Nutrition, 57*(11), 2286–2295. DOI: 10.1080/10408398.2016.1257481

Masekoameng, M., & Molotja, M. (2023). Contribution of indigenous food preservation and processing practices to food security of rural households in Sekhukhune District of Limpopo Province, South Africa. *African Journal of Food, Agriculture, Nutrition and Development, 23*(122), 24115–24135. DOI: 10.18697/ajfand.122.21705

Masipa, T. (2017). The Impact of Climate Change on Food Security in South Africa: Current Realities and Challenges Ahead. *Jàambá, 9*(1). Advance online publication. DOI: 10.4102/jamba.v9i1.411

Massoud, R., Sharifan, A., Khosravi-Darani, K., & Asadi, G. (2020). Cadmium Bioremoval by Saccharomyces cerevisiae in Milk. *Journal of Medical Microbiology and Infectious Diseases, 8*(1), 29–33. DOI: 10.29252/JoMMID.8.1.29

Masumuzzaman, M., Evivie, S. E., Ogwu, M. C., Bailiang, L., Jin-Cheng, D., Wan, L., Guicheng, H., Fei, L., & Song, W. (2021). Genomic and *in vitro* properties of the dairy *Streptococcus thermophilus* SMQ-301 strain against selected pathogens. *Food & Function, 12*(8), 7017–7028. DOI: 10.1039/D0FO02951C

Matheron, A., Guerault, M. N., Vazquez, R., Cheron, M., Brossard, D., & Crauste-Manciet, S. (2020). Microbiological stability tests with simulated broth preparations and integrity testing for sterile standard cytotoxic preparations. *Pharmaceutical Technology in Hospital Pharmacy, 5*(1). Advance online publication. DOI: 10.1515/pthp-2020-0002

Ma, X., Cai, D., Chen, Q., Zhu, Z., Zhang, S., Wang, Z., Hu, Z., Shen, H., & Meng, Z. (2024). Hunting Metabolic Biomarkers for Exposure to Per- and Polyfluoroalkyl Substances: A Review. *Metabolites, 14*(7), 392. DOI: 10.3390/metabo14070392 PMID: 39057715

Ma, Y. (2024). Study on food locking and packaging by nanomaterials. *Applied and Computational Engineering, 56*(1), 180–185. DOI: 10.54254/2755-2721/56/20240655

McCarter, J., Gavin, M., Baereleo, S., & Love, M. (2014). The Challenges of Maintaining Indigenous Ecological Knowledge. *Ecology and Society, 19*(3), art39. Advance online publication. DOI: 10.5751/ES-06741-190339

McClements, D. J. (2020). Nanotechnology approaches for improving the healthiness and sustainability of the modern food supply. *ACS Omega, 5*(46), 29623–29630. DOI: 10.1021/acsomega.0c04050 PMID: 33251398

Medina, Á., Akbar, A., Baazeem, A., Rodríguez, A., & Magan, N. (2017). Climate change, food security, and mycotoxins: Do we know enough? *Fungal Biology Reviews, 31*(3), 143–154. DOI: 10.1016/j.fbr.2017.04.002

Meghwar, P., Saeed, S. M. G., Forte, L., Smaoui, S., Khalid, N. I., De Palo, P., & Maggiolino, A. (2024). Electrolyzed Water: A Promising Strategy for Improving Food Quality and Safety of Fruits, Vegetables, and Meat. *Journal of Food Quality, 2024*(1), 3272823. DOI: 10.1155/2024/3272823

Mercer, J., Kelman, I., Alfthan, B., & Kurvits, T. (2012). Ecosystem-based adaptation to climate change in Caribbean Small Island Developing States: Integrating Local and External Knowledge. *Sustainability (Basel), 4*(8), 1908–1932. DOI: 10.3390/su4081908

Mercer, J., Kelman, I., Taranis, L., & Suchet-Pearson, S. (2009). Framework for integrating indigenous and scientific knowledge for disaster risk reduction. *Disasters*, *34*(1), 214–239. DOI: 10.1111/j.1467-7717.2009.01126.x

Methuselah, J. (2024). Digital twin technology for smart manufacturing. *Journal of Technology and Systems*, *6*(4), 52–65. DOI: 10.47941/jts.2143

Michnik, K., Thompson, S., and Beardy, B. (2021). Moving your body, soul, and heart to share and harvest food. Canadian Food Studies / La Revue Canadienne Des Études Sur L Alimentation, 8(2). DOI: 10.15353/cfs-rcea.v8i2.446

Mihai, F.-C., Gündoğdu, S., Khan, F. R., Olivelli, A., Markley, L. A., & Van Emmerik, T. (2022). Plastic pollution in marine and freshwater environments: Abundance, sources, and mitigation. In *Emerging Contaminants in the Environment* (pp. 241–274). Elsevier. DOI: 10.1016/B978-0-323-85160-2.00016-0

Milia, A., Buonocunto, F., Mali, M., Giandomenico, S., Leo, A., Spada, L., Ferraro, L., & Giordano, L. (2023). Natural vs. Anthropogenic Metals for Reconstructing the Source-to-Sink Pathway (Naples Bay, Eastern Tyrrhenian Sea). *Frontiers in Earth Science (Lausanne)*, *11*, 1225658. Advance online publication. DOI: 10.3389/feart.2023.1225658

Miller, F. A., Silva, C. L., & Brandao, T. R. (2013). A review on ozone-based treatments for fruit and vegetables preservation. *Food Engineering Reviews*, *5*(2), 77–106. DOI: 10.1007/s12393-013-9064-5

Mills, J., Gage, K., & Khan, A. (2010). Potential influence of climate change on vector-borne and zoonotic diseases: A review and proposed research plan. *Environmental Health Perspectives*, *118*(11), 1507–1514. DOI: 10.1289/ehp.0901389

Min, K., & Hong, W. (2021). The effect of food sustainability and the food safety climate on the job stress, job satisfaction, and job commitment of kitchen staff. *Sustainability (Basel)*, *13*(12), 6813. DOI: 10.3390/su13126813

Misra, N. N., Tiwari, B. K., Raghavarao, K. S. M. S., & Cullen, P. J. (2011). Nonthermal plasma inactivation of food-borne pathogens. *Food Engineering Reviews*, *3*(3), 159–170. DOI: 10.1007/s12393-011-9041-9

Misra, S., Mandliya, S., & Mishra, H. N. (2025). Evaluation of Stability of Coencapsulated Probiotics, γ-Aminobutyric Acid (GABA), and Physicochemical Properties of Functional Bar Matrix during Storage and Digestion Using a Chemometric Approach. *ACS Food Science & Technology*, *5*(6), 2371–2380. DOI: 10.1021/acsfoodscitech.5c00151

Mlalila, N., Kadam, D. M., Swai, H., & Hilonga, A. (2016). Transformation of food packaging from passive to innovative via nanotechnology: Concepts and critiques. *Journal of Food Science and Technology*, *53*(9), 3395–3407. DOI: 10.1007/s13197-016-2325-6 PMID: 27777446

Mmbando, G. S. (2023). The legal aspect of the current use of genetically modified organisms in Kenya, Tanzania, and Uganda. *GM Crops and Food: Biotechnology in Agriculture and the Food Chain*, *14*(1), 1–12. DOI: 10.1080/21645698.2023.2208999

Modonesi, C., & Gusmeroli, F. (2018). Evidence-based science or science-based evidence? The GM crops between false myths and ecological systems. *Organisms. The Journal of Biological Sciences*, *2*(1), 63–72. DOI: 10.13133/2532-5876_3.13

Mo, G., Sun, Y., Huang, S., Gao, F., Huang, J., Lai, L., Wang, G., & Yan, Y. (2024). Reconfigurable digital twin from the perspective of smart manufacturing systems in Industry 4.0/5.0. *Journal of Physics: Conference Series*, *2832*(1), 012007. DOI: 10.1088/1742-6596/2832/1/012007

Mohamad, A. A. (2019). Physical properties of quasi-solid-state polymer electrolytes for dye-sensitised solar cells: A characterisation review. *Solar Energy*. DOI: 10.1016/j.foodchem.2016.07.004

Mohamad, A., Rizwan, M., Keasberry, N. A., & Ahmed, M. U. (2019). Fabrication of label-free electrochemical food biosensor for the sensitive detection of ovalbumin on nanocomposite-modified graphene electrode. *Biointerface Research in Applied Chemistry*, *9*(6), 4655–4662. DOI: 10.33263/BRIAC96.655662

Mohamad, A., Rizwan, M., Keasberry, N. A., Nguyen, A. S., Dai Lam, T., & Ahmed, M. U. (2020). Gold-microrods/Pd-nanoparticles/polyaniline-nanocomposite-interface as a peroxidase-mimic for sensitive detection of tropomyosin. *Biosensors & Bioelectronics*, *155*, 112108. DOI: 10.1016/j.bios.2020.112108 PMID: 32217333

Mohanty, L., Singh, A., Pandey, A., Kumar, R., Ramesh, G., Swamy, G., Pandey, S. K., & Singh, B. (2024). Harnessing nanotechnology for eco-friendly crop enhancement and sustainable agriculture. *Journal of Experimental Agriculture International*, *46*(7), 154–167. DOI: 10.9734/jeai/2024/v46i72568

Mohapatra, D., Mishra, S., Giri, S., & Kar, A. (2013). Application of hurdles for extending the shelf life of fresh fruits. *Trends in Post-Harvest Technology*, *1*(1), 37–54.

Morais, S., Tortajada-Genaro, L. A., Maquieira, A., & Martinez, M. A. (2020). Biosensors for food allergy detection according to specific IgE levels in serum. *Trends in Analytical Chemistry*, *127*, 115904. DOI: 10.1016/j.trac.2020.115904

Morales-Díaz, A. B., Ortega-Ortíz, H., Juárez-Maldonado, A., Cadenas-Pliego, G., González-Morales, S., & Benavides-Mendoza, A. (2017). Application of nanoelements in plant nutrition and its impact in ecosystems. *Advances in Natural Sciences Nanoscience and Nanotechnology*, *8*(1), 013001. DOI: 10.1088/2043-6254/8/1/013001

Moresis, A., Restivo, L., Bromilow, S., Flik, G., Rosati, G., Scorrano, F., Tsoory, M., O'Connor, E. C., Gaburro, S., & Bannach-Brown, A. (2024). A minimal metadata set (MNMS) to repurpose nonclinical in vivo data for biomedical research. *Lab Animal*, *53*(3), 67–79. DOI: 10.1038/s41684-024-01335-0

Mosharaf, M. K., Gomes, R. L., Cook, S., Alam, M. S., & Rasmusssen, A. (2024). Wastewater reuse and pharmaceutical pollution in agriculture: Uptake, transport, accumulation and metabolism of pharmaceutical pollutants within plants. *Chemosphere*, *364*, 143055. DOI: 10.1016/j.chemosphere.2024.143055 PMID: 39127189

Mostafa, I., Abbas, H. A., Ashour, M. L., Yasri, A., El-Shazly, A. M., Wink, M., & Sobeh, M. (2020). Polyphenols from Salix tetrasperma Impair Virulence and Inhibit Quorum Sensing of Pseudomonas aeruginosa. *Molecules (Basel, Switzerland)*, *25*(6), 1341. DOI: 10.3390/molecules25061341 PMID: 32187997

Moulton, A., & Schramm, P. (2017). Climate Change and Public Health Surveillance: Toward a Comprehensive Strategy. *Journal of Public Health Management and Practice*, *23*(6), 618–626. DOI: 10.1097/PHH.0000000000000550

Muharremi, H., Raka, L., Spahiu, J., Tershnjaku, I., & Topi, D. (2022). Investigation of aflatoxin M1 in baby milk and aflatoxin B1 in infant cereals marketed in Kosovo. *Journal of Food Processing and Preservation*, *46*(6). Advance online publication. DOI: 10.1111/jfpp.16285

Mukherjee, A., Gómez-Sala, B., O'Connor, E. M., Kenny, J. G., & Cotter, P. D. (2022). Global regulatory frameworks for fermented foods: A review. *Frontiers in Nutrition*, *9*, 902642. DOI: 10.3389/fnut.2022.902642 PMID: 35719144

Muller, M. (2018). Promoting or protecting traditional knowledges? tensions in the resurgence of indigenous food practices on vancouver island. *International Indigenous Policy Journal*, *9*(4). Advance online publication. DOI: 10.18584/iipj.2018.9.4.4

Muñoz, I., Rodríguez, C., Gillet, D., & Moerschbacher, B. (2018). Life cycle assessment of chitosan production in India and Europe. *The International Journal of Life Cycle Assessment*, *23*, 1151–1160. DOI: 10.1007/s11367-017-1290-2

Munoz-Muriedas, J. (2021). Large scale meta-analysis of preclinical toxicity data for target characterisation and hypotheses generation. *PLoS One, 16*(6), e0252533. DOI: 10.1371/journal.pone.0252533

Murayi, J. A., Evenson, E., Britton, C., Gehred, A., & Goday, P. S. (2025). Clinical effects of pediatric commercial food-based formulas: A systematic review. *Journal of Pediatric Gastroenterology and Nutrition, 80*(3), 501–509. DOI: 10.1002/jpn3.12450

Nabi, B. G., Mukhtar, K., Arshad, R. N., Radicetti, E., Tedeschi, P., Shahbaz, M. U., Walayat, N., Nawaz, A., Inam-Ur-Raheem, M., & Aadil, R. M. (2021). High-pressure processing for sustainable food supply. *Sustainability (Basel), 13*(24), 13908. DOI: 10.3390/su132413908

Nagaraj, K., Radha, S., Deepa, C. G., Raja, K., Umapathy, V., Badgujar, N. P., Parekh, N. M., Manimegalai, T., Devi, L. A., & Uthra, C. (2025). Photocatalytic advancements and applications of titanium dioxide (TiO_2): Progress in biomedical, environmental, and energy sustainability. *Next Research, 2*(1), 100180. DOI: 10.1016/j.nexres.2025.100180

Naik, G., Haider, S. Z., Bhandari, U., Lohani, H., & Chauhan, N. (2021). Comparative Analysis of In vitro Antimicrobial and Antioxidant Potential of Cinnamomum tamala Extract and their Essential Oils of Two Different Chemotypes. *Agricultural Science Digest - A Research Journal.* DOI: 10.18805/ag.D-5187

Nang'Ayo, F., Simiyu-Wafukho, S., & Oikeh, S. O. (2014). Regulatory challenges for GM crops in developing economies: The African experience. *Transgenic Research, 23*(6), 1049–1055. DOI: 10.1007/s11248-014-9805-0

Narayanan, A., Peter, B., & Keellings, D. (2024). A Climate Extremes Resilience Index for the Conterminous United States. *Weather, Climate, and Society, 16*(1), 87–103. DOI: 10.1175/WCAS-D-23-0008.1

Naseri, I., & Yourdkhani, M. (2022). Rapid and energy-efficient frontal curing of multifunctional composites using integrated nanostructured heaters. *ACS Applied Materials & Interfaces, 14*(44), 50215–50224. DOI: 10.1021/acsami.2c15415 PMID: 36226889

Nasir-Naeem, K. O., Akande, S. A., Garba, R., & Akolade, J. O. (2021). Assessment of milling as an alternative to chemical additives in processing of jute mallow (Corchorus olitorius). *AROC in Food and Nutrition, 1*(2), 10–17. DOI: 10.53858/arocfn01021017

Nawaz, N., Wen, S., Wang, F., Nawaz, S., Raza, J., Iftikhar, M., & Usman, M. (2022). Lysozyme and its application as antibacterial agent in food industry. *Molecules (Basel, Switzerland), 27*(19), 6305. DOI: 10.3390/molecules27196305 PMID: 36234848

Ndung'u, J. K., Nguta, J. M., Mapenay, I. M., & Moriasi, G. A. (2024). A Comprehensive Review of Ethnomedicinal Uses, Phytochemistry, Pharmacology, and Toxicity of Prunus africana (Hook. F.) Kalkman from Africa. *Scientifica, 2024*, 1–17. DOI: 10.1155/2024/8862996 PMID: 38654751

Neufeld, H., Richmond, C., & Centre, S. (2017). Impacts of place and social spaces on traditional food systems in southwestern ontario. *International Journal of Indigenous Health, 12*(1), 93–115. DOI: 10.18357/ijih112201716903

Nielsen, K., Clayton, S., Stern, P., Dietz, T., Capstick, S., & Whitmarsh, L. (2021). How Psychology Can Help Limit Climate Change. *The American Psychologist, 76*(1), 130–144. DOI: 10.1037/amp0000624

Niewczas-Dobrowolska, M., Sikora, T., & Prusak, A. (2019). Ethics and trust while food safety crisis. *Proceedings on Engineering Sciences, 1*(2), 257–260. DOI: 10.24874/PES01.02.022

Nikolaou, A. (2013). Pharmaceuticals and related compounds as emerging pollutants in water: Analytical aspects. *Global NEST Journal, 15*(1), 1–12. DOI: 10.30955/gnj.000969

Nilsson, M., Griggs, D., & Visbeck, M. (2016). Policy: Map the interactions between Sustainable Development Goals. *Nature, 534*(7607), 320–322. DOI: 10.1038/534320a

Nizzy, A. M., & Ogwu, M. C. (2024). Valorization of cassava processing by-products into biofuel for a sustainable environment. In *Plant Biology, Sustainability and Climate Change, Sustainable Cassava.* Academic Press. DOI: 10.1016/B978-0-443-21747-0.00010-2

Nji, Q., Babalola, O., Ekwomadu, T., Nleya, N., & Mwanza, M. (2022). Six main contributing factors to high levels of mycotoxin contamination in african foods. *Toxins, 14*(5), 318. DOI: 10.3390/toxins14050318

Nji, Q., Babalola, O., & Mwanza, M. (2022). Aflatoxins in maize: Can their occurrence be effectively managed in Africa in the face of climate change and food insecurity? *Toxins, 14*(8), 574. DOI: 10.3390/toxins14080574

Nji, Q., & Mwanza, M. (2024). Three-year multi-mycotoxin analysis of south african commercial maize from three provinces. *Frontiers in Fungal Biology, 5*, 1426782. Advance online publication. DOI: 10.3389/ffunb.2024.1426782

Nleya, N., Adetunji, M., & Mwanza, M. (2018). Current Status of Mycotoxin Contamination of Food Commodities in Zimbabwe. *Toxins, 10*(5), 89. DOI: 10.3390/toxins10050089

Nnamani, C., Ajayi, S., Oselebe, H., Atkinson, C., Igboabuchi, A., & Ezigbo, E. (2017). Sphenostylis stenocarpa (ex. A. Rich.) harms., a fading genetic resource in a changing climate: Prerequisite for conservation and sustainability. *Plants, 6*(3), 30. DOI: 10.3390/plants6030030

Nuttavuthisit, K., & Thøgersen, J. (2015). The importance of consumer trust for the emergence of a market for green products: The case of organic food. *Journal of Business Ethics, 140*(2), 323–337. DOI: 10.1007/s10551-015-2690-5

Nwizugbo, K. C., Ogwu, M. C., Eriyamremu, G. E., & Ahana, C. M. (2023). Alterations in energy metabolism, total protein, uric and nucleic acids in African sharptooth catfish (*Clarias gariepinus* Burchell.) exposed to crude oil and fractions. *Chemosphere, 316*, 137778. DOI: 10.1016/j.chemosphere.2023.137778 PMID: 36640975

Obahiagbon, E. G., & Ogwu, M. C. (2023). Consumer Perception and Demand for Sustainable Herbal Medicine Products and Market. In *Herbal Medicine Phytochemistry.* Springer. DOI: 10.1007/978-3-031-21973-3_65-1

Obahiagbon, E. G., & Ogwu, M. C. (2024). Organic Food Preservatives: The Shift Towards Natural Alternatives and Sustainability in the Global South's Markets. In Ogwu, M. C., Izah, S. C., & Ntuli, N. R. (Eds.), *Food Safety and Quality in the Global South* (pp. 299–329). Springer. DOI: 10.1007/978-981-97-2428-4_10

Obi, F. O., Ugwuishiwu, B. O., & Nwakaire, J. N. (2016). Agricultural waste concept, generation, utilization and management. *Nigerian Journal of Technology, 35*(4), 957–964. DOI: 10.4314/njt.v35i4.34

Odubo, T. C., & Izah, S. C. (2024). Safety considerations of trace metals in locally produced nutritive food-drinks consumed in Yenagoa metropolis, Nigeria. *Biological Trace Element Research, 203*(8), 4408–4419. Advance online publication. DOI: 10.1007/s12011-024-04488-8

Oelofse, S., & Nahman, A. (2012). Estimating the magnitude of food waste generated in South Africa. *Waste Management & Research, 31*(1), 80–86. DOI: 10.1177/0734242X12457117

Ofosu, G., Dittmann, A., Sarpong, D., & Botchie, D. (2020). Socio-economic and environmental implications of artisanal and small-scale mining (ASM) on agriculture and livelihoods. *Environmental Science & Policy, 106*, 210–220. DOI: 10.1016/j.envsci.2020.02.005

Ogamba, E. N., Charles, E. E., & Izah, S. C. (2021). Distributions, pollution evaluation and health risk of selected heavy metal in surface water of Taylor creek, Bayelsa State, Nigeria. *Toxicology and Environmental Health Sciences, 13*(2), 109–121. DOI: 10.1007/s13530-020-00076-0

Ogwu, M. C. (2023). Local Food Crops in Africa: Sustainable Utilization, Threats, and Traditional Storage Strategies. In *Sustainable Utilization and Conservation of Africa's Biological Resources and Environment.* Springer. DOI: 10.1007/978-981-19-6974-4_13

Ogwu, M. C., & Kulkarni, S. (2024). Sustainable cassava: An overview. In *Plant Biology, Sustainability and Climate Change, Sustainable Cassava.* Academic Press. DOI: 10.1016/B978-0-443-21747-0.00024-2

Ogwu, M. C., & Osawaru, M. E. (2022). Traditional Methods of Plant Conservation for Sustainable Utilization and Development. In *Biodiversity in Africa: Potentials, Threats and Conservation.* Springer. DOI: 10.1007/978-981-19-3326-4_17

Ogwu, M. C., & Osawaru, M. E. (2023). Disease Outbreaks in Ex-Situ Plant Conservation and Potential Management Strategies. In *Sustainable Utilization and Conservation of Africa's Biological Resources and Environment.* Springer. DOI: 10.1007/978-981-19-6974-4_18

Ogwu, M. C., Iyiola, A. O., & Izah, S. C. (2023). Overview of African Biological Resources and Environment. In *Sustainable Utilization and Conservation of Africa's Biological Resources and Environment.* Springer. DOI: 10.1007/978-981-19-6974-4_1

Ogwu, M. C., Izah, S. C., & Iyiola, A. O. (2022). An Overview of the Potentials, Threats and Conservation of Biodiversity in Africa. In *Biodiversity in Africa: Potentials, Threats and Conservation.* Springer. DOI: 10.1007/978-981-19-3326-4_1

Ogwu, M. C., Osawaru, M. E., Obadiaru, O., & Itsepopo, I. C. (2016). Nutritional composition of some uncommon indigenous spices and vegetables. *International Conference on Natural Resource Development and Utilization (4th Edition): Positioning small and medium enterprises (SMEs) for optimal utilization of available resources and economic growth.* Raw Materials Research and Development Council (Federal Ministry of Science and Technology), Maitama, Abuja Nigeria.

Ogwu, M.C., Onah, T.P., Kosoe, E.A., Mumuni, E., Akolgo-Azupogo, Helen, Obahiagbon, E.G., Izah, S.C., & Imarhiagbe, O. (2024b). Factors Driving the Acceptance of Genetically Modified Food Crops in Ghana. *Food Safety and Health.* DOI: 10.1002/fsh3.12031

Ogwu, M. C. (2019a). Lifelong Consumption of Plant-Based GM Foods: Is It Safe? In Papadopoulou, P., Misseyanni, A., & Marouli, C. (Eds.), *Environmental Exposures and Human Health Challenges* (pp. 158–176). IGI Global. DOI: 10.4018/978-1-5225-7635-8.ch008

Ogwu, M. C. (2019b). Understanding the Composition of Food Waste: An "-Omics" Approach to Food Waste Management. In Gunjal, A. P., Waghmode, M. S., Patil, N. N., & Bhatt, P. (Eds.), *Global Initiatives for Waste Reduction and Cutting Food Loss* (pp. 212–236). IGI Global. DOI: 10.4018/978-1-5225-7706-5.ch011

Ogwu, M. C., Ahana, C. M., & Osawaru, M. E. (2018). Sustainable food production in Nigeria: A case study for Bambara groundnut (*Vigna subterranean* (L.) Verdc. Fabaceae). *Journal of Energy and Natural Resources Management, 1*, 68–77.

Ogwu, M. C., Osawaru, M. E., & Owie, M. O. (2023). Effects of Storage at Room Temperature on the Food Components of Three Cocoyam Species (*Colocasia esculenta, Xanthosoma atrovirens, and X. sagittifolium*). *Food Studies, 13*(2), 59–83. DOI: 10.18848/2160-1933/CGP/v13i02/59-83

Ohimain, E. I., & Izah, S. C. (2015). Physicochemical characteristics and microbial population of palm oil sold in major markets in Yenagoa metropolis, Bayelsa States, Nigeria. *Journal of Environmental Treatment Techniques, 3*(3), 143–147.

Ohoro, C. R., Adeniji, A. O., Okoh, A. I., & Okoh, O. O. (2019). Distribution and Chemical Analysis of Pharmaceuticals and Personal Care Products (PPCPs) in the Environmental Systems: A Review. *International Journal of Environmental Research and Public Health, 16*(17), 3026. DOI: 10.3390/ijerph16173026 PMID: 31438569

Ojha, K. S., Kerry, J. P., Duffy, G., Beresford, T., & Tiwari, B. K. (2015). Technological advances for enhancing quality and safety of fermented meat products. *Trends in Food Science & Technology, 44*(1), 105–116. DOI: 10.1016/j.tifs.2015.03.010

Okechukwuyem Ojji, S. (2024). Emerging Technology Integration—Artificial Intelligence (AI) and Machine Learning (ML) for Predictive Analysis for Safety and Toxicity Assessment in Environmental Toxicology. *International Journal of Scientific Research and Management, 12*(05), 1182–1195. DOI: 10.18535/ijsrm/v12i05.ec03

Oladipo, O. H., Ibrahim, R. R., Adeboye, S. E., & Kuiper, H. (2020). Readiness of the Nigerian public for the introduction of genetically modified crops into the food market. *African Journal of Biotechnology*, *19*(7), 426–438. DOI: 10.5897/AJB2020.17136

Oldenkamp, R., Benestad, R., Hader, J., Mentzel, S., Nathan, R., Madsen, A., & Moe, S. (2023). Incorporating climate projections in the environmental risk assessment of pesticides in aquatic ecosystems. *Integrated Environmental Assessment and Management*, *20*(2), 384–400. DOI: 10.1002/ieam.4849

Oliveira, M., Tiwari, B. K., & Duffy, G. (2020). Emerging technologies for aerial decontamination of food storage environments to eliminate microbial cross-contamination. *Foods*, *9*(12), 1779. DOI: 10.3390/foods9121779 PMID: 33266230

Omari, R., Bellingrath-Kimura, S., Addo, E., Oikawa, Y., & Fujii, Y. (2018). Exploring farmers' indigenous knowledge of soil quality and fertility management practices in selected farming communities of the guinea savannah agro-ecological zone of ghana. *Sustainability (Basel)*, *10*(4), 1034. DOI: 10.3390/su10041034

Omorotionmwan, F. O.-O., Ogwu, H. I., & Ogwu, M. C. (2019). Antibacterial characterization and bacteria composition of pineapple (*Ananas comosus* [Linn.] Merr.) peel and pulp. *Food and Health*, *5*(1), 1–11. DOI: 10.3153/FH19001

Omotayo, O., Omotayo, A., Mwanza, M., & Babalola, O. (2019). Prevalence of mycotoxins and their consequences on human health. *Toxicological Research*, *35*(1), 1–7. DOI: 10.5487/TR.2019.35.1.001

Ongu, I., Olayide, P., Alexandersson, E., Mugwanya Zawedde, B., & Eriksson, D. (2023). Biosafety regulatory frameworks in Kenya, Nigeria, Uganda and Sweden and their potential impact on international R&D collaborations. *GM Crops and Food: Biotechnology in Agriculture and the Food Chain*, *14*(1), 1–17. DOI: 10.1080/21645698.2023.2194221

Onoabhagbe, O., Ogwu, M. C., & Ikhajiagbe, B. (2024). Germination characteristics of *Sorghum bicolor* (L.) Moench. under different pH regimes after chemo-priming. *Vegetos*, *37*(5), 1876–1886. Advance online publication. DOI: 10.1007/s42535-024-00909-0

Onomu, A. (2023). Pitfalls and potential pathways to commercialization of indigenous food crops, fruits, and vegetables in Africa. *Asian Journal of Agriculture and Rural Development*, *13*(1), 25–38. DOI: 10.55493/5005.v13i1.4716

Onwudiegwu, C., & Izah, S. C. (2025). From Science to Policy: Safeguarding Food Security and Public Health through Strengthened Soil Pollution Regulations. *ES General.*, *7*, 1363. DOI: 10.30919/esg1363

Onyeaka, H., Miri, T., Hart, A., Anumudu, C., & Nwabor, O. F. (2023). Application of ultrasound technology in food processing with emphasis on bacterial spores. *Food Reviews International*, *39*(7), 3663–3675. DOI: 10.1080/87559129.2021.2013255

Onyeaka, H., Passaretti, P., Miri, T., & Al-Sharify, Z. T. (2022). The safety of nanomaterials in food production and packaging. *Current Research in Food Science*, *5*, 763–774. DOI: 10.1016/j.crfs.2022.04.005 PMID: 35520272

Oprea, I., Fărca , A. C., Leopold, L. F., Diaconeasa, Z., Coman, C., & Socaci, S. A. (2022). Nano-encapsulation of citrus essential oils: Methods and applications of interest for the food sector. *Polymers*, *14*(21), 4505. DOI: 10.3390/polym14214505 PMID: 36365499

Oriekhoe, O. I., Ilugbusi, B. S., & Adisa, O. (2024). Ensuring global food safety: Integrating blockchain technology into food supply chains. *Engineering Science & Technology Journal*, *5*(3), 811–820. DOI: 10.51594/estj.v5i3.905

Orozco, D., Alarcon-Rojo, A. D., Chavez-Mendoza, C., Luna, L., Carrillo-Lopez, L. M., & Ronquillo, O. (2021). Fat replacement by pecan nut and oregano oil and their impact on the physicochemical properties and consumer acceptability of frankfurters. *Animal Bioscience*, *34*(10), 1674–1683. DOI: 10.5713/ab.20.0622 PMID: 33332948

Orutugu,, L.A., & Izah,, S.C., & Aseibai. (2015). Microbiological quality of Kunu drink sold in some major markets of Yenagoa Metropolis, Nigeria. *Continental Journal of Biomedical Science.*, *9*(1), 9–16.

Osawaru, M. E., & Ogwu, M. C. (2020). Survey of plant and plant products in local markets within Benin City and environs. In Filho, L. W., Ogugu, N., Ayal, D., Adelake, L., & da Silva, I. (Eds.), *African Handbook of Climate Change Adaptation* (pp. 1–24). DOI: 10.1007/978-3-030-42091-8_159-1

Osawaru, M. E., Ogwu, M. C., & Aigbefue, D. (2014). Survey of ornamental gardens in five local government areas of Southern Edo State Nigeria. *The Bioscientist*, *2*(1), 87–102.

Ovuru, K. F., Izah, S. C., Ogidi, O. I., Imarhiagbe, O., & Ogwu, M. C. (2024). Slaughterhouse facilities in developing nations: Sanitation and hygiene practices, microbial contaminants and sustainable management system. *Food Science and Biotechnology*, *33*(3), 519–537. DOI: 10.1007/s10068-023-01406-x

Palla, U., & Iwunwa, M. G. C. (2025). Leveraging predictive models to enhance infection control and reduce cross-contamination in public health settings. *International Journal of Research Publication and Reviews*, *6*(1), 407–422. DOI: 10.55248/gengpi.6.0125.0212

Pal, N., Sharma, P., Kumawat, M., Singh, S., Verma, V., Tiwari, R. R., Sarma, D. K., Nagpal, R., & Kumar, M. (2024). Phage therapy: An alternative treatment modality for MDR bacterial infections. *Infectious Diseases*, *56*(10), 785–817. DOI: 10.1080/23744235.2024.2379492 PMID: 39017931

Pankaj, S. K., Wan, Z., & Keener, K. M. (2018). Effects of cold plasma on food quality: A review. *Foods*, *7*(1), 4. DOI: 10.3390/foods7010004 PMID: 29301243

Pan, M., Xie, X., Liu, K., Yang, J., Hong, L., & Wang, S. (2020). Fluorescent carbon quantum dots—Synthesis, functionalization and sensing application in food analysis. *Nanomaterials (Basel, Switzerland)*, *10*(5), 930. DOI: 10.3390/nano10050930 PMID: 32403325

Parhusip, A. J. N., Wijaya, T. A., & Irawati, W. (2021). *Reduction of Heavy Metals in Squid, Green Mussels, White Shrimp Using Combination of Saccharomyces cerevisiae and Candida albicans*. DOI: 10.2991/absr.k.210621.072

Parise, I. (2018). A brief review of global climate change and the public health consequences. *Australian Journal of General Practice*, *47*(7), 451–456. DOI: 10.31128/AJGP-11-17-4412

Park, Y. W., Kim, S. M., Lee, J. Y., & Jang, W. (2015). Application of biosensors in smart packaging. *Molecular & Cellular Toxicology*, *09*(11), 277–285. DOI: 10.1007/s13273-015-0027-1

Pasecnaja, E., Bartkevics, V., & Zacs, D. (2022). Occurrence of selected per- and polyfluorinated alkyl substances (PFASs) in food available on the European market – A review on levels and human exposure assessment. *Chemosphere*, *287*, 132378. DOI: 10.1016/j.chemosphere.2021.132378 PMID: 34592212

Patial, R., & Sobti, R. C. (2024). Exploring the impact of meta-analysis in scientific research: A review. *Medinformatics*, *00*(00), 1–11. DOI: 10.47852/bonviewMEDIN42022447

Patil, A., Somvanshi, S. B., Sawale, M., Desai, K., Ozadali, F., Benyathiar, P., & Mishra, D. K. (2025). Advances in Microbubble Technologies for Food Sanitization: A Comprehensive Review. *Comprehensive Reviews in Food Science and Food Safety*, *24*(4), e70230. DOI: 10.1111/1541-4337.70230 PMID: 40661042

Peelman, N., Ragaert, P., De Meulenaer, B., Adons, D., Peeters, R., Cardon, L., & Devlieghere, F. (2013). Application of bioplastics for food packaging. *Trends in Food Science & Technology*, *32*(2), 128–141. DOI: 10.1016/j.tifs.2013.06.003

Perrone, G., Ferrara, M., Medina, Á., Pascale, M., & Magan, N. (2020). Toxigenic fungi and mycotoxins in a climate change scenario: Ecology, genomics, distribution, prediction, and prevention of the risk. *Microorganisms*, *8*(10), 1496. DOI: 10.3390/microorganisms8101496

Persaud, J., Wannamaker, K., Stark, K., Lambert, C., Harrison, C., & Keller, N. (2025). Decolonizing education: Advancing indigenous student success through culturally responsive practices in Ontario. *Alternative*, *21*(2), 264–273. DOI: 10.1177/11771801251340657

Peters, R., Brandhoff, P., Weigel, S., Marvin, H., Bouwmeester, H., Aschberger, K., Rauscher, H., Amenta, V., Arena, M., Botelho Moniz, F., Gottardo, S., & Mech, A. (2014). Inventory of nanotechnology applications in the agricultural, feed and food sector. *EFSA Supporting Publications*, *11*(7). Advance online publication. DOI: 10.2903/sp.efsa.2014.EN-621

Phillipps, B., Skinner, K., Parker, B., & Neufeld, H. (2022). An intersectionality-based policy analysis examining the complexities of access to wild game and fish for urban indigenous women in northwestern Ontario. *Frontiers in Communication*, *6*, 762083. Advance online publication. DOI: 10.3389/fcomm.2021.762083

Pickson, R., & Boateng, E. (2021). Climate Change: A Friend or Foe to Food Security in Africa? *Environment, Development and Sustainability*, *24*(3), 4387–4412. DOI: 10.1007/s10668-021-01621-8

Pilnik, M., Argentim, T., Kinupp, V., Haverroth, M., & Ming, L. (2023). Traditional botanical knowledge: Food plants from the Huni Kuĩ indigenous people, Acre, Western Brazilian Amazon. Rodriguésia, 74. https://doi.org/DOI: 10.1590/2175-7860202374016

Pironti, C., Ricciardi, M., Motta, O., Miele, Y., Proto, A., & Montano, L. (2021). Microplastics in the Environment: Intake through the Food Web, Human Exposure and Toxicological Effects. *Toxics*, *9*(9), 224. DOI: 10.3390/toxics9090224 PMID: 34564375

Pirozzi, A., Pataro, G., Donsì, F., & Ferrari, G. (2021). Edible coating and pulsed light to increase the shelf life of food products. *Food Engineering Reviews*, *13*(3), 544–569. DOI: 10.1007/s12393-020-09245-w

Piva, E., Fais, P., Ioime, P., Forcato, M., Viel, G., Cecchetto, G., & Pascali, J. P. (2023). Per- and polyfluoroalkyl substances (PFAS) presence in food: Comparison among fresh, frozen and ready-to-eat vegetables. *Food Chemistry*, *410*, 135415. DOI: 10.1016/j.foodchem.2023.135415 PMID: 36652797

Popken, L., Griffin, P., Coté, C., & Angel, É. (2023). Indigenous Food Sovereignty through Resurgent Self-Governance: Centering Nuuchah-nulth Principles in Sea Otter Management in Canada. *Ecology and Society*, *28*(2), art12. Advance online publication. DOI: 10.5751/ES-13702-280212

Prasad, R., Bhattacharyya, A., & Nguyen, Q. D. (2017). Nanotechnology in sustainable agriculture: Recent developments, challenges, and perspectives. *Frontiers in Microbiology*, *8*, 1014. Advance online publication. DOI: 10.3389/fmicb.2017.01014 PMID: 28676790

Pratiwi, R., Ramadhanti, S. P., Amatulloh, A., Megantara, S., & Subra, L. (2023). Recent Advances in the Determination of Veterinary Drug Residues in Food. *Foods*, *12*(18), 3422. DOI: 10.3390/foods12183422 PMID: 37761131

Principal Food Safety Regulations. (2020). In International Finance Corporation, Food Safety Handbook: A Practical Guide for Building a Robust Food Safety Management System (pp. 23–54). The World Bank. DOI: 10.1596/978-1-4648-1548-5_ch2

Prodanović, R., Nedić, S., Vujanac, I., Bojkovski, J., Nedić, S., Jovanović, L., Kirovski, D., & Borozan, S. (2023). Dietary Supplementation of Chestnut Tannins in Prepartum Dairy Cows Improves Antioxidant Defense Mechanisms Interacting with Thyroid Status. *Metabolites*, *13*(3), 334. DOI: 10.3390/metabo13030334 PMID: 36984774

Prosser, R. S., & Sibley, P. K. (2015). Human health risk assessment of pharmaceuticals and personal care products in plant tissue due to biosolids and manure amendments, and wastewater irrigation. *Environment International*, *75*, 223–233. DOI: 10.1016/j.envint.2014.11.020 PMID: 25486094

Prüst, M., Meijer, J., & Westerink, R. H. S. (2020). The plastic brain: Neurotoxicity of micro- and nanoplastics. *Particle and Fibre Toxicology*, *17*(1), 24. DOI: 10.1186/s12989-020-00358-y PMID: 32513186

Puckowski, A., Cwięk, W., Mioduszewska, K., Stepnowski, P., & Białk-Bielińska, A. (2021). Sorption of pharmaceuticals on the surface of microplastics. *Chemosphere*, *263*, 127976. DOI: 10.1016/j.chemosphere.2020.127976 PMID: 32835979

Pujiono, C., Taufiq Mukti, A., & Hastuti Satyantini, W. (2021). Study of Pb Heavy Metal Pollution Level on Tannin Content of Seaweed (Kappaphycus alvarezii) in Bluto and Talango Sea Waters, Sumenep, East Java. *Journal of Aquaculture and Fish Health, 10*(1), 1. DOI: 10.20473/jafh.v10i1.17088

Purohit, S. R., Sharma, V., Kumari, M., Muthukumarappan, K., & Kane-Potaka, J. (Eds.). (2024). *Future Crops and Processing Technologies for Sustainability and Nutritional Security* (1st ed.). CRC Press. DOI: 10.1201/9781003280170

Putri, L. A., Prabowo, Y. D., Dewi, D. M. M., Mumtazah, Z., Adila, F. P., Fadillah, G., Amrillah, T., Triyana, K., Nugroho, F. A. A., & Wasisto, H. S. (2024). Review of noble metal nanoparticle-based colorimetric sensors for food safety monitoring. *ACS Applied Nano Materials, 7*(17), 19821–19853. DOI: 10.1021/acsanm.4c04327

Puza, B. (2015). *Bayesian methods for statistical analysis.* DOI: 10.26530/OAP-EN_611011

Rabiepour, A., Zahmatkesh, F., & Babakhani, A. (2024). Preservation techniques to increase the shelf life of seafood products: An overview. *Journal of Food Engineering and Technology, 13*(1), 1–24. DOI: 10.32732/jfet.2024.13.1.1

Raja, I. S., Vedhanayagam, M., Preeth, D. R., Kim, C., Lee, J. H., & Han, D. W. (2021). Development of two-dimensional nanomaterials based electrochemical biosensors on enhancing the analysis of food toxicants. *International Journal of Molecular Sciences, 22*(6), 3277. DOI: 10.3390/ijms22063277 PMID: 33806998

Rajeshkumar, S., & Malarkodi, C. (2014). In vitro antibacterial activity and mechanism of silver nanoparticles against foodborne pathogens. *Bioinorganic Chemistry and Applications, 2014*, 1–10. DOI: 10.1155/2014/581890 PMID: 25313307

Ramazanidoroh, F., Hosseininezhad, M., Shahrampour, D., & Wu, X. (2024). Edible packaging as a functional carrier of prebiotics, probiotics, and postbiotics to boost food safety, quality, and shelf life. *Probiotics and Antimicrobial Proteins, 16*(4), 1327–1347. DOI: 10.1007/s12602-023-10110-5 PMID: 37389789

Rameshrad, M., Razavi, B. M., & Hosseinzadeh, H. (2017). Protective effects of green tea and its main constituents against natural and chemical toxins: A comprehensive review. *Food and Chemical Toxicology, 100*, 115–137. DOI: 10.1016/j.fct.2016.11.035 PMID: 27915048

Ramírez Carnero, A., Lestido-Cardama, A., Vazquez Loureiro, P., Barbosa-Pereira, L., Rodríguez Bernaldo De Quirós, A., & Sendón, R. (2021). Presence of Perfluoroalkyl and Polyfluoroalkyl Substances (PFAS) in Food Contact Materials (FCM) and Its Migration to Food. *Foods, 10*(7), 1443. DOI: 10.3390/foods10071443 PMID: 34206351

Ramírez-Morales, D., Masís-Mora, M., Montiel-Mora, J. R., Cambronero-Heinrichs, J. C., Pérez-Rojas, G., Tormo-Budowski, R., Méndez-Rivera, M., Briceño-Guevara, S., Gutiérrez-Quirós, J. A., Arias-Mora, V., Brenes-Alfaro, L., Beita-Sandí, W., & Rodríguez-Rodríguez, C. E. (2021). Multi-residue analysis of pharmaceuticals in water samples by liquid chromatography- mass spectrometry: Quality assessment and application to the risk assessment of urban-influenced surface waters in a metropolitan area of Central America. *Process Safety and Environmental Protection, 153*, 289–300. DOI: 10.1016/j.psep.2021.07.025

Ramos, C. V., Samelo, J., Martins, P. A., & Moreno, M. J. (2022). Protein Quantification in Complex Matrices. *Journal of Chemical Education, 99*(3), 1488–1496. DOI: 10.1021/acs.jchemed.2c00109

Rana, M. J., Riaz, U., Iqbal, R. K., Deeba, F., Iqbal, N., & Ali, E. (2023). Silver nanoparticles and their potential antimicrobial applications in food industry. *The Journal of Microbiology and Molecular Genetics, 4*(1), 59–84. DOI: 10.52700/jmmg.v4i1.81

Ranjbaran, M., Carciofi, B. A., & Datta, A. K. (2021). Engineering modeling frameworks for microbial food safety at various scales. *Comprehensive Reviews in Food Science and Food Safety, 20*(5), 4213–4249. DOI: 10.1111/1541-4337.12818

Ranta, J., Siekkinen, K. M., Nuotio, L., Laukkanen, R., Hellström, S., Korkeala, H., & Maijala, R. (2010). Causal hidden variable model of pathogenic contamination from pig to pork. *Statistical Modelling, 10*(1), 69–87. DOI: 10.1177/1471082X0801000105

Rao, A. (2022). Industry-funded research and bias in food science. *Quantitative Marketing and Economics, 20*(1), 39–67. DOI: 10.1007/s11129-021-09244-z

Rashidi, L., & Khosravi-Darani, K. (2011). The applications of nanotechnology in food industry. *Critical Reviews in Food Science and Nutrition, 51*(8), 723–730. DOI: 10.1080/10408391003785417 PMID: 21838555

Rastogi, S., Kumari, V., Sharma, V., & Ahmad, F. J. (2021). Gold nanoparticle-based sensors in food safety applications. *Food Analytical Methods, 15*(2), 468–484. DOI: 10.1007/s12161-021-02131-z

Ravichandran, R. (2010). Nanotechnology applications in food and food processing: Innovative green approaches, opportunities and uncertainties for global market. *International Journal of Green Nanotechnology Physics and Chemistry, 1*(2), 72–P96. DOI: 10.1080/19430871003684440

Rebezov, M., Farhan Jahangir Chughtai, M., Mehmood, T., Khaliq, A., Tanweer, S., Semenova, A., Khayrullin, M., Dydykin, A., Burlankov, S., Thiruvengadam, M., Shariati, M. A., & Lorenzo, J. M. (2021). Novel techniques for microbiological safety in meat and fish industries. *Applied Sciences (Basel, Switzerland), 12*(1), 319. DOI: 10.3390/app12010319

Rebezov, M., Saeed, K., Khaliq, A., Rahman, S. J. U., Sameed, N., Semenova, A., Khayrullin, M., Dydykin, A., Abramov, Y., Thiruvengadam, M., Shariati, M. A., Bangar, S. P., & Lorenzo, J. M. (2022). Application of electrolyzed water in the food industry: A review. *Applied Sciences (Basel, Switzerland), 12*(13), 6639. DOI: 10.3390/app12136639

Rehman, A. U., Crimi, M., & Andreescu, S. (2023). Current and emerging analytical techniques for the determination of PFAS in environmental samples. *Trends in Environmental Analytical Chemistry, 37*, e00198. DOI: 10.1016/j.teac.2023.e00198

Ren, Zhang, Li, Jian, Zhao, & Song. (2021). Development of a pulse-induced electrochemical biosensor based on gluconamide for Gram-negative bacteria detection. *Microchimica Acta*, 11, 188:399. .DOI: 10.1016/j.bioelechem.2022.108226

Rendueles, C., Duarte, A. C., Escobedo, S., Fernández, L., Rodríguez, A., García, P., & Martínez, B. (2022). Combined use of bacteriocins and bacteriophages as food biopreservatives. A review. *International Journal of Food Microbiology, 368*, 109611. DOI: 10.1016/j.ijfoodmicro.2022.109611 PMID: 35272082

Revelou, P. K., Tsakali, E., Batrinou, A., & Strati, I. F. (2025). Applications of Machine Learning in Food Safety and HACCP Monitoring of Animal-Source Foods. *Foods, 14*(6), 922. DOI: 10.3390/foods14060922 PMID: 40231903

Rhodes, L., Smith, K., Murray, J., Nishimura, T., & Finch, S. (2020). Ciguatera fish poisoning: The risk from an Aotearoa/New Zealand perspective. *Toxins, 12*(1), 50. DOI: 10.3390/toxins12010050

Richard, G., Izah, S. C., & Ogwu, M. C. (2022). Implications of Artisanal Crude Oil Refining on Sustainable Food Production in the Niger Delta Region of Nigeria. *Journal of Environmental Bioremediation and Toxicology, 5*(2), 69–77. DOI: 10.54987/jebat.v5i2.775

Rincón, A., Santamaría-Pérez, B., Rabasa, S., Coince, A., Marçais, B., & Buée, M. (2015). Compartmentalized and contrasted response of ectomycorrhizal and soil fungal communities of Scots pine forests along elevation gradients in France and Spain. *Environmental Microbiology*, *17*(8), 3009–3024. DOI: 10.1111/1462-2920.12894

Rizzo, L., Manaia, C., Merlin, C., Schwartz, T., Dagot, C., Ploy, M. C., Michael, I., & Fatta-Kassinos, D. (2013). Urban wastewater treatment plants as hotspots for antibiotic resistant bacteria and genes spread into the environment: A review. *The Science of the Total Environment*, *447*, 345–360. DOI: 10.1016/j.scitotenv.2013.01.032 PMID: 23396083

Robin, T. (2019). Our Hands at Work: Indigenous Food Sovereignty in Western Canada. *Journal of Agriculture, Food Systems, and Community Development*, 1–15. DOI: 10.5304/jafscd.2019.09B.007

Robin, T., Dennis, M., & Hart, M. (2020). Feeding indigenous people in Canada. *International Social Work*, *65*(4), 652–662. DOI: 10.1177/0020872820916218

Robin, T., & Hart, M. (2025). Cree food knowledge and being well. *International Journal of Environmental Research and Public Health*, *22*(2), 181. DOI: 10.3390/ijerph22020181

Roche, H., & Tidou, A. (2008). First ecotoxicological assessment assay in a hydroelectric reservoir: The Lake Taabo (côte d'Ivoire). *Bulletin of Environmental Contamination and Toxicology*, *82*(3), 322–326. DOI: 10.1007/s00128-008-9572-9

Rochman, C. M., Hoh, E., Hentschel, B. T., & Kaye, S. (2013). Long-Term Field Measurement of Sorption of Organic Contaminants to Five Types of Plastic Pellets: Implications for Plastic Marine Debris. *Environmental Science & Technology*, *130109073312009*. Advance online publication. DOI: 10.1021/es303700s PMID: 23270427

Rodríguez-González, A., Zanin, M., & Menasalvas, E. (2019). Public Health and Epidemiology Informatics: Can Artificial Intelligence Help Address Future Global Challenges? An overview of antimicrobial resistance and impact of climate change in disease epidemiology. *Yearbook of Medical Informatics*, *28*(01), 224–231. DOI: 10.1055/s-0039-1677910

Rodsamran, P., & Sothornvit, R. (2019). Lime peel pectin integrated with coconut water and lime peel extract as a new bioactive film sachet to retard soybean oil oxidation. *Food Hydrocolloids*, *97*, 105173. DOI: 10.1016/j.foodhyd.2019.105173

Romani, A., Simone, G., Campo, M., Moncini, L., & Bernini, R. (2021). Sweet chestnut standardized fractions from sustainable circular process and green tea extract: In vitro inhibitory activity against phytopathogenic fungi for innovative applications in green agriculture. *PLoS One*, *16*(2), e0247298. DOI: 10.1371/journal.pone.0247298 PMID: 33617600

Rong, G., Ma, F., Qian, Y., Zhang, Z., Chen, S., & Sun, Y. (2024). Exploring the critical years for interdisciplinary citations. *Journal of the Association for Information Science and Technology*, *75*(12), 1398–1412. DOI: 10.1002/asi.24940

Rosman, N., Salleh, W. N. W., Mohamed, M. A., Jaafar, J., Ismail, A. F., & Harun, Z. (2018). Hybrid membrane filtration-advanced oxidation processes for removal of pharmaceutical residue. *Journal of Colloid and Interface Science*, *532*, 236–260. DOI: 10.1016/j.jcis.2018.07.118 PMID: 30092507

Rossi, M., Cubadda, F., Dini, L., Terranova, M., Aureli, F., Sorbo, A., & Passeri, D. (2014). Scientific basis of nanotechnology, implications for the food sector and future trends. *Trends in Food Science & Technology*, *40*(2), 127–148. DOI: 10.1016/j.tifs.2014.09.004

Roth, K., Imran, Z., Liu, W., & Petriello, M. C. (2020). Diet as an Exposure Source and Mediator of Per- and Polyfluoroalkyl Substance (PFAS) Toxicity. *Frontiers in Toxicology*, *2*, 601149. DOI: 10.3389/ftox.2020.601149 PMID: 35296120

Rouhi, M., Razavi, S. H., & Mousavi, S. M. (2017). Optimization of crosslinked poly (vinyl alcohol) nanocomposite films for mechanical properties. *Materials Science and Engineering C*, *71*, 1052–1063. DOI: 10.1016/j.msec.2016.11.135 PMID: 27987659

Roy, S. J. W. R., & Rhim, J.-W. (2022). Genipin-crosslinked gelatin/chitosan-based functional films incorporated with rosemary essential oil and quercetin. *Materials (Basel)*, *15*(11), 3769. DOI: 10.3390/ma15113769 PMID: 35683069

Rubio-Armendáriz, C., Alejandro-Vega, S., Paz-Montelongo, S., Gutiérrez-Fernández, Á. J., Carrascosa-Iruzubieta, C. J., & Hardisson-de La Torre, A. (2022). Microplastics as Emerging Food Contaminants: A Challenge for Food Safety. *International Journal of Environmental Research and Public Health*, *19*(3), 1174. DOI: 10.3390/ijerph19031174 PMID: 35162198

Ryu, H., Li, B., De Guise, S., McCutcheon, J., & Lei, Y. (2021). Recent progress in the detection of emerging contaminants PFASs. *Journal of Hazardous Materials*, *408*, 124437. DOI: 10.1016/j.jhazmat.2020.124437 PMID: 33162244

Sabola, G. (2024). Climate Change Impacts on Agricultural Trade and Food Security in Emerging Economies: A Case Study of Southern Africa. *Discover Agriculture, 2*(1), 12. Advance online publication. DOI: 10.1007/s44279-024-00026-1

Sabraoui, T., Khider, T., Nasser, B., Eddoha, R., Moujahid, A., Benbachir, M., & Essamadi, A. (2020). Determination of Punicalagins Content, Metal Chelating, and Antioxidant Properties of Edible Pomegranate (*Punica granatum L*) Peels and Seeds Grown in Morocco. *International Journal of Food Sciences, 2020*, 1–8. DOI: 10.1155/2020/8885889 PMID: 33029486

Sabui, P., Mallick, S., Singh, K. R., Natarajan, A., Verma, R., Singh, J., & Singh, R. P. (2022). Potentialities of fluorescent carbon nanomaterials as sensor for food analysis. *Luminescence, 38*(7), 1047–1063. DOI: 10.1002/bio.4406 PMID: 36355396

Sadikiel Mmbando, G. (2024). The adoption of genetically modified crops in Africa: The public's current perception, the regulatory obstacles, and ethical challenges. *GM Crops and Food: Biotechnology in Agriculture and the Food Chain, 15*(1), 185–199. DOI: 10.1080/21645698.2024.2345401

Saeedi, M. (2024). How microplastics interact with food chain: A short overview of fate and impacts. *Journal of Food Science and Technology, 61*(3), 403–413. DOI: 10.1007/s13197-023-05720-4 PMID: 37360257

Sahu, R., Singh, A. K., Kumar, A., Singh, K., & Kumar, P. (2023). Bacteriophages concept and applications: A review on phage therapy. *Current Pharmaceutical Biotechnology, 24*(10), 1245–1264. DOI: 10.2174/13892010246662211041124 57 PMID: 36336808

Salih, A. M., Al-Qurainy, F., Khan, S., Nadeem, M., Tarroum, M., & Shaikhaldein, H. O. (2022). Biogenic silver nanoparticles improve bioactive compounds in medicinal plant Juniperus procera in vitro. *Frontiers in Plant Science, 13*, 962112. Advance online publication. DOI: 10.3389/fpls.2022.962112 PMID: 36226285

Saltzman, L., Hansel, T., & Ferreira, R. (2023). Climate change and health equity: Expanding our reach using technology. *PLOS Climate, 2*(7), e0000242. DOI: 10.1371/journal.pclm.0000242

Samarajeewa, U. (2023). Polycyclic aromatic hydrocarbons and food safety: A review. *Journal of the National Science Foundation of Sri Lanka, 51*(2). Advance online publication. DOI: 10.4038/jnsfsr.v51i2.11396

Sampson, D., Cely-Santos, M., Gemmill-Herren, B., Babin, N., Bernhart, A., Kerr, R., & Wittman, H. (2021). Food sovereignty and rights-based approaches strengthen food security and nutrition across the globe: A systematic review. *Frontiers in Sustainable Food Systems, 5*, 686492. Advance online publication. DOI: 10.3389/fsufs.2021.686492

Samut, H., Namli, Ş., Özdemir, F., Çömlekçioğlu, N., & Soyer, Y. (2023). Climate change and food safety: Temperature impact on the attachment of Escherichia coli pathogroups on cress leaf. *Journal of Food Safety, 43*(5), e13059. Advance online publication. DOI: 10.1111/jfs.13059

Sandhu, R., Chaudhary, N., Shams, R., & Dash, K. K. (2025). Genetically modified crops and sustainable development: Navigating challenges and opportunities. *Food Science and Biotechnology, 34*(2), 307–323. DOI: 10.1007/s10068-024-01669-y

Sani, M. A., Azizi-Lalabadi, M., Tavassoli, M., Mohammadi, K., & McClements, D. J. (2021). Recent advances in the development of smart and active biodegradable packaging materials. *Nanomaterials (Basel, Switzerland), 11*(5), 1331. DOI: 10.3390/nano11051331 PMID: 34070054

Santos, V. P., Marques, N. S., Maia, P. C., Lima, M. A. B. D., Franco, L. D., & Campos-Takaki, G. M. D. (2020). Seafood waste as attractive source of chitin and chitosan production and their applications. *International Journal of Molecular Sciences, 21*(12), 4290. DOI: 10.3390/ijms21124290 PMID: 32560250

Sanyang, M. L., Sapuan, S. M., Jawaid, M., Ishak, M. R., & Sahari, J. (2016). Effect of plasticizer type and concentration on physical properties of biodegradable films based on sugar palm (Arenga pinnata) starch for food packaging. *Journal of Food Science and Technology, 53*(1), 326–336. DOI: 10.1007/s13197-015-2009-7 PMID: 26787952

Sapuppo, A., & Uga, E. (2024). PFAS e salute: Quali rischi per l'uomo? *QUADERNI ACP, 31*(5), 218. DOI: 10.53141/QACP.2024.218-222

Saputra, A., Gavahian, M., Purnomo, E. H., Saraswati, S., Hasanah, U., & Boing Sitanggang, A. (2025). High-pressure processing as an emerging technology for enhancing dairy product safety and quality: A critical review of latest advances, sustainability impact, limitations, and prospects. *Critical Reviews in Food Science and Nutrition*, 1–25. DOI: 10.1080/10408398.2025.2499623 PMID: 40326610

Sarabaegi, M., & Roushani, M. (2018). Rapid and sensitive determination of Pseudomonas aeruginosa by using a glassy carbon electrode modified with gold nanoparticles and aptamer-imprinted polydopamine. *Microchemical Journal, 168*, 106388. DOI: 10.1016/j.microc.2021.106388

Sarmah, A. K., Meyer, M. T., & Boxall, A. B. A. (2006). A global perspective on the use, sales, exposure pathways, occurrence, fate and effects of veterinary antibiotics (VAs) in the environment. *Chemosphere*, *65*(5), 725–759. DOI: 10.1016/j.chemosphere.2006.03.026 PMID: 16677683

Sawatzky, A., Cunsolo, A., Jones-Bitton, A., Middleton, J., & Harper, S. (2018). Responding to climate and environmental change impacts on human health via integrated surveillance in the circumpolar north: A systematic realist review. *International Journal of Environmental Research and Public Health*, *15*(12), 2706. DOI: 10.3390/ijerph15122706

Sawyer, W. E., & Izah, S. C. (2024). *Unmasking food adulteration: Public health challenges, impacts and mitigation strategies*. ES General.

Sawyer, W. E., & Izah, S. C. (2024). *Unmasking Food Adulteration: Public Health Challenges, Impacts and Mitigation Strategies. ES General.*

Sawyer, W. E., Nabebe, G., & Izah, S. C. (2024). Public health implications of pesticide residues in food: Risks, regulations, and interventions. *Greener Journal of Biomedical and Health Sciences*, *7*(1), 31–46. DOI: 10.15580/gjbhs.2024.1.102024144

Scallan, E., Hoekstra, R. M., Angulo, F. J., Tauxe, R. V., Widdowson, M. A., Roy, S. L., Jones, J. L., & Griffin, P. M. (2011). Foodborne illness acquired in the United States—Major pathogens. *Emerging Infectious Diseases*, *17*(1), 7–15. DOI: 10.3201/eid1701.P11101 PMID: 21192848

Schaider, L. A., Balan, S. A., Blum, A., Andrews, D. Q., Strynar, M. J., Dickinson, M. E., Lunderberg, D. M., Lang, J. R., & Peaslee, G. F. (2017). Fluorinated Compounds in U.S. Fast Food Packaging. *Environmental Science & Technology Letters*, *4*(3), 105–111. DOI: 10.1021/acs.estlett.6b00435 PMID: 30148183

Scharadin, B., Zanocco, C., & Chistolini, J. (2023). Food retail environments, extreme weather, and their overlap: Exploratory analysis and recommendations for U.S. food policy. *PLoS One*, *18*(11), e0289282. DOI: 10.1371/journal.pone.0289282

Schattman, R., Grubinger, V., McKaeg, L., & Nelson, K. (2018). Whole farm water use: a survey of vegetable producers in New England states. DOI: 10.32747/2018.6938606.ch

Schiemann, J., Dietz-Pfeilstetter, A., Hartung, F., Kohl, C., Romeis, J., & Sprink, T. (2019). Risk assessment and regulation of plants modified by modern biotechniques: Current status and future challenges. *Annual Review of Plant Biology*, *70*(1), 699–726. DOI: 10.1146/annurev-arplant-050718-100025

Schleich, B., Anwer, N., Mathieu, L., & Wartzack, S. (2017). Shaping the digital twin for design and production engineering. *CIRP Annals*, *66*(1), 141–144. DOI: 10.1016/j.cirp.2017.04.040

Schmitt, H., & Sullivan, D. (2022). Communities living with chronic environmental contamination: Leveraging interdisciplinarity to address environmental justice issues. *Translational Issues in Psychological Science*, *8*(4), 473–488. DOI: 10.1037/tps0000338

Scholes, C. (2019). Educational modules for increasing indigenous australian students' involvement in chemistry. *Journal of Chemical Education*, *96*(9), 1914–1921. DOI: 10.1021/acs.jchemed.9b00207

Schrenk, D., Bignami, M., Bodin, L., Chipman, J. K., del Mazo, J., Grasl-Kraupp, B., Hogstrand, C., Hoogenboom, L. R., Leblanc, J.-C., Nebbia, C. S., Nielsen, E., Ntzani, E., Petersen, A., Sand, S., Vleminckx, C., Wallace, H., Barregård, L., Ceccatelli, S., Cravedi, J.-P., & Schwerdtle, T.EFSA Panel on Contaminants in the Food Chain (EFSA CONTAM Panel). (2020). Risk to human health related to the presence of perfluoroalkyl substances in food. *EFSA Journal*, *18*(9). Advance online publication. DOI: 10.2903/j.efsa.2020.6223 PMID: 32994824

Schrenk, D., Bignami, M., Bodin, L., Chipman, J. K., del Mazo, J., Grasl-Kraupp, B., Hoogenboom, L. R., Leblanc, J.-C., Nebbia, C. S., Nielsen, E., Ntzani, E., Petersen, A., Sand, S., Schwerdtle, T., Vleminckx, C., Wallace, H., Rose, M., Cottrill, B., Lundebye, A. K., & Hogstrand, C.EFSA GMO Panel on Contaminants in the Food Chain (CONTAM). (2022). Decontamination process for dioxins and dioxin-like PCBs from fish oil and vegetable oils and fats by a physical process with activated carbon. *EFSA Journal*, *20*(9), e07524. DOI: 10.2903/j.efsa.2022.7524

Schueler, V., Kuemmerle, T., & Schröder, H. (2011). Impacts of Surface Gold Mining on Land Use Systems in Western Ghana. *Ambio*, *40*(5), 528–539. DOI: 10.1007/s13280-011-0141-9

Schwabl, P., Köppel, S., Königshofer, P., Bucsics, T., Trauner, M., Reiberger, T., & Liebmann, B. (2019). Detection of Various Microplastics in Human Stool: A Prospective Case Series. *Annals of Internal Medicine*, *171*(7), 453–457. DOI: 10.7326/M19-0618 PMID: 31476765

Schwiesow, M., Elde, N., & Hilbert, Z. (2024). Distinct routes to thermotolerance in the fungal pathogen Cryptococcus neoformans. https://doi.org/DOI: 10.1101/2024.04.08.588590

Sęczyk, Ł., Gawlik-Dziki, U., & Świeca, M. (2021). Influence of Phenolic-Food Matrix Interactions on In Vitro Bioaccessibility of Selected Phenolic Compounds and Nutrients Digestibility in Fortified White Bean Paste. *Antioxidants, 10*(11), 1825. DOI: 10.3390/antiox10111825 PMID: 34829697

Seiyaboh, E. I., Kigigha, L. T., Alagoa, C. T., & Izah, S. C. (2018). Microbial quality of palm oil sold in Amassoma, Bayelsa State, Nigeria. *International Journal of Public Health and Safety, 3*, 153.

Seiyaboh, E. I., Kigigha, L. T., Alagoa, C. T., & Izah, S. C. (2018b). Microbial Quality of Palm Oil Sold in Amassoma, Bayelsa State, Nigeria. *Int J Pub Health Safe, 3*, 153.

Seiyaboh, E. I., Kigigha, L. T., Aruwayor, S. W., & Izah, S. C. (2018). Level of selected heavy metals in liver and muscles of cow meat sold in Yenagoa metropolis, Bayelsa State, Nigeria. *International Journal of Public Health and Safety, 3*, 154.

Seiyaboh, E. I., Kigigha, L. T., Aruwayor, S. W., & Izah, S. C. (2018a). Level of Selected Heavy Metals in Liver and Muscles of Cow Meat Sold in Yenagoa Metropolis, Bayelsa State, Nigeria. *International Journal of Public Health and Safety, 3*, 154.

Sekhon, B. S. (2014). Nanotechnology in agri-food production: An overview. *Nanotechnology, Science and Applications, 31*, 31. Advance online publication. DOI: 10.2147/NSA.S39406 PMID: 24966671

Sekton, B. S. (2010). Food nanotechnology – an overview. *Nanotechnology, Science and Applications, 1*, 1. Advance online publication. DOI: 10.2147/NSA.S8677

Semenza, J., Herbst, S., Rechenburg, A., Suk, J., Höser, C., Schreiber, C., & Kistemann, T. (2012). Climate Change Impact Assessment of Food- and Waterborne Diseases. *Critical Reviews in Environmental Science and Technology, 42*(8), 857–890. DOI: 10.1080/10643389.2010.534706

Sendhil, R., Nyika, J., Yadav, S., Mackolil, J., Workie, E., Ragupathy, R., & Ramasundaram, P. (2022). Genetically modified foods: Bibliometric analysis on consumer perception and preference. *GM Crops and Food: Biotechnology in Agriculture and the Food Chain, 13*(1), 65–85. DOI: 10.1080/21645698.2022.2038525

Serna-Escolano, V., Serrano, M., Valero, D., Isabel Rodríguez-López, M., Gabaldón, J. A., Castillo, S., Valverde, J. M., Zapata, P. J., Guillén, F., & Martínez-Romero, D. (2020). Thymol Encapsulated into HP-β-Cyclodextrin as an Alternative to Synthetic Fungicides to Induce Lemon Resistance against Sour Rot Decay. *Molecules (Basel, Switzerland), 25*(18), 4348. DOI: 10.3390/molecules25184348 PMID: 32971952

Servin, A. L. (2004). Antagonistic activities of lactobacilli and bifidobacteria against microbial pathogens. *FEMS Microbiology Reviews, 28*(4), 405–440. DOI: 10.1016/j.femsre.2004.01.003 PMID: 15374659

Shaala, L. A., Asfour, H. Z., Youssef, D. T., Żółtowska-Aksamitowska, S., Wysokowski, M., Tsurkan, M., & Ehrlich, H. (2019). New source of 3D chitin scaffolds: The Red Sea demosponge Pseudoceratina arabica (Pseudoceratinidae, Verongiida). *Marine Drugs, 17*(2), 92. DOI: 10.3390/md17020092 PMID: 30717221

Shafiq, M., Anjum, S., Hano, C., Anjum, I., & Abbasi, B. H. (2020). An overview of the applications of nanomaterials and nanodevices in the food industry. *Foods, 9*(2), 148. DOI: 10.3390/foods9020148 PMID: 32028580

Shafique, B., Ranjha, M. M. A. N., Murtaza, M. A., Walayat, N., Nawaz, A., Khalid, W., Mahmood, S., Nadeem, M., Manzoor, M. F., Ameer, K., Aadil, R. M., & Ibrahim, S. A. (2022). Recent trends and applications of nanoencapsulated bacteriocins against microbes in food quality and safety. *Microorganisms, 11*(1), 85. DOI: 10.3390/microorganisms11010085 PMID: 36677377

Shanmugasundaram, R., Lourenco, J., Al Hakeem, W., Dycus, M. M., & Applegate, T. J. (2023). Subclinical doses of dietary fumonisins and deoxynivalenol cause cecal microbiota dysbiosis in broiler chickens challenged with Clostridium perfringens. *Frontiers in Microbiology, 14*, 1106604. Advance online publication. DOI: 10.3389/fmicb.2023.1106604 PMID: 37082176

Sharma, M., Vidhya, C. S., Sunitha, N. H., Sachan, P., Singh, B., Santhosh, K., & Shameena, S. (2024). Emerging food processing and preservation approaches for nutrition and health. *European Journal of Nutrition & Food Safety, 16*(1), 112–127. DOI: 10.9734/ejnfs/2024/v16i11382

Sheikhzadeh, E., Hamsaz, M. C., Turner, A. P., Jager, E. W., & Beni, V. (2016). Label-free impedimetric biosensor for *Salmonella Typhimurium* detection based on poly [pyrrole-co-3-carboxyl-pyrrole] copolymer supported aptamer. *Biosensors & Bioelectronics, 15*(80), 194–200. DOI: 10.1016/j.bios.2016.01.057 PMID: 26836649

Shen, C., Yin, X. C., Jiao, B. Y., Li, J., Jia, P., Zhang, X. W., Cheng, X.-H., Ren, J.-X., Lan, H.-D., Hou, W.-B., Fang, M., Li, X., Fei, Y.-T., Robinson, N., & Liu, J. P. (2022). Evaluation of adverse effects/events of genetically modified food consumption: A systematic review of animal and human studies. *Environmental Sciences Europe, 34*(1), 1–33. DOI: 10.1186/s12302-021-00578-9

Shen, Y., Wang, L., Ding, Y., Liu, S., Li, Y., Zhou, Z., & Liang, Y. (2024). Trends in the Analysis and Exploration of per- and Polyfluoroalkyl Substances (PFAS) in Environmental Matrices: A Review. *Critical Reviews in Analytical Chemistry*, *54*(8), 3171–3195. DOI: 10.1080/10408347.2023.2231535 PMID: 37410543

Sherriff, S., Kalucy, D., Tong, A., Naqvi, N., Nixon, J., Eades, S., Ingram, T., Slater, K., Dickson, M., Lee, A., & Muthayya, S. (2022). Murradambirra dhangaang (make food secure): Aboriginal community and stakeholder perspectives on food insecurity in urban and regional Australia. *BMC Public Health*, *22*(1), 1066. Advance online publication. DOI: 10.1186/s12889-022-13202-z

Shih, Y., Chen, J., Chen, Y., Yang, P., Kuo, Y., Chen, T., & Hsu, B. (2021). Impact of heavy precipitation events on pathogen occurrence in estuarine areas of the Puzi River in Taiwan. *PLoS One*, *16*(8), e0256266. DOI: 10.1371/journal.pone.0256266

Shit, S. C., & Shah, P. M. (2014). Edible polymers: Challenges and opportunities. *Journal of Polymers*, *2014*, 1–13. DOI: 10.1155/2014/427259

Siddiqui, S. A., Singh, S., Bahmid, N. A., & Sasidharan, A. (2024). Applying innovative technological interventions in the preservation and packaging of fresh seafood products to minimize spoilage-A systematic review and meta-analysis. *Heliyon*, *10*(8), e29066. DOI: 10.1016/j.heliyon.2024.e29066

Sidikjan, N., Eziz, M., Xin-guo, L., & Wang, Y. (2022). Spatial distribution, contamination levels, and health risks of trace elements in topsoil along an urbanization gradient in Urumqi, China. *Sustainability (Basel)*, *14*(19), 12646. DOI: 10.3390/su141912646

Siedlok, F., & Hibbert, P. (2013). The organization of interdisciplinary research: Modes, drivers and barriers. *International Journal of Management Reviews*, *16*(2), 194–210. DOI: 10.1111/ijmr.12016

Sievert, K., Lawrence, M., Naika, A., & Baker, P. (2019). Processed Foods and Nutrition Transition in the Pacific: Regional Trends, Patterns, and Food System Drivers. *Nutrients*, *11*(6), 1328. DOI: 10.3390/nu11061328

Sikora, F., Ochmian, I., & Sobolewska, M. (2022). Quality and technological properties of flour with the addition of aesculus hippocastanum and castanea sativa. *Acta Universitatis Cibiniensis. Series E: Food Technology*, *26*(1), 43–54. DOI: 10.2478/aucft-2022-0004

Silbergeld, E. K., Graham, J., & Price, L. B. (2008). Industrial Food Animal Production, Antimicrobial Resistance, and Human Health. *Annual Review of Public Health*, *29*(1), 151–169. DOI: 10.1146/annurev.publhealth.29.020907.090904 PMID: 18348709

Silva, F. V., & Gibbs, P. A. (2012). Thermal pasteurization requirements for the inactivation of Salmonella in foods. *Food Research International*, *45*(2), 695–699. DOI: 10.1016/j.foodres.2011.06.018

Sim, S., & Wiwanitkit, V. (2021). Food contamination, food safety and covid-19 outbreak. *Warasan Wichai Witthayasat Kanphaet*, *35*(5), 463–466. DOI: 10.1108/JHR-01-2021-0014

Singh, A., Patel, A., & Minocha, N. (2025). Nano innovation: Enhancing food packaging through nanotechnology. *Current Nanoscience*, *21*(6), 957–971. DOI: 10.2174/0115734137314364240920052006

Singh, S., Gaikwad, K. K., & Lee, Y. S. (2018). Anthocyanin-A natural dye for smart food packaging systems. *Korean Journal of Packaging Science & Technology*, *24*(3), 167–180. DOI: 10.20909/kopast.2018.24.3.167

Siracusa, V., Rocculi, P., Romani, S., & Dalla Rosa, M. (2008). Biodegradable polymers for food packaging: A review. *Trends in Food Science & Technology*, *19*(12), 634–643. DOI: 10.1016/j.tifs.2008.07.003

Siya, A., Mafigiri, R., Migisha, R., & Kading, R. (2021). Uganda mountain community health system perspectives and capacities towards emerging infectious disease surveillance... https://doi.org/DOI: 10.20944/preprints202105.0691.v1

Smith, B., & Fazil, A. (2019). How will climate change impact microbial foodborne disease in Canada? *Canada Communicable Disease Report = Relevé des Maladies Transmissibles au Canada*, *45*(4), 108–113. DOI: 10.14745/ccdr.v45i04a05

Smith, B., Meadows, S., Meyers, R., Parmley, E., & Fazil, A. (2019). Seasonality and Zoonotic Foodborne Pathogens in Canada: Relationships between Climate and Campylobacter, E. coli, and Salmonella in Meat Products. *Epidemiology and Infection*, *147*, e190. Advance online publication. DOI: 10.1017/S0950268819000797

Smith, J., Ross, K., & Whiley, H. (2016). Australian food safety policy changes from a "command and control" to an "outcomes-based" approach: Reflection on the effectiveness of its implementation. *International Journal of Environmental Research and Public Health*, *13*(12), 1218. DOI: 10.3390/ijerph13121218

Smith, M., Love, D. C., Rochman, C. M., & Neff, R. A. (2018). Microplastics in Seafood and the Implications for Human Health. *Current Environmental Health Reports*, *5*(3), 375–386. DOI: 10.1007/s40572-018-0206-z PMID: 30116998

Smout, F., Skerratt, L., Butler, J., Johnson, C., Congdon, B., & Thompson, R. (2017). The hookworm Ancylostoma ceylanicum: An emerging public health risk in Australian tropical rainforests and Indigenous communities. *One Health, 3*, 66–69. DOI: 10.1016/j.onehlt.2017.04.002

Smyth, S. J. (2017). Genetically modified crops, regulatory delays, and international trade. *Food and Energy Security*, *6*(2), 78–86. DOI: 10.1002/fes3.100

Soares, G., Hedges, J., Sethi, S., Poirier, B., & Jamieson, L. (2023). From biocolonialism to emancipation: Considerations on ethical and culturally respectful omics research with indigenous Australians. *Medicine, Health Care, and Philosophy*, *26*(3), 487–496. DOI: 10.1007/s11019-023-10151-1

Sobhan, A., Hossain, A., Wei, L., Muthukumarappan, K., & Ahmed, M. (2025). IoT-Enabled Biosensors in Food Packaging: A Breakthrough in Food Safety for Monitoring Risks in Real Time. *Foods*, *14*(8), 1403. DOI: 10.3390/foods14081403 PMID: 40282803

Sohail, M., Urooj, Z., Noreen, S., Baig, M. M. F. A., Zhang, X., & Li, B. (2023). Micro- and nanoplastics: Contamination routes of food products and critical interpretation of detection strategies. *The Science of the Total Environment*, *891*, 164596. DOI: 10.1016/j.scitotenv.2023.164596 PMID: 37279810

Sonawane, S. K., Patil, S. P., & Arya, S. S. (2018). Nanotechnology enrolment in food and food safety. *Journal of Microbiology, Biotechnology and Food Sciences*, *8*(3), 893–900. DOI: 10.15414/jmbfs.2018-19.8.3.893-900

Song, S., Huang, T., Ma, J., Mao, X., Gao, H., Zhao, Y., Jiang, W., Lian, L., Chen, B., Liu, L., & Wang, Z. (2022). Assessing safety of market-sold fresh fish: Tracking fish origins and toxic chemical origins. *Environmental Science & Technology*, *56*(13), 9505–9514. DOI: 10.1021/acs.est.2c00495

Soni, D. K., Ahmad, R., & Dubey, S. K. (2018). Biosensor for the detection of Listeria monocytogenes: Emerging trends. *Critical Reviews in Microbiology*, *44*(5), 590–608. DOI: 10.1080/1040841X.2018.1473331 PMID: 29790396

Soni, K. A., & Nannapaneni, R. (2010). Bacteriophage significantly reduces Listeria monocytogenes on raw salmon fillet tissue. *Journal of Food Protection*, *73*(1), 32–38. DOI: 10.4315/0362-028X-73.1.32 PMID: 20051201

Sorrentino, A., Gorrasi, G., & Vittoria, V. (2007). Potential perspectives of bionanocomposites for food packaging applications. *Trends in Food Science & Technology*, *18*(2), 84–95. DOI: 10.1016/j.tifs.2006.09.004

Spence, G. T., Steinsaltz, D., & Fanshawe, T. R. (2016). A Bayesian approach to sequential meta-analysis. *Statistics in Medicine*, *35*(29), 5356–5375. DOI: 10.1002/sim.7052

Spencer, J., Maxwell, D., Erickson, K., Wall, D., Nicholas-Figueroa, L., Pratt, K., & Shultz, G. (2021). Cultural relevance in chemistry education: Snow chemistry and the iñupiaq community. *Journal of Chemical Education*, *99*(1), 363–372. DOI: 10.1021/acs.jchemed.1c00480

Spiegel, M., Fels-Klerx, H., & Marvin, H. (2012). *Effects of Climate Change on Food Safety Hazards in the Dairy Production Chain. Food Research International*, *46*(1), 201–208. DOI: 10.1016/j.foodres.2011.12.011

Stecconi, T., Tavoloni, T., Stramenga, A., Bacchiocchi, S., Barola, C., Dubbini, A., Galarini, R., Moretti, S., Sagratini, G., & Piersanti, A. (2024). A LC-MS/MS procedure for the analysis of 19 perfluoroalkyl substances in food fulfilling recent EU regulations requests. *Talanta*, *266*, 125054. DOI: 10.1016/j.talanta.2023.125054 PMID: 37611367

Stefanova, R., Vasilev, N. V., & Spassov, S. L. (2010). Irradiation of food, current legislation framework, and detection of irradiated foods. *Food Analytical Methods*, *3*(3), 225–252. DOI: 10.1007/s12161-009-9118-8

Steffens, K., Larsbo, M., Moeys, J., Kjellström, E., Jarvis, N., & Lewan, E. (2014). Modelling pesticide leaching under climate change: Parameter vs. climate input uncertainty. *Hydrology and Earth System Sciences*, *18*(2), 479–491. DOI: 10.5194/hess-18-479-2014

Stoma, M., & Dudziak, A. (2022). Eastern Poland Consumer Awareness of Innovative Active and Intelligent Packaging in the Food Industry: Exploratory Studies. *Sustainability (Basel)*, *14*(20), 13691. DOI: 10.3390/su142013691

Subedi, J., Houston, J., & Sherman–Morris, K. (2018). Interdisciplinary research as an iterative process to build disaster systems knowledge. *Risk Analysis*, *41*(7), 1072–1077. DOI: 10.1111/risa.13244

Suderman, N., Isa, M. I. N., & Sarbon, N. M. (2018). The effect of plasticizers on the functional properties of biodegradable gelatin-based film: A review. *Food Bioscience*, *24*, 111–119. DOI: 10.1016/j.fbio.2018.06.006

Sugg, M. M., Ryan, S., Spurlock, T., Duncan, S., Hege, A., Ogwu, M. C., Tyson, J., Shay, E., Renwick, K., Hendren, C., Roy, M., & Runkle, J. D. (2023). Climate Change and Health in Rural Mountain Environments: Summary of a Workshop on Knowledge Gaps, barriers, and Opportunities for Action. *GeoJournal.* DOI: 10.1007/s10708-023-10916-4

Sultan, K., Ali, T., Fahmy, N., & El-Shibiny, A. (2019). Using millimeter-waves for rapid detection of pathogenic bacteria in food based on bacteriophage. *Engineering Reports*, *1*(1), e12026. Advance online publication. DOI: 10.1002/eng2.12026

Sundararajan, N., Habeebsheriff, H. S., Dhanabalan, K., Cong, V. H., Wong, L. S., Rajamani, R., & Dhar, B. K. (2023). Mitigating global challenges: Harnessing green synthesized nanomaterials for sustainable crop production systems. *Global Challenges (Hoboken, NJ)*, *8*(1), 2300187. Advance online publication. DOI: 10.1002/gch2.202300187 PMID: 38223890

Sunderland, E. M., Hu, X. C., Dassuncao, C., Tokranov, A. K., Wagner, C. C., & Allen, J. G. (2019). A review of the pathways of human exposure to poly- and perfluoroalkyl substances (PFASs) and present understanding of health effects. *Journal of Exposure Science & Environmental Epidemiology*, *29*(2), 131–147. DOI: 10.1038/s41370-018-0094-1 PMID: 30470793

Sun, L., Jiang, Y., Pan, R., Li, M., Wang, R., Chen, S., Fu, S., & Man, C. (2018). A novel, simple and low-cost paper-based analytical device for colorimetric detection of Cronobacter spp. *Analytica Chimica Acta*, *7*(1036), 80–88. DOI: 10.1016/j.aca.2018.05.061 PMID: 30253840

Sun, T., Wang, S., Ji, C., Li, F., & Wu, H. (2022). Microplastics aggravate the bioaccumulation and toxicity of coexisting contaminants in aquatic organisms: A synergistic health hazard. *Journal of Hazardous Materials*, *424*, 127533. DOI: 10.1016/j.jhazmat.2021.127533 PMID: 34879523

Sun, X., Jia, M., Ji, J., Guan, L., Zhang, Y., Tang, L., & Li, Z. (2015). Enzymatic amplification detection of peanut allergen Ara h1 using a stem-loop DNA biosensor modified with a chitosan-mutiwalled carbon nanotube nanocomposite and spongy gold film. *Talanta*, *131*, 521–527. DOI: 10.1016/j.talanta.2014.07.078 PMID: 25281135

Suvarna, V., Nair, A., Mallya, R., Khan, T., & Omri, A. (2022). Antimicrobial nanomaterials for food packaging. *Antibiotics (Basel, Switzerland)*, *11*(6), 729. DOI: 10.3390/antibiotics11060729 PMID: 35740136

Suwardi, A., Navia, Z., Harmawan, T., Syamsuardi, S., & Mukhtar, E. (2020). Ethnobotany and Conservation of Indigenous Edible Fruit Plants in South Aceh, Indonesia. *Biodiversitas (Surakarta), 21*(5). Advance online publication. DOI: 10.13057/biodiv/d210511

Swiderska, K., Argumedo, A., Wekesa, C., Ndalilo, L., Song, Y., Rastogi, A., & Ryan, P. (2022). Indigenous Peoples' Food Systems and Biocultural Heritage: Addressing Indigenous Priorities through Decolonial and Interdisciplinary Research Approaches. *Sustainability (Basel), 14*(18), 11311. DOI: 10.3390/su141811311

Szelenberger, R., Cichoń, N., Zajaczkowski, W., & Bijak, M. (2024). Application of biosensors for the detection of mycotoxins for the improvement of food safety. *Toxins, 16*(6), 249. DOI: 10.3390/toxins16060249 PMID: 38922144

Szewczuk, M. A., Zych, S., Oster, N., & Karakulska, J. (2023). Activity of Patchouli and Tea Tree Essential Oils against Staphylococci Isolated from Pyoderma in Dogs and Their Synergistic Potential with Gentamicin and Enrofloxacin. *Animals (Basel), 13*(8), 1279. DOI: 10.3390/ani13081279 PMID: 37106842

Tagliapietra, B. L., Felisberto, M. H. F., Sanches, E. A., Campelo, P. H., & Clerici, M. T. P. S. (2021). Non-conventional starch sources. *Current Opinion in Food Science, 39*, 93–102. DOI: 10.1016/j.cofs.2020.11.011

Taha, A. A., & Omar, M. M. (2024). Review article: nanoparticle applications in plant nutrition: a comprehensive review and future perspectives. *Journal of Soil Sciences and Agricultural Engineering, 0*(0), 25–30. DOI: 10.21608/jssae.2024.271443.1218

Taha, H., Nasraween, M., Khader, Y., Al Omari, L., Al Nsour, M., Al Shdaifat, A., & Berggren, V. (2022). Microbial load of chicken shawerma and the handlers' compliance with food safety practices in Jordan. *Arab Journal of Nutrition and Exercise, 6*(1). Advance online publication. DOI: 10.18502/ajne.v6i1.10063

Tahir, A., Shabir Ahmad, R., Imran, M., Ahmad, M. H., Kamran Khan, M., Muhammad, N., Nisa, M. U., Tahir Nadeem, M., Yasmin, A., Tahir, H. S., Zulifqar, A., & Javed, M. (2021). Recent approaches for utilization of food components as nano-encapsulation: A review. *International Journal of Food Properties, 24*(1), 1074–1096. DOI: 10.1080/10942912.2021.1953067

Tamborlin, L., Sumere, B. R., de Souza, M. C., Pestana, N. F., Aguiar, A. C., Eberlin, M. N., Simabuco, F. M., Rostagno, M. A., & Luchessi, A. D. (2020). Characterization of pomegranate peel extracts obtained using different solvents and their effects on cell cycle and apoptosis in leukemia cells. *Food Science & Nutrition, 8*(10), 5483–5496. DOI: 10.1002/fsn3.1831 PMID: 33133551

Tarigholizadeh, S., Sushkova, S., Rajput, V. D., Ranjan, A., Arora, J., Dudnikova, T., Barbashev, A., Mandzhieva, S., Minkina, T., & Wong, M. H. (2023). Transfer and degradation of PAHs in the soil–plant system: A review. *Journal of Agricultural and Food Chemistry*, *72*(1), 46–64. DOI: 10.1021/acs.jafc.3c05589

Tegegne, H., & Phyo, H. (2017). Food safety knowledge, attitude, and practices of meat handler in abattoir and retail meat shops of Jigjiga town, Ethiopia. DOI: 10.15167/2421-4248/jpmh2017.58.4.737

Temkin, A. M., Hocevar, B. A., Andrews, D. Q., Naidenko, O. V., & Kamendulis, L. M. (2020). Application of the Key Characteristics of Carcinogens to Per and Polyfluoroalkyl Substances. *International Journal of Environmental Research and Public Health*, *17*(5), 1668. DOI: 10.3390/ijerph17051668 PMID: 32143379

Tesfay, S. (2024). What are the impacts of climate change on sustainable food production, food demand, and population numbers in sub-Saharan Africa? A systematic review. Food journal, 2(2). DOI: 10.59411/sc2by231

Thakur, A. (2025). Innovation Through Collaboration Leveraging Knowledge Sharing for Global R&D Success. In *Research and Development Practices in Innovation* (pp. 127–164). IGI Global Scientific Publishing. DOI: 10.4018/979-8-3373-0812-8.ch006

Tharanathan, R. N. (2003). Biodegradable films and composite coatings: Past, present and future. *Trends in Food Science & Technology*, *14*(3), 71–78. DOI: 10.1016/S0924-2244(02)00280-7

The Codex Alimentarius and the European Legislation on Food Additives. (2017). In P. Laganà, E. Avventuroso, G. Romano, M. E. Gioffré, P. Patanè, S. Parisi, U. Moscato, & S. Delia, SpringerBriefs in Molecular Science (pp. 23–32). Springer International Publishing. DOI: 10.1007/978-3-319-57042-6_2

Then, C., Miyazaki, J., & Bauer-Panskus, A. (2022). Deficiencies in the risk assessment of genetically engineered Bt cowpea approved for cultivation in Nigeria: A critical review. *Plants*, *11*(3), 380. DOI: 10.3390/plants11030380

Thiruvengadam, M., Rajakumar, G., & Chung, I. M. (2018). Nanotechnology: current uses and future applications in the food industry. *Biotech*, *8*(1), 74.

Thomann, E. (2018). Food safety policy: Transnational, hybrid, wicked. In *Oxford research encyclopedia of politics*. DOI: 10.1093/acrefore/9780190228637.013.540

Thompson, L., & Darwish, W. (2019). Environmental chemical contaminants in food: Review of a global problem. *Journal of Toxicology*, *2019*, 1–14. DOI: 10.1155/2019/2345283

Thow, A., Heywood, P., Schultz, J., Quested, C., Jan, S., & Colagiuri, S. (2011). Trade and the nutrition transition: Strengthening policy for health in the Pacific. *Ecology of Food and Nutrition*, *50*(1), 18–42. DOI: 10.1080/03670244.2010.524104

Tirado, C., Clarke, R., Jaykus, L., McQuatters-Gollop, A., & Frank, J. (2010). Climate change and food safety: A review. *Food Research International*, *43*(7), 1745–1765. DOI: 10.1016/j.foodres.2010.07.003

Tolve, R., Bianchi, F., Lomuscio, E., Sportiello, L., & Simonato, B. (2022). Current advantages in the application of microencapsulation in functional bread development. *Foods*, *12*(1), 96. DOI: 10.3390/foods12010096 PMID: 36613312

Tomczyk, Ł., Stępień, Ł., Urbaniak, M., Szablewski, T., Cegielska-Radziejewska, R., & Stuper-Szablewska, K. (2018). Characterisation of the mycobiota on the shell surface of table eggs acquired from different egg-laying hen breeding systems. *Toxins*, *10*(7), 293. DOI: 10.3390/toxins10070293 PMID: 30012982

Tseng, W. B., Hsieh, M. M., Chen, C. H., Chiu, T. C., & Tseng, W. L. (2020). Functionalized gold nanoparticles for sensing of pesticides: A review. *Yao Wu Shi Pin Fen Xi*, *28*(4), 522–539. DOI: 10.38212/2224-6614.1092 PMID: 35696144

Tsitsos, A., Economou, V., Chouliara, E., Koutouzidou, G., Arsenos, G., & Ambrosiadis, I. (2023). Effect of Chitosan and Alginate-Based Edible Membranes with Oregano Essential Oil and Olive Oil in the Microbiological, Physicochemical and Organoleptic Characteristics of Mutton. *Microorganisms*, *11*(2), 507. DOI: 10.3390/microorganisms11020507 PMID: 36838470

Tudi, M., Ruan, H., Wang, L., Lyu, J., Sadler, R., Connell, D., & Phung, D. (2021). Agriculture development, pesticide application and its impact on the environment. *International Journal of Environmental Research and Public Health*, *18*(3), 1112. DOI: 10.3390/ijerph18031112

Tuominen, I., Ballardini, R., Mähönen, J., & Pihlajarinne, T. (2023). Protecting and accessing indigenous peoples' digital cultural heritage through sustainable governance and IPR structures - the case of Sámi culture. *Arctic Review on Law and Politics*, *14*. Advance online publication. DOI: 10.23865/arctic.v14.5809

Turhan, E. U., Polat, S., Erginkaya, Z., & Konuray, G. (2022). Investigation of synergistic antibacterial effect of organic acids and ultrasound against pathogen biofilms on lettuce. *Food Bioscience*, *47*, 101643. DOI: 10.1016/j.fbio.2022.101643

Uğur, F. A. (2023). *Food Irradiation: A Promising Technology for Enhancing Food Safety, Quality, and Sustainability*. BIDGE Publications.

Ullah, M. (2024). NGOs' role in sustaining indigenous knowledge in rural Bangladesh: Agriculture, healthcare, and disaster management. *South Asian Journal of Social Sciences and Humanities*, *5*(1), 79–102. DOI: 10.48165/sajssh.2024.5106

UNESCO. (2017). Communication for Development: Harnessing information for sustainable development. Retrieved from https://en.unesco.org/themes/communication

United Nations. (2015). The 17 Goals - Sustainable Development. Available at: https://sdgs.un.org/goals

Upadhyay, P., Zubair, M., Roopesh, M. S., & Ullah, A. (2024). An overview of advanced antimicrobial food packaging: Emphasizing antimicrobial agents and polymer-based films. *Polymers*, *16*(14), 2007. DOI: 10.3390/polym16142007 PMID: 39065324

Urango, A. C. M., Strieder, M. M., Silva, E. K., & Meireles, M. A. A. (2022). Impact of thermosonication processing on food quality and safety: A review. *Food and Bioprocess Technology*, *15*(8), 1700–1728. DOI: 10.1007/s11947-022-02760-0

Usha, T., Middha, S. K., Hemavathi, K. N., Hemalatha, P., Harini, R., Afreen, Z., Dwivedi, N., Akshaya, K. A., Goyal, A. K., & Baliga, M. S. (2025). Innovations in nanotechnology for food production and packaging: A review. *Current Nutrition and Food Science*, *21*(5), 559–572. DOI: 10.2174/0115734013297685240820043042

Uzoekwe, S. A., Izah, S. C., & Aigberua, A. O. (2021). Environmental and human health risk of heavy metals in atmospheric particulate matter (PM_{10}) around gas flaring vicinity in Bayelsa State, Nigeria. *Toxicology and Environmental Health Sciences*, *13*(4), 323–335. DOI: 10.1007/s13530-021-00085-7

Vaičiulienė, G., Bakutis, B., Jovaišienė, J., Falkauskas, R., Gerulis, G., & Baliukonienė, V. (2020). Origanum vulgareandthymus vulgarisextract usability to improve silage hygienic quality and reduce mycotoxin concentrations. *Journal of Microbiology and Biotechnology*, *30*(8), 1149–1155. DOI: 10.4014/jmb.2003.03010 PMID: 32522971

Vaiserman, A., Koliada, A., Zabuga, O., & Socol, Y. (2018). Health impacts of low-dose ionizing radiation: Current scientific debates and regulatory issues. *Dose-Response*, *16*(3), 1559325818796331. DOI: 10.1177/1559325818796331 PMID: 30263019

Vaisocherová-Lísalová, H., Víšová, I., Ermini, M. L., Špringer, T., Song, X. C., Mrázek, J., Lamačová, J.Jr, Lynn, N. S., Šedivák, P., & Homola, J. (2016). Low-fouling surface plasmon resonance biosensor for multi-step detection of foodborne bacterial pathogens in complex food samples. *Biosensors & Bioelectronics*, *15*(80), 84–90. DOI: 10.1016/j.bios.2016.01.040 PMID: 26807521

Valdés, A., Burgos, N., Jiménez, A., & Garrigós, M. C. (2015). Natural pectin polysaccharides as edible coatings. *Coatings*, *5*(4), 865–886. DOI: 10.3390/coatings5040865

Van Asselt, E. D., Rietra, R. P. J. J., Römkens, P. F. A. M., & Van Der Fels-Klerx, H. J. (2011). Perfluorooctane sulphonate (PFOS) throughout the food production chain. *Food Chemistry*, *128*(1), 1–6. DOI: 10.1016/j.foodchem.2011.03.032 PMID: 25214321

Van Boeckel, T. P., Brower, C., Gilbert, M., Grenfell, B. T., Levin, S. A., Robinson, T. P., Teillant, A., & Laxminarayan, R. (2015). Global trends in antimicrobial use in food animals. *Proceedings of the National Academy of Sciences of the United States of America*, *112*(18), 5649–5654. DOI: 10.1073/pnas.1503141112 PMID: 25792457

Van Der Voet, H., De Boer, W. J., Kruisselbrink, J. W., Goedhart, P. W., Van Der Heijden, G. W. A. M., Kennedy, M. C., Boon, P. E., & Van Klaveren, J. D. (2015). The MCRA model for probabilistic single-compound and cumulative risk assessment of pesticides. *Food and Chemical Toxicology*, *79*, 5–12. DOI: 10.1016/j.fct.2014.10.014 PMID: 25455888

Van Dijk, H., Houghton, J., Van Kleef, E., Van Der Lans, I., Rowe, G., & Frewer, L. (2008). Consumer responses to communication about food risk management. *Appetite*, *50*(2–3), 340–352. DOI: 10.1016/j.appet.2007.08.011 PMID: 17945386

Van Raamsdonk, L. W. D., Van Der Zande, M., Koelmans, A. A., Hoogenboom, R. L. A. P., Peters, R. J. B., Groot, M. J., Peijnenburg, A. A. C. M., & Weesepoel, Y. J. A. (2020). Current Insights into Monitoring, Bioaccumulation, and Potential Health Effects of Microplastics Present in the Food Chain. *Foods*, *9*(1), 72. DOI: 10.3390/foods9010072 PMID: 31936455

Varma, R., & Vasudevan, S. (2020). Extraction, Characterization, and Antimicrobial Activity of Chitosan from Horse Mussel *Modiolus modiolus. ACS Omega*, *5*(32), 20224–20230. DOI: 10.1021/acsomega.0c01903 PMID: 32832775

Varzakas, T., & Αντωνιάδου, M. (2024). A holistic approach for ethics and sustainability in the food chain: The gateway to oral and systemic health. *Foods*, *13*(8), 1224. DOI: 10.3390/foods13081224

Verlicchi, P., Al Aukidy, M., & Zambello, E. (2012). Occurrence of pharmaceutical compounds in urban wastewater: Removal, mass load and environmental risk after a secondary treatment—A review. *The Science of the Total Environment*, *429*, 123–155. DOI: 10.1016/j.scitotenv.2012.04.028 PMID: 22583809

Verma, S., Singh, A., Pradhan, S., & Kushuwaha, M. (2024). Impact of climate change on agriculture: A review. *International Journal of Environment and Climate Change, 14*(3), 615–620. DOI: 10.9734/ijecc/2024/v14i34069

Vieira, T. M., Moldão-Martins, M., & Alves, V. D. (2021). Composite Coatings of Chitosan and Alginate Emulsions with Olive Oil to Enhance Postharvest Quality and Shelf Life of Fresh Figs (Ficus carica L. cv. 'Pingo De Mel'). *Foods, 10*(4), 718. DOI: 10.3390/foods10040718 PMID: 33805309

Vieira, Y., Lima, E. C., Foletto, E. L., & Dotto, G. L. (2021). Microplastics physicochemical properties, specific adsorption modeling and their interaction with pharmaceuticals and other emerging contaminants. *The Science of the Total Environment, 753*, 141981. DOI: 10.1016/j.scitotenv.2020.141981 PMID: 32911167

Vijay Rakesh Reddy, S., Sudhakar Rao, D. V., Sharma, R. R., Preethi, P., & Pandiselvam, R. (2022). Role of ozone in post-harvest disinfection and processing of horticultural crops: A review. *Ozone Science and Engineering, 44*(1), 127–146. DOI: 10.1080/01919512.2021.1994367

Vijayakumar Sreelatha, H., Patel, S., & Nagarajan, P. (2024). The Key Concepts of Animal Models. In V. Sreelatha, H., Patel, S. & Nagarajan, P. (Eds.), *Animal models in research: Principles and practice* (pp. 3–16). Singapore: Springer Nature Singapore. DOI: 10.1007/978-981-97-0048-6_1

Vijayan, D., Ludwig, D., Rybak, C., Kaechele, H., Hoffmann, H., Schönfeldt, H., Mbwana, H. A., Rivero, C. V., & Löhr, K. (2022). Indigenous knowledge in food system transformations. *Communications Earth & Environment, 3*(1), 213. Advance online publication. DOI: 10.1038/s43247-022-00543-1

Vu, T. P., Gumus-Bonacina, C. E., Corradini, M. G., He, L., McClements, D. J., & Decker, E. A. (2022). Role of Solid Fat Content in Oxidative Stability of Low-Moisture Cracker Systems. *Antioxidants, 11*(11), 2139. DOI: 10.3390/antiox11112139 PMID: 36358510

Wagh, R. V., Priyadarshi, R., Khan, A., Riahi, Z., Packialakshmi, J. S., Kumar, P., Rindhe, S. N., & Rhim, J. W. (2025). The role of active packaging in the defense against foodborne pathogens with particular attention to bacteriophages. *Microorganisms, 13*(2), 401. DOI: 10.3390/microorganisms13020401 PMID: 40005767

Wana, H., & Kudama, G. (2022). Do traditional food commodities transform food security? evidence from indigenous knowledge in West part of Ethiopia. DOI: 10.21203/rs.3.rs-1687653/v1

Wang, Wang, Yan, Luan, Wu, & Bian. (2021). Rapid, sensitive and label-free detection of pathogenic bacteria using a bacteria-imprinted conducting polymer film-based electrochemical sensor. *Talanta*, 1, 226:122135.

Wang, B., Hong, J., Liu, C., Zhu, L., & Jiang, L. (2021). An electrochemical molecularly imprinted polymer sensor for rapid β-Lactoglobulin detection. *Sensors (Basel)*, *21*(24), 8240. DOI: 10.3390/s21248240 PMID: 34960338

Wang, J., & Chen, T. (2016). The spread model of food safety risk under the supply-demand disturbance. *SpringerPlus*, *5*(1), 1765. Advance online publication. DOI: 10.1186/s40064-016-2881-2

Wang, J., & Wu, Z. (2023). Minimal processing of produce using a combination of UV-C irradiation and ultrasound-assisted washing. *Lebensmittel-Wissenschaft + Technologie*, *182*, 114901. DOI: 10.1016/j.lwt.2023.114901

Wang, S. L., & Nguyen, V. B. (2019). Production of potent antidiabetic compounds from shrimp head powder via Paenibacillus conversion. *Process Biochemistry (Barking, London, England)*, *76*, 18–24. DOI: 10.1016/j.procbio.2018.11.004

Wang, W., Chen, W., Lv, M., Zou, R., Wang, D., Hou, F., & Liu, D. (2018). Applications of power ultrasound in oriented modification and degradation of pectin: A review. *Journal of Food Engineering*, *234*, 98–107. DOI: 10.1016/j.jfoodeng.2018.04.016

Wang, X., Bouzembrak, Y., Lansink, A. O., & Van Der Fels-Klerx, H. J. (2021). Application of machine learning to the monitoring and prediction of food safety: A review. *Comprehensive Reviews in Food Science and Food Safety*, *21*(1), 416–434. DOI: 10.1111/1541-4337.12868

Wang, X., Xu, Y., Liang, M., Gao, J., Wang, J., Chen, S., & Cheng, J. (2022). Chinese consumers' trust in food safety surveillance sampling of commonly consumed foods. *Foods*, *11*(13), 1971. DOI: 10.3390/foods11131971

Wang, Y., Li, L., Li, H., Peng, Y., & Fu, L. (2022). A fluorometric sandwich biosensor based on rationally imprinted magnetic particles and aptamer modified carbon dots for the detection of tropomyosin in seafood products. *Food Control*, *132*, 108552. DOI: 10.1016/j.foodcont.2021.108552

Wang, Y., Li, Y., Yang, S., Li, C., Li, L., Gao, S., & Wu, Z. (2023). Mechanism of ozone treatment in delayed softening of fresh-cut kiwifruit during storage. *Postharvest Biology and Technology*, *204*, 112469. DOI: 10.1016/j.postharvbio.2023.112469

Wanihsuksombat, C., Hongtrakul, V., & Suppakul, P. (2010). Development and characterization of a prototype of a lactic acid–based time–temperature indicator for monitoring food product quality. *Journal of Food Engineering*, *100*(3), 427–434. DOI: 10.1016/j.jfoodeng.2010.04.027

Wanyama, R., Gödecke, T., Chege, C., & Qaim, M. (2019). How important are supermarkets for the diets of the urban poor in Africa? *Food Security*, *11*(6), 1339–1353. DOI: 10.1007/s12571-019-00974-3

Waterhouse, B., Hartford, A., Mukhopadhyay, S., Ferguson, R., & Hendrickson, B. A. (2022). Using the Bayesian detection of potential risk using inference on blinded safety data (BDRIBS) method to support the decision to refer an event for unblinded evaluation. *Pharmaceutical Statistics*, *21*(2), 372–385. DOI: 10.1002/pst.2175

Weiler, A., Hergesheimer, C., Brisbois, B., Wittman, H., Yassi, A., & Spiegel, J. (2014). Food sovereignty, food security, and health equity: A meta-narrative mapping exercise. *Health Policy and Planning*, *30*(8), 1078–1092. DOI: 10.1093/heapol/czu109

Wei, Y., Lan, G., Wu, Z., Chen, B., Quan, F., Li, M., Sun, S., & Du, H. (2022). Phyllosphere fungal communities of rubber trees exhibited biogeographical patterns, but not bacteria. *Environmental Microbiology*, *24*(8), 3777–3790. DOI: 10.1111/1462-2920.15894

Whyte, K. (2018). Food sovereignty, justice, and indigenous peoples. DOI: 10.1093/oxfordhb/9780199372263.013.34

Williams, L., Jones, M., & Wright, P. (2023). Decolonising food regulatory frameworks: Importance of recognising traditional culture when assessing dietary safety of traditional foods. *The Proceedings of the Nutrition Society*, *82*(4), 454–467. DOI: 10.1017/S0029665123003026

Williams, M. S., Ebel, E. D., & Vose, D. (2010). Framework for microbial food-safety risk assessments amenable to Bayesian modeling. *Risk Analysis*, *31*(4), 548–565. DOI: 10.1111/j.1539-6924.2010.01532.x

Williams, P., Sikutshwa, L., & Shackleton, S. (2020). Acknowledging indigenous and local knowledge to facilitate collaboration in landscape approaches—Lessons from a systematic review. *Land (Basel)*, *9*(9), 331. DOI: 10.3390/land9090331

Wilson, P. D. C.L. (Ed.). (2007). Intelligent and Active Packaging for Fruits and Vegetables (1st ed.). CRC Press. DOI: 10.1201/9781420008678

Woesik, R., Woesik, K., Woesik, L., & Woesik, S. (2013). Effects of ocean acidification on the dissolution rates of reef-coral skeletons. *PeerJ*, *1*, e208. DOI: 10.7717/peerj.208

World Health Organization. (2017). Pharmaceuticals in drinking-water: Public health and environment. https://www.who.int/publications/i/item/9789241512753

Wu, B., Gao, C., Chen, L., Buscot, F., Goldmann, K., Purahong, W., Ji, N.-N., Wang, Y.-L., Lü, P.-P., Li, X.-C., & Guo, L. (2018). Host phylogeny is a major determinant of Fagaceae-associated ectomycorrhizal fungal community assembly at a regional scale. *Frontiers in Microbiology*, *9*, 2409. Advance online publication. DOI: 10.3389/fmicb.2018.02409

Wudil, A., Usman, M., Rosak-Szyrocka, J., Pilař, L., & Boye, M. (2022). Reversing years for global food security: A review of the food security situation in sub-Saharan Africa (SSA). *International Journal of Environmental Research and Public Health*, *19*(22), 14836. DOI: 10.3390/ijerph192214836

Wu, F., Bhatnagar, D., Bui-Klimke, T., Carbone, I., Hellmich, R., Munkvold, G., Paul, P., Payne, G., & Takle, E. (2011). Climate change impacts on mycotoxin risks in us maize. *World Mycotoxin Journal*, *4*(1), 79–93. DOI: 10.3920/WMJ2010.1246

Wu, F., & Güçlü, H. (2012). Aflatoxin regulations in a network of global maize trade. *PLoS One*, *7*(9), e45151. DOI: 10.1371/journal.pone.0045151

Wu, X., Ernst, F., Conkle, J. L., & Gan, J. (2013). Comparative uptake and translocation of pharmaceutical and personal care products (PPCPs) by common vegetables. *Environment International*, *60*, 15–22. DOI: 10.1016/j.envint.2013.07.015 PMID: 23973619

Wysokowski, M., Bazhenov, V. V., Tsurkan, M. V., Galli, R., Stelling, A. L., Stöcker, H., & Ehrlich, H. (2013). Isolation and identification of chitin in three-dimensional skeleton of Aplysina fistularis marine sponge. *International Journal of Biological Macromolecules*, *62*, 94–100. DOI: 10.1016/j.ijbiomac.2013.08.039 PMID: 23994783

Xiang, L., Li, Y.-W., Yu, P.-F., Feng, N.-X., Zhao, H.-M., Li, H., Cai, Q.-Y., Mo, C.-H., & Li, Q. X. (2019). Food Safety Concerns: Crop Breeding as a Potential Strategy to Address Issues Associated with the Recently Lowered Reference Doses for Perfluorooctanoic Acid and Perfluorooctane sulfonate. Journal of Agricultural and Food Chemistry. DOI: 10.1021/acs.jafc.9b04625

Xie, L., Guo, C., Yang, L., & He, Y. (2024). Harnessing gold nanomaterials for advanced multicolor colorimetric biosensors in food hazards detection. *Yao Wu Shi Pin Fen Xi*, *32*(3), 274–295. DOI: 10.38212/2224-6614.3511 PMID: 39636771

Xing, Y., Zhou, Y., Zhang, X., Lin, X., Li, J., Liu, P., Lee, H. K., & Huang, Z. (2023). The sources and bioaccumulation of per- and polyfluoroalkyl substances in animal-derived foods and the potential risk of dietary intake. *The Science of the Total Environment, 905*, 167313. DOI: 10.1016/j.scitotenv.2023.167313 PMID: 37742961

Xiong, J., Zhang, H., Qin, L., Zhang, S., Cao, J., & Jiang, H. (2022). Magnetic fluorescent quantum dots nanocomposites in food contaminants analysis: Current challenges and opportunities. *International Journal of Molecular Sciences, 23*(8), 4088. DOI: 10.3390/ijms23084088 PMID: 35456904

Xu, M. L., Gao, Y., Wang, X., Han, X. X., & Zhao, B. (2021). Comprehensive strategy for sample preparation for the analysis of food contaminants and residues by GC–MS/MS: A review of recent research trends. *Foods, 10*(10), 2473. DOI: 10.3390/foods10102473 PMID: 34681522

Xu, X., Zhou, X. D., & Wu, C. D. (2012). Tea catechin epigallocatechin gallate inhibits Streptococcus mutans biofilm formation by suppressing gtf genes. *Archives of Oral Biology, 57*(6), 678–683. DOI: 10.1016/j.archoralbio.2011.10.021 PMID: 22169220

Xu, Y. X., Kim, K. M., Hanna, M. A., & Nag, D. (2005). Chitosan–starch composite film: Preparation and characterization. *Industrial Crops and Products, 21*(2), 185–192. DOI: 10.1016/j.indcrop.2004.03.002

Yadav, A., Bhagat, R., & Yadav, V. (2019). Does coal mining affect household food security? An Empirical Study of Dhanbad, India. *Agriculture & Food Security, 8*(1), 7. Advance online publication. DOI: 10.1186/s40066-018-0244-7

Yahaya, U., Suleiman, R. A., Hussaini, Y., Hamidu, S. T., Odey, B. O., & Adaaja, B. O. (2024). Environmental risk and biosafety of genetically modified plants. *Nigerian Journal of Biotechnology, 41*(2), 104–108. DOI: 10.4314/njb.v41i2.9

Yan, D., Li, Y., Liu, Y., Li, N., Zhang, X., & Yan, C. (2021). Antimicrobial Properties of Chitosan and Chitosan Derivatives in the Treatment of Enteric Infections. *Molecules (Basel, Switzerland), 26*(23), 7136. DOI: 10.3390/molecules26237136 PMID: 34885715

Yang, A., Zheng, Y., Long, C., Chen, H., Liu, B., Li, X., Yuan, J., & Cheng, F. (2014). Fluorescent immunosorbent assay for the detection of alpha lactalbumin in dairy products with monoclonal antibody bioconjugated with CdSe/ZnS quantum dots. *Food Chemistry, 150*, 73–79. DOI: 10.1016/j.foodchem.2013.10.137 PMID: 24360421

Yang, P., Rao, L., Zhao, L., Wu, X., Wang, Y., & Liao, X. (2021). High pressure processing combined with selected hurdles: Enhancement in the inactivation of vegetative microorganisms. *Comprehensive Reviews in Food Science and Food Safety*, *20*(2), 1800–1828. DOI: 10.1111/1541-4337.12724 PMID: 33594773

Yang, Y., Shi, Y., Kerfahi, D., Ogwu, M. C., Wang, J., Dong, K., Takahashi, K., Moroenyane, I., & Adams, J. M. (2021). Elevation-related climate trends dominate fungal co-occurrence network structure and the abundance of keystone taxa on Mt. Norikura, Japan. *The Science of the Total Environment*, *799*, 149368. DOI: 10.1016/j.scitotenv.2021.149368

Yan, M. R., Hsieh, S., & Ricacho, N. (2022). Innovative food packaging, food quality and safety, and consumer perspectives. *Processes (Basel, Switzerland)*, *10*(4), 747. DOI: 10.3390/pr10040747

Yao, L., Zhu, S., Liu, W., Manzoor, Z., Nisar, M. F., & Li, M. (2021). A comprehensive review of pharmacological and Analytical Aspects of Acacetin. *Natural Resources for Human Health*, *1*(1), 8–18. DOI: 10.53365/nrfhh/141018

Yaun, B. R., Sumner, S. S., Eifert, J. D., & Marcy, J. E. (2004). Inhibition of pathogens on fresh produce by ultraviolet energy. *International Journal of Food Microbiology*, *90*(1), 1–8. DOI: 10.1016/S0168-1605(03)00158-2 PMID: 14672825

Yiridomoh, G. (2021). "Illegal" gold mining operations in Ghana: Implication for climate-smart agriculture in northwestern Ghana. *Frontiers in Sustainable Food Systems*, *5*, 745317. Advance online publication. DOI: 10.3389/fsufs.2021.745317

Yoon, J. H., Oh, M. S., & Lee, S. Y. (2024). Effectiveness of organic acids for inactivating pathogenic bacteria inoculated in laboratory media and foods: An updated minireview. *Food Science and Biotechnology*, *33*(12), 2715–2728. DOI: 10.1007/s10068-024-01618-9 PMID: 39184989

Young, L., Shukla, S., & Wilson, T. (2024). Indigenous values and perspectives for strengthening food security and sovereignty: Learning from a community-based case study of misko-ziibiing (bloodvein river First Nation), Manitoba, Canada. *Frontiers in Sustainable Food Systems*, *8*, 1321231. Advance online publication. DOI: 10.3389/fsufs.2024.1321231

Yu, H., Park, J. Y., Kwon, C. W., Hong, S. C., Park, K. M., & Chang, P. S. (2018). An overview of nanotechnology in food science: Preparative methods, practical applications, and safety. *Journal of Chemistry*, *2018*, 1–10. DOI: 10.1155/2018/5427978

Yu, J., Hennessy, D. A., Tack, J., & Wu, F. (2022). Climate change will increase aflatoxin presence in US Corn. *Environmental Research Letters, 17*(5), 054017. DOI: 10.1088/1748-9326/ac6435

Zai, B., McReavy, S., Hogan, G., Ng, V., Papadopoulos, A., Young, I., & Grant, L. (2024). Identifying the Impact of Climate Variables on Biological and Chemical Contaminants in Preharvest Foods and Their Associated Food Safety Risks: A Scoping Review Protocol. *BMJ Open, 14*(8), e083749. DOI: 10.1136/bmjopen-2023-083749

Zajác, P., Čapla, J., & Čurlej, J. (2025). Microplastic contamination of food. *Scifood, 19*, 1–16. DOI: 10.5219/scifood.1

Zanni, S., Cammalleri, V., D'Agostino, L., Protano, C., & Vitali, M. (2024). Occurrence of pharmaceutical residues in drinking water: A systematic review. *Environmental Science and Pollution Research International, 32*(16), 10436–10463. DOI: 10.1007/s11356-024-34544-8 PMID: 39103588

Zapico, A., Ruiz-Saavedra, S., Gómez-Martín, M., de Los Reyes-Gavilán, C. G., & González, S. (2022). Pilot study for the dietary assessment of xenobiotics derived from food processing in an adult Spanish sample. *Foods, 11*(3), 470. DOI: 10.3390/foods11030470

Zárate-Moreno, J. C., Escobar-Sierra, D. M., & Ríos-Estepa, R. (2023). Development and Evaluation of Chitosan-Based Food Coatings for Exotic Fruit Preservation. *Biotech, 12*(1), 20. DOI: 10.3390/biotech12010020 PMID: 36810447

Zhanbolat, A. A., & Tungyshbaeva, U. O. (2024). Current strategies for the use of active packaging in food products. *Bulletin of Shakarim University Technical Sciences*, 1(13), 108–115. DOI: 10.53360/2788-7995-2024-1(13)-14

Zhang, Zhou, Li, Fan, Li, Lu, Wen, & Ren. (2023). Recent advances of fluorescent sensors for bacteria detection-A review. *Talanta,* 3(1), 254:124133. DOI: 10.1016/j.talanta.2022.124133

Zhang, J., Zou, X., Zhai, X., Huang, X., Jiang, C., & Holmes, M. (2019). Preparation of an intelligent pH film based on biodegradable polymers and roselle anthocyanins for monitoring pork freshness. *Food Chemistry, 272*, 306–312. DOI: 10.1016/j.foodchem.2018.08.041 PMID: 30309548

Zhang, K., Perussello, C. A., Milosavljević, V., Cullen, P. J., Sun, D. W., & Tiwari, B. K. (2019). Diagnostics of plasma reactive species and induced chemistry of plasma treated foods. *Critical Reviews in Food Science and Nutrition, 59*(5), 812–825. DOI: 10.1080/10408398.2018.1564731 PMID: 30676057

Zhang, L., Huang, Y., Zhou, Y., Buckley, T., & Wang, H. H. (2013). Antibiotic Administration Routes Significantly Influence the Levels of Antibiotic Resistance in Gut Microbiota. *Antimicrobial Agents and Chemotherapy*, *57*(8), 3659–3666. DOI: 10.1128/AAC.00670-13 PMID: 23689712

Zhang, Q., Xu, E. G., Li, J., Chen, Q., Ma, L., Zeng, E. Y., & Shi, H. (2020). A Review of Microplastics in Table Salt, Drinking Water, and Air: Direct Human Exposure. *Environmental Science & Technology*, *54*(7), 3740–3751. DOI: 10.1021/acs.est.9b04535 PMID: 32119774

Zhang, T., Ma, L., & Liu, J. (2023). Influence of Structural Characteristics of Interdisciplinary Knowledge Networks on College Students' Innovation Ability. *International Journal of Emerging Technologies in Learning*, *18*(03), 51–66. DOI: 10.3991/ijet.v18i03.38053

Zhang, X., Khan, I. M., Ji, H., Wang, Z., Tian, H., Cao, W., & Mi, W. (2020). A label-free fluorescent aptasensor for detection of staphylococcal enterotoxin a based on aptamer-functionalized silver nanoclusters. *Polymers*, *12*(1), 152. DOI: 10.3390/polym12010152 PMID: 31936075

Zhang, Y., Moran, M., Nearing, M., Campos, G., Huete, A., Buda, A., & Starks, P. (2013). Extreme precipitation patterns and reductions of terrestrial ecosystem production across biomes. *Journal of Geophysical Research. Biogeosciences*, *118*(1), 148–157. DOI: 10.1029/2012JG002136

Zhao, G., Du, J., Chen, W., Pan, M., & Chen, D. (2019). Preparation and thermostability of cellulose nanocrystals and nanofibrils from two sources of biomass: Rice straw and poplar wood. *Cellulose (London, England)*, *26*(16), 8625–8643. DOI: 10.1007/s10570-019-02683-8

Zhong, Y., Godwin, P., Jin, Y., & Xiao, H. (2020). Biodegradable polymers and green-based antimicrobial packaging materials: A mini-review. *Advanced Industrial and Engineering Polymer Research.*, *3*(1), 27–35. DOI: 10.1016/j.aiepr.2019.11.002

Zhong, Z., Gao, R., Chen, Q., & Jia, L. (2020). Dual-aptamers labeled polydopamine-polyethyleneimine copolymer dots assisted engineering a fluorescence biosensor for sensitive detection of Pseudomonas aeruginosa in food samples. *Spectrochimica Acta. Part A: Molecular and Biomolecular Spectroscopy*, *224*, 117417. DOI: 10.1016/j.saa.2019.117417 PMID: 31362188

Zhou, J., Lv, X., Jia, J., Din, Z. U., Cai, S., He, J., Xie, F., & Cai, J. (2022). Nanomaterials-based electrochemiluminescence biosensors for food analysis: Recent developments and future directions. *Biosensors (Basel)*, *12*(11), 1046. DOI: 10.3390/bios12111046 PMID: 36421164

Zhou, S., Yu, B., & Zhang, Y. (2023). Global concurrent climate extremes exacerbated by anthropogenic climate change. *Science Advances*, *9*(10), eabo1638. Advance online publication. DOI: 10.1126/sciadv.abo1638

Zhu, Y., Li, C., Cui, H., & Lin, L. (2020). Feasibility of cold plasma for the control of biofilms in food industry. *Trends in Food Science & Technology*, *99*, 142–151. DOI: 10.1016/j.tifs.2020.03.001

Zhu, Y., & Wang, D. (2016). Rapid Detection of Enterobacter Sakazakii in milk Powder using amino modified chitosan immunomagnetic beads. *International Journal of Biological Macromolecules*, *1*(93), 615–622. DOI: 10.1016/j.ijbiomac.2016.09.024 PMID: 27616695

Zia, S., Khan, M. R., Shabbir, M. A., Aslam Maan, A., Khan, M. K. I., Nadeem, M., Khalil, A. A., Din, A., & Aadil, R. M. (2022). An inclusive overview of advanced thermal and nonthermal extraction techniques for bioactive compounds in food and food-related matrices. *Food Reviews International*, *38*(6), 1166–1196. DOI: 10.1080/87559129.2020.1772283

Zinsstag, J., Crump, L., Schelling, E., Hattendorf, J., Maidane, Y., Ali, K., Muhummed, A., Umer, A. A., Aliyi, F., Nooh, F., Abdikadir, M. I., Ali, S. M., Hartinger, S., Mäusezahl, D., de White, M. B. G., Cordon-Rosales, C., Castillo, D. A., McCracken, J., Abakar, F., & Cissé, G. (2018). Climate change and one health. *FEMS Microbiology Letters*, *365*(11). Advance online publication. DOI: 10.1093/femsle/fny085

Ziska, L., Crimmins, A., Auclair, A., DeGrasse, S., Garofalo, J., Khan, A., … Walls, I. (2016). Ch. 7: food safety, nutrition, and distribution. The impacts of climate change on human health in the United States: a scientific assessment. DOI: 10.7930/J0ZP4417

Zolfaghari, A., Bazargani-Gilani, B., & Aghajani, N. (2023). Edible film based on corn zein containing dill extract and essential oil/β-cyclodextrin inclusion complex: Shelf life enhancement of common carp fillet. *Food Science & Nutrition*, *11*(7), 4275–4288. DOI: 10.1002/fsn3.3353 PMID: 37457154

Zornes, G., Curl, J., & Holmer, M. (2022). PFAS Complicate Potable Reuse and Advanced Treatment. *Journal - American Water Works Association*, *114*(2), 72–74. DOI: 10.1002/awwa.1871

Zorraquín-Peña, I., Cueva, C., Bartolomé, B., & Moreno-Arribas, M. V. (2020). Silver nanoparticles against foodborne bacteria. effects at intestinal level and health limitations. *Microorganisms*, *8*(1), 132. DOI: 10.3390/microorganisms8010132 PMID: 31963508

Zoungranan, Y., Lynda, E., Dobi-Brice, K. K., Tchirioua, E., Bakary, C., & Yannick, D. D. (2020). Influence of natural factors on the biodegradation of simple and composite bioplastics based on cassava starch and corn starch. *Journal of Environmental Chemical Engineering*, *8*(5), 104396. DOI: 10.1016/j.jece.2020.104396

Żukowska, G., Durczyńska, Z., Roszkowski, S., Myszura-Dymek, M., & Bik-Małodzińśka, M. (2024). Possibilities of using bio-based nanomaterials in sustainable agriculture. *Journal of Ecological Engineering*, *25*(3), 313–322. DOI: 10.12911/22998993/182966

About the Contributors

Matthew Chidozie Ogwu is an Assistant Professor of Integrated Ecology and Sustainable Development at Appalachian State University, USA. He is an accomplished interdisciplinary scholar with transdisciplinary expertise spanning One Health, EcoHealth, and sustainability science. His research is deeply rooted in the assessment of coupled human and natural systems and socio-ecological dynamics, with a strong focus on resilience, environmental justice, and sustainable development. Dr. Ogwu has earned numerous prestigious awards, competitive research grants, and international scholarships in recognition of his academic excellence and impactful contributions. He is widely published and serves as a board member and peer reviewer for several high-impact journals across environmental science, public health, and ecological disciplines. Passionate about science-policy integration and community-engaged scholarship, Dr. Ogwu consistently lends his expertise to international collaborative projects and volunteers his time to promote education, health equity, and environmental sustainability worldwide. His leadership and mentoring have inspired students, early-career researchers, and community stakeholders in various global regions.

Sylvester Izah is a distinguished faculty member at Bayelsa Medical University in Yenagoa, Nigeria. Dr. Izah holds dual Ph.D. in Public Health, and Applied Microbiology and Environmental Health, an M.Sc. in Applied Microbiology, and a B.Sc. in Biological Sciences from Niger Delta University, Nigeria. In addition to his teaching role, he holds the position of Assistant Director of Academic Planning, Research, and Innovations at the university. Dr. Izah is a licensed Environmental Health Specialist in Nigeria. His expertise spans various disciplines, and he has a particular focus on Sustainable Human-Environmental Health Interactions. This broad area of study involves the assessment and management of air, soil, and water quality, as well as toxicology, hygiene, sanitation, food science, waste management, biodiversity, and sustainability. Dr. Izah's academic prowess is highlighted by an impressive research and publication portfolio, which includes more than 300 peer-reviewed publications. His work extends to the publication of edited books with renowned publishers such as Springer and Elsevier, contributions to book chapters in publications by Elsevier, Springer, and Taylor and Francis, as well as numerous journal articles.

* * *

Kholis Ernawati, S.Si, S.Ag, M.Kes., is a scholar in Environmental Health, Public Health, and Digital Health, with interdisciplinary interests in environmental science, tech-based healthcare services, and Islamic perspectives on health. She is a faculty member at the Faculty of Medicine, Universitas YARSI, Indonesia. She is known for contributing to health promotion strategies, digital health innovation, and integrating Islamic ethics into public health education. Dr. Ernawati holds a B.Sc. in Chemistry and an M.P.H. from Universitas Airlangga, a Doctorate in Environmental Health (cum laude) from Universitas Indonesia, and a B.A. in Islamic Studies (cum laude) from STIU Darul Hikmah Bekasi. She has authored over 130 publications, with 1,250+ citations, an H-index of 25, and a SINTA score of 1,558. Her research appears in international journals (Q2–Q4), covering dengue prevention, COVID-19, and big data in health surveillance. Dr. Ernawati work fosters technology-driven, culturally sensitive, and ethically grounded health practices, advancing sustainable public health in Indonesia and beyond.

Abubakar Tijjani Ibrahim, PhD, is a Lecturer at the Department of Mass Communication, Kano State Polytechnic, Nigeria, where he currently serves as the Head of Department. In addition, he held various administrative positions including Acting Head of department, Chief Examination Officer and Deputy Chief Examination Officer. Dr. Abubakar Ibrahim also served in a number of committees including ASUP National Conference Committee, Examination Misconduct Committee among others. His research interest focuses on fake news, misinformation, social media and news content credibility, and media literacy.

Adamkolo Mohammed Ibrahim, PhD, is a Lecturer at the Department of Mass Communication, University of Maiduguri, Nigeria. He is an accomplished researcher and peer reviewer of more than a decade. He reviews quite a lot of local and international journals, notably Computers in Human Behavior (Elsevier), SAGE Open (SAGE) and Journal of Systems and Information Technology (JSIT) (Emerald). He has published many research articles and book chapters and attended several local and international conferences. His research interests include social TV, social media, development communication, health communication, development communication, mass communication and journalism, ICT adoption and computer-mediated communication.

Meenambiga S. S. completed 6 years of teaching experience and 4 years of research experience in the field of Biotechnology. Passionate about teaching and have trained many young students in the field of Mycology, Natural Product Research, Nanoparticle synthesis, Drug Discovery from Plants and fungal sources and in-silico analysis of drug targets using modern Bio-informatics tools. Skilled in scientific writing covering the following aspects: Proposal submission, research articles suitable for national and international publications, drafting, proof reading and editing research thesis and expertise in critically reviewing the research manuscript for indexed journals. Member of Board of Studies for B.Tech Biotechnology curriculum in VISTAS. Guided more than 20 undergraduate students projects and served as Principal Investigator in the student's project funded by Tamilnadu State Council for Science and Technology. Published more than 30 research and review articles in national and international indexed journal and has volunteered as reviewer for many reputed journals. Life member of Mycological Society of India, BRSI, IARA, IAENG and many other professional organizations.

Mohammed Bashir Tijjani, PhD, is a Professor of Veterinary Medicine at the Department of Veterinary Pharmacology and Toxicology, Faculty of Veterinary Medicine, University of Maiduguri, Nigeria. His research interests include Pharmacology, Chemotherapy, Clinical Pharmacology and Toxicology, Herbal Medicine and Toxicology of Poisonous Plants and Vertebrate Toxicology.

Index

I

K

M

N

P

R

S

T

www.ingramcontent.com/pod-product-compliance
Lightning Source LLC
LaVergne TN
LVHW062326300926
849161LV00007B/1595